NOBEL LECTURES

CHEMISTRY

1996–2000

NOBEL LECTURES

INCLUDING PRESENTATION SPEECHES AND LAUREATES' BIOGRAPHIES

PHYSICS
CHEMISTRY
PHYSIOLOGY OR MEDICINE
LITERATURE
PEACE
ECONOMIC SCIENCES

Nobel Lectures

Including Presentation Speeches and Laureates' Biographies

Chemistry

1996–2000

Editor

Ingmar Grenthe
Royal Institute of Technology,
Stockholm

World Scientific
New Jersey • London • Singapore • Hong Kong

Published for the Nobel Foundation in 2003 by

World Scientific Publishing Co. Pte. Ltd.
5 Toh Tuck Link, Singapore 596224
USA office: Suite 202, 1060 Main Street, River Edge, NJ 07661
UK office: 57 Shelton Street, Covent Garden, London WC2H 9HE

NOBEL LECTURES IN CHEMISTRY (1996–2000)

ISBN 981-02-4958-6
ISBN 981-02-4959-4 (pbk)

Printed by FuIsland Offset Printing (S) Pte Ltd, Singapore

FOREWORD

Since 1901 the Nobel Foundation has published annually "Les Prix Nobel" with reports from the Nobel award ceremonies in Stockholm and Oslo as well as the biographies and Nobel lectures of the laureates. In order to make the lectures available for people with special interests in the different prize fields the Foundation gave Elsevier Publishing Company the right to publish in English the lectures for 1901–1970, which were published in 1964–1972 through the following volumes:

Physics 1901–1970	4 vols.
Chemistry 1901–1970	4 vols.
Physiology or Medicine 1901–1970	4 vols.
Literature 1901–1967	1 vol.
Peace 1901–1970	3 vols.

Thereafter, and until the year 2000, the Nobel Foundation has given World Scientific Publishing Company the right to bring the series up to date and also publish the Prize lectures in Economics from the year 1969. The Nobel Foundation is very pleased that the intellectural and spiritual message to the world laid down in the laureates' lectures, thanks to the efforts of World Scientific, will reach new readers all over the world.

Bengt Samuelsson
Chairman of the Board

Michael Sohlman
Executive Director

Stockholm, August 2001

PREFACE

This volume describes the exciting developments and fundamental discoveries in chemistry that were rewarded with Nobel Prizes from 1996 to 2000. They range from the apparently simple, the discovery of a new stable modification of carbon, to the complex, the understanding of how living cells can store and make use of energy. They cover important experimental and theoretical developments that have increased our understanding of the most fundamental question in chemistry, how chemical bonds are formed and broken. They demonstrate how the discovery that an insulating organic polymer could be made electrically conducting has led to prolific developments, not only in fundamental chemistry and physics, but also to important applications.

Sir Issac Newton said that he could see far because he was standing on the shoulders of giants. This is true also for the scientists of today, but equally important is the courage to question established truths; there was a new form of carbon, all organic polymers were not insulators, it was possible to obtain detailed insight into the dynamics of the chemical bond and to give a detailed quantitative basis for the heuristic bonding concepts "ionic" and "covalent" that we all have met in the introductory courses in chemistry. The lectures also describe how the borders between the traditional classifications of natural science have gradually been blurred; the "giants" are no longer confined to a specific discipline. The lectures demonstrate the importance of technological developments, the laser, advanced analytical tools and methods for structure determination of macromolecules, but also of faster and cheaper computers.

This volume gives a vivid picture of the activity surrounding important sections of the research frontiers in chemistry. By reading it I hope the reader will get as much pleasure and insight as I have had.*

Ingmar Grenthe
Editor

*The material in this volume reproduces the original in *Les Prix Nobel* from 1996 to 2000. The laureates have been consulted in order to correct errors and update references, and these are given as errata at the end of the lectures.

CONTENTS

NOBEL LECTURES

CHEMISTRY

1996–2000

Chemistry 1996

ROBERT F. CURL, Jr., HAROLD W. KROTO and RICHARD E. SMALLEY

"for their discovery of fullerenes"

THE NOBEL PRIZE IN CHEMISTRY

Speech by Professor Lennart Eberson of the Royal Swedish Academy of Sciences.
Translation of the Swedish text.

Your Majesties, Your Royal Highness, Ladies and Gentlemen,

We like to think that everything worth knowing about the chemical elements is already known, and that carbon–one of our most thoroughly researched elements–could not possibly yield further important discoveries. Carbon has been known since prehistoric times as soot, coal and charcoal. By the late 18th century, graphite and diamonds had been shown to be different forms of the element carbon. We employ carbon in countless ways: the large-scale burning of coal as a fuel; the use of coke in steel production; the use of graphite in lubricants, pencils, brake linings, etc. That rare form of carbon known as a diamond has many applications, aside from its aesthetic function. An ordinary automobile tire contains 3 kilograms of carbon black, and activated carbon is highly useful in a wide variety of fields. Carbon is the basis of life processes; it is extremely important to all of us.

It was therefore a first-class scientific sensation when this year's Laureates in Chemistry – Robert Curl, Harold Kroto and Richard Smalley – together with graduate students James Heath and Sean O'Brien, reported in 1985 that they had discovered a new, stable form of carbon in which sixty carbon atoms are arranged in a closed shell. They named this new carbon molecule *buckminsterfullerene* after the American architect R. Buckminster Fuller, inventor of the "geodesic dome", a building perhaps internationally best known from the United States pavilion at the 1967 world's fair in Montreal. To understand how the carbon atoms in buckminsterfullerene are connected to each other, we need only recall the pattern on the surface of a soccer ball, or European football. This ball is stitched together from 12 black pentagons and 20 white hexagons, in such a way that no pentagon comes into contact with another pentagon. The result is a highly symmetrical structure with sixty corners. If we now imagine that we place a carbon atom at each one of these 60 corners, we know how buckminsterfullerene looks. Although it is 300 million times smaller than a soccer ball!

The discovery of buckminsterfullerene–or C_{60}–was made by using an advanced instrument, in which a laser vaporized a very small quantity of carbon in one five billionth of a second. When the hot carbon gas condensed, it formed clusters containing different numbers of carbon atoms. The one with 60 carbon atoms was the most common. Many of these various carbon molecules were shown to have the same stability as C_{60} and were therefore also assumed to be closed; the collective name for such clusters was *fullerenes*. It was also possible to produce fullerenes that enclosed a metal atom inside, for example potassium or cesium.

The problems with these experiments was that fullerenes were available in such small quantities that their postulated structure could not be rigorously verified. The years from 1985 to 1990 were filled with scientific disputes, in which the stubbornness, ingenuity and enthusiasm of the fullerenes' discoverers kept their hypothesis alive despite rather severe criticism. Only in 1990 were physicists Donald Huffman and Wolfgang Krätschmer able to produce gram-sized quantities of C_{60} using a method that could be quickly and inexpensively duplicated in any laboratory. This made it possible to apply the whole battery of structural determination methods and show that C_{60} really had the structure its discoverers had hypothesized. Chemists now quickly went to work studying the chemistry of fullerenes. They were able to try out various applications of the chemistry and physics of fullerenes.

So why are the fullerenes so interesting? To understand this, we must look at the structure of other forms of carbon. Graphite consists of carbon atoms attached together in very large flat networks that are piled on top of each other, whereas diamonds consist of carbon atoms bound into endless three-dimensional networks. Both are examples of what we usually call polymers. The chemistry that can be done using these forms of carbon is rather limited –and not entirely inexpensive, in the case of diamonds! A fullerene, on the other hand, has a closed, low-molecular structure that can be chemically processed and modified in an almost infinite number of ways.

This year's Nobel Prize in Chemistry has implications for all the natural sciences. The seeds of the discovery were sowed by a desire to understand the behavior of carbon in red giant stars and interstellar gas clouds. The discovery of fullerenes has expanded our knowledge and changed our thinking in chemistry and physics. It has given us new hypotheses on the occurrence of carbon in the universe. It has also led us to discover small quantities of fullerenes in geological formations. Fullerenes are probably present in much larger amounts on earth than previously believed. It has been shown that most sooty flames contain small quantities of fullerenes. Think of this the next time you light a candle!

The symmetry concept has played an important role in the history of ideas and the natural sciences. Ideas of symmetry dominate many important theories and comprise a strong driving force behind scientific thinking. We are fascinated by the beautiful structure of C_{60}–and this feeling has existed ever since humans began to reflect on natural phenomena. In the *Timaeus* dialogue, Plato described his theory of the four elementary particles for fire, earth, air and water:

"And next we have to determine what are the four most beautiful bodies which are unlike one another, and of which some are capable of resolution into one another; for having discovered thus much, we shall know the true origin of earth and fire and of the proportionate and intermediate elements. And then we shall not be willing to allow that there are any distinct kinds of visible bodies fairer than these..."

He went on to describe four of the five regular polyhedrons–the tetrahedron (fire), the cube (earth), the octahedron (air) and the icosahedron (water). The dodecahedron represents the cosmos because it is closest to that most perfect of forms, the sphere. Plato would certainly have found the structure of C_{60}–an expanded dodecahedron, which is about as close to a sphere as you can get–to be an unusually beautiful body.

Professors Curl, Kroto and Smalley,

You have been awarded the 1996 Nobel Prize in Chemistry for your discovery of a new form of the element carbon, the fullerenes. It is a privilege and a great pleasure for me to congratulate you on behalf of the Royal Swedish Academy of Sciences, and I now ask you to receive your Nobel Prizes from the hands of His Majesty the King.

Robert Curl

ROBERT F. CURL JR.

I was born in Alice, Texas on August 23, 1933. My father was a Methodist minister, and my mother was what we then called a housewife. I have a sister, Mary, who is some years my elder. In those days, Methodist ministers moved often, and as a child I lived in a succession of mostly small towns in south Texas: Alice, Brady, San Antonio, Kingsville, Del Rio, Brownsville, McAllen, Austin, then back to San Antonio. During this time the church hierarchy recognized that my father was an able administrator capable of organizing people to get things done and gifted at resolving conflicts. From the time I was about nine my father was no longer pastor of a church but rather a supervisor of church activities over a district. This was a great relief to me as I was spared being the center of judgmental attention as the "preacher's kid."

By the time I reached adulthood my father was universally revered as a fair, kind, and gentle man with an acute mind. His most enduring monument will be the San Antonio Medical Center as he worked hard and effectively to start the Methodist Hospital there, which really started the center.

When I was nine years old, my parents gave me a chemistry set. Within a week, I had decided to become a chemist and never wavered from that choice. As I grew my interest in chemistry grew more intense, if not more sophisticated. Of course there was no chemistry in the school program until high school.

I was not a particularly distinguished student as a child. My grades were good but obtained more by steady work than any brilliance on my part. I vividly remember my father telling me that one of my elementary school teachers had told him that I was not brilliant but I was a steady hard worker. Somehow the further I progressed in school, the easier it became to do well.

It was a great delight when I finally got to study chemistry in high school. My teacher, Mrs. Lorena Davis, saw that I was keenly interested and did her best to foster and nourish that interest. As only one year of chemistry was offered then, I had no formal course in the subject to take my final year in high school. Mrs. Davis offered me special projects to satisfy my appetite for chemistry. I remember most constructing a Cottrell Precipitator. I was shocked to see Mrs. Davis, who didn't smoke, light up a cigarette and blow smoke into the precipitator to demonstrate that it worked.

When it came time to choose a college, I got interested in Rice Institute. It had an excellent reputation as being a good school for a dedicated student. I was also impressed by how well its football team was doing. My parents loved my choice because at that time Rice charged no tuition, and they would have been hard pressed to send me to a university that did. While my father held the highest administrative office, not counting the Bishop, in the Southwest Texas Conference, he did not make much money.

At that time there was a high failure rate at Rice. With no tuition, students were expected to prove themselves worthy or make way for someone else. However, I was ready for the challenge that Rice presented and prospered academically. Socially, my fellow students were ready for the challenge that I presented and worked hard to convert a rather straight-laced, serious boy into someone they could stand to be around.

By a quirk of fate, the most colorful professors I encountered in my first years taught subjects other than chemistry. I liked my first and second year chemistry professors (in fact I later developed a closer relation with my second year professor, John T. Smith), but they were not particularly colorful. It was not until my third year when I had John E. Kilpatrick for Physical Chemistry and George Holmes Richter for Organic Chemistry that the chemistry department began to pull ahead in the colorfulness race. John Kilpatrick came to class, sat in the middle of the table in front, lit a cigarette, took an enormous drag, and began to speak. No smoke came out! Richter enlivened his lectures by describing the pharmacological effects of various organic chemicals. Richter was a fine teacher of Organic Chemistry, but that was of little use to me since I had an almost unnatural aversion to Organic Chemistry. Kilpatrick was the most welcoming to students of any person I ever encountered with absolutely no regard for the amount of time he spent with a student. This, happily for him, made the time he devoted self-limiting, because I would think about whether I had an hour or two to spare before dropping by to see him.

The most impressive chemistry teacher I had was Richard Turner, whom I first encountered in a senior Natural Products course. (The curriculum was cleverly constructed so that it was impossible to avoid a second encounter with Organic Chemistry.) It was his enthusiastic discussion of barriers to internal rotation and the pioneering work of Kenneth Pitzer in the area that made me resolve to go to University of California, Berkeley, and work with Pitzer. This is a decision I have never regretted.

While I was at Berkeley, Pitzer was the Dean of the College of Chemistry and a very busy man. Nevertheless he was always completely accessible to his graduate students, and always genuinely delighted to see me when I interrupted his work. When our conversation reached its conclusion, he graciously got me out of his office. I was grateful for this as well because at the time, as you can see from my comments about visiting John Kilpatrick, I had trouble with leave-takings. I think that I received an excellent education in how to do research from Pitzer. The most important work I did at Berkeley was on Pitzer's extension of the Theory of Corresponding States. Over the years, I have remained in contact with Ken and Jean Pitzer. Indeed, we were able to collaborate again in research some years later when he was president of Rice University.

My years at Berkeley were some of the happiest of my life primarily because it was during this time that I met and married my wife, Jonel. Our union seemed pre-ordained when we discovered that our ancestors came from the tiny town of Center Point, Texas (pop. 300).

At that time, there seemed to be an unwritten rule that Pitzer's students should do experiment as well as theory. This suited me, because I had always been interested in experiments. Pitzer suggested that I investigate the matrix isolation infrared spectrum of disiloxane in order to establish whether the SiO-Si bond was linear or bent. If I had tried to do these experiments involving liquid hydrogen without help, I believe there is a good chance an explosion would have resulted. However, a fellow student, Dolphus Milligan, helped me tremendously with these experiments and with his aid I was able to collect the necessary data, which indicated that Si-O-Si is somewhat bent from linearity.

Pitzer was able to help me get a post-doctoral position with E. Bright Wilson at Harvard. At that time, Wilson had developed a method for measuring barriers to internal rotation using microwave spectroscopy and I was still interested in internal rotation barriers. It seemed a perfect situation.

I enjoyed Harvard scientifically. Wilson's personality was very different from Pitzer's. Although he was born in Tennessee, he personified the New England virtues of upright integrity and serious concern about all aspects of life. His disapproval of superstition in all forms was well-known; none of us would dare mention in his presence the gremlins we all suspected inhabited his microwave spectrometer. Wilson above all was a fine, decent, caring person who wanted the best for his students.

The atmosphere in Mallinkrodt Laboratory at Harvard was somewhat different from that of Lewis Hall at Berkeley. Perhaps it was because the graduate system and expectations for graduate students were different. At that time, a student was expected to complete his Ph.D. at Berkeley in three years while at Harvard it took many students five or even more years. Compared with the laid-back Berkeley graduate students of my day, Harvard students seemed intense and often eccentric. The big exception was Dudley Herschbach, who was modest, relaxed, and friendly, and the most brilliant intellect I had encountered in someone my own age.

In those days faculty hiring was done with few formalities. Somewhat out-of-the-blue, I got an offer to come back to Rice as an Assistant Professor. The prospect of returning to a warm climate and familiar surroundings full of many happy memories was delightful and with no negotiations I happily accepted.

I inherited George Bird's graduate students and his microwave spectrometer, which was more sensitive than Wilson's. Of these two strokes of good luck, Bird's students proved the greater treasure. My very first student was Jim Kinsey. He accomplished so much in the first year that I was at Rice that he graduated. The work we did together on the microwave spectrum of ClO_2 and the treatment of fine and hyperfine structure set me up for a productive period of studying the spectra of stable free radicals.

I have remained at Rice from 1958 until today. In my professional and research career, I have played a variety of roles and worked in several areas of

Physical Chemistry, too varied to describe further. A great deal of my research has been collaborative involving other principals both at Rice and elsewhere. I have enjoyed quite a few very pleasant research associations over the years. Outside Rice I have collaborated with C. A. Coulson, Roger Kewley, Takeshi Oka, Ken Evenson, John Brown, Eizi Hirota, Shuji Saito, Anthony Merer, Wolfgang Urban, Harry Kroto and Leon Phillips. Among the Rice Faculty, I have enjoyed collaborations with John Kilpatrick, Frank Tittel (for the last 25 years), Phil Brooks, Rick Smalley, Graham Glass and Bruce Weisman. The Nobel Prize in Chemistry was awarded to Rick Smalley, Harry Kroto, and myself for the fruits of one of these collaborations, the discovery of the fullerenes.

I must point out that we do not claim this discovery is ours alone. James Heath and Sean O'Brien, who were graduate students at the time, have equal claim to this discovery. Both Jim and Sean were equal participants in the scientific discussions that directed the course of this work and actually did most of the experiments. The early experiments that Sean and Jim did not do were carried out by Yuan Liu and Qing-Ling Zhang. At an early stage, Frank Tittel became involved in this work. At a later stage, F. D. Weiss and J. L. Elkind did the shrink wrap experiments, which were among the strongest evidence for the fullerene hypothesis.

DAWN OF THE FULLERENES: EXPERIMENT AND CONJECTURE

Nobel Lecture, December 7, 1996

by

ROBERT F. CURL JR.

Chemistry Department and Rice Quantum Institute,
Rice University, Houston, Texas 77005, USA

CONJECTURE

Several individuals in widely separated parts of the world had envisioned the class of carbon cage compounds we now know as the fullerenes, in particular C_{60}, long before we began our work. The earliest reference in the area appears to be the somewhat whimsical proposal by Jones[1] that giant carbon cage molecules, which we would now call giant fullerenes, might be synthesized by introducing defects in the graphitic sheet to allow them curve and close. He believed that such molecules should exhibit unique properties such as having a very low density. He realized some time later that the required defects would be pentagons.[2]

Apparently the first person to imagine the truncated icosahedron isomer of C_{60} shown in Fig. 1 was Osawa.[3,4] Osawa conceived of the C_{60} structure while meditating on the structure of corannulene ($C_{20}H_{10}$), which has a cent-

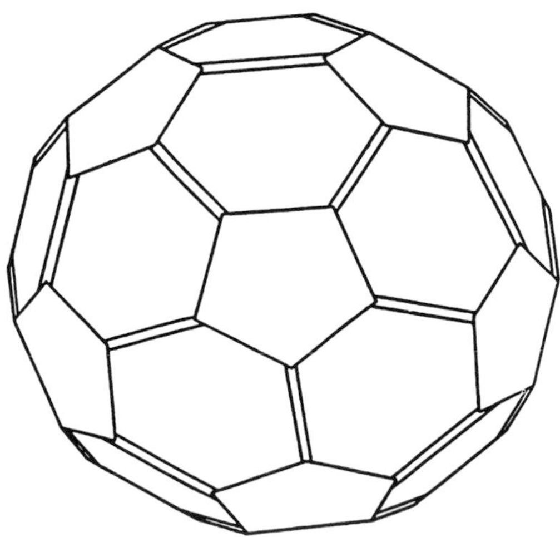

Fig. 1. Truncated icosahedron C_{60} with dominant Kekulé structure.

ral pentagon of carbon atoms surrounded by five hexagons, when he glanced at his son's soccer ball (football) and recognized the same pattern. Shortly thereafter a Hückel treatment of the π electrons was carried out in the Soviet Union.[5] and repeated[6] independently some years later in the USA as part of a larger article. The Hückel calculation was repeated again by Haymet[7] in his discussion of the stability of the molecule.

The synthetic organic chemist, Orville Chapman, took the challenge of the synthesis of truncated icosahedron C_{60} seriously in the early 1980's, obtained funding from the National Science Foundation for this purpose, and set to work with several graduate students on its total synthesis.[8] This project is indeed a tremendous challenge to the conventional methods of organic synthesis, and to date no such total synthesis of C_{60} has been completed.

These conjectures concerning C_{60} were based upon good chemical intuition backed up by approximate quantum chemical calculations. The conclusion of these conjectures was that truncated icosahedron C_{60} would be a chemically stable compound that, once prepared, could be handled much as any common substance.

EXPERIMENT

When we started our experiments on carbon clusters in late August 1985, we were completely ignorant of the conjectures just related. The purposes of our carbon cluster project were to determine whether the sort of carbon chain compounds such as HC_7N found[9,10] by radioastronomy in the interstellar medium could be synthesized by mixing carbon vapor with a suitable reagent such as ammonia and to find the conditions needed for a study of the low temperature electronic spectra of carbon chain compounds using resonance-enhanced two photon ionization. The spectroscopic work on carbon chains was motivated by the proposal made by Douglas[11] that electronic absorptions of long-chain carbon molecules, C_n (n=5–15), are the source of the diffuse interstellar bands.[12] The work on formation of carbon chain compounds was ultimately published,[13,14] but the carbon chain spectroscopy was never really begun.

I won't review in detail here the experiments we carried out in August and September of 1985 which resulted in our proposal[15] that truncated icosahedron C_{60} (buckminsterfullerene) is formed spontaneously in condensing carbon vapor. Accounts of this have been given[16,17] by two of the five people involved and in two books on the discovery of the fullerenes.[8,18] The authors of these books consulted all five of us and, in my opinion, made the best effort possible to use the recollections of those involved to recreate these events.

However, in order to understand what we did it is necessary to learn something about the apparatus[19–21] invented by Richard Smalley to investigate compounds and clusters formed from refractory elements. He and his students used it to investigate the high resolution electronic spectra of a number of metal dimers,[20–23] copper trimer,[24] and SiC_2.[25] In 1984, Frank Tittel,

Smalley, and I with our students began investigating semiconductor clusters using this apparatus.

The heart of the experiment is the laser vaporization supersonic molecular beam source. The source went through several design variations primarily to accommodate the physical form of the available sample material to be vaporized. For these experiments, it was a disk vaporization source because the semiconductor samples that we had been investigating were more readily available in sheet form. Fig. 2 shows a cross-section of this pulsed molecular beam source. In operation, the solenoid-actuated pulsed valve was fired to release through the 1 mm orifice a pulse of He gas over the sample lasting somewhat less than a millisecond. The backing pressure could be as high as 10 atm. At some point during the gas pulse usually near its middle, the Q-switched frequency-doubled vaporization laser was fired generating a 5 nsec long pulse of green light (532 nm) with an energy of roughly 30–40 mJ. This laser was focused onto the rotating-translating graphite disk (to avoid digging pits into the sample) vaporizing a plume of carbon vapor into the gas stream. Multiphoton ionization and the subsequent heating of the resulting plasma limited the amount of material vaporized in a single shot and insured that the species initially contained in it were atoms or very small molecules such as C_2 and C_3.

The material vaporized was caught up in the helium gas flow, was mixed with it and cooled by it. The cooling vapor then began to condense into clusters. The extent of clustering could be varied by changing the backing pressure, the timing of the firing of the vaporization laser with respect to the center of the gas pulse, and by varying the length and geometry of the channel downstream from the vaporization point. In the configuration shown in Fig. 2, an integration cup has been added to the end of the gas channel to provide more time for clustering and reaction before supersonic expansion.

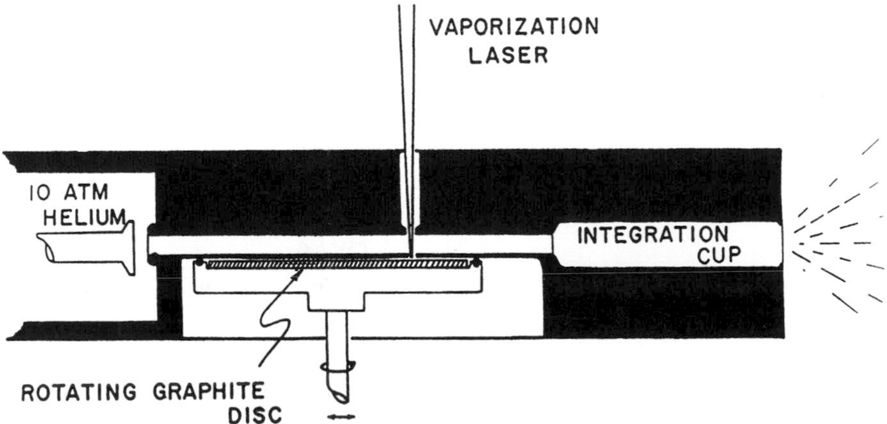

Fig. 2. Laser vaporization source for producing molecular beams of clusters of refractory materials. The integrating cup can be removed. For the carbon experiments, the target is a graphite disk which is rotated slowly to provide fresh vaporization surface. The vaporization laser, a 5 nsec pulse at 532 nm of about 30–40 mJ, is focused onto the surface of the graphite. The pulsed nozzle passes helium over this vaporization zone.

After clustering, the gas pulse was expanded supersonically through a nozzle into a large vacuum chamber (see Fig. 3). Because this expansion is essentially an adiabatic reversible expansion, the temperature of the species in the gas drops from somewhat above room temperature to a few K. After a few dozen expansion nozzle diameters, collisions between particles in the expansion jet cease resulting in gas stream with a narrow, highly directional velocity distribution. The resulting jet of cold clusters can be skimmed into a molecular beam and interrogated by mass spectrometry. Mass spectrometric detection labels the species by mass, which is a particularly important consideration in the study of clusters where a wide distribution of cluster sizes is always found. In addition, mass spectrometric detection provides high sensitivity and permits extensive control of the trajectory of the cluster ions. Consequently, a variety of methods for manipulating and probing the cluster ions were developed and used in this work. These more sophisticated methods will be described later at the appropriate point. For the present, we focus on the simplest mass spectrometry.

For mass spectrometric detection, the skimmer at the end of the large chamber forms a molecular beam from the portion of the jet moving directly away from the nozzle. This beam is passed through a differential pumping chamber and another skimmer and thence between the plates of the ion extraction field. In the most usual experiments, this field is a DC field and ions are produced by a pulsed ionization laser (normally an ArF 193 nm excimer laser, 6.4 eV, pulse length about 10 nsec). Once produced the ions are accelerated by the DC field into the drift tube of a time-of-flight mass spectrometer. Because all ions of unit charge receive the same energy, ions of greater

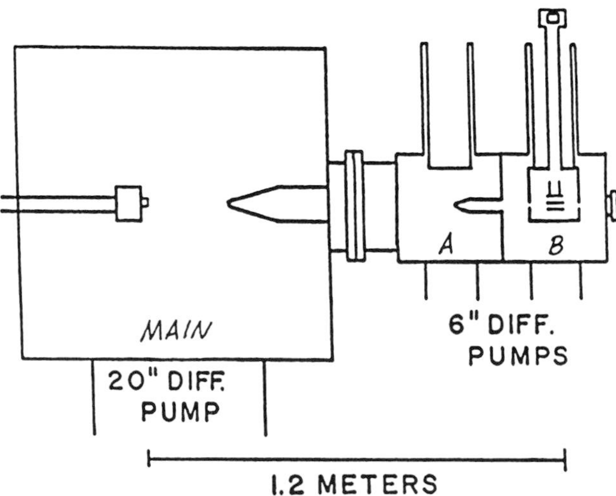

Fig. 3. Molecular beam photoionization time-of-flight mass spectrometer.

mass reach a final velocity which is less than ions of lesser mass. Therefore, the mass of the ion in the acceleration region is determined by its time of arrival at the ion detector. By plotting ion detector signal vs arrival time, a mass spectrum of the cluster distribution is obtained. Often these cluster distributions are quite colorful exhibiting "magic number" cluster sizes where a peak is more prominent than its neighbors by a factor of perhaps 2. Fig. 4 shows a cluster distribution for carbon similar to that obtained previously by Rohlfing, Cox and Kaldor[26] in a essentially the same apparatus. In Rohlfing, Cox, and Kaldor distribution, C_{11}^+, C_{15}^+, C_{19}^+, and C_{60}^+ might be called magic numbers. The discovery of the fullerenes began when we found that under the right conditions, C_{60}^+, could become a super magic number far more prominent than its neighbors.

When the carbon system was investigated in September 1985, major fluctuations in the prominence of the C_{60} peak compelled us to examine this mass region more carefully under a variety of clustering conditions. It was found that the relative prominence of the C_{60} mass peak could varied from roughly twice the intensity of its nearest neighbors to about 50 times the intensity of its nearest neighbors. This led us to propose[15] that the very prominent C_{60} cluster we had observed has a closed, highly symmetric, carbon cage structure in the form of a truncated icosahedron. Since our inspiration to look for a spherical cage structure came from the geodesic domes of R. Buckminster Fuller, we dubbed this molecule, buckminsterfullerene.

The buckminsterfullerene proposal rested on the single experimental observation that carbon vapor condensation conditions could be found where the intensity of the mass spectrum peak of the C_{60} in the carbon cluster beam

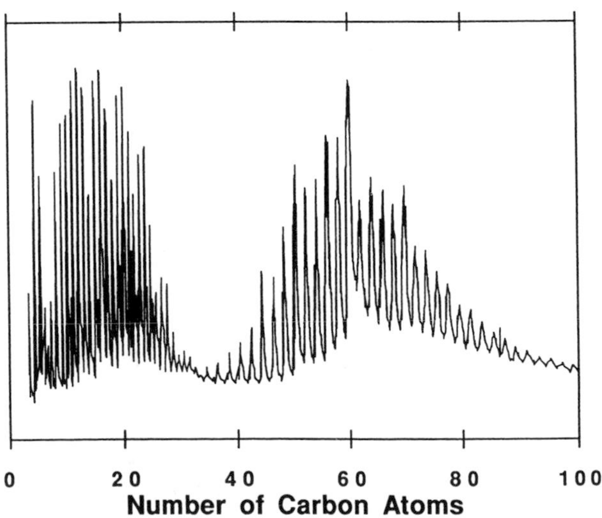

Fig. 4. Carbon cluster distribution observed under mild clustering conditions. This distribution is similar to that obtained by Rohlfing, Cox, and Kaldor.[26]

was many times the intensity of any of its near neighbors in mass as shown at the top of Fig. 5. Was this proposal a lucky guess or is this single observation taken in context sufficient to prove that the prominent C_{60} peak in Fig. 5 is the truncated icosahedron isomer shown in Fig. 1? Our claim has always been that the situation is much closer to proof than conjecture.

Note in Fig. 5 that the relative prominence of C_{60} depends upon the clustering conditions. The C_{60} peak becomes more prominent when more time is given for high temperature (room temperature and above) collisions between the carbon clusters. This immediately indicates that whatever isomer(s) of C_{60} are responsible for its prominence must be "survivors" that are relatively impervious to chemical attack.

There are probably millions of plausible isomers of C_{60} which differ in chemical connectivity. Most of these millions of C_{60} isomers will be obviously chemically reactive with dangling carbon bonds and thus unable to survive

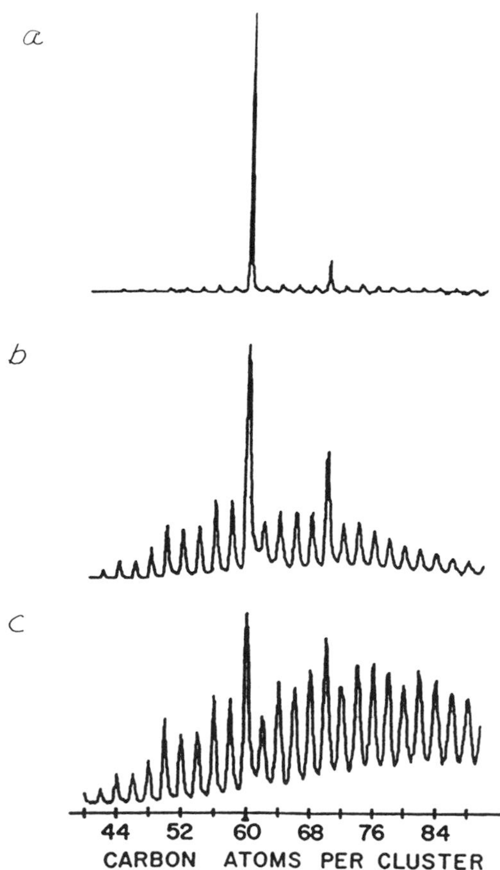

Fig. 5. Mass spectrum of the C_{60} mass region showing C_{60} growing prominent. Panel c corresponds to a situation where the helium backing pressure is low and there is no integration cup. Panel b was obtained with full helium backing pressure with vaporization in the center of the gas pulse, but no integration cup. Panel a has the integration cup in addition to full helium backing pressure.

chemical attack. The role of chemical attack is manifested in the fact that in all three distributions in Fig. 5 only even carbon clusters are observed in contrast to observations not shown here but shown in Fig. 4 of the region below 25 carbon atoms where both even and odd clusters are observed with comparable intensities. It is possible to find clustering conditions where both even and odd carbon clusters are observed with comparable intensity near C_{60}; these conditions correspond to much less time for chemical reaction than for any of the mass spectra shown in Fig. 5. Fig. 6 "control" shows a distribution informediate between Rohlfing, Cox and Kaldor and those observed with nearly equal intensity odd clusters. Thus even in the bottom panel of Fig. 5 the observation of only clusters with even carbon numbers is evidence that already all the clusters in the region must have some special structures that are less susceptible to chemical attack than a typical dangling bond isomer. The obvious explanation from our present viewpoint is that these clusters are all closed carbon cage structures (fullerenes) also. In September 1985, we recognized, without the fullerene concept, that the even cluster distribution probably reflected isomers of reduced reactivity compared with the odd clusters.

Thus we believed that the very prominent C_{60} peak in the top panel of Fig. 5 could only be explained by a single isomer of C_{60} remarkably impervious to chemical attack. A readily imaginable alternative explanation would be in terms of a C_{60} isomer that is much easier to photoionize than its neighbors by the 6.4 eV ArF ionization laser employed. However this explanation ignores the clear increase of the prominence in the C_{60} signal when more time is allowed for chemical reaction. Thus an explanation for the prominence of C_{60} based on its easier photoionization does not take into account the obvious reduced chemical reactivity of C_{60} compared with its neighbors. Further evidence against an explanation based upon photoionization efficiency was obtained in later experiments[27] which demonstrated a similar prominence of C_{60}^+ upon photoionization with a 7.9 eV F_2 excimer laser.

The truncated icosahedron form of C_{60} is clearly a special structure which should be chemically very stable. It has no dangling bonds with the valences of every atom satisfied. The pattern of double and single bonds depicted in Fig. 1 is just one of 12,500 possible Kekulé structures[28] (but it has proved to be the dominant one). By symmetry, every atom is equivalent so that there is no specific point of chemical attack. Strain is introduced in curving the intrinsically planar system of double bonds into a spherical shape, but the strain is symmetrically and uniformly distributed over the molecule thereby again avoiding a weak point for chemical attack. There is obviously no other structure with this high degree of symmetry, and very little reason to fear that another structure could be found that offers this unique combination of advantages. For these reasons, and in spite of the fact that it seems very counterintuitive for this high symmetry, low entropy molecule to form out of the chaos of carbon vapor condensing at high temperature, we have never really considered the assignment of the prominent C_{60} peak in the mass spectrum to the buckminsterfullerene structure to be a guess.

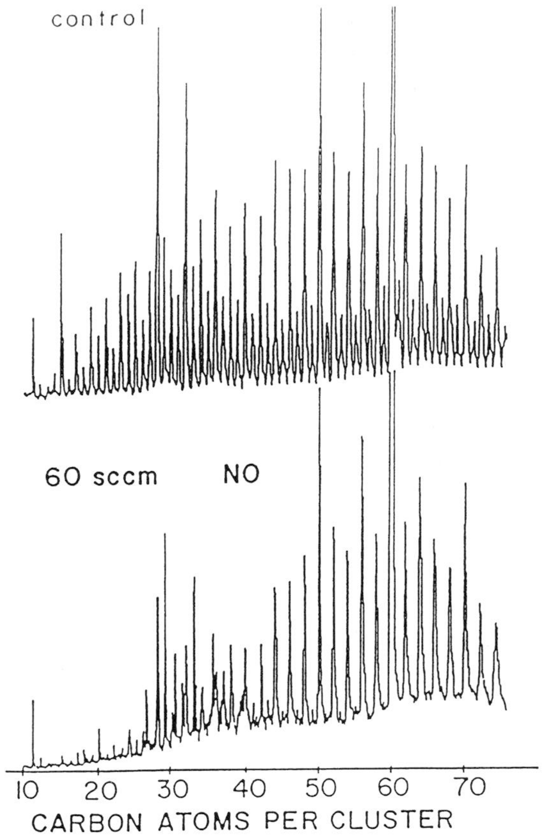

Fig. 6. Reaction of carbon clusters with NO. The upper control mass spectrum was obtained without added NO. In the lower spectrum roughly 1 Torr of NO was added to the gas stream in the fast-flow reaction tube. It is clear that the odd clusters react with NO. The various products of these reactions are not resolved, but contribute to the elevated baseline seen in this spectrum.

In science, more proof is always demanded. Later experiments[27,29] demonstrated that the special prominence of C_{60} was not the result of some special preference for the $C_{60}{}^{+}$ ion upon photofragmentation and that C_{60} can be made specially prominent in both the residual cations and anions (a residual ion being an ion that is formed in the vaporization plasma and survives the expansion process).

CONJECTURE – THE FULLERENE HYPOTHESIS

We soon learned from Haymet's paper[7] of Euler's rule[30] stating that a solid figure with any even number, *n*, of 24 or more vertices could be constructed with 12 pentagons and $(n\text{-}20)/2$ hexagons. This immediately provides an explanation in terms of such carbon cage molecules of the even cluster distribution that appears at carbon numbers above 30 in the mass spectrum, as these molecules would have no dangling bonds and would thus be relatively unreactive. These spheroidal carbon cage carbon molecules consisting only of pentagons and hexagons were given the generic name of fullerenes.

However, in contrast with the truncated icosahedron explanation for the prominent C_{60} peak, this conclusion has always seemed to me to be much more conjecture, however plausible. The next few years of our lives were devoted to testing experimentally this fullerene hypothesis and finding that it passed every test.

EXPERIMENT – REACTIVITY AND PHOTOFRAGMENTATION

It is possible to inject chemical reagent gases into the cluster stream prior to the expansion and then to observe reaction product ions in the mass spectrum.[31,32] A reaction tube was added to the end of the cluster source and various reagents, such as NO, SO_2, NH_3, H_2, CO and O_2 were injected into the gas stream.[33] It was possible to obtain a mass distribution without added reagent that showed both odd and even carbon number peaks with the odd carbon number peaks about half the intensity of the even ones. When a reagent such as NO or SO_2 was added, the odd carbon number peaks disappeared, but the even carbon peaks with 40 atoms or more remained unreactive as would be expected if they were fullerenes having no dangling bonds. The distributions observed in this experiment are shown in Fig. 6. Note that the odd clusters which are believed to have dangling chemical bonds are much more reactive than the even clusters.

A series of photofragmentation experiments were carried out on the carbon cluster ions hypothesized to be fullerenes using a tandem time-of-flight mass spectrometer.[34] The apparatus is shown in Fig. 7. In these experiments,

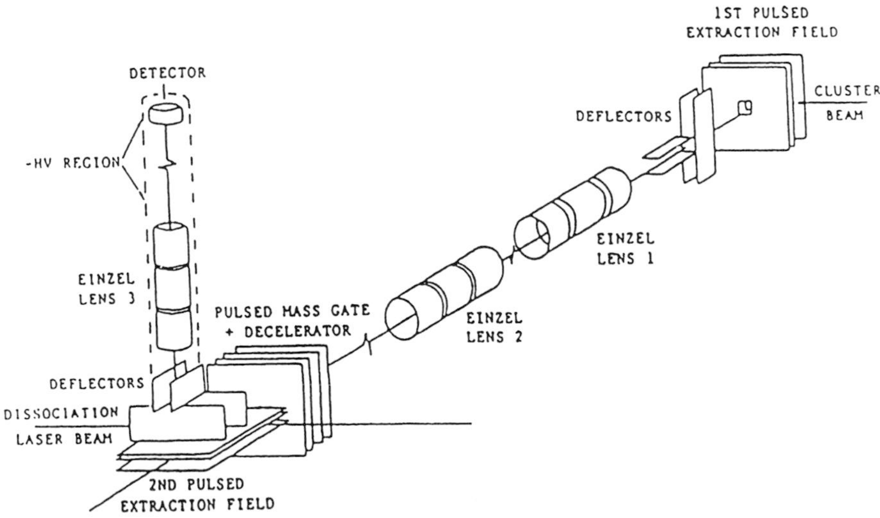

Fig. 7. Tandem time-of-flight mass spectrometer. The molecular beam containing residual ions from the vaporization plasma enters the extraction region of the primary mass spectrometer where a 2000 V pulse is applied across the grids. The deflectors remove the molecular beam velocity. The einzel lenses focus the ion beam. Ions are selected by the mass gate and then fragmented by the laser and analyzed by the second time-of-flight mass spectrometer.

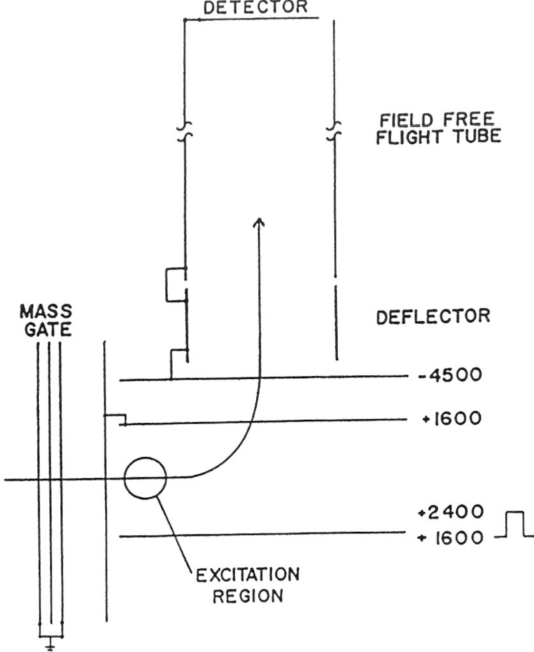

Fig. 8. Detail of the photofragmentation region of the tandem TOF apparatus showing the mass gate, laser excitation region, extraction optics and flight tube.

a single carbon cluster ion was mass selected and then interrogated by a photofragmentation laser as shown in the apparatus detail, Fig. 8. It was then accelerated into a second time-of-flight drift region and the mass of its ionic fragments determined. The photofragmentation pattern of C_{60}^+ is shown in Fig. 9. As can be seen, the fragmentation pattern corresponds to the loss of an even number of carbon atoms down to C_{32}^+ where it changes abruptly to produce ions containing about 20 atoms. We believe that C_{60}^+ is losing an even number of carbon atoms in a few step process with the fullerene cage of the ion reclosing upon loss of the neutral even-number fragment. The abrupt change in pattern at C_{32}^+ takes place because the strained small fullerene can no longer close upon carbon loss and instead a large neutral fragment is shaken off when the strain energy is suddenly released upon opening the cage.

The energies of many of the fullerenes have been calculated at the STO-3G/SCF level of theory by Scuseria.[35] and are shown in Fig. 10. It should be noted that for fullerenes of size C_{28} and larger there is more than one fullerene structure for a given number of carbon atoms.[36] For example, there are [37,38] 1812 fullerene isomers (i.e. cages containing 12 pentagons and 20 hexagons) of C_{60} alone. The energies plotted in Fig. 10 are for the lowest energy fullerene isomers found. Truncated icosahedron C_{60} and D_{5h} symmetry C_{70}, which are respectively the lowest energy fullerene isomers of 60 and 70 carbon atoms, are local minima in the energy curve, and this can be strikingly seen in these energetics. We label the buckminsterfullerene isomer of C_{60}, C_{60}^{BF}.

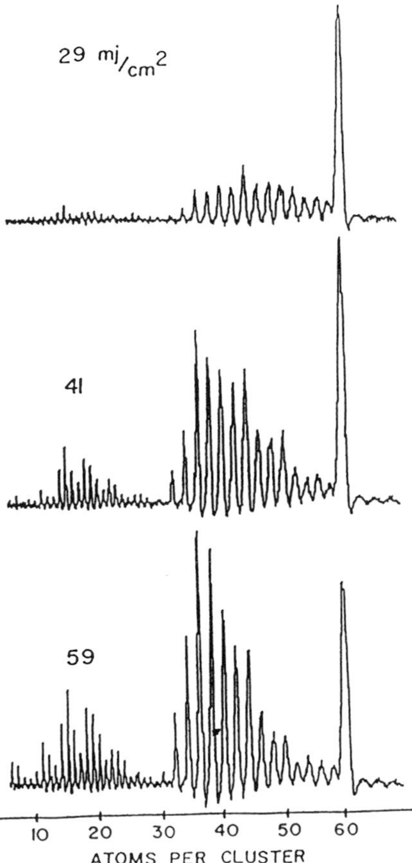

Fig. 9. Photofragmentation pattern of C_{60}^+ showing loss of even number of carbon atoms and break off at C_{32}^+.

When these energies are combined with the bond dissociation energy of C_2 [39] of 6.21 eV, the overall energy change in fragmentation can be calculated. Thus the overall energy change in

$$C_{62} \rightarrow C_{60}^{BF} + C_2 \tag{1}$$

is ΔE= 0.7 eV while for the loss from C_{60}^{BF}

$$C_{60}^{BF} \rightarrow C_{58} + C_2 \tag{2}$$

ΔE = 11.2 eV. The activation barrier for the fragmentation of C_{60} cannot be less than the fragmentation energy of 11.2 eV. In order to have substantial fragmentation of C_{60}^{BF+} in the few μsec available in Fig. 9, I estimate that about 100 eV must be deposited into the C_{60}^{BF} ion.

If the activation barriers to fragmentation follow the energetics, one would expect that the special stability of C_{60} would be apparent in the fragmentation pattern. Fig. 11 shows fragmentation of some larger clusters when the sample is irradiated just before the acceleration region of the second TOF as shown in Fig. 9. Only a few μsec are available for fragmentation after irradiation and

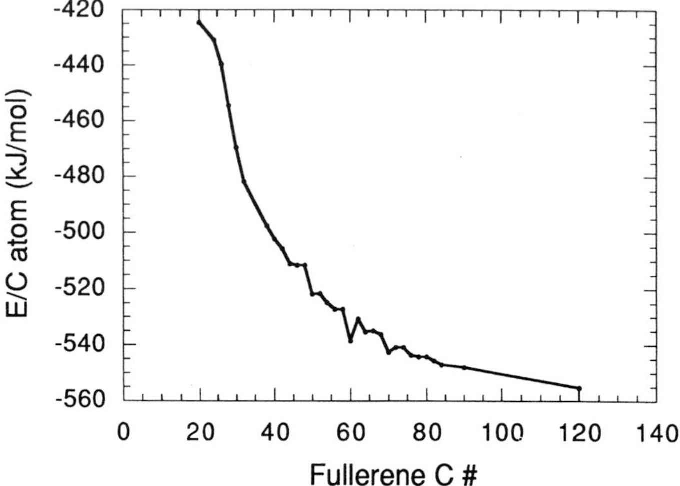

Fig. 10. Energy per carbon atom relative to atomic C as a function of fullerence cluster size at the STO-3G/SCF level at the MM3 optimized geometry for the lowest energy fullerence structure for each size. This figure is based upon the calculations of R. L. Murry, D. L. Strout, and G. E. Scuseria.[35]

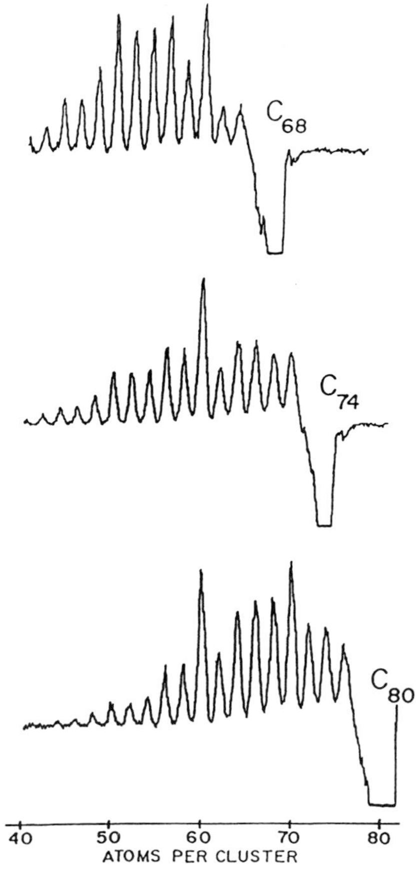

Fig. 11. ArF (15 mJ cm^{-2}) fragmented large even carbon clusters. There is no difference in the fragmentation pattern of the large clusters. C_{60} and C_{70} are slightly favored.

before acceleration and analysis, and C_{60} and C_{70} are only slightly special fragments. However, if the irradiation is carried out in the *first* TOF extraction region just before extraction and the ions which have the right time-of-flight for the initial species of Fig. 12 are permitted through the mass gate, the pattern is much different. Here the sequence is that the large ion is irradiated and accelerated before it fragments; it then has time to fragment all the way to the second extraction region which corresponds to a time of about 120 μsec.

In Fig. 12, it is clearly seen that C_{60} is quite prominent. The original ion is less energetic than is the case of the short-term fragmentation and thus the fragmentation pattern is much more sensitive to the fragmentation energetics. The relationship between the long time fragmentations of Fig. 12 where $C_{60}{}^{+}$ is very prominent to the short time fragmentations of Fig. 11 where $C_{60}{}^{+}$ is less prominent can be explained if one assumes that the activation barriers to ring rearrangements on the surface of the fullerene ions are much less than the activation barrier to fragmentation of a typical fullerene ion.

Assume that a fragmentation which leads to a $C_{60}{}^{+}$ fullerene takes place. It is unlikely that this C_{60} ion has the buckminsterfullerene structure. Probably several rerrangements of this ion must take place before the especially stable $C_{60}{}^{BF}$ structure is found. When long time fragmentation is investigated, there is sufficient time for a number of lower activation energy ring rearrangement processes to take place before fragmentation because the energy deposited in the original ion is relatively small compared with the short time fragmentation energy. These ring rearrangements find the low energy buckminsterfullerene structure which is very hard to fragment. With the short time fragmentation, far more energy is deposited in the original ion and fragmentation occurs before the low energy buckminsterfullerene structure is found.

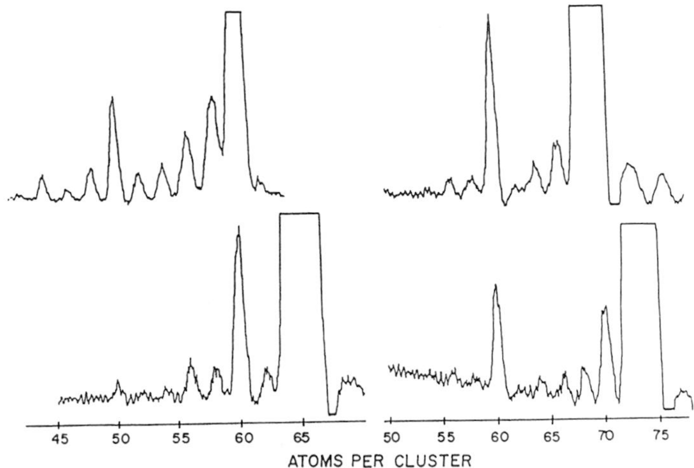

Fig. 12. Metastable TOF mass spectra for 60, 66, 70, and 74 atom clusters. The clusters were irradiated 1 μsec before the first extraction pulse with 15 mJ cm^{-2} ArF. The ions were mass gated at the time appropriate to the parent ions listed above and analyzed in the second TOF mass spectrometer. The travel time to the second mass spectrometer was approximately 120 μsec. The special prominence of C_{60} and C_{70} is clear.

Figs. 9, 11, and 12 taken together provide striking evidence that cations being examined are structurally related to each other and to C_{60}^{BF}. It appears that fragmentation takes place with the preservation of the cage structure until the break-off point at C_{32} is reached.

CONJECTURE – THE EXISTENCE OF ENDOHEDRAL COMPLEXES

The fullerenes are hollow. Buckminsterfullerene has a cavity almost 4 Å in diameter that is capable of holding any atom of the periodic table. It seemed to us that it might be possible to introduce a foreign atom into the central cavity to produce an endohedral adduct. We recognized that bulk samples of such materials, if they could be obtained, might have many unusual and potentially useful properties. For this narrative, the important point is that such an endrohedral atom would be difficult to dislodge.

EXPERIMENT – ENDOHEDRAL METALLOFULLERENES AND "SHRINK WRAPPING"

Success in forming adducts with a single lanthanum atom was almost immediately achieved.[40] In these experiments, a low-density graphite disk was soaked in a water solution of $LaCl_3$, dried and used as target for laser vaporization. The mass spectrum at low ionization laser fluence showed many peaks from both pure carbon and carbon-lanthanum adducts, but when the ionization laser power was turned up somewhat so that the least stable species would photofragment, all bare cluster peaks except for C_{60} and C_{70} disappeared, but clusters with one lanthanum atom at every even carbon number remained. There were no clusters remaining with more than one La atom. Thus under laser fluences capable of destroying the less stable carbon clusters, one and only one lanthanum atom stuck. This is a strong indication that the lanthanum atom is inside the cage.

It was found that endohedral metallofullerenes containing the alkali atoms K and Cs could be readily formed. This led to a unique way to test the fullerene hypothesis by "shrink-wrapping". A series of photofragmentation experiments were carried out[41] in a fourier transform ion cyclotron resonance cell on $C_{60}K^+$ and $C_{60}Cs^+$. A supersonic beam of cluster ions is prepared as described above and injected into the ion cyclotron resonance cell where they can be trapped for several minutes. By applying a range of rf frequencies to the cell, the orbits of almost all ions but the desired one, $C_{60}K^+$ (or $C_{60}Cs^+$), can be excited thereby driving the unwanted ions from the cell. Photofragmentation experiments can then be carried out on the remaining ions.

From our previous experiments on photofragmentation and $C_{60}La^+$, we expected that at low laser fluences the ions would lose C_2, C_4, C_6 while retaining the metal. If the metal is in the cage as proposed, the cage will become increasingly strained upon loss of neutral carbon because it is shrinking down upon the resistant metal core. A point will be reached where the cage will

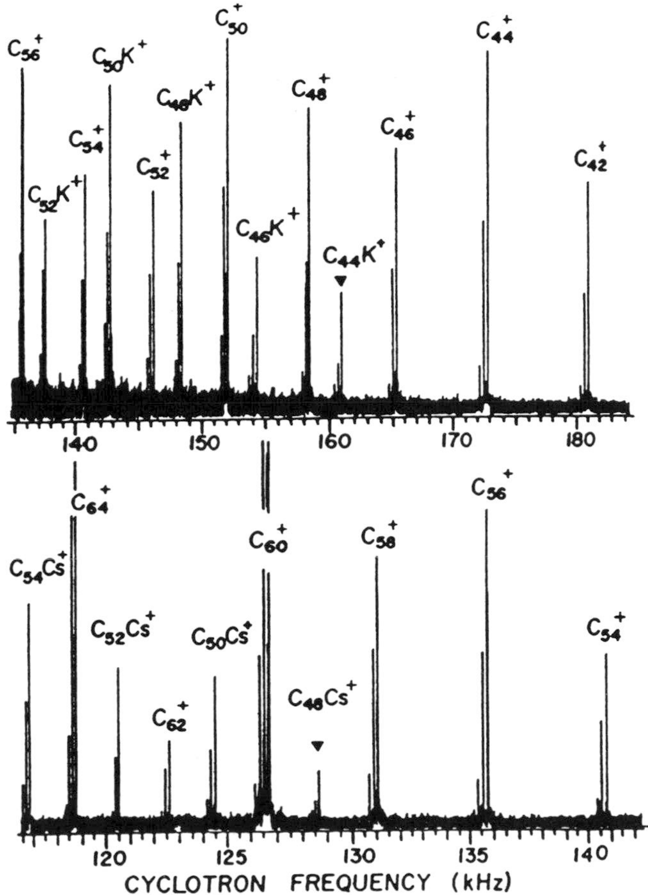

Fig. 13. The low mass portion of the fragment ions produced by intense laser excitation of $C_{60}K^+$ (panel A) and $C_{60}Cs^+$ (panel B) in the FT-ICR trap. The clusters containing carbon only seen in these spectra arise from fragmentation of empty fullerene ions simultaneously trapped with the metal species.

break releasing the metal. This point will depend upon the size of the metal ion and will therefore be reached for larger clusters in the case of $C_{60}Cs^+$ than in the case of $C_{60}K^+$. Furthermore, the cage breaking point can be roughly estimated from the van der Waals radii of the alkali ion and the carbon atoms. Fig. 13 shows the photofragmentation results. The even carbon loss breaks off for $C_{60}K^+$ at $C_{44}K^+$ and for $C_{60}Cs^+$ at $C_{48}Cs^+$ which agrees well with predictions from the van der Waals radii. We can conceive of no explanation for these observations other than that we were observing fragmentation of endohedral fullerene complexes.

CONJECTURE – C_{70} HAS D_{5H} SYMMETRY

In almost all mass spectrometer carbon cluster distributions where C_{60}^+ (or C_{60}^-) is prominent, C_{70}^+ (or C_{70}^-) is usually the next most prominent ion in the C_{40}–C_{90} mass range. Therefore it is likely to have a somewhat special structure. A plausible guess for the structure of C_{70} seemed to be[40] one in which a band of five hexagons was added around the equator of C_{60}. To form such a structure (see Fig. 14), choose an orientation of C_{60} so that pentagons are at the north and south poles; then cut the C-C bonds connecting the northern and southern hemispheres, separate them, rotate one hemisphere by 1/10 turn with respect to the other and add a string of 10 carbon atoms at the equator to rejoin the two hemispheres.

Both the structures of C_{60}, buckminsterfullerene, and D_{5h} C_{70} have no abutting pentagons. Haymet suggested[7] that such connected rings would be destabilizing. Shortly thereafter, Schmalz *et al.*[42] pointed out that abutting pentagons necessarily involve destabilizing, antiaromatic eight atom conjugated π electron circuits around the ring making structures with abutting pentagons less stable. Both Kroto[43] and Schmalz *et al.*[44] proposed that D_{5h} C_{70} is the smallest cage structure larger than C_{60}^{BF} without abutting pentagons. Both reported that they had made a diligent, but not exhaustive, search of structures between C_{60} and C_{70} for isolated pentagon cages. Subsequently, Liu *et al.* were able to prove[45] this conjecture. Kroto[43] showed that the "magic number" mass spectrometer peaks containing fewer than 60 carbon atoms corresponded to structures with the minimum number of abutting pentagons. These considerations became the "isolated pentagon rule" which states that in the stable fullerenes the pentagons are isolated.

Although it was strongly supported by these theoretical considerations, the conjecture that the structure of C_{70} has D_{5h} symmetry could never be verified experimentally by molecular beam mass spectrometry. The proof of this structure had to wait until the production[46] of macroscopic mixtures of C_{60} and C_{70} permitted the separation of C_{60} and C_{70} and the observation of the ^{13}C NMR spectrum of C_{70}.[47]

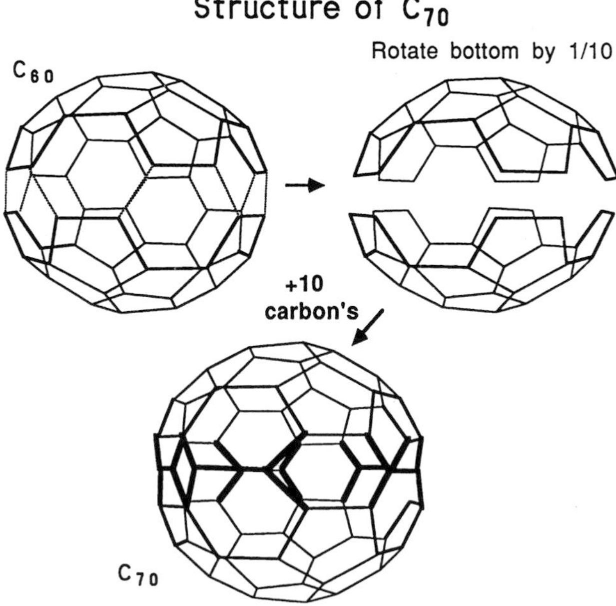

Fig. 14. Relationship of the proposed structure of C_{70} to that of C_{60} (buckminsterfullerene).

CONJECTURE – C_{60} MIGHT BE THE CARRIER OF THE DIFFUSE INTERSTELLAR BANDS

Not all conjectures can be expected to be correct. When we wrote our initial paper[15] on C_{60}, we were inspired with wildly imaginative thoughts about the potential significance of formation of this remarkable molecule in all areas of chemistry and astrophysics. Since we had been thinking of carbon chain molecules as possible carriers of the diffuse interstellar bands (dib's),[11] it was natural to consider C_{60} as offering a potential explanation of the diffuse interstellar bands, and we proposed that as a possibility.

C_{60} seemed an attractive candidate for several reasons. First, any molecular carrier of the dib's must be large enough so that it is not dissociated by absorbing photons of energies of up to 13.6 eV. In a large molecule such as C_{60}, when a photon is absorbed internal conversion rapidly brings the molecule back to its ground electronic state. If there is more energy available in the photon than is needed for unimolecular dissociation, then dissociation competes with vibrational reradiation of the excess energy in the infrared. With a dissociation threshold of 11 eV and so many vibrational modes, unimolecular decay even with 13.6 eV of excitation is a very slow process and loses completely to infrared reradiation. Second, there are not many dib's known, and therefore, any proposed carrier must give only a few bands and should not be one of a large class of equally attractive candidates. The high symmetry of C_{60} suggests only a few bands, and, while there is a fullerene family, C_{60} is often uniquely prominent. Any carrier should consist of the more cosmically abundant elements. Lastly, the diffuseness of the bands could come from mixing of the spectroscopically active excited electronic state with ground or lower state levels. Subsequent to our original proposal, it became obvious that C_{60} would be likely to be photoionized or to react with H atoms so that $C_{60}{}^+$ and slightly hydrogenated derivatives of C_{60} would be more attractive.[48-50]

Eventually, it proved possible[51] to obtain a portion of the visible electronic spectra of C_{60} and C_{70} using resonant two-photon ionization, and these spectra demonstrated conclusively that neither neutral C_{60} nor neutral C_{70} have absorptions that correspond to the known diffuse interstellar lines in the same region. These observations alone do not rule out $C_{60}{}^+$ or perhaps protonated C_{60} derivatives as carriers of the diffuse interstellar bands; however, the matrix isolation spectrum of $C_{60}{}^+$ seems to rule it out[52] as a diffuse interstellar band carrier.

CONJECTURE – SOOT IS FORMED FROM SPIRALING ICOSAHEDRAL CARBON SHELLS

The control of soot formation is of enormous practical value; consequently, the nature of soot and the processes involved in its formation have been extensively studied, and this remains an active research field. Since soot consists primarily of carbon, we naturally thought about whether the chemistry involved in fullerene formation might be also applicable to soot formation. The combustion environment principally differs from condensing carbon vapor in that large quantities of hydrogen are present in combustion. Indeed, almost as many H atoms as C atoms are present in combustion soot. Combustion soot is considered to be regions of layered sheets of large polycyclic aromatic hydrocarbons joined by disorganized regions.

It seemed to us that the fullerenes were forming in a process where small carbon clusters such as C_2 and C_3 were adding to a growing, curving sheet of five- and six-membered rings. Curvature and ultimate closure into fullerenes was brought about by some sort of ring rearrangement process that let the carbon cluster find its lowest energy, least reactive forms. The growth of polycyclic hydrocarbons (PAH) in soot formation was thought[53] to be by the creation of a reactive carbon atom on the periphery by H atom abstraction from a peripheral CH bond followed by acetylene addition to the reactive center, followed by ring closure with H atom elimination. The similarity between the process we imagined for fullerene formation and the PAH growth process in combustion soot formation seemed striking to us. The main difference seemed likely to be that hydrogen on the edges might interfere with the fullerene closure processes leading to imperfect cages. Perhaps a soot particle was the result of such imperfect growth and would resemble a spiraling fullerenic shell. Thus we proposed that a soot particle would be based upon a spiraling icosahedral shell similar to the structure shown in Fig. 15.

This suggestion was met with disfavor by some members of the soot community.[54–56] However, it did result in searches for fullerenes in flames by some of the leading combustion and soot scientists. Thus Homann[57] found C_{50}^+, C_{60}^+, and C_{70}^+ prominently in hydrocarbon flames. He suggested[58] that soot particles might be supports for fullerene growth. More recently he has proposed more elaborate and detailed models for fullerene formation[59] and

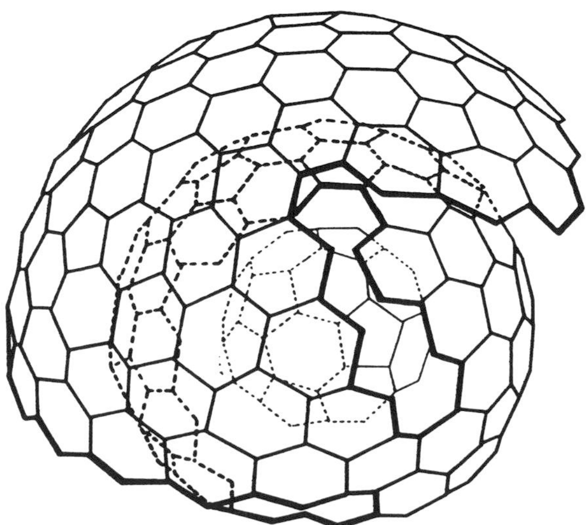

Fig. 15. Spiraling icosahedral shell model of a growing soot particle. It has almost completed its second shell of growth.

its relation to soot.[60] After C_{60} and C_{70} became available in bulk, soot scientists began looking for these species in combustion soots with success.[61,62] Under optimal conditions the yield of C_{60} plus C_{70} was as much as 20% of the soot and up to 0.5% of the carbon fed,[63] making flames a practical preparation scheme for fullerenes. Quite recently, electron microscope examination[64] of nanostructures in flame soots has revealed the whole range of fullerenic structures: fullerenes, nanotubes, and the nested fullerenes resembling Russian dolls. These observations are strong evidence that carbon has a strong propensity to form these structures in chemically quite different environments where elemental carbon forms are produced.

Thus some very special chemistry does not appear to be needed for fullerene formation. Howard has proposed[65] a detailed kinetic scheme for the formation of C_{60} and C_{70} in flames which parallels the standard mechanism for the formation of polyaromatic hydrocarbons, which are considered to be the soot precursors. Nevertheless, fullerene formation and soot formation remain to him clearly distinct processes.

The soot community remains united in the belief that, while soot might produce fullerenes as an offshoot, soot formation is a different process from fullerene formation and that soot is primarily composed of PAH's and that there is no evidence for spiraling icosahedral shells. It should be noted that it is known[66] that more than half of the soluble PAH inventory of flames contain five-membered rings and therefore curvature. Whether the curvature introduced by such five membered rings plays any role in the formation of soot particles remains an interesting open question.

CONCLUSION

In science, conjecture drives both experiment and theory for it is only by forming conjectures (hypotheses) that we can make the direction of our experiments and theories informed. If such and such is true, then I should be able to do this experiment and look for this particular result or I should be able to find this theoretical formulation. Conversely, experiment and theory drive conjecture. One makes a startling observation or has a sudden insight and begins to speculate on its significance and implications and to draw possible conclusions (conjecture).

However, not all conjectures are equally valid or useful. Thus the conjecture that C_{60} might be related to the diffuse interstellar bands was and remains wildly speculative. It has been one of my favorite speculations,[50] because it has many things going for it: carbon is injected into the medium and C_{60} forms spontaneously in condensing carbon under the right conditions, C_{60} is unique and its compounds are limited in number and only a few species should be involved in the diffuse interstellar bands, C_{60}^+ could survive the UV present in the interstellar medium, the diffuseness of the bands could come from mixing of the spectroscopically active excited electronic state with ground or lower state levels. However, conjectures should be judged by the accumulation of evidence that supports and contradicts them. To date, there has been plentiful contradictory evidence for the C_{60} connection to the diffuse interstellar bands and no supporting evidence. Primarily, this conjecture has the fatal defect that it has stimulated little productive science.

The conjecture that soot may consist of spiraling spheroidal shells of carbon belongs in a slightly different category. It was based upon a hypothesis, which was somewhat vague at the time, about how fullerenes may form. Thus, it was not a wild idea in that there was some support for the speculation. In my opinion, the conjecture that soot consists of spiraling spheroidal shells is probably wrong. However, I think it likely that there is some more subtle connection between the curvature introduced by five-membered rings and soot formation. Regardless of its validity, this conjecture has turned out to be extremely valuable, because it got the soot community, in some cases somewhat grumpily, thinking about the formation of fullerenes and other carbon morphologies in flames. As a result, fullerenes were found in flames and soot. It appears[64] that all the new carbon morphologies can be produced in flames.

Turning to the central theme, the conjecture of Jones[1,2] that carbon cage compounds might have interesting properties, and the conjecture of Osawa[3] and several others that C_{60} would be a stable, chemically interesting molecule are examples of conjectures which are correct and valuable, but which by themselves cannot be made useful. There was no way, or no easy way (remember Chapman), to proceed to further work based on the conjecture.

On the other hand, the conjecture that a new whole class of carbon cage compounds, the fullerenes, are formed spontaneously in condensing carbon vapor has led to sweeping consequences. At the time, this hypothesis seemed to be the only logical explanation of the observed carbon mass spectra distributions, but it was not self-evident. As we have seen, we tested this conjecture in a variety of experiments which always provided evidence supporting the conjecture. This pattern of repeated confirmation of expected consequences is what is expected for a correct hypothesis. In the long run, the fullerene hypothesis has proved to be spectacularly correct and it has provided the basis for a whole new branch of organic chemistry.

Finally, I believe that the conjecture that started it all, namely that truncated icosahedron C_{60} forms spontaneously in condensing carbon, scarcely belongs in the category of conjecture. The three mass spectra in Fig. 5 when coupled with the conditions under which they were obtained demand that the species responsible for the prominent peak at C_{60} must be singularly different and chemically relatively unreactive. The human mind can conceive of no other isomer of C_{60} that better fits this requirement.

BIBLIOGRAPHY

1 D. E. H. Jones, *New Scientist* **32,** 245 (1966).

2 D. E. H. Jones, in *The Inventions of Daedalus* (W. H. Freeman, Oxford and San Francisco, 1982), pp. 118–119.

3 E. Osawa, *Kagaku (Kyoto)* **25,** 854–863 (1970).

4 Z. Yoshida and E. Osawa, in *Aromaticity* (Kagakudojin, Kyoto, 1971), pp. 174–178.

5 D. A. Bochvar and E. G. Gal'pern, *Dokl. Akad. Nauk SSSR* **209,** 610–612 (1973).

6 R. A. Davidson, *Theor. Chim. Acta* **58,** 193–235 (1981).
7 A. D. J. Haymet, *J. Am. Chem. Soc.* **108,** 319–321 (1986).
8 J. Baggott, *Perfect Symmetry: The Accidental Discovery of Buckminsterfullerene* (Oxford University Press, Oxford, 1994).
9 H. W. Kroto, C. Kirby, D. R. M. Walton, L. W. Avery, N. W. Broten, J. M. McCleod, and T. Oka, *Astrophys. J.* **219,** L133–L138 (1978).
10 L. W. Avery, in *Interstellar Molecules,* edited by B. H. Andrew (D. Reidel, Hingham, MA, 1979), pp. 47–55.
11 A. E. Douglas, *Nature* **269,** 130–132 (1977).
12 E. Herbig, *Astrophys. J.* **196,** 129–60 (1975).
13 J. R. Heath, Q. Zhang, S. C. O'Brien, R. F. Curl, H. W. Kroto, and R. E. Smalley, *J. Am. Chem. Soc.* **109,** 359–63 (1987).
14 H. W. Kroto, J. R. Heath, S. C. O'Brien, R. F. Curl, and R. E. Smalley, *Astrophys. J.* **314,** 352–5 (1987).
15 H. W. Kroto, J. R. Heath, S. C. O'Brien, R. F. Curl, and R. E. Smalley, *Nature* **318,** 162–163 (1985).
16 R. E. Smalley, *The Sciences* **31,** 22–30 (1991).
17 H. W. Kroto, *Angew. Chem.* **3l,** 111–129 (1992).
18 H. Aldersey–Williams, *The Most Beautiful Molecule: An Adventure in Chemistry* (Aurum Press, London, 1995).
19 T. G. Dietz, M. A. Duncan, D. E. Powers, and R. E. Smalley, *J. Chem. Phys.* **74,** 6511-2 (1981).
20 D. E. Powers, S. G. Hansen, M. E. Geusic, A. C. Puiu, J. B. Hopkins, T. G. Dietz, M. A. Duncan, P.R.R.Langridge-Smith, and R. E. Smalley, *J. Phys. Chem.* **86,** 2556–60 (1982).
21 J. B. Hopkins, P. R. R. Langridge-Smith, M. D. Morse, and R. E. Smalley, *J. Chem. Phys.* **78,** 1627–1637 (1983).
22 D. L. Michalopoulos, M. E. Geusic, S. G. Hansen, D. E. Powers, and R. E. Smalley, *J. Phys. Chem.* **86,** 3914–3916 (1982).
23 P. R. R. Langridge-Smith, M. D. Morse, R. E. Smalley, and A. J. Merer, *J. Chem. Phys.* **80,** 593–600 (1984).
24 M. D. Morse, J. B. Hopkins, P. R. R. Langridge-Smith, and R. E. Smalley, *J. Chem. Phys.* **79,** 5316–5328 (1983).
25 D. L. Michalopolous, M. E. Geusic, P. R. R. Langridge-Smith, and R. E. Smalley, *J. Chem. Phys.* **80,** 3556–3560 (1984).
26 E. A. Rohlfing, D. M. Cox, and A. Kaldor, *J. Chem. Phys.* **81,** 3322–3330 (1984).
27 Y. Liu, S. C. O'Brien, Q. Zhang, J. R. Heath, F. K. Tittel, R. F. Curl, H. W. Kroto, and R. E. Smalley, *Chem. Phys. Lett.* **126,** 215–217 (1986).
28 D. J. Klein, T. G. Schmalz, T. G. Hite, and W. A. Seitz, *J. Am. Chem. Soc* **108,** 1301–2 (1986).
29 S. C. O'Brien, J. R. Heath, H. W. Kroto, R. F. Curl, and R. E. Smalley, *Chem. Phys. Lett.* **132,** 99–102 (1986).
30 L. Euler, *Novi corumentarii academie Petropolitanae* **4,** 109 (1752/3) (1758).
31 M. E. Geusic, M. D. Morse, S. C. O'Brien, and R. E. Smalley, *Rev. Sci. Inst.* **56,** 2123–30 (1985).
32 M. D. Morse, M. E. Geusic, J. R. Heath, and R. E. Smalley, *J. Chem. Phys.* **83,** 2293–2304 (1985).
33 Q. L. Zhang, S. C. O'Brien, J. R. Heath, Y. Liu, R. F. Curl, H. W. Kroto, and R. E. Smalley, *J. Phys. Chem.* **90,** 525–8 (1986).
34 S. C. O'Brien, J. R. Heath, R. F. Curl, and R. E. Smalley, *J. Chem. Phys.* **88,** 220–230 (1988).
35 R. L. Murry, D. L. Strout, and G. E. Scuseria, *Int. J. Mass Spect. and Ion Proc.* **138,** 113–31 (1994).
36 P. W. Fowler and D. E. Manolopoulos, *An Atlas of Fullerenes* (Clarendon Press, Oxford, 1995).
37 D. E. Manolopoulos, *Chem. Phys. Lett* **192,** 330 (1992).

38 X. Liu, T. G. Schmalz, and D. J. Klein, *Chem. Phys. Lett.* **192,** 331 (1992).
39 K.-P. Huber and G. H. Herzberg, *Constants of Diatomic Molecules* (van Nostrand, New York, 1979).
40 J. R. Heath, S. C. O'Brien, Q. Zhang, Y. Liu, R. F. Curl, H. W. Kroto, F. K. Tittel, and R. E. Smalley, *J. Am. Chem. Soc.* **107,** 7779–80 (1985).
41 F. D. Weiss, J. L. Elkind, S. C. O'Brien, R. F. Curl, and R. E. Smalley, *J. Am. Chem. Soc.* **110,** 4464–5 (1988).
42 T. G. Schmalz, W. A. Seitz, D. J. Klein, and G. E. Hite, *Chem. Phys. Lett.* **130,** 203–7 (1986).
43 H. W. Kroto, *Nature* **329,** 529–31 (1987).
44 T. G. Schmalz, W. A. Seitz, D. J. Klein, and G. E. Hite, *J. Am. Chem. Soc.* **110,** 1113–27 (1988).
45 X. Liu, D. J. Klein, T. G. Schmalz, and W. A. Seitz, *J. Compt. Chem.* **12,** 1252–1259 (1991).
46 W. Krätschmer, L. D. Lamb, K. Fostiropoulos, and D. R. Huffman, *Nature* **347,** 354–8 (1990).
47 R. Taylor, J. P. Hare, A. K. Abdul-Sada, and H. W. Kroto, *J. Chem. Soc. Chem. Comm.* **20,** 1423–5 (1990).
48 A. Léger, L. d'Hendecourt, L. Vertraete, and W. Schmidt, *Astr. Astrophys.* **203,** 145–8 (1988).
49 A. Léger, L. d'Hendecourt, L. Vertraete, and W. Schmidt, in *Quasicrystals, Networks, and Molecules of Fivefold Symmetry*, edited by I. Hargattai (VCH, New York, 1990) pp. 247–255.
50 R. F. Curl, in *Buckminsterfullerenes*, edited by W. E. Billups and M. A. Ciufolini (VCH Publishers Inc, New York, 1993), pp. 1–20.
51 R. E. Haufler, Y. Chai, L. P. F. Chibante, M. R. Fraelich, R. B. Weisman, R. F. Curl, and R. E. Smalley, *J. Chem. Phys.* **95,** 2197–2199 (1991).
52 J. Fulara, M. Jakobi, and J. P. Maier, *Chem. Phys. Lett.* **211,** 227–234 (1993).
53 M. Frenklach, D. W. Clary, J. W. C. Gardiner, and S. E. Stein, in *Twentieth Symposium (International) on Combustion* (The Combustion Institute, 1984), pp. 887–901.
54 M. Frenklach and L. B. Ebert, *J. Phys. Chem.* **92,** 561–563 (1988).
55 L. B. Ebert, *Science* **247,** 1468–1471 (1990).
56 R. M. Baum, *Chem. Eng'g. News* February 5, 1990, pp. 30–32.
57 P. Gerhardt, S. Löffler, and K. H. Homann, *Chem. Phys. Lett.* **137,** 306–10 (1987).
58 P. Gerhardt, S. Löffler, and K. H. Homann, in *Twenty-Second Symposium (International) on Combustion* (The Combustion Institute, 1989), pp. 395–401.
59 T. Baum, S. Löffler, P. Weilmünster, and K. H. Homann, in *ACS Div. Fuel Chem. Prepr.*, Vol. 36 (1991), pp. 1533.
60 T. Baum, S. Löffler, P. Löffler, P. Weilmünster, and K. H. Homann, *Ber. Bunsen-Ges. Phys. Chem.* **96,** 841–857 (1992).
61 J. B. Howard, J. T. McKinnon, Y. Makarovsky, A. Lafleur, and M. E. Johnson, *Nature* **352,** 139–141 (1991).
62 J. T. McKinnon, W. L. Bell, and R. B. Barkley, *Combust. Flame* **88,** 102–112 (1992).
63 J. B. Howard, A. L. Lafleur, Y. Makarovsky, S. Mitra, C. J. Pope, and T. K. Yadav, *Carbon* **30,** 1183–1201 (1992).
64 K. D. Chowdhury, J. B. Howard, and J. B. VanderSande, *J. Mater. Res.* **11,** 341–7 (1996).
65 C. J. Pope, J. A. Marr, and J. B. Howard, *J. Phys. Chem.* **97,** 11001–13 (1993).
66 F. W. Lam, J. P. Longwell, and J. B. Howard, in *Twenty-Third Symposium (International) on Combustion* (The Combustion Institute, 1991), pp. 1477–1484.

Photograph by Nicholas Sinclair

HAROLD W. KROTO

I was the kid with the funny name in my form. That is one of the earliest memories I have of school (except for being forced to finish school dinners). Other kids had typical Lancashire names such as Chadderton, Entwistle, Fairhurst, Higginbottom, Mottershead and Thistlethwaite though I must admit that there were the odd Smith, Jones and Brown. My name at that time was Krotoschiner (my father changed it to Kroto in 1955 so it is now occasionally thought, by some, to be Japanese). I felt as though I must have come from outer space–or maybe they did! I now realise that I had made a continual subconscious effort to blend as best I could into the environment by making my behaviour as identical as possible to that of the other kids. This was not easy–indeed it was almost impossible with a couple of somewhat eccentric parents (in particular an extrovertly gregarious mother) who were born in Berlin and came to Britain as refugees in their late 30's.

Bolton is a once prosperous but then (the fifties) decaying northern English town which is rightfully proud of its legendary contributions to the industrial revolution–the likes of Samuel Crompton and Richard Arkwright were Boltonians. Indeed we lived in Arkwright St and I shall always remember walking to school each morning past the windows of cotton mills through which I could see the vast rows of massive looms and spinning frames operated by women who had been working from at least six o'clock in the morning, if not earlier.

My efforts to merge into the background meant, among other things such as fighting (literally) for survival, speaking only English (all real Englishmen expect others to speak English)–though I allowed myself to absorb just enough German to understand what my parents were saying about me when they spoke German. One specific memory was that when I did particularly poorly at French one year my Father gave me a very large French dictionary for my birthday–was I pleased!!!

My name seems to have its origins in Silesia where my father's family originated and there is a town in Poland now called Krotoszyn (then Krotoschin). My father's family came from Bojanowo and set up a shop in Berlin where my father was born in 1900. The original family house, which was then a shop, still exists in the main square in Bojanowo. I have an old photograph which shows the sign "I. Krotoschiner" in gothic characters emblazened over the window. I visited the town recently and, apart from cars rather than horse-drawn carts and the sign, little has changed–the Hotel Centralny is now the Restauracja Centralny and the aerials on the roofs are still there!

My father, who originally wanted to be a dress designer but somehow ended up running a small business printing faces and other images on toy balloons, had to leave Berlin in 1937 and my mother (who was not Jewish) followed a few months later. I always felt that my parents had a really raw deal, as did almost everyone born in Europe at the turn of the Century. The First World War took place while they were teenagers, then the Depression struck and Hitler came to power while they were young adults. They had to leave their home country and then the Second World War broke out and they had to leave their home again. When my father was 45 he had to find a new profession, when he was 55 he set up his business again and when he was 65 he realised I was not going to take it over. He sold the business and retired in his early 70's.

I do not know how my father managed to catch the train to take him over the border into Holland in 1937. For as long as I knew him he was always late for everything; he invariably missed every train or bus he was supposed to catch. He told me that this was because he was called up in 1917 to go to the Front but arrived at the station just as the train was pulling out. When he asked the station master what he should do, he was told to go home. From then on he decided to make a point of missing trains and buses, but seems to have made one exception, in 1937. My parents managed to set up their small business again in London but the effort was, of course, shortlived due to the outbreak of the War in September 1939. I was born in Wisbech (a very small town in Cambridgeshire to which my mother was evacuated) on Oct 7th 1939 in the first month of the War so I was a war baby. My father was interned on the Isle of Man because he was considered to be an enemy alien; my mother (who was also an alien, but presumably assumed not to be an enemy one) was moved (with me–when I was about one year old) from London to Bolton in 1940. After the war my father became an apprentice engineer and because he was so good with his hands he managed to get a job as a fully qualified toolmaker at an engineering company in months rather than years.

In 1955, with help from friends in England and Germany from before the war, he set up his own small factory again, this time to make balloons as well as print them. I spent much of my school holidays working at the factory. I was called upon to fill in everywhere, from mixing latex dyes to repairing the machinery and replacing workers on the production line. I only now realise what an outstanding training ground this had been for the development of the problem solving skills needed by a research scientist. I am also sure that what I was doing then would contravene present-day health and safety at work regulations. I would have been considered too young and inexperienced to do the sort of maintenance work that I was often called upon to do. I did the stocktaking twice-a-year using a set of old scales with sets of individual gram weights (weighing balloons 10 at-a-time to obtain their average weights), my head, log tables and a sliderule to determine total numbers of various types of balloons. No paradise of microprocessor controlled balances then. After each stocktaking session I invariably felt that I never wanted to see another balloon as long as I lived.

My parents had lost almost everything and we lived in a very poor part of

Bolton. However they did everything they could to get me the best education they could. As far as they were concerned this meant getting me into Bolton School, a school with exceptional facilities and teachers. As a consequence of misguided politically motivated educational policies this school has become an independent school and it bothers me that, were I today in the same financial position as my parents had been when I was a child, I would not be able to send my children to this school. Though I did not like exams or homework any more than other kids, I did like school and spent as much time as I could there. At first I particularly enjoyed art, geography, gymnastics and woodwork. At home I spent much of the time by myself in a large front room which was my private world. As time went by it filled up with junk and in particular I had a Meccano set with which I "played" endlessly. Meccano which was invented by Frank Hornby around 1900, is called Erector Set in the US. New toys (mainly Lego) have led to the extinction of Meccano and this has been a major disaster as far as the education of our young engineers and scientists is concerned. Lego is a technically trivial plaything and kids love it partly because it is so simple and partly because it is seductively coloured. However it is only a toy, whereas Meccano is a real engineering kit and it teaches one skill which I consider to be the most important that anyone can acquire: This is the sensitive touch needed to thread a nut on a bolt and tighten them with a screwdriver and spanner just enough that they stay locked, but not so tightly that the thread is stripped or they cannot be unscrewed. On those occasions (usually during a party at your house) when the handbasin tap is closed so tightly that you cannot turn it back on, you know the last person to use the washroom never had a Meccano set.

At no point do I ever remember taking religion very seriously or even feeling that the biblical stories were any different from fairy stories. Certainly none of it made any sense. By comparison the world in which I lived, though I might not always understand it in all aspects, always made a lot of sense. Nor did it make much sense that my friends were having a good time in a coffee bar on Saturday mornings while I was in schul singing in a language I could not understand. Once while my father and I were fasting, I remember my mother having some warm croissants–and did they smell good! I decided to have one too–ostensibly a heinous crime. I waited for a 10 ton "Monty Python" weight to fall on my head! It didn't. Some would see this lack of retribution as proof of a merciful God (or that I was not really Jewish because my mother wasn't), but I drew the logical (Occam's razor) conclusion that there was "nothing" there. There are serious problems confronting society and a "humanitarian" God would not have allowed the unaccountable atrocities carried out in the name of any philosophy, religious or otherwise, to happen to anyone let alone to his/her/its chosen people. The desperate need we have for such organisations as Amnesty International has become, for me, one of the pieces of incontrovertible evidence that no divine *(mystical)* creator (other than the simple Laws of Nature) exists.

The illogical excuses, involving concepts such as free will(!), convoluted into confusing arguments by clerics and other self-appointed guardians of universal morality, have always seemed to me to be just so much fancy (or actually clumsy) footwork devised to explain why the fascinating and beautifully elegant world I live in operates exactly the way one would expect it to in the absence of a mystical power. Of course the excuses have been honed and polished over millenia to retain a hold over those unwilling or unable to accept that, as a Croatian friend of mine once neatly put it, "When you've had it you've had it".

The humanitarian philosophies that have been developed (sometimes under some religious banner and invariably in the face of religious opposition) are human inventions, as the name implies–and our species deserves the credit. I am a devout atheist–nothing else makes any sense to me and I must admit to being bewildered by those, who in the face of what appears so obvious, still believe in a mystical creator. However I can see that the promise of infinite immortality is a more palatable proposition than the absolute certainty of finite mortality which those of us who are subject to free thought (as opposed to free will) have to look forward to and many may not have the strength of character to accept it.

[After all this, I have ended up a supporter of ideologies which advocate the right of the individual to speak, think and write in freedom and safety (surely the bedrock of a civilised society). I have very serious personal problems when confronted by individuals, organisations and regimes which do not accept that these freedoms are fundamental human rights. I feel one must oppose those who claim that the "good" of the community must come before that of the individual–this claim is invariably used to justify oppression by the state. Furthermore there has never been any consensus on what the "good" of the community actually consists of, whereas for individuals there is little difficulty. Thus I am a supporter of Amnesty International, a humanist and an atheist. I believe in a secular, democratic society in which women and men have total equality, and individuals can pursue their lives as they wish, free of constraints–religious or otherwise. I feel that the difficult ethical and social problems which invariably arise must be solved, as best they can, by discussion and am opposed to the crude simplistic application of dogmatic rules invented in past millennia and ascribed to a plethora of mystical creators–or the latest invention; a single creator masquerading under a plethora of pseudonyms. Organisations which seek political influence by co-ordinated effort disturb me and thus I believe religious and related pressure groups which operate in this way are acting antidemocratically and should play no part in politics. I also have problems with those who preach racist and related ideologies which seem almost indistinguishable from nationalism, patriotism and religious conviction.]

My art teacher, Mr Higginson, would give me special tuition at lunch times or after school was over. My father made me finish all my homework and I had to stay up until it was not only complete but passed his inspection–midnight if necessary. As time progressed, for reasons which I am not sure I understand, I gravitated towards chemistry, physics and maths (in that order) and these became my specialist subjects in the 6th form. I was keen on sport, and in school I concentrated on gymnastics whilst outside school I played as much tennis as I could. I patterned my backhand (and my haircut) on that of Dick Savitt and my service on that of Neil Fraser. At one time I remember wanting to be Wimbledon champion but decided that this goal was going to be a bit hard to achieve as I seemed to be having too much difficulty winning.

I started to develop an unhealthy interest in chemistry during enjoyable lessons with Dr. Wilf Jary who fascinated me most with his ability, when using a gas blowpipe to melt lead, to blow continuously without apparently stopping to breath in. I, like almost all chemists I know, was also attracted by the smells and bangs that endowed chemistry with that slight but charismatic element of danger which is now banned from the classroom. I agree with those of us who feel that the wimpish chemistry training that schools are now forced to adopt is one possible reason that chemistry is no longer attracting as many talented and adventurous youngsters as it once did. If the decline in hands-on science education is not redressed, I doubt that we shall survive the 21st century. I became ever more fascinated by chemistry–particularly organic chemistry–and was encouraged by the sixth form chemistry teacher (Harry Heaney, now Professor at Loughborough) to go to Sheffield University because he reckoned it had, at the time, the best chemistry department in the UK (and perhaps anywhere)–a friendly interview with the amazing Tommy Stephens (compared with a most forbidding experience at Nottingham) settled it.

I was born during the war so I just escaped military service. As all the normal places at Oxbridge were already assigned for the next two years to re-emerging national servicemen, I needed to achieve scholarship level to get to Cambridge. This turned out to be a bit difficult as I had been assigned a college with an examination syllabus orthogonal to the one that I had studied. Ian McKellen, the actor, who was in the same year at school, only seems to have needed to remember his lines from his part as Henry V in the school play!

The first day that I arrived in Sheffield, I walked past a building which had a nameplate saying it was the Department of Architecture and was bemused–did people do that at University? I had somehow missed this possibility because general careers advice was non-existent at that time. With hindsight I am sure that with the advice available today I would have done something like architecture which would have conflated my art and technology interests. At Sheffield I did as much as I could. Initially I lived with a family in Hillsborough, near to the Sheffield Wednesday football ground and occasionally watched them–very occasionally as I *am* a Bolton Wanderers supporter. I played as much tennis as I could which helped to get me a room in a hall of residence (Crewe Hall). I played for the university tennis team and we got to the UAU (Universities Athletics Union) final twice–the team would probably have been champions without me–which they were in 1964. I wanted to continue with some form of art, which was really my passion, and became art editor of "Arrows" (the student magazine which we published each term), specialising in designing the magazine's covers and the screenprinted advertising posters. Whilst a research student I won a *Sunday Times* bookjacket design competition–the first important (national) prize I was to get for a very long time. Later my cover design for the departmental teaching and research brochure *"Chemistry at Sussex"* was featured in "Modern Publicity" (an international annual of the best in professional graphic design)–I consider this to be one of my best publications.

In the 1960s almost everybody could play the guitar well enough to play and sing two or three songs at a party so I had a go at that too and learned just enough chords (about half-a-dozen) to play some simple songs at local student folk clubs. I also decided that I should do some administration in the Students' Union and from secretary of the tennis team I somehow ended up as President of the Athletics Council. During my last year at University (1963–64) I spent some 2–3 hours of each day attending to administration in the sports office in the Union. That year's involvement in embryonic politics was enough to last a lifetime. I managed to do enough chemistry in between the tennis, some snooker and football, designing covers and posters for "Arrows", painting murals as backdrops for balls and trying to play the guitar, to get a first class honours BSc degree (1958-61) and a PhD (1961–64) as well as some job offers. I also got married.

I had been keen on organic chemistry when I arrived at Sussex (at the behest of Harry Heaney I had bought Fieser and Fieser's Organic Chemistry and read much of it while at school–it was a good read), but as the university course progressed I started to get interested in quantum mechanics and when I was introduced to spectroscopy (by Richard Dixon, who was to become Professor at Bristol) I was hooked. It was fascinating to see spectroscopic band patterns which showed that molecules could count. I had a problem as I really liked organic chemistry (I guess I really liked drawing hexagons) but in the end I decided to do a PhD in the Spectroscopy of Free Radicals produced by Flash Photolysis–with Richard Dixon. George Porter was Professor of Physical Chemistry at that time so there was a lot of flashing going on at Sheffield.

In 1964 I had several job offers but Marg(aret) and I decided that we wanted to live abroad for a while and Richard Dixon had inveigled an attractive offer of a postdoctoral position for me from Don Ramsay at the National Research Council in Ottawa. In 1964 Marg and I left Liverpool, on the Empress of Canada, for Montreal and then went on to Ottawa by train. I arrived at the famous No. 100, Sussex Drive, NRC, Ottawa, where Gerhard Herzberg (GH) had created the mecca of spectroscopy with his colleagues Alec Douglas, Cec Costain, Don Ramsay, Boris Stoicheff and others. At the time NRC was the only national research facility worldwide that was recognised as a genuine success. I suspect that this was because the legendary Steacie had left researchers to do the science they wanted; now unfortunately–as almost everywhere else–administrators decide what should be done. I remember easily making friends with all the other postdocs who congregated each morning and afternoon in the historical room 1057–the spectroscopy tea/coffee area. The atmosphere was, in retrospect, quite exhilarating and many there, including: Reg Colin, Cec Costain, Fokke Creutzberg, Alec Douglas, Werner Goetz, Jon Hougen, Takeshi Oka and Jim Watson and their families became our lifelong close friends. As I look back I realise that Cec Costain, Jon Hougen, Takeshi Oka and Jim Watson were to exert enormous direct and indirect influence on my scientific development. I gradually learned to recognise who was good at what and what (if anything) I was good at. To paraphrase Clint Eastwood "A (scientist's) gotta know his limitations"–and in this somewhat daunting company I learned mine. Although I knew that my level of knowledge and understanding was limited when I arrived, I was never made to feel inferior. This encouraging atmosphere was, in my opinion, the most important quality of the laboratory and permeated down directly from GH, Alec and Cec–it was a fantastic, free environment. The philosphy seemed to be to make state-of-the-art equipment available and let budding young scientists loose to do almost whatever they wanted. Present research funding policies appear to me to be opposed to this type of intellectual environment. I have severe doubts about policies (in the UK and elsewhere) which concentrate on "relevance" and fund only those with foresight when it is obvious that many (including me) haven't got much. There are as many ways to do science as there are scientists and thus when funds are scarce good scientists have to be supported even if they do not know where their studies are leading. Though it seems obvious (at least to me) that unexpected discoveries must be intrinsically more important than predictable (applied) advances it is now more difficult than ever before to obtain support for more non-strategic research.

In 1965 after a further year of flash photolysis/spectroscopy in Don Ramsay's laboratory, where I discovered a singlet-singlet electronic transition of the NCN radical and worked on pyridine which turned out to have a non-planar excited state (still to be fully published!), I transferred to Cec Costain's laboratory because I had developed a fascination for microwave spectroscopy. There I worked on the rotational spectrum of NCN_3. Sometimes Takeshi Oka would be on the next spectrometer–working next to someone with such an exceptional blend of theoretical and experimental expertise did not help to alleviate the occasional sense of inadequacy. I really learned quantum mechanics (as did we all) from an intensive course that Jon Hougen gave at Carlton University. Whenever I was in difficulty theoretically (which was most of the time) Jim Watson helped me out–when he was not busy helping everyone else out. Gradually I realised that many in the field were stronger at physics than chemistry and in retrospect I subconsciously recognised that there might be a niche for me in spectrocopy research if I could exploit my relatively strong chemistry backgound.

In 1966, after two years at NRC, John Murrell (who had taught me quantum chemistry at Sheffield) offered me a postdoctoral position at Sussex. We were quite keen to live in the US, however, and I managed to get a postdoctoral position at Bell Labs (Murray Hill) with Yoh Han Pao (later Professor at Case Western) to carry out studies of liquid phase interactions by laser Raman spectroscopy. David Santry (now Professor at McMaster) was also working with Yoh Han at that time and each evening Dave and I carried out CNDO theoretical calculations on the electronic transitions of small molecules and radicals. I learned programming (Fortran) from Dave who threw me in at the deep end by showing me how to modify and correct the programs and then left me to see if I could do it myself.

During the year I received another letter from John Murrell to say that the position that had been available at Sussex the previous year was still available but would not be so for much longer. Thus Marg, Stephen (who had been born in Ottawa) and I came back to the UK–my annual salary dropped from $14000 to £1400, ouch! Marg had to find part-time employment as soon as possible although pregnant with our second son, David (we were poorer–but we were happier .... !!!). I was just about to start writing off for some positions back in the US and had just located the address of Buckminster Fuller's research group (I was interested in the way that predesigned urban sub-structures might be welded into an efficient large urban complex) when John Murrell offered me a permanent lectureship at Sussex which I accepted.

I remember thinking I would give myself five years to make a go of research and teaching and if it was not working out I would re-train to do graphic design (my first love) or go into scientific educational TV (I had had an interview with the BBC before we went to Canada). I started to build up a microwave laboratory to probe unstable molecules and Michael Lappert encouraged me to use his photoelectron spectrometer to carry out work independently.

By 1970 I had carried out research in the electronic spectroscopy of gas phase free radicals and rotational microwave spectrocopy, I had built He-Ne and argon ion lasers to study intermolecular interactions in liquids, carried out theoretical calculations and learned to write programs. At Sussex I carried on liquid phase Raman studies, rebuilt a flash photolysis machine and built a microwave spectrometer and started to do photoelectron spectroscopy. I had applied for a Hewlett Packard microwave spectrometer and SERC, in its infinite wisdom, decided to place the equipment at Reading (where my co-applicant, a theoretician (!), worked) so requiring me and my group (the experimentalists) to travel each month to Reading to make our measurements! However by 1974, after three further attempts to get my own spectrometer (with help in consolidating my proposal from David Whiffen), the SERC finally gave in and I got one of my own at Sussex. The first molecule we studied was the carbon chain species HC_5N–to which the start of my role in the discovery of C_{60} can be traced directly.

The discovery of C_{60} in 1985 caused me to shelve my dream of setting up a studio specialising in scientific graphic design (I had been doing graphics semiprofessionally for years and it was clear that the computer was starting to develop real potential as an artistically creative device). That was the downside of our discovery. I decided to probe the consequences of the C_{60} concept. In 1990 when the material was finally extracted by Krätschmer, Lamb, Fostiropoulos and Huffman, I and my colleagues Roger Taylor and David Walton, decided to exploit the synthetic chemistry and materials science implications. I began to realise that I might never fulfill my graphics aspirations. In 1991 I was fortunate enough to be awarded a Royal Society Research Professorship which enables me to concentrate on research by allowing me to do essentially no teaching. However I like teaching so I continue to do some. I have discovered that since I stopped teaching 1st and 2nd-year students, home-grown graduate students are few and far between.

In 1995, together with Patrick Reams a BBC producer, I inaugurated the Vega Science Trust to create science films of sufficiently high quality for network television broadcast (BBC2 and BBC Prime). Our films not only reflect the excitement of scientific discovery but also the intrinsic concepts and principles without which fundamental understanding is impossible. The Trust also seeks to preserve our scientific cultural heritage by recording scientists who have not only made outstanding contributions but also are outstanding communicators. The trust, whose activitities are coordinated by Gill Watson, has now made some 20 films of Royal Institution (London) Discourses archival programmes and interviews.

I have been asked many questions about our Nobel Prize and have many conflicting thoughts about it. I have particular regrets about the fact that the contributions of our student co-workers Jim Heath, and Sean O'Brien as well as Yuan Liu receive such disparate recognition relative to that accorded to ours (e.g. Bob, Rick and me). I also have regrets with regard to the general recognition accorded to the amazing breakthrough that Wolfgang Krätschmer and Don Huffman made with their students Kostas Fostiropoulos and Lowell Lamb in extracting C_{60} using the carbon arc technique and which did so much to ignite the explosive growth of Fullerene Science. I have heard some scientists say that young scientists need prizes such as the Nobel Prize as an incentive. Maybe some do, but I don't. I never dreamed of winning the Nobel Prize–indeed I was very happy with my scientific work prior to the discovery of C_{60} in 1985. The creation of the first molecules with carbon/phosphorus double bonds and the discovery of the carbon chains in space seemed (to me) like nice contributions and even if I did not do anything else as significant I would have felt quite successful as a scientist. A youngster recently asked what advice I would give to a child who wanted to be where I am now. One thing I would not advise is to do science with the aim of winning any prizes let alone the Nobel Prize that seems like a recipe for eventual disillusionment for a lot of people. [Over the years I have given many lectures for public understanding of science and some of my greatest satisfaction has come in conversations with school children, teachers, lay people, retired research workers who have often exhibited a fascination for science as a cultural activity and a deep and understanding of the way nature works.] I believe competition is to be avoided as much as possible. In fact this view applies to any interest–I thus have a problem with sport which is inherently competitive. My advice is to do something which interests you or which you enjoy (though I am not sure about the definition of enjoyment) and do it to the absolute best of your ability. If it interests you, however mundane it might seem on the surface, still explore it because something unexpected often turns up just when you least expect it. With this recipe, whatever your limitations, you will almost certainly still do better than anyone else. Having chosen something worth doing, *never give up and try not to let anyone down.*

SYMMETRY, SPACE, STARS and C_{60}

Nobel Lecture, December 7, 1996

by

HAROLD W. KROTO

School of Chemistry, Physics and Environmental Science, University of Sussex, Brighton, BN1 9QJ, Great Britain

INTRODUCTION

A Summary of the Key Phases in the Birth of the Fullerenes

The story of the discovery of C_{60} Buckminsterfullerene, Fig. 1, and the birth of Fullerene Science consists of several disparate strands which came together over ten days in September 1985. During this period of feverish activity, working with Jim Heath, Sean O'Brien, Yuan Liu, Bob Curl and Rick Smalley at Rice University in Texas, evidence was found that a C_{60} molecule self-assembled spontaneously from a hot nucleating carbon plasma [1]. The molecule had however a prehistory: The earliest paper which describes the C_{60} molecule is to be found in Kagaku (in Japanese) where Eiji Osawa in 1970 sugges-

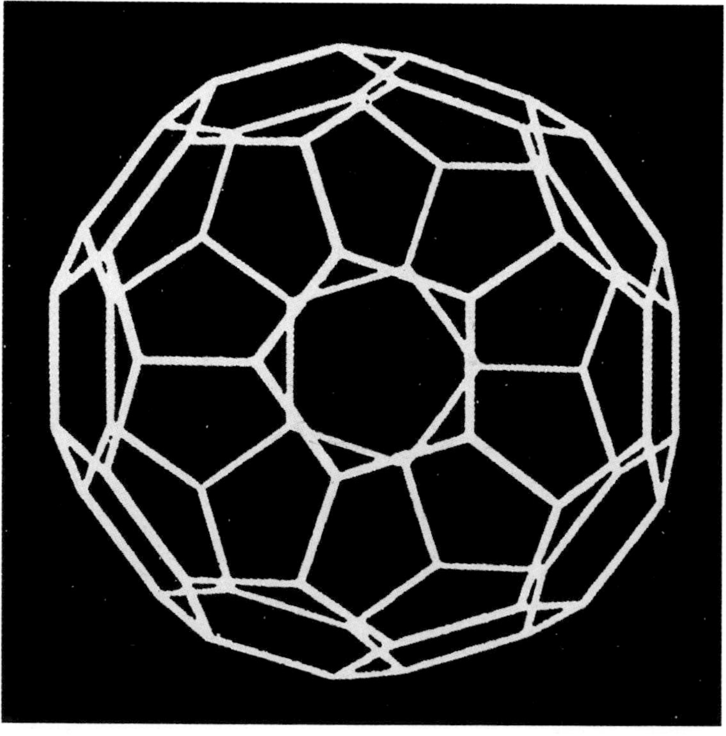

Figure 1. C_{60} Buckminsterfullerene [1].

ted that it should be stable [2] and the following year discussed its possible aromatic properties in more detail in a book with Yoshida [3]. Bochvar and Gal'pern published a theoretical study in 1972 [4,5]. Somewhat earlier in 1966 David Jones had conjectured that if pentagonal disclinations could be introduced among the hexagons in a graphene sheet, the sheet would close into a hollow balloon [6,7]. We however did not know about this prehistory and none of us had ever heard of such a molecule or thought about it.

In retrospect, as far as I was concerned, I seem to have been fascinated by various peculiar aspects of carbon chemistry for much of my research career. With the perspective of hindsight my involvement can be split into five phases: At Sussex, in the late 1960's, David Walton and I initiated a synthetic/spectroscopic study (Phase I) which led directly–in the mid-1970's–to some unexpected discoveries (with colleagues at Sussex and NRC Ottawa, Canada) of long carbon chain molecules in the interstellar medium (Phase II). At about this time Don Huffman and Wolfgang Krätschmer were developing laboratory experiments which focused on the composition of the particulate carbon material which occupies the same interstellar regions–indeed the agents which caused the scattering in the dark regions and observed some intriguing new unexplainable features in the 220 nm region of their laboratory data. Also at about this time Richard Smalley was developing a range of advanced chemical physics techniques which culminated in an ingenious apparatus which enabled small clusters of *refractory* materials to be created and probed for the first time. This was the pulsed supersonic nozzle–laser vaporisation apparatus [8]. It is my view that the *raison d'être* of this machine was to discover C_{60}.

It was the marriage of ideas resulting from the discoveries of carbon chains in space with the recognition that the cluster beam technique might offer laboratory support for their circumstellar origin that in 1985 resulted in the serendipitous discovery that the C_{60} molecule not only existed but that it self-assembled (Phase III) [1,9]. After this revelation it still remained for unequivocal proof of the soccerball structure to be obtained (Phase IV). Several groups made important contributions: Theoreticians, who had a field day and came up with some results which ultimately proved crucial for the extraction [9,10]; the Rice group (initially in collaboration with Sussex) [9,11,12]; The Sussex Group [9,13]; The IBM group [14] and last but not least; the Heidelberg/Tucson group which succeeded in the first extraction of macroscopic quantities [15–17]. It was this fantastic breakthrough that has made the explosive growth of Fullerene Science possible (Phase V). The advance made by Krätschmer and Huffman and their students is one of the most beautiful pieces of fundamental science.

In any complex story, the events can only be understood by reference to the complete set of personal subjective accounts of the participants–even purportedly objective accounts must be treated with skepticism [18]. In this case *all* the scientists–including research students–played crucial roles in the story. This article describes parts of the story to which Sussex researchers made contributions. There have now been numerous accounts ranging from the personal articles [19–22] as well as books by Baggott [23], Aldersey-Williams [24] Krätschmer and Schuster [25], Dettman [26] and Ball [27]. An excellent text focusing on the chemistry is that by Hirsch [28]. Several compendia have been published–some edited by or with Sussex researchers include one with a historical perspective [29], a second from the viewpoint of carbon materials and physics [30], and a third which concentrates on fullerene chemistry [31]. It should also be recognised that the C_{60} led to a further amazing advance–this was the discovery by Iijima that carbon nanotubes also form at the same time as C_{60} [32].

The story of C_{60} also cannot be recounted without reference to its beauty which results from the incredible symmetry. Another important aspect of the molecule's aura lies in the name buckminsterfullerene [33] and the direct association it has with the geodesic domes designed by Buckminster Fuller [34–36]. It invests this elegant molecule with a charisma that has fascinated scientists, delighted lay people and has infected children with a new enthusiasm for science and in particular it has given chemistry a new lease of life [37].

PROLOGUE

Symmetry, the Key to the Theory of Everything

Symmetry appears to be fundamental to our perception of the physical world and it also plays a major role in our attempts to explain everything about it. As far as structural symmetry is concerned it goes back to ancient times, as indicated by the (pre-)Platonic structures exhibited in the Ashmolean Museum in Oxford [38,39]. The most famous examples are of course to be found in "The Timaeus" where in the section relating to "The Elements" Plato says: "In the first place it is clear to everyone (!) that fire, earth, water and air are bodies and all bodies are solids" (!!) [40]. Plato goes on to discuss chemistry in terms of these elements and associates them with the four Platonic solids (only four at that time–until Hippasus discovered the fifth–the dodecahedron).

Although this may at first sight seem like a somewhat naive philosophy it indicates a very deep understanding of the way Nature actually functions. A curiously close contemporary analogue of this ancient assumption is to be found in the historially important article by Van Vleck [41] on which almost all of modern molecular spectroscopy is based: Van Vleck says "Practically every-one (!) knows that the components of total angular momentum (NB the angular momentum operator is usually denoted by the symbol J and the associated quantum number by j) of the molecule relative to the axes X,Y,Z fixed in space satisfy the commutation relation of the form

$$J_X J_Y - J_Y J_X = iJ_Z \tag{1}$$

Klein discovered the rather surprising fact that when total angular momentum is referred to axes mounted in the molecule which we will denote by x,y,z the sign of i in the commutation relation is reversed i.e.

$$J_x J_y - J_y J_x = -iJ_z \text{ ”} \tag{2}$$

Does practically everyone know this?–I wondered whether to check this claim out by asking everyone on the main street in Brighton whether they did. I hardly knew–or more accurately–really understood the first relation, let alone the second. However I did know that angular momentum was quantised and governed by the fundamental relations

$$\langle j | J^2 | j \rangle = \hbar^2 j(j+1) \tag{3}$$

$$M_J = -j \ldots +j \tag{4}$$

which means that J has $2j+1$ possible orientations, and

$$\Delta j = 0, \pm 1 \tag{5}$$

which indicates that when a transition occurs, j may only change by one unit or on occasion remain unchanged.

I knew that the row structure of Mendeleef's periodic table could be explained on the basis of these quantum properties of angular momentum, in particular the fact that a system with angular momentum j possesses $2j+1$ quantised orientations. I decided I ought to understand these and under the tutelage of Jon Hougen and Jim Watson I dived into the "Theory of Atomic Spectra" of Condon and Shortley [42]. There can surely be no more elegant printed page in the Sciences or the Arts than page 60 of this book, Fig. 2, where Dirac's proof of the $\Delta j = \pm 1$ selection rule (it is non-committal on 0) is presented–a perfect example of elegant and powerful use of Roman typeface. One only has to look at this page to know that it must be important. As far as I was concerned it planted the desire to understand the embedded theory–the effort was (for me) considerable, but worth it and remains one of the most intellectually rewarding tasks I have undertaken. Having finally grasped the implications of these relations, I ended up writing a book called "Molecular Rotation Spectra" [43] which deals with the way the angular momentum relations govern the rotational dynamics of molecules.

The geometric pattern of the periodic table, Fig. 3, is an implicit result of the angular momentum description of atoms and electrons (together with a few odds and ends (!) such as the Pauli exclusion principle). Any intelligent life on another planet that had developed an understanding of chemistry would almost certainly recognise it. The electron angular momentum wavefunctions are described in terms of spherical harmonics–known by mathematicians long before 1925 when Quantum Mechanics was created and thus these symmetric mathematical functions lend an elegant beauty to the abstract description of rotational/orbital motion in atoms and molecules. Thus we discover that symmetry principles underpin the elegant quantum mechanical description of atoms and molecules in an abstract picture in which statics and dynamics are paradoxically conflated in a way that often leaves us hovering on the boundary between abstract mathematical understanding and literal physical misunderstanding. It is the existence of similar abstract aspects of symmetry that has invested the C_{60} molecule with a charismatic quality that few other molecules possess.

Another important fact which follows immediately from $12^2 6$ is that the scalar product of any two vectors* which satisfy this commutation rule with respect to $\boldsymbol{J}$ will commute with J_x, J_y, J_z, and hence with $\boldsymbol{J}^2$:

$$[\boldsymbol{J}, \boldsymbol{T}_1 \cdot \boldsymbol{T}_2] = 0, \quad [\boldsymbol{J}^2, \boldsymbol{T}_1 \cdot \boldsymbol{T}_2] = 0. \tag{4}$$

This is independent of whether $\boldsymbol{T}_1$ commutes with $\boldsymbol{T}_2$ or not.

We shall now consider the problem of obtaining the matrices of T_x, T_y, T_z in a representation in which $\boldsymbol{J}^2$, J_z, and a set A of observables which commute with $\boldsymbol{J}$ are diagonal. We shall first obtain a selection rule on j, i.e. a condition on $j'-j$ necessary for the non-vanishing of a matrix component connecting the states j and j'. This we may do by a method outlined by Dirac (p. 158).

Using the relation $12^2 6$, we find that

$$[\boldsymbol{J}^2, \boldsymbol{T}] = \boldsymbol{J}\cdot[\boldsymbol{J}, \boldsymbol{T}] - [\boldsymbol{T}, \boldsymbol{J}]\cdot\boldsymbol{J} = -i\hbar(\boldsymbol{J}\cdot\boldsymbol{T}\times\mathfrak{J} - \boldsymbol{T}\times\mathfrak{J}\cdot\boldsymbol{J})$$
$$= -i\hbar(\boldsymbol{J}\times\boldsymbol{T} - \boldsymbol{T}\times\boldsymbol{J}) = -2i\hbar(\boldsymbol{J}\times\boldsymbol{T} - i\hbar\boldsymbol{T}).$$

From this we have

$$[\boldsymbol{J}^2, [\boldsymbol{J}^2, \boldsymbol{T}]] = -2i\hbar[\boldsymbol{J}^2, (\boldsymbol{J}\times\boldsymbol{T} - i\hbar\boldsymbol{T})] = -2i\hbar\{\boldsymbol{J}\times[\boldsymbol{J}^2, \boldsymbol{T}] - i\hbar[\boldsymbol{J}^2, \boldsymbol{T}]\}$$
$$= -2i\hbar\{-2i\hbar\boldsymbol{J}\times(\boldsymbol{J}\times\boldsymbol{T} - i\hbar\boldsymbol{T}) - i\hbar(\boldsymbol{J}^2\boldsymbol{T} - \boldsymbol{T}\boldsymbol{J}^2)\}$$
$$= 2\hbar^2(\boldsymbol{J}^2\boldsymbol{T} + \boldsymbol{T}\boldsymbol{J}^2) - 4\hbar^2\boldsymbol{J}(\boldsymbol{J}\cdot\boldsymbol{T}),$$

using $12^2 11$a to expand $\boldsymbol{J}\times(\boldsymbol{J}\times\boldsymbol{T})$. But

$$[\boldsymbol{J}^2, [\boldsymbol{J}^2, \boldsymbol{T}]] \equiv [\boldsymbol{J}^2, (\boldsymbol{J}^2\boldsymbol{T} - \boldsymbol{T}\boldsymbol{J}^2)] \equiv \boldsymbol{J}^4\boldsymbol{T} - 2\boldsymbol{J}^2\boldsymbol{T}\boldsymbol{J}^2 + \boldsymbol{T}\boldsymbol{J}^4.$$

Hence

$$\boldsymbol{J}^4\boldsymbol{T} - 2\boldsymbol{J}^2\boldsymbol{T}\boldsymbol{J}^2 + \boldsymbol{T}\boldsymbol{J}^4 = 2\hbar^2(\boldsymbol{J}^2\boldsymbol{T} + \boldsymbol{T}\boldsymbol{J}^2) - 4\hbar^2\boldsymbol{J}(\boldsymbol{J}\cdot\boldsymbol{T}). \tag{5}$$

Take the matrix component of this equation referring to the states $\alpha\, j\, m$ and $\alpha'\, j'\, m'$, where $j' \neq j$. Since $\boldsymbol{J}\cdot\boldsymbol{T}$ commutes with $\boldsymbol{J}$ [by (4)], this component will vanish for the last term in the equation. From the rest we obtain:

$$\hbar^4[j^2(j+1)^2 - 2j(j+1)j'(j'+1) + j'^2(j'+1)^2](\alpha j m|\boldsymbol{T}|\alpha' j' m')$$
$$= 2\hbar^4[j(j+1) + j'(j'+1)](\alpha j m|\boldsymbol{T}|\alpha' j' m').$$

The bracket on the left is

$$[j(j+1) - j'(j'+1)]^2 = (j-j')^2(j+j'+1)^2,$$

while

$$2[j(j+1) + j'(j'+1)] = (j+j'+1)^2 + (j-j')^2 - 1.$$

Hence

$$[(j-j')^2(j+j'+1)^2 - (j+j'+1)^2 - (j-j')^2 + 1](\alpha j m|\boldsymbol{T}|\alpha' j' m') = 0,$$

or

$$[(j+j'+1)^2 - 1][(j-j')^2 - 1](\alpha j m|\boldsymbol{T}|\alpha' j' m') = 0. \qquad (j \neq j')$$

In order to obtain a non-vanishing matrix component one of the brackets must vanish. The first cannot since $j' \neq j$ and $j, j' \geqslant 0$. The second vanishes only when $j'-j = \pm 1$. Hence for a non-vanishing matrix component we must have

$$j' - j = 0, \pm 1. \tag{6}$$

* An important special instance of this is the square of such a vector.

Figure 2. Page 60 of Condon and Shortley's book [42] in which Dirac's elegant derivation of the fundamental selection rule on j, the angular momentum quantum number, is derived (published by permission of Cambridge Univ Press). These relations not only describe a beautiful example of the way Nature behaves but also are an exquisite example of man's mathematical invention to describe the behaviour, as well as an elegant example of typeface which visually evokes a powerful feeling for the underlying abstract concept. Buried deep in this derivation are fundamental laws of symmetry which govern the way in which light interacts with matter. These relations govern the origin of the sunlight falling on this page, the modulation of that light by the page and the print on it, as well as the way in which the receptors in your eye detect the reflected light and transforms it into perceived images in your brain. By the way, it may be worth noting that not only does this paper consist mainly of carbon but so also does the ink.

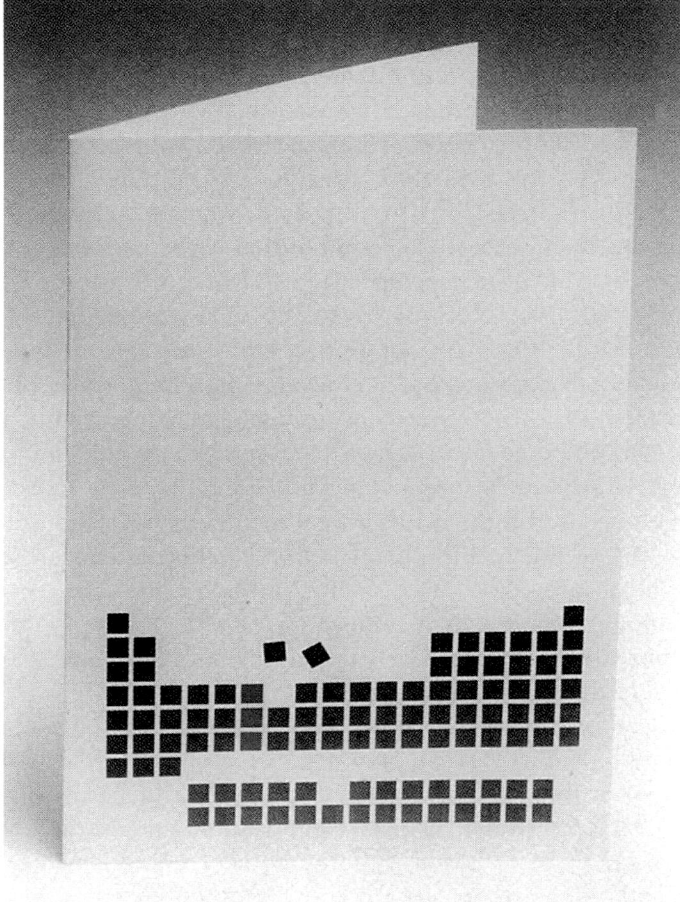

Figure 3. This item was designed by the author for a conference. It is left to the reader to deduce the topic of the conference and also the nature of item depicted. The answer to the latter may be in the Hitchhiker's Guide to the Galaxy.

PHASE I

Carbon, Still Crazy After All These Years

Carbon is really peculiar–in fact, the fact that there is enough carbon around in the Universe to allow enough biology to occur and result in the Human Race at all is due to peculiarities in the nuclear chemistry of carbon. The "molecular" chemistry of carbon is no less peculiar. Even if we discount the whole of Organic Chemistry, the Inorganic Chemistry of carbon is also amazing. I worked with carbon containing species right from the start when I was a graduate student with Richard Dixon at Sheffield. My first (successful) research project was a study of the electronic spectrum of the diatomic free radical CBr in which we also explored the intrinsic nature of the bond that the carbon atom formed with halogens [44]. Even more interesting was a preliminary (aborted) study of the fascinating molecule carbon suboxide OCCCO. The most interesting aspect of this molecule was the fact that it did not seem to know whether it wanted to be a linear or bent. The triatomic molecule C_3 was no less odd, exhibiting similar structural schizophrenia. The identification of C_3 in comets and flames, in the elegant historical study by Alec Douglas [45], was a landmark in spectroscopy which added to the ubiquitous charismatic attraction of carbon species on earth as well as in space.

When I arrived at Sussex in 1967, after two years at NRC (Ottawa) and a year at Bell Labs (Murray Hill) I set up a microwave spectroscopy research programme which aimed at the creation of new molecules which contained multiple bonds between carbon and various second or third row atoms [46] e.g. sulphur (C=S, C=Se), phosphorus (C=P, C=P) and Si (C=Si). At this time the so-called double bond rule–Multiple bonds involving non-first-row elements should not exist–appeared to have some validity. However the stability of CS_2 indicated that the embargo was not total. Most importantly, as far as I was concerned, I became fascinated (while a postdoc at NRC in Canada) by the molecule HC≡P which Gier had made in 1961 [47]. I became convinced that the existence of this molecule (which incidently was produced spontaneously by a carbon arc in phosphine (!)) implied the existence of derivatives such as ClC≡P and $CH_3C{\equiv}P$. I also wondered whether these species could be used as synthons to create organic phosphorus cycloaddition products (analogues of the nitrogen heterocyclic compounds) as well as metal complexes which parallel those of the metal-nitrile complexes.

The first new molecules that we created possessed carbon to sulphur double bonds [46]. Numerous compounds such as $CH_3CH{=}S$ (thioacetaldehyde), $CH_2{=}C{=}S$ (thioketene) and $CH_2{=}CH\text{-}CH{=}S$ (thioacrolein) were produced (with Barry Landsberg, Krini Georgiou and Roger Suffolk). They were created by thermolysis of custom-synthesised precursors and analysed by both microwave and photoelectron spectroscopy. After the success with sulphur, it seemed that the same techniques might also work for phosphorus and, encouraged by Gier's HC≡P result of some dozen years previously, we tried it–and it worked like a dream [46]!

With my Sussex colleague John Nixon, together with Nigel Simmons and Nick Westwood, Osamu Ohashi, Keichi Ohno and James Burkett St Laurent we produced a whole range of these species. With Colin Kirby and Terry Cooper we also produced some elegant boron-sulphur XB=S analogues [46]. $CH_2{=}PH$ was my favourite as it was the first (and simplest) molecule with a carbon-phosphorus double bond. Our first paper, on $CH_2{=}PH$, $CH_2{=}PCl$ and $CF_2{=}PH$ [48] and the paper by Becker, also on the creation of C=P containing species [49], appeared almost simultaneously. Our successful creation of $CH_3C{\equiv}P$ [50] was almost equally exciting as it was the first analogue of HC≡P. It still seems amazing that Gier's 1961 result had not been recognised sooner by anyone else as a clear sign that a whole new class of compounds just waiting to be discovered. The creation and structural determination of species such as $CH_2{=}PH$ and $CH_2{=}PCl$, Fig 4 as well as $CH_3C{\equiv}P$ and other related species, were (for me) some of our most exciting advances [46]. Since these papers were published these compounds, which we call phosphaethenes and phosphaethynes, have become the basis of some most fruitful fields of new

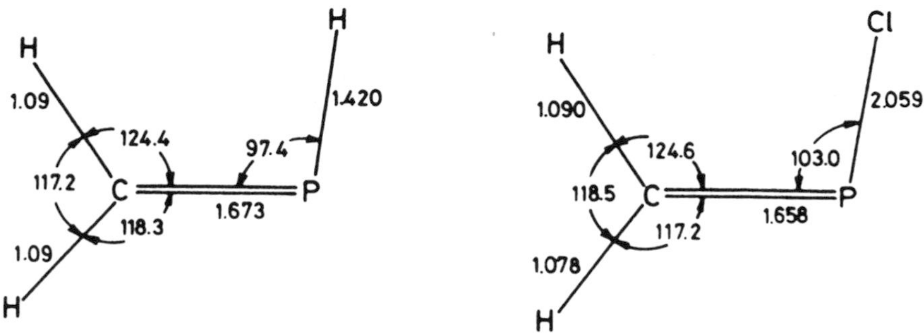

Figure 4. The structures of the first phosphaalkynes created, CH_2=PH and CH_2=PCl, as determined by microwave spectroscopy [48,46].

chemistry. However we never got to grips with the carbon-silicon double bond–a problem solved by others–as our interests led in other directions.

During this time (ca 1970) my colleague David Walton was perfecting techniques for synthesising very long linear chains of carbon atoms [51–53]. These molecules appeared like perfect test-beds for quantum mechanical study of the dynamics of bending and rotation in simple linear systems. In 1972, as luck would have it, David and I proposed a project for the Chemistry by Thesis (CT) course at Sussex which aimed at probing such structural and dynamic behavior of moderately long carbon chains. The CT course had been initiated by Colin Eaborn [54] as a new approach to undergraduate teaching in which a BSc degree could be obtained on the basis of a thesis which the student would write on a research topic carried out under the supervision of two tutors from different chemistry disciplines (e.g. synthetic chemistry and spectroscopy). The course recognised that the traditional examination approach to the assessment of a student's ability–and in particular future capability–was an artificial and often unsatisfactory guide.

The course was unique to Sussex and produced a long stream of highly accomplished graduates with remarkably few who did not do well on the course. It was one of the numerous outstanding success stories of Chemistry at Sussex. I highlight "was" because the course has now been legislated out of existence by the combined efforts of Health and Safety Executive officials, lawyers and misguided university administrators. This is an archetypal example of how a disparate mix of uncorrelated antagonists–each focused on individual goals with no perception of the true overall consequences can achieve negative results. A similar mix of advisors and officials in the secondary education sector has all but eliminated any real chemistry from the UK school curriculum. Today, students with any real chemical facility are virtually an extinct species.

Andrew Alexander was the outstanding undergraduate student catalyst of the research project which aimed at the synthesis of a range of polyynes in particular HC_5N and their analysis by ir, nmr and in particular microwave spectroscopy [55,46]. In fact it was Alex's keenness in the first place that got the project going and his enthusiasm for the work, in the second place which played a key role in its success. The start of the Sussex role in the story of C_{60} is really to be found in this project. It was the epitome of basic research and in these times when the mania for applied research is rampant, it is hard to imagine one less likely to gain support. There is little doubt in my mind that, had this project not been initiated, we at Sussex would not have been involved in the discovery of C_{60}.

HC_5N was the first molecule to be studied at Sussex on the newly arrived Hewlett Packard Microwave Spectrometer in 1974. This superb instrument enabled the rotational spectra of molecules to be measured with such ease (relative to the home-built instruments that existed in most microwave laboratories at the time). The machine enabled us to concentrate almost all our attention on the demanding chemistry involved in producing new species without worrying about how to optimise the detection parameters. There is no doubt that much of the work we did during the 1970's would have been essentially impossible without this amazing instrument.

PHASE II

A Tale of Cold Black Giant Clouds and Warm Red Giant Stars

About the same time that we were creating the long chain carbon molecules in the laboratory at Sussex, a veritable Pandora's Box of molecules had been opened up by Townes and co-workers [56] who discovered ammonia in Orion. After this discovery microwavers and radioastronomers joined forces and showed that the vast dark clouds which lie between the stars, Fig. 5, harbour scores of molecules (methanol, carbon monoxide, formaldehyde, ethanol, hydrogen cyanide, formic acid, formamide etc)–all the molecules the Galaxy could possibly need for a primordial pot of prebiotic soup [57]. Furthermore they were all identified by the detection of their microwave rotational emission spectra using radiotelescopes. Cyanoethyne (or cyanoacetylene to us older guys) HC_3N was also detected [58] and when we made HC_5N I began to wonder whether it might also be present in interstellar space. It seemed like a very long shot at the time because an abundance *rule-of-thumb* had materialised, based on a rough correlation of available empirical data, that suggested that an additional carbon atom would reduce the abundance by about a factor of ten, and two extra atoms implied that HC_5N should be down by a factor of 100 from HC_3N and this meant that it would be undetectable. However my view is never to assume anything is right until it has been tested experimentally because, as often as not, assumptions are wrong and even when they are not something unexpected and interesting often happens.

However, the strong bond between laboratory spectroscopy and astrophysics, which goes back at least to Newton's time and perhaps even to antiquity, was again giving birth to yet another major new field–in this case *interstellar astrochemistry.* Molecular radioastronomy was the crucial medium in uncovering the hordes of molecules–indeed it has now shown that the molecules play a crucial role in the collapse of the clouds to form stars. As gravitational attraction causes the interstellar gas clouds to collapse and heat up, rotational tran-

Figure 5. The dark clouds in Taurus from Barnards 1927 Survey Atlas of Selected Regions of the Milky Way, (ed. E B Frost and R Calvert, Carnegie Institute of Washington 1927). Heiles Cloud 2 is the dark area in the left-hand bottom corner.

sitions (mainly of CO) leak radio energy out allowing further cloud collapse to occur, ultimately to such high pressures and temperatures that new stars and planets can form.

I wrote to my friend Takeshi Oka (we had both been postdoctoral fellows at NRC during the "Golden Years" when NRC had been the outstanding National Research facility) and asked him whether he was interested in searching for HC_5N in space–he turned out to be as keen as was I. Together with NRC astronomers, Lorne Avery, Norm Broten and John McLeod, the molecule was detected in the giant molecular cloud SgrB2 towards the galactic centre [59]. The discovery was really a surprise because the molecule turned out to be much more abundant than expected–one more example of why one should never let a theoretical assumption deter one from carrying out an experiment.

The next question was clear–was HC_7N present? David Walton worked out a synthesis and Colin Kirby set about the demanding task of synthesising it and measuring the rotational frequencies [60]. After making it in the laboratory and searching for it in a dark cloud in Taurus, Fig. 5, using the 46 meter radio dish in Algonquin Park, Canada we found that it too was present, Fig. 6 [61,62,57]–it scarcely seemed possible. This family of molecules was making total nonsense of the abundance *rule-of-thumb* which almost all other families

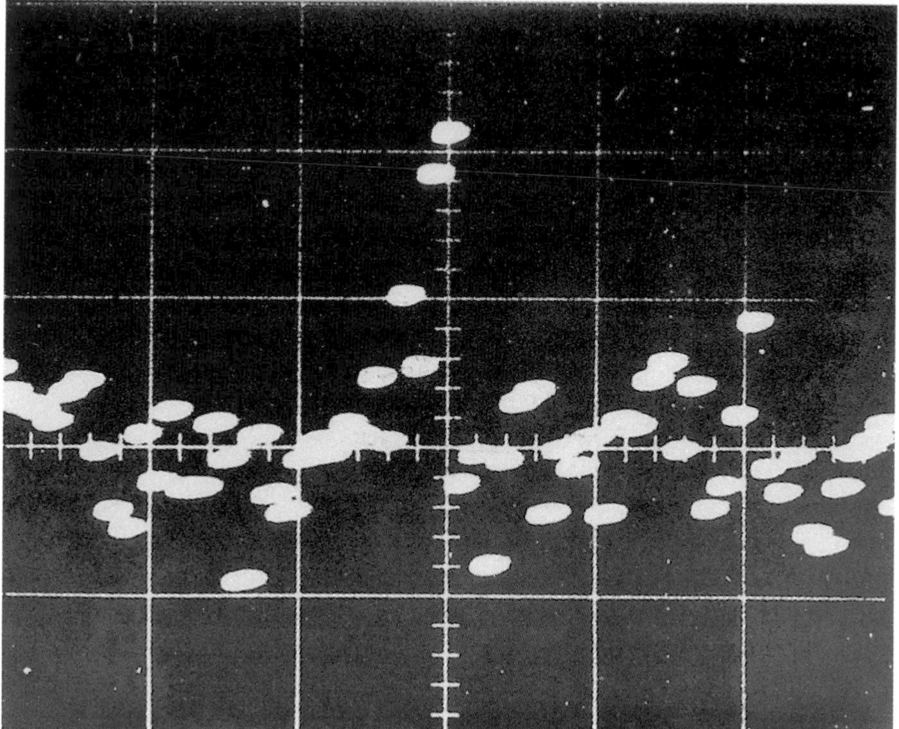

Figure 6. A photograph of the first oscilloscope trace of the radio signal due to interstellar HC_7N in Heiles's Cloud 2 in the constellation of Taurus (Fig. 5).

of molecules appeared to follow. Then, just as we at Sussex were about to attempt the daunting task of synthesising HC_9N, Takeshi found a clever way of estimating its rotational frequencies by extrapolation from the known frequencies of the set HC_nN (n=1,3,5,7) [63]; and we detected it as well [64]. These were then the longest and heaviest molecules conclusively detected in space–and still remain so nearly 20 years later. From these discoveries it became clear that there was a problem: The ion-molecule reaction theories of Klemperer, Herbst, Dalgarno and Black [65,66], which accounted almost perfectly for almost all the other species observed were just not able to account for the high abundances of such long carbon chains. It seemed to me that some alternative source was needed and fortunately a possible answer started to appear. As time progressed cool red giant carbon stars stars, such as the fascinating object IRC+10216 [68], were found to be pumping the chain molecules out into the ISM.

PHASE III

Ten Days in September 1985

During the early 1980's Robert Curl visited my laboratory at Sussex and invited me to come to Rice University in Houston at some time in the future. During Easter 1984, just after attending the Conference on Molecular Structure that Jim Boggs organised biennially in Austin, I took up the invitation and visited Rice. Apart from this superb conference another good reason for going to Texas is that there are excellent Half-Priced Book Stores in Austin, Houston and Dallas. When I arrived in Houston, Bob was enthusing about a very recent result that Rick Smalley and his group had just obtained. They had shown that SiC_2 was triangular [68]–a fascinating result. The way in which the Si had avoided forming a classical double bond in this species, at a stroke, explained why we had had so much difficulty observing molecules with Si=C bonds. It also fitted in nicely with the flexibility of C_3 and C_3O_2 which I had studied years before. I went over to Rick's laboratory where he enthusiastically described how his cluster beam machine worked. He explained how a pulsed laser, focussed on a metal disc vaporised material producing atoms which were swept up by pulses of helium which caused clustering and concomitant cooling. Then by, expansion through a nozzle into a vacuum, they are cooled further and by skimming into a pulsed beam which is analysed by a time-of-flight mass spectrometer the cluster mass distributions can be ascertained. As Rick described the mode of operation, it reminded me of previous work on carbon clusters, particularly by Hintenberger and coworkers [69] and I began to wonder whether, by substituting the metal target disc by a graphite one, the plasma produced by the laser might simulate the circumstellar shell of cool red giant carbon stars such as IRC+10216 and produce the carbon chains. This would give strong support for the stellar solution to the chain problem and provide an important alternative mechanism to the ion-molecule schemes.

As that day wore on, I thought more-and-more about the idea and became more-and-more convinced that Rick's apparatus was the key to proving a circumstellar source of the chains. Furthermore the detection of the spectrum of SiC_2 suggested that the apparatus might be able to confirm the conjecture of Alec Douglas [70] that the carbon chains might be carriers of the Diffuse Interstellar Bands (DIBs) [71]. The DIBs are interstellar spectroscopic features which have puzzled astronomers and spectroscopists since the 1930's. That night I described these ideas to Bob who was keen to collaborate–particularly on the second much more difficult DIB problem. In the event, in August 1985, Bob called me at my home in the UK to say that my experiments were to be carried out at Rice imminently and asked whether I was interested in coming to Houston. I didn't need to be asked twice–I dropped everything and within three days I was in Houston.

As soon as I arrived I gave a concentrated 2–3 hour seminar on all aspects of interstellar molecules and their spectra. I met Jim Heath, Yuan Liu and Sean O'Brien, the key students with whom I was to work intimately for the amazing ten-day period from 1st to 10th Sept 1985. Jim Heath and I struck an instant rapport partly because we are both addicted to books and enjoy browsing in bookstores. So after long stints on the apparatus (Ap2 as Rick affectionately called his second generation cluster beam machine) Jim and I would occasionally play hooky to visit Houston book stores. We would invariably end up, often accompanied by Carmen (Jim's wife) and Sean, late in the night at the House of Pies–a Houston 24-hour coffee bar which served outstanding Dutch apple pie. Apparently (according to a recent TV programme on the history of the PC) the founders of COMPAQ hatched their company at the House of Pies.

The experiments began on Sept 1st, and the good news is that they were successful. Almost immediately we detected the linear molecules with 5 to 9 carbon atoms that we had observed in space [72] and thus we got the evidence needed to confirm the idea that the chains could easily have originated in red giant stars. The bad news is that an interloper was present Fig. 7. As the results emerged, our attention was quickly attracted by the precocious antics of this uninvited guest. It had 60 carbon atoms and was accompanied by a more diminutive but still fairly prominent partner with 70 carbon atoms. These characters actually had already appeared in a publication some twelve months previously by Rohlfing, Cox and Kaldor in 1984 [73] and another group, Bloomfield et al. had also been probing their properties [74]. I had actually read the Exxon paper in some detail at the time it appeared because it described almost exactly the experiments I had proposed to the Rice group a few months previously. I had not however been moved to rub 60 brain cells together to divine what might be going on and apparently neither had anyone else. The Exxon group had made the discovery that a new family of car-

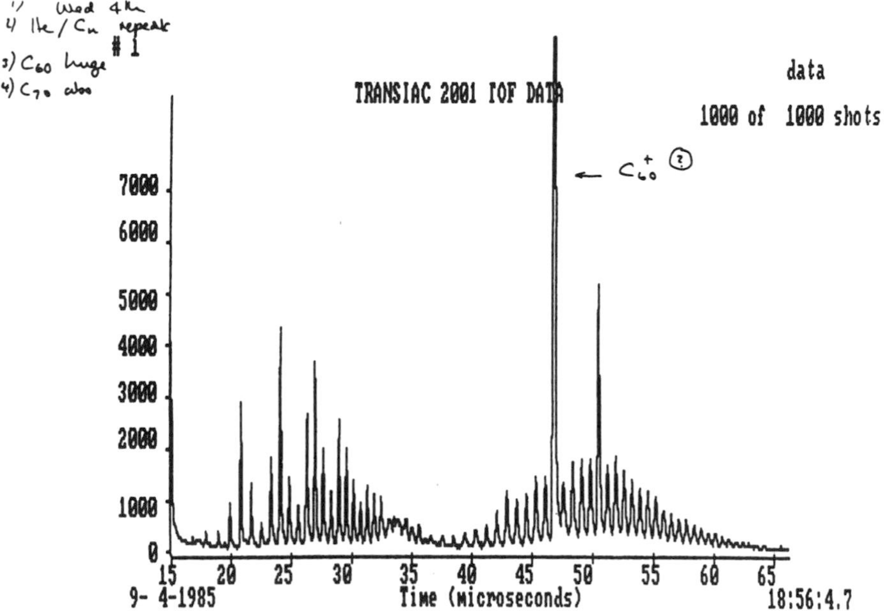

Figure 7. Time-of-flight mass spectrum, annotated by the author, of carbon clusters produced on Wed 4th Sept 1985 the day on which the dominance of the C_{60} signal was first recorded (see Fig 8a).

bon clusters with more than 30 atoms existed which appeared only to be even numbered. Though C_{60} was definitely more prominent than its neighbours, it was not sufficiently prominent to attract any particular attention, and was certainly not dominant as it was after Jim and Sean had finally finished with it.

The strongest peak in the 30–100 atomic mass range of the mass spectrum was usually the peak for C_{60} and it was used to tee up the experimental parameters for these larger species. During these studies, conditions were discovered which gave a distribution in which C_{60} was completely off-scale Fig. 7. This observation was made on Wednesday Sept 4th and our reactions were written up in the lab notebook by the graduate students Fig. 8. Two days later, on Friday, a group meeting took place at which Jim and Sean offered to work over the weekend to optimise conditions. Sean worked that evening and Jim took over the next day and worked all weekend Fig. 8. By Sunday evening

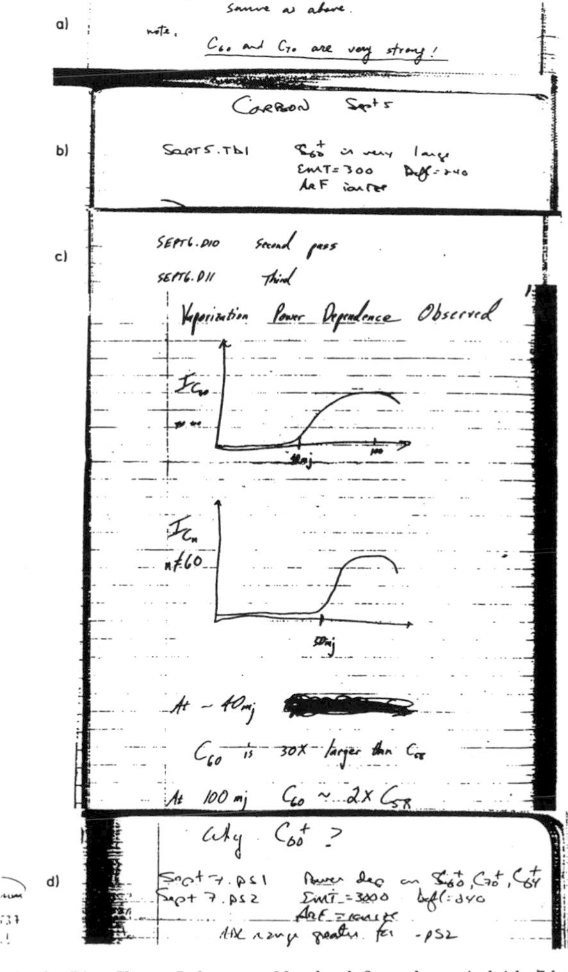

Figure 8. Entries in the Rice Cluster Laboratory Notebook from the period 4th–7th Sept 1985 when key experiments were carried out at Rice University. a) Sept 4th; b) Sept 5th; c) Sept 6th; d) Sept 7th. Entries by Heath, Liu and O'Brien.

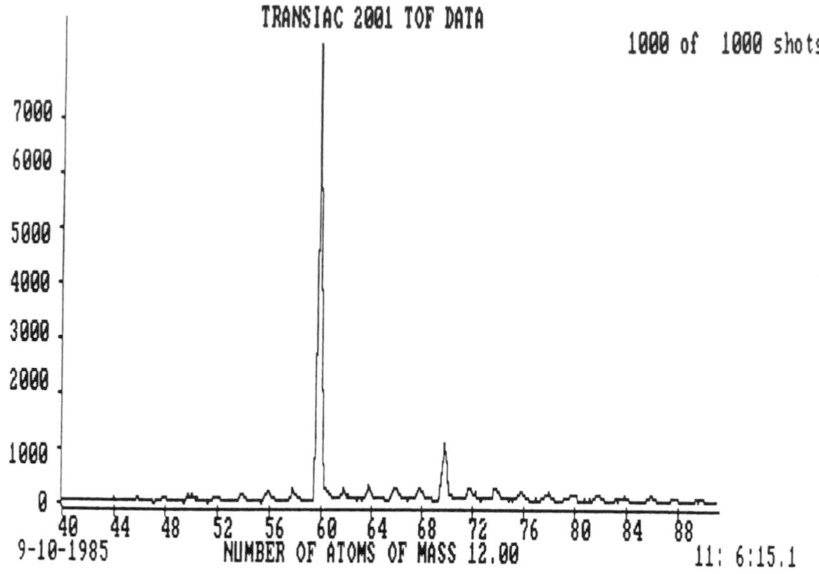

Figure 9. Time-of-flight mass spectrum produced by Heath of carbon clusters under the optimum conditions for the observation of a dominant C_{60} signal.

(8th Sept) the result was the spectrum shown in Fig. 9 which exhibited almost nothing but the C_{60}; with C_{70} as a somewhat smaller but still prominent chaperone.

As C_{60} and its sidekick became the prime focus of our attentions, I began to call these peculiar wadges of carbon the Lone Ranger and Tonto or sometimes Don Quixote and Sancho Panza. Our deliberations came to a climax during discussions on Monday 9th. C_{60} appeared to be really quite unreactive, a behaviour difficult to reconcile with a flat hexagonal graphene sheet–the most obvious first thought that would occur to anyone–which must have some 20 or more dangling bonds. After all, the chains had ends (two) and they added on two H atoms; thus a 60 atom flat hexagonal graphene sheet should add on ca 20 or more H atoms, but it did not. As we sought a solution, we discussed many possible structures that might explain the perplexing results.

Gradually a consensus developed that what might have happened was that flat hexagonal sheets had either formed or ablated from the surface of the graphite disc and closed into a cage, so eliminating the reactive edge. The closed hexagonal cage idea reminded me of a visit with my family to Expo67 in Montreal where Buckminster Fuller's dome had dominated the horizon. It particularly called to mind an image of the dome in an issue of Graphis magazine [75] devoted to the exhibition, Fig. 10. Rick went to the Rice Library and extracted a book about Buckminster Fuller and his inventions [35]. I also recalled a "stardome" skymap, Fig. 11a, which I had built for my children some years before. I remembered that the stardome had not only hexagons but also pentagons and wondered, as Bob and I went home for lunch, whether I should call my wife at home back in England to check whether or not it

Figure 10. Photograph taken by Michel Proulx of the Geodesic Dome designed by Buckminster Fuller for the US exhibit at Montreal EXPO67 [75]. One of the pentagons necessary for closure is discernable in this photograph. The variation in the strut-lengths of the hexagons in the vicinity of the pentagon, necessary to achieve a relatively smooth round surface, can also be discerned.

had 60 vertices. As I was scheduled to leave for the UK the next day, I invited the team out for dinner at a Mexican Restaurant to celebrate the exciting discovery. Needless to say we spent all the time at the restaurant trying to solve the puzzle.

That evening/night Rick experimented with sheets of hexagons, Jim together with Carmen experimented with toothpicks and jelly beans and Bob and I discussed the stardome solution again. Rick could not make any progress until he remembered our discussion of the pentagons in the stardome [19]. Rick discovered that the structure started to curve into a saucer shape as soon as pentagons were included among the hexagons and finally closed as he added a twelfth pentagon. The next morning, when he revealed his paper model, Fig. 11b, I remember being ecstatic. It was beautiful and looked just like the stardome, Fig. 11a, as I remembered it. It is of course a truncated icosahedron and the fact that it turned out to be a football too was most appropriate. After all, the whole discovery story is an archetypal example of team effort. I remember thinking that the molecule was so beautiful that it just had to be right–and anyway even if it were not, everybody would surely love it, which they did–eventually! My suggestion that we call the molecule Buckminsterfullerene (the -ene ending fitted perfectly) was, after some discussion, accepted and we sent off the paper to Nature–the date of receipt was 13th Sept (NB–the experiments had started on 1st Sept).

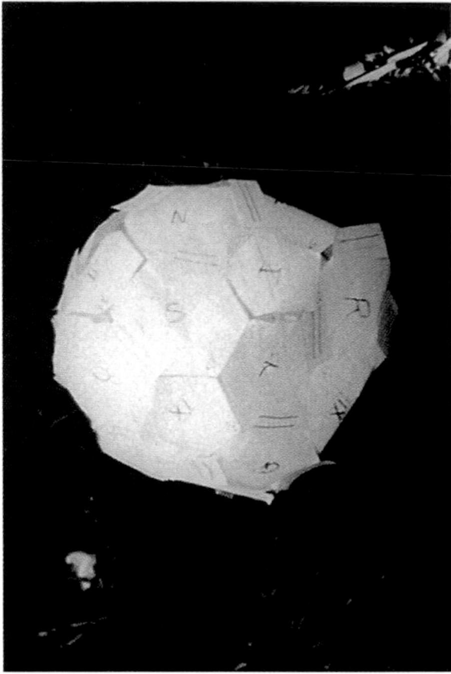

Figure 11. The two card models which played key roles in the positing of the truncated icosahedral structure of C_{60}. a) (left) Stardome map of the sky (Buckminster Fuller patented t-icosahedral, and other polyhedral, world map projections); b) (right) protype C_{60} made by Smalley. Both models are truncated icosahedra with 60 vertices, 12 pentagonal and 20 hexagonal faces.

PHASE IV

Little Fullerenes, Giant Fullerenes, Red Solutions, and One-line Solutions, or–How to Play Football Without Knowing the Rules

We decided to probe the consequences of the C_{60} discovery and experiments were initiated at Rice [76]. Jim Heath, Sean O'Brien, Yuan Liu and Qing Ling Zhang were among the most heavily involved students. A series of experiments was also initiated at Sussex: With Tony Stace, a cluster beam machine was constructed by a highly dedicated set of students (Simon Balm, Richard Hallett and Wahab Allaf) and we thus started to probe cluster behaviour in general. With the aid of support from British Gas (via Steve Wood) we also started to probe the implications of our discovery for combustion.

One of the most charismatic things about C_{60}, and one which continually lay simmering in the back of my mind, was the fact that all the carbon atoms in C_{60} are equivalent and so the ^{13}C nmr spectrum should consist of but a single elegant line. However, how could one make enough to measure it?. That was a daunting task and I must admit that whenever my thoughts drifted onto this subject they invariably ended with the conclusion that one day some young, ingenious, upstart, synthetic organic chemist would surely observe it first. I certainly never entertained much hope of observing the line at Sussex and so it was not really an objective. It is important to understand how impossible such a task appeared at that time. My attitude to the research was to probe those aspects of the C_{60} discovery which particularly puzzled me and let the research flow whichever way it would.

A few days after the discovery I arrived back in Europe and announced our discovery and our conjectured structure at a conference in Riccione. Julie August, a former student now working at Nottingham, told Martyn Poliakoff of the discovery and he sent me a zerox copy of the idea that David Jones had had in 1966 [6,7]. Writing under the pseudonym of Daedalus in the Ariadne column of the New Scientist David had proposed the amazingly imaginative idea that large carbon balloons might be feasible. We also discovered that C_{60} itself had already been proposed by Osawa in a Japanese article 15 years earlier [2] and in a book with Yoshida [3]. Not only that, Bochvar and Gal'pern had even carried out a Hückel calculation and published the pattern of the molecular orbitals [4,5]. We also discovered that Orville Chapman at UCLA had initiated a programme aimed at the synthesis of the molecule (private communication). In retrospect it is rather peculiar that this charismatic molecule had not attracted the attention of the chemistry community earlier. Jones, in his articles [6,7] introduced me to D'arcy Thompson's book on Growth and Form [77] and pointed out that Euler's Law indicated that no sheet of hexagons could close. If however 12 pentagons were introduced into a hexagonal sheet of any size it would close. Almost everybody (who knows any organic chemistry!) knows that unsaturated molecules with adjacent pentagonal rings are extremely unstable. Thus from Jones' article it became immediately obvious that as 5x12=60, C_{60} had to be the smallest cage able to close without abutting pentagons. Thus the secret of C_{60}'s stability was beautifully simple. It lay in the accommodation of the two requirements: Euler's 12-pentagon closure principle and the chemical stability conferred by pentagon non-adjacency. I consider these two points to be the "2+2" of Fullerene stability. Schatz, Seitz, Klein and Hite, using circuit theory, placed these principles on a firm theoretical basis [78]. The C_{70} peak was, as mentioned previously, also prominent and an elegant and a plausible structure was found by Rick in which two C_{30} hemispherical halves were separated by a ring of ten extra C atoms was proposed in the paper describing the crucially important discovery, which Jim Heath made, that atoms (such as La) could be put inside the cage [79]. This discovery (and its refinements), that endohedral complexes could form, was perhaps the strongest *empirical* observation supporting the cage proposal prior to extraction in 1990.

During some moments of musing over the above two stability criteria I began to wonder about the identity of the next cage (after C_{60}) which might be able to close without abutting pentagons. I started to play around with a model of C_{60} by adding atoms and before I had made much headway it suddenly struck me that perhaps closure could not occur again until C_{70}, Fig. 12–if that were the case it would explain–at a stroke–the prominence of C_{70}. That seemed to me a most unexpected and certainly not a very obvious result.

Furthermore a subtle circular argument appeared–if it were true, then closure demanded that C_{70} be stable and that would be, for me, by far the most convincing support for the fullerene proposal and there was not a hope-in-hell of an alternative explanation of *both* the 60 *and* 70 magic numbers. I knew from that moment on that one day we would be proven right.

A call to the Galveston group to see whether they could prove the conjecture was in order. To my absolute delight Tom Schmalz told me that they had already shown that cages with 62, 64 and 66 atoms could not close without abutting pentagons and that the analysis on 68 atom cages was in mid-stream at that very moment. Thus the Pentagon Isolation Rule, which governs general fullerene stability, was born [80,81]. Another thought struck me during the phonecall with Schmalz. I remembered that C_{50} was often a magic number in some of our experiments and suggested that they might check whether the 50 atom cage might be the smallest cage to close without triplets of abutting pentagons. Schmalz *et al.* showed this to be the case too [81]. I also wondered about a result that Sean had obtained [82]. He had found that laser irradia-

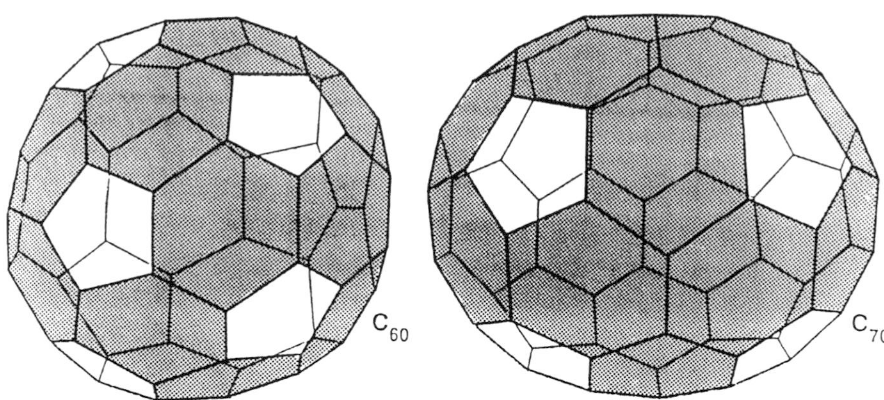

Figure 12. For species with less than 72 atoms these two structures are the only ones which can be constructed without abutting pentagons. Thus on the basis of the pentagon isolation principle [80,81] and geodesic considerations, C_{60} and C_{70} are predicted to be the first and second fullerene magic numbers respectively. This result provided the simplest and most convincing circumstantial evidence in support of the closed cage concept prior to extraction.

tion of C_{60} caused it to fragment by reducing the size of the clusters more-or-less sequentially by even numbers from 60 to 58 to 56 etc. .... all the way down to 32. After 32 further irradiation blew the cluster into small fragments. I tried to push the pentagon multiplet isolation idea to quartets and in the hope that it would explain the C_{32} result but to my amazement and delight an elegant C_{28} structure [80] formed in my hands, Fig. 13. This reminded me of a result that had puzzled us for some time–in some of our runs the 28 carbon atom signal sometimes rivalled that of C_{60}. It also struck me that this species should be a sort of superatom cluster analogue of the carbon atom with effective tetravalency suggesting that an elegant tetrahedral $C_{28}H_4$ derivative,

Fig. 13, might actually be a stable molecule [80]. Indeed in some Exxon data [83], I found just the mass spectrum distribution which confirmed that the isolation principle could be generalised to include various sizes of multiplets and account for magic numbers of smaller cages down to C_{20} [80,13]. It not only exhibited magic numbers at 60, 50 and 28 but it stopped abruptly at 24. From Patrick Fowler I had learned that no cage could be constructed with 22 atoms [84]. Thus there appeared to be "semistable" fullerenes (at least in beams) down to C_{20} which were predictable, Fig. 14, on the basis of the cage closure concept. What more "proof" could one possibly want for the whole fullerene concept? None.

Then one day I decided that we should build our own Buckminster Fuller domes, or rather molecular models of the giant fullerenes [85] and ordered 10 000 carbon atoms (molecular model atoms and bonds). Ken McKay, armed with Goldberg's paper [86] and Coxeter's book [87] set about building C_{240}, C_{540} and later C_{960} and C_{1500} with icosahedral symmetry. When Ken came in with the model of C_{540} it was beautiful but I could not quite understand its shape–the model was not round like Buckminster Fuller's Montreal

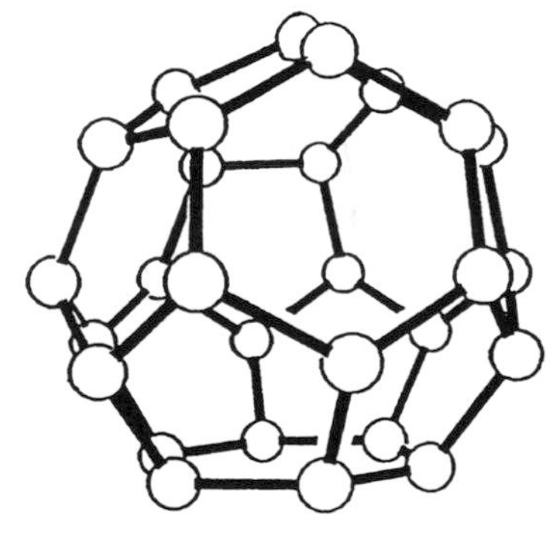

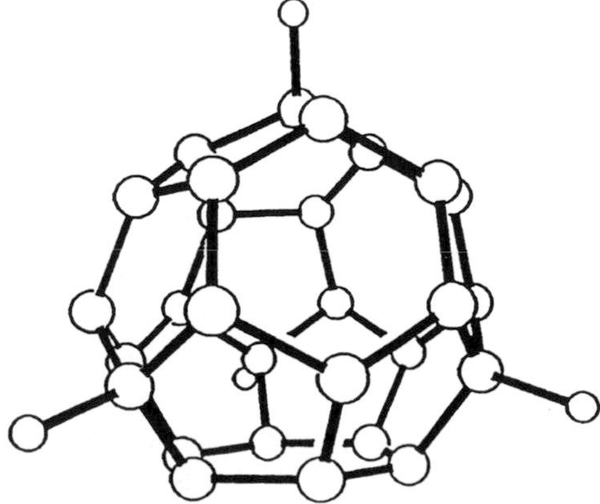

Figure 13. The tetrahedral small fullerene C_{28} (above) and the tetrahydrogenated derivative, $C_{28}H_4$ (below), which is expected to exhibit some stability [80].

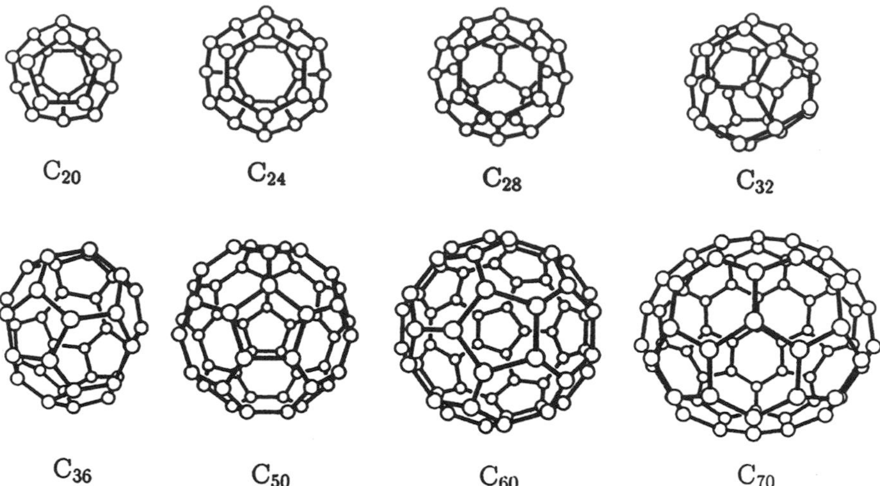

Figure 14. A set of fullerenes between C_{20-70} some of which are magic and others which are predicted to be "semi"-magic [80,81].

dome but had clear icosahedral tendencies, Fig. 15 [90]. Indeed Ken's model had cusps focused at the 12 pentagons and from a distance had a definite polygonal outline. Then we realised that the structure explained some very interesting results that Sumio Iijima had obtained in 1980 [89]. He had observed concentric shell onion-like carbon particles by transmission electron microscopy. Superficially many of Iijima's particles appeared to be round, as one might assume if they were nanoscale geodesic domes. However because of this assumption I had glossed over some crucial subtle features in the

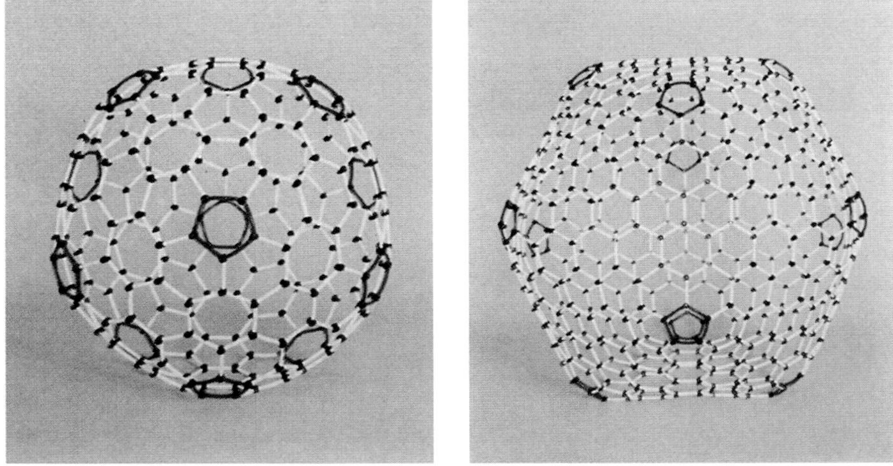

Figure 15. Photographs of molecular models of the giant fullerenes C_{240} (left) and C_{540} (right). The models turned out to possess icosahedral shapes rather than perfectly spheroidal ones similar to the geodesic domes. Each of these structures can be considered as an initially flat hexagonal network which has warped into a closed monosurface (of icosahedral symmetry) by 12 pentagonal disclinations which give rise to the cusps. These structures provided the first explanation of faceting in closed carbon species from giant fullerenes to nanotubes.

micrographs–the outlines were not quite as round as they superficially appeared–they were actually polygonal [88,13]. In fact as we looked more carefully at the infrastructure of Buckminster Fuller's domes we realised that the strut lengths in the vicinity of the pentagons had been adjusted to give them a smooth spheroidal shape, Fig. 10. Thus the C_{60} discovery had led to an elegant explanation of a previously unexplained result. It showed that a cage with an icosahedral shape could be constructed entirely from sp^2 carbon atoms. Thus 12 pentagonal disclinations–a minimal number of defects by comparison with those found in any normal crystals–would convert a perfectly flat graphene sheet of any size into a closed cage.

There is a rather nice lesson to be drawn from this advance: There had been no premeditated research aim, the original reason for constructing the giant fullerene model was solely for the intrinsic interest and pleasure of building an elegant structure–and yet it revealed an unexpected important result. The result had important consequences–for instance it was the first study of the way that the inclusion of pentagons influences the curvature of a graphene sheet–a result that was to find particular use later when elongated giant fullerenes (nanotubes) were discovered. On occasion I would wonder about our cage proposal, re-assess each and every observation: 60 and 70 should be magic numbers; 50, 28 were also explainable magic numbers; there was no clear evidence of C_{22}; the icosahedral shape of the giant fullerenes could be explained; Jim had observed metal complexes that behaved as though they were endohedral [79]; Sean had found that C_2 units were ejected [82] and there were other supporting experimental results [12,19]. It all fitted perfectly–there was no way we could be wrong.

I thought we should probe the Iijima result and see whether C_{60} might reveal some clues about the round particles in general. This was based on the idea we had proposed [90] that the mechanism of C_{60} production might be related to soot formation. We were able to acquire an old carbon arc evaporating unit and drill a hole in the base plate to admit He. Ken McKay monitored the structure of the carbon deposit by scanning electron microscopy as a function of He pressure. The idea was that as the He pressure increased C_{60} formation might be initiated and this would be accompanied by the creation of round carbon particles. A change in the deposit did occur as the pressure rose above 50 microns [20]. The next step seemed obvious–to monitor the possible production of C_{60}. However I made a fatal error. I assumed that if C_{60} were created in the arc, it would only be in minuscule quantities and only the most sensitive of detection techniques available (mass spectrometry) would work. I then tried, for the next two years (unsuccessfully), to obtain the financial support for quadrupole mass spectrometer.

I was still trying, with Geoff Cloke, to get a mass spectrometer when we learned of a study by Krätschmer, Fostiropoulos and Huffman who had detected four infrared bands, Fig. 16, (exactly as predicted theoretically [91-93]) in a carbon deposit made in exactly the same way as ours [16]. If this were correct then their deposit (and ours also!!!) must contain at least 1% C_{60}. I could not believe it–and yet there were four sharp resonant frequencies in the pure carbon deposit–what on earth could they be. I decided to check out this incredible (in its literal meaning) result and proposed it as a third year project for Amit Sarkar (an undergraduate) to work with Jonathan Hare (graduate student) to see whether they could repeat it–and they did!!! [20]. It is difficult, today, to explain just how difficult it was to accept the Krätschmer *et al.* observations but it is important, for an understanding of how science advances, to try. How could C_{60} have remained undetected until the end of the 20th century if it could be produced in such a yield by this seemingly simple approach?

We then tried to see whether the deposit gave a mass spectrum and Ala'a Abdul Sada obtained a 720 mass signal, Figs. 17 and 18a–it still seemed impossible. I was still very suspicious, especially as I knew that C_{60} can be formed readily during the mass spectrometric sampling process. The mass spectrometer then broke down (Fig. 18a)! Over the years we had often discussed the likely physical properties of C_{60}–some thought it would be a high melting point solid, others thought it would be a liquid or even a gas. On a Friday evening 3rd August Jonathan decided, with the unblinkered optimism that only the young possess, to test whether the species would dissolve in benzene, Fig. 18b, and amazingly produced a red solution which he placed on my desk on

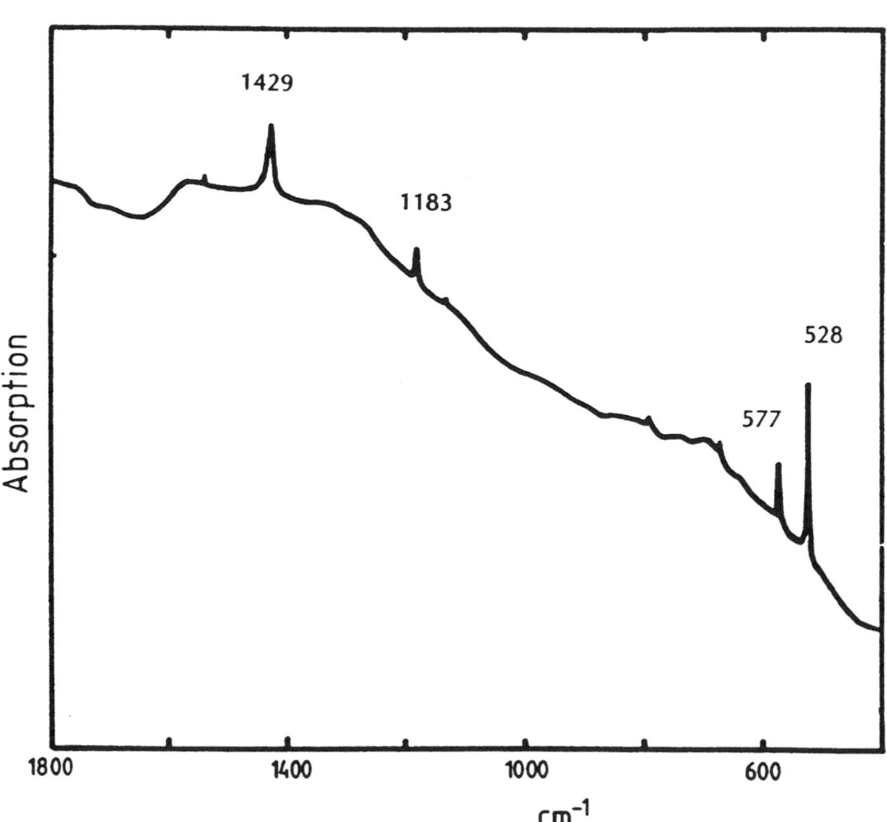

Figure 16. The four historic IR bands observed by Krätschmer, Lamb, Fostiropoulos and Huffman, from [16].

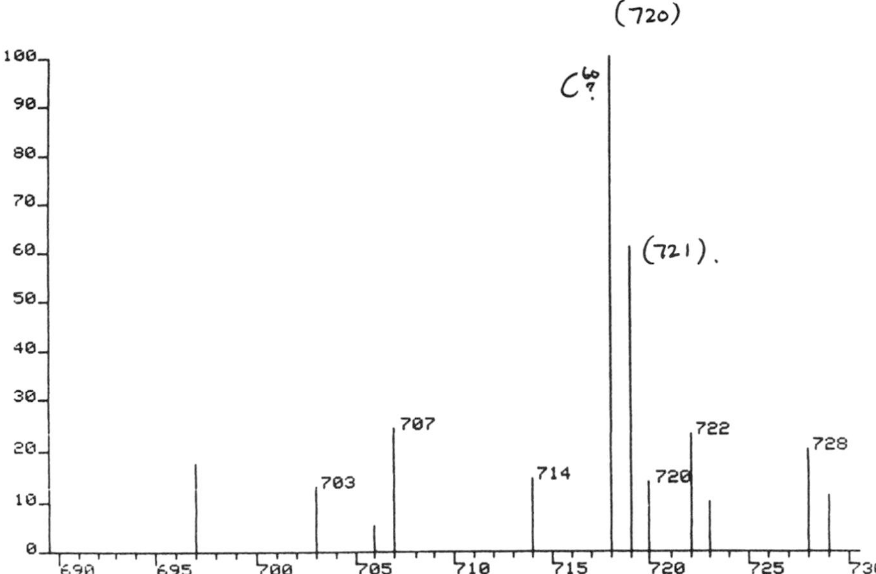

Figure 17. Fast atom bombardment (FAB) mass spectrum of a deposit of arc-processed soot obtained on 23rd July 1990 at Sussex by Ala'a Abdul-Sada. The machine calibration is out by 2 amu however the isotope pattern was convincing as the peaks are close to the intensity ratio 1.0:0.66:0.22 as expected for $0^{12}C_{60}{:}^{12}C_{59}{}^{13}C{:}^{12}C_{58}{}^{13}C_2$.

Monday morning (6th), Figs. 19 and 18c. Jonathan seemed to have no difficulty in accepting that it was C_{60} as is indicated by the fact that he called it $C_{60}{}^{B}$ (Fig.18) which was our shorthand notation for Buckminsterfullerene. On the Thursday some of the solution was concentrated to see whether a deposit could be obtained which would yield a stronger 720 mass spectrum–but it didn't, Fig.18d. This was exactly the right experiment and actually a rather difficult one for technical reasons and would almost certainly have worked the next time–*had there been one(!)*.

The next day, disaster struck–I had a call from *Nature*; Philip Ball asked whether I would referee a new paper by Krätschmer *et al.* When the ms arrived by fax at around 12 noon it was a bombshell. Krätschmer *et al.* had also obtained a red (!!!!) solution and what is more their solution had yielded crystals, Fig. 20, which x-ray analysis conclusively indicated consisted of arrays of spheroidal molecules exactly the right size to be C_{60}. There was absolutely no doubt about it–Wolfgang Krätschmer, Lowell Lamb, Kostas Fostiropoulos and Don Huffman had done it [15]. It was only at that moment that I suddenly realised that we had been in a race and–we had been pipped at the post. I considered committing suicide–but decided to go for lunch instead. Over lunch I read and re-read this fantastic manuscript and I doubt whether I shall ever have an experience like it again. It was a very strange mix of alternating feelings: First, exhilaration from being proven right, and then disappointment from having come so close to proving it ourselves. Immediately after lunch I called back to Philip Ball and told him to accept the paper immediately.

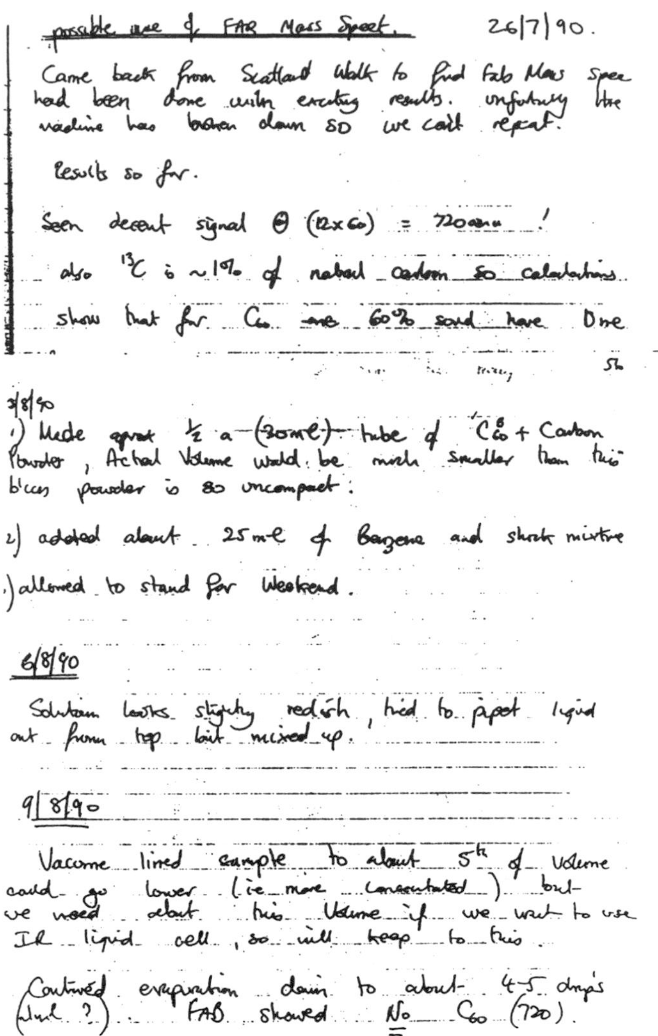

possible use of FAB Mass Spect. 26/7/90.

Came back from Scotland Walk to find Fab Mass Spec had been done with exciting results. unfortunately the machine has broken down so we can't repeat.

Results so far.

Seen decent signal ⊖ (12x60) = 720amu !

also ^{13}C is ~1% of natural carbon so calculations show that for C_{60} ~ 60% should have One

3/8/90

1) Made approx ½ a (30ml) tube of C_{60} + Carbon Powder, Actual Volume would be much smaller than this b'cos powder is so uncompact.

2) added about 25ml of Benzene and shook mixture

) allowed to stand for Weekend.

6/8/90

Solution looks slightly reddish, tried to pipet liquid out from top but mixed up.

9/8/90

Vacume lined sample to about 5th of Volume could go lower (ie more concentrated) but we need about this Volume if we want to use IR liquid cell, so will keep to this.

Continued evaporation down to about 4-5 drops (dual ?) FAB showed No C_{60} (720).

Figure 18. Entries by Jonathan Hare in his laboratory notebook: a) 26/7/90 b) 3/8/90 c) 6/8/90 d) 9/8/90.

But what to do now, it was difficult to think at all let alone make any sensible decisions. However as the afternoon wore on it gradually dawned on me that although almost everything was lost–it was not quite all. Krätschmer *et al.* had analysed their material by x-ray crystallography and not by nmr. So, the much-coveted single nmr line was still waiting to be observed. Thus I realised that there was just a single crumb of comfort left, but what a tasty morsel it would be and we should be able to observe it. (The nmr shift should also indicate just how aromatic the molecule was or was not. Of course the meaning of aromaticity for a species that cannot undergo substitution is debatable!) After all Jonathan had extracted the precious red solution on the Monday, four days before the extraction ms arrived on the Friday and we also had the confirmatory 720 amu mass spectrum. So we already had material in our hands and we were the only ones apart from Krätschmer and colleagues who

Figure 19. Original "reddish" extract obtained by Jonathan Hare on Monday 6th Sept 1990.

had. We could also take immense intellectual satisfaction at having extracted the red solution before the arrival of the paper of Krätschmer and colleagues. I suppose that that should be enough for a scientist–but of course it never is.

In 1982 Krätschmer and Huffman had observed puzzling UV absorption features while studying carbon dust produced by a carbon arc to simulate interstellar dust in the laboratory and measure the optical spectrum. After the publication of our discovery and theoretical predictions of the UV spectrum such as those by Rosén and co-workers [94,95] and others [9,10] they conjectured that the features might be due to C_{60}. They then set about testing their idea by searching for the four tell-tale IR lines which theory suggested the molecule should exhibit [91–93], Fig. 16. When they found them they went on to make C_{60} from pure ^{13}C and show that the isotope shifts were perfectly consistent with their conjecture [17]. They then extracted crystals, Fig. 20 and proved their conjecture conclusively. I look upon the conjecture that they had made C_{60} in 1982 together with the decision to test it by analysing their deposit by IR as one of the most prescient pieces of science I have ever come across and as time has progressed I have come to appreciate it more

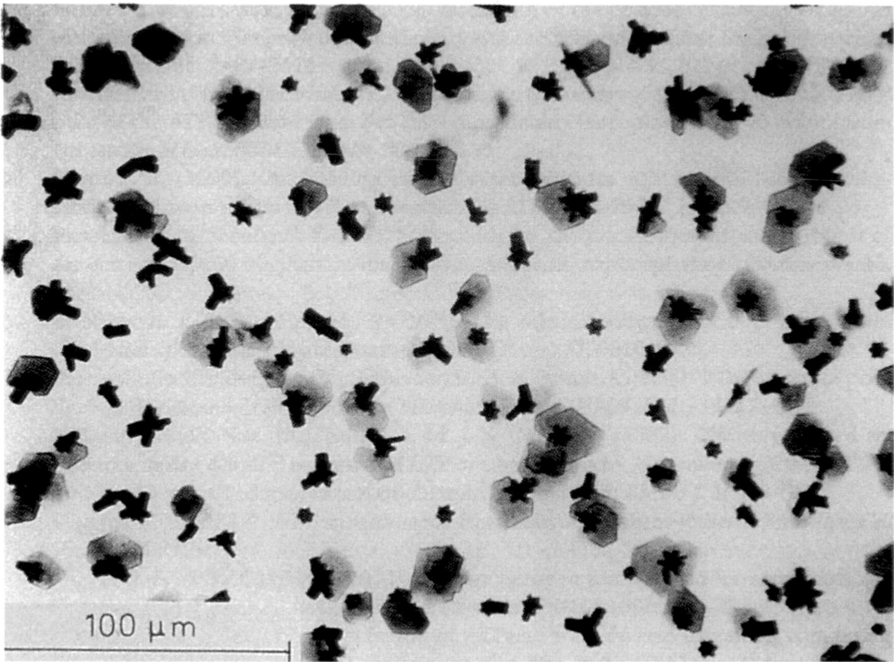

Figure 20. The astounding photograph of crystals of pure carbon which Krätschmer, Lamb, Fostiropoulos and Huffman extracted during their breakthrough in extraction [15]. Photograph supplied by Wolfgang Krätschmer.

and more. The role played by spectroscopy–in both the initial UV observation and the crucial tell-tale IR fingerprinting is an archetypal example of the way spectroscopy can be used to solve important problems, in this case an historically significant one as well. To my mind there is no more amazing discovery than that there is a form of pure carbon that is *soluble* and the image of the carbon crystals of Krätschmer *et al.,* shown in Fig. 21 has to be one of the most sublime images of 20th century chemistry.

During the tea-break that afternoon I showed the paper to my colleagues. Roger Taylor offered to help us to get the one-line solution and immediately tried to extract as much material as he could from Jonathan who was somewhat reluctant to part with the small amount of the precious material that he had. Roger then made a key discovery, he found that the red solution could be chromatographically resolved into two components, one was red and the other a beautiful delicate magenta. (Chromatography has since become the accepted way to separate all the members of the fullerene family.)

Would nmr analysis show the single line? Tony Avent who operated the nmr spectrometer produced a two(?)-line spectrum from the magenta material, Fig. 21. One was a magnificent strong line, but it was benzene–quite a well known compound. The second was a puny little blip in the base line–but Tony reckoned it was C_{60}–and he was right. To us it was the most beautiful little blip one could ever imagine. After further refinement the single line was observed as a strong signal Fig. 22a. The icing on the cake was the red material which gave five nmr lines–exactly what was expected for C_{70} [96], Fig. 22b.

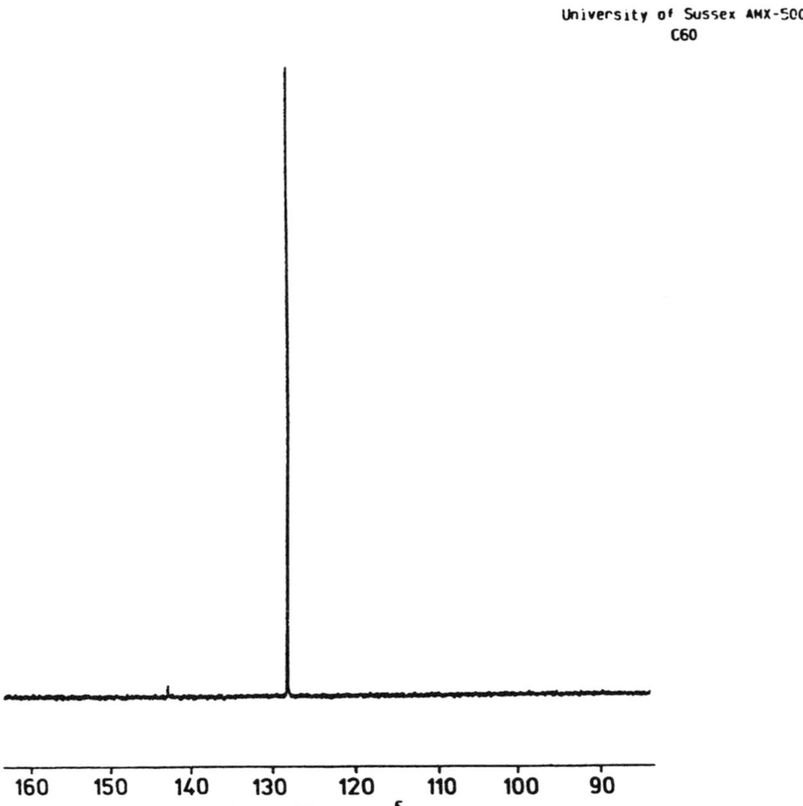

Figure 21. The first NMR trace in which the C_{60} resonance (at 143 ppm) was first identified (just!). The strong line at 128 ppm is (rather appropriately) the resonance of benzene.

Thus in summary, Krätschmer, Lamb, Fostiropoulos and Huffman had indeed extracted C_{60} and at Sussex we had come a very close second. However Jonathan had extracted a red solution and Ala'a had detected a mass spectrum independently and before the Heidelberg/Tucson (*Nature*) ms had arrived. After this Roger had discovered the chromatographic separation of the fullerenes. Tony had detected the much coveted single line (Meijer and Bethune at IBM were also close and with Johnson also observed a single line nmr spectrum [97]). As a certain football manager, noted for his "eloquence", in adversity would say–"The lads dun good!"

So in September 1990 some 20 years after the molecule was conceived by Osawa and five years after we had discovered that it could self-assemble, Krätschmer *et al.* had extracted it and Fullerene Science was well and truly on its way.

PHASE V

C_{60} Buckminsterfullerene, Not Just a Pretty Molecule

After the extraction breakthrough in 1990, the field of Fullerene research exploded and now some thousand papers a year are published in the field.

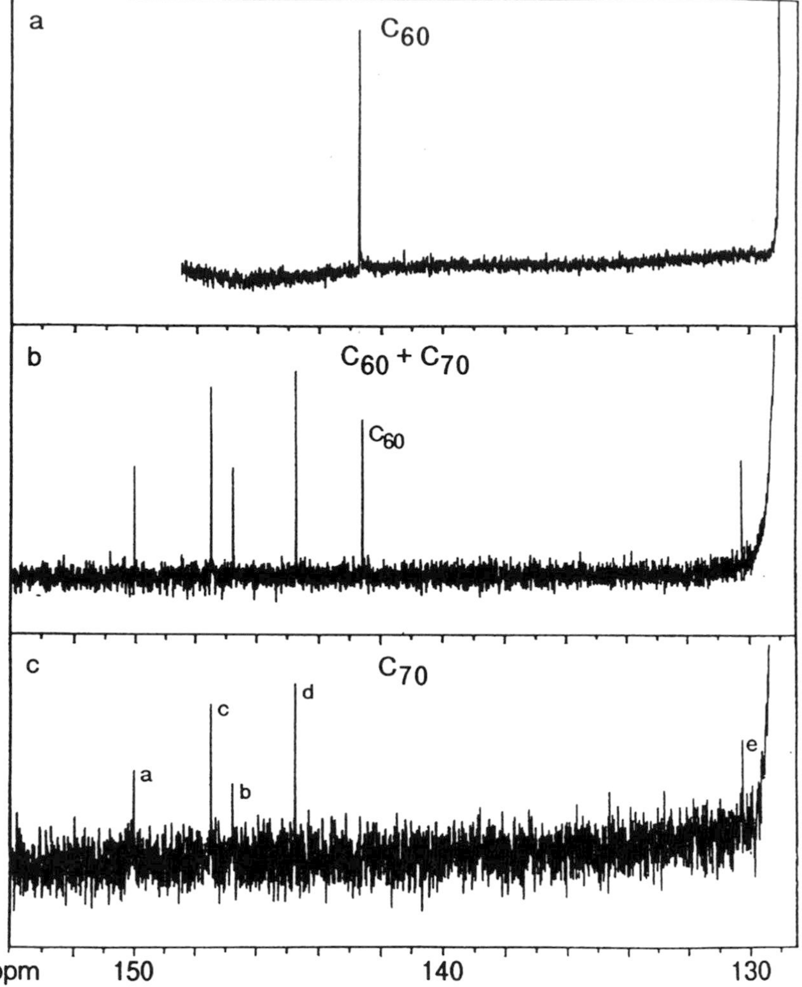

Figure 22. ^{13}C NMR spectrum obtained from chromatographically purified samples of soluble material extracted from arc-processed graphite. a) spectrum of purified C_{60} (magenta fraction) b) mixed sample c) spectrum of purified C_{70} (red fraction) showing five-lines as expected for the symmetric egg-shaped molecule, Fig. 12.

Several new research projects, designed to explore the chemical and physical behaviour of the fullerenes were launched at Sussex. With Roger Taylor and David Walton fullerene synthetic chemistry became a primary aim [98]. Roger Taylor and others pioneered the extraction of higher fullerenes C_{76}, C_{78}... etc [31]. Spectroscopic studies were carried out with John Dennis and Jonathan Hare and also in collaboration with Sydney Leach and colleagues in Meudon/Orsay. Halogenation studies have turned out to be particularly important and Paul Birkett was able to show that $C_{60}Cl_6$, Fig. 23, can form [98]. These observations which may be considerd to be the fullerene equivalent of *ortho, meta* and *para* direction in benzene have been followed by the creation of some key acylated derivatives. Adam Darwish has probed a wide range of reactions in particular hydrogenation and hydroxylation reactions.

Interesting mixed crystalline compounds involving ferrocene and phosphorus have been made and structurally characterised by Jonathan Crane, Wyn Locke and Peter Hitchcock [98]. Further chemical properties of C_{60} as a synthon, including for instance cycloaddition and metal complexation reactions, have been investigated by Mohammed Meidine, Brian O'Donovan and Martin van Wijnkoop and solubilisation methods have been developed by Huang Shaoming. Colin Crowley and Andy Caffyn have been investigating thermolytic routes to C_{60}. Fragmentation behaviour is also being investigated with Perdita Barran, Steve Firth and Tony Stace (in collaboration with Eleanor Cambell of the Max Born Institute in Berlin). These studies followed on from the work of Simon Balm, Richard Hallet, Ken McKay and Wahab Allaf. Cold fluorescence studies are being carried out with Reg Colin, Pierre Coheur, Steve Firth, Michele Carlier (in Brussels) and Eleanor Cambell (in Berlin).

A solid state chemistry programme has been initiated by Kosmas Prassides in which fascinating solid state dynamics and phase behaviour [99] as well as superconducting properties of the metal intercalation complexes have been probed [100]. With Fred Wudl's group, the elegant azafullerenes are under detailed investigation [101].

On the nanotechnology front some fascinating advances have been made [102]. Collaboration with Morinobu Endo of Shinshu University in Japan has opened up several new avenues in the creation of nanostructures by pyrolytic

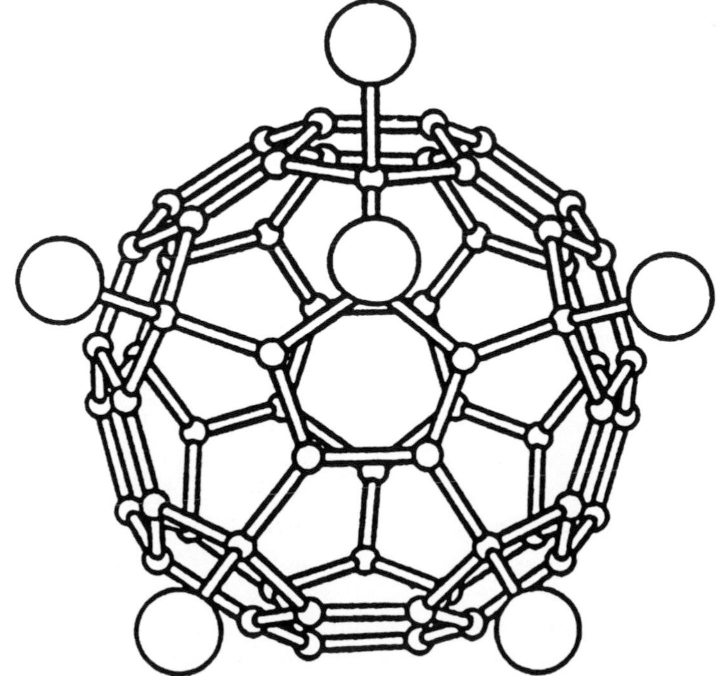

Figure 23. The structure of $C_{60}Cl_6$ [98].

methods and possible nanotube growth mechanisms. With Mauricio Terrones, Wen Kuang Hsu and Jon Hare, studies of the fascinating way that nanostructure creation is governed by metal cluster catalysts are being investigated (with Umberto Terrones in Mexico). Related investigations of these promising routes to nanoscale materials are under way with Thomas Muller, Doug Reid and Nicole Grobert. Wen Kuang Hsu has observed the amazing result that nanotubes can be obtained by electrolysis [101]. Just as exciting as the new carbon structures are the boron nitride nanotubes. Some of this nanotechnology work is being carried out in collaboration with Tony Cheetham and Xing Ping Zhang at UCSB and Laurence Dunne at the University of the South Bank. All in all the way ahead looks exciting as is exemplified by the nanostructure shown in Fig. 24, which appears very much like an extra which has escaped from the set of the film Alien II.

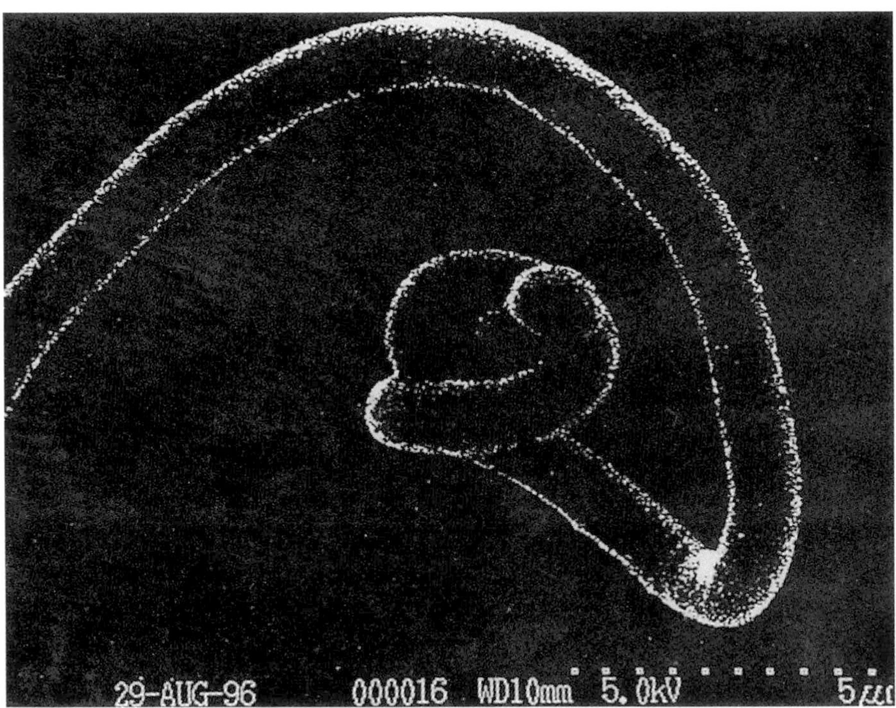

Figure 24. Amazing carbon nanostructure created by thermolysis.

EPILOGUE

The Cosmic and Microcosmic Charisma of the Soccerball

One cannot end any account of the discovery of C_{60} without noting that it has the same structure as a football (or soccerball) and that this led to its original conception. Thus it seems fitting to reproduce in Fig. 25 the image from the book by Osawa and Yoshida [2,3] and point out that much of the world now plays football (or soccer). Furthermore a large number of researchers are studying the peculiar behaviour of the ball on a microcosmic scale. Last but not least we should note that, perhaps, the molecules most delightful property lies in the inherent charisma [37] which arises from its elegantly simple and highly symmetric structure that is quite unlike any other. It is this charisma that has stimulated delight and fascination for chemistry in young and old alike. Perhaps no image captures the most important quality of C_{60} better than the photograph depicted in Fig. 26. It says it all.

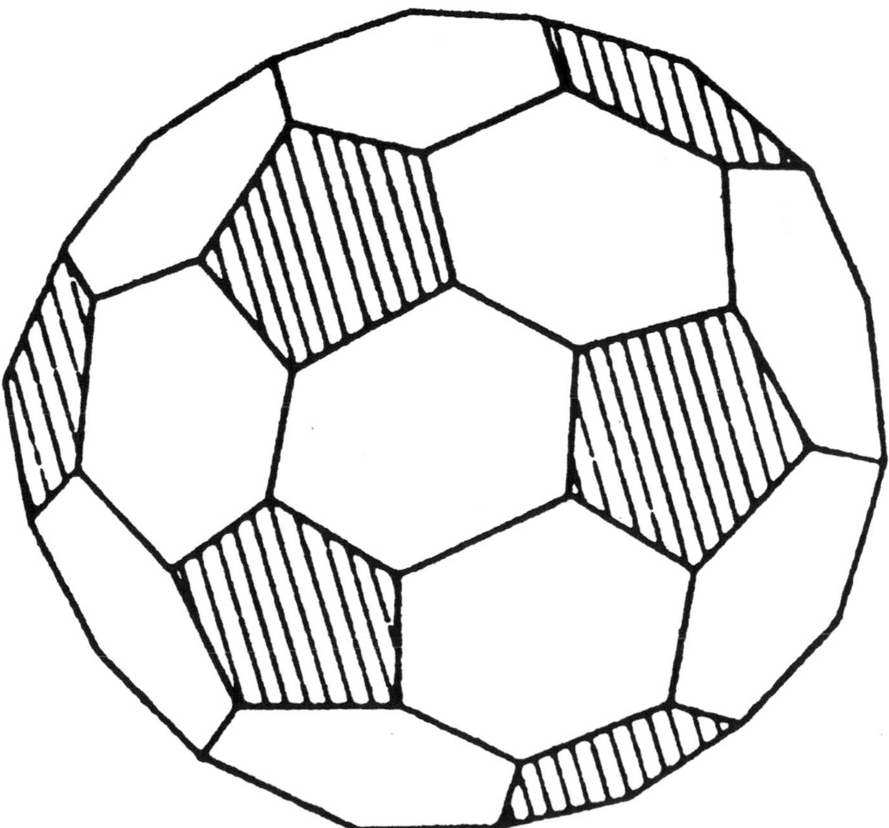

Figure 25. The first image of C_{60} ever published (Osawa and Yoshida [2,3]–(taken from ref 3).

Figure 26. This photograph of Ellis captures perfectly the delight that C_{60} can stimulate in youngsters of all ages (photographer–G. E. Watson, published with permission).

ACKNOWLEDGEMENTS

There are many people to whom I owe deep debts of gratitude in addition to the students, colleagues and friends who appear in the above text. I thank in particular those students who were not directly involved with C_{60} related studies because much of their work formed a basis to which many ideas may be tenuously traced. The support over the years has come from several sources, mainly the The School of Chemistry and Molecular Sciences of the University of Sussex, SRC, SERC, EPSRC, NATO, The Royal Society, BP (John Cadogan), ICI/Zeneca (Peter Doyle, Mike Hutchinson, Neil Winterton), British Gas and BOC (Steve Wood) and Alfred Bader.

REFERENCES

1. Kroto, H. W.; Heath, J. R.; O'Brien, S. C.; Curl, R. F.; Smalley, R. E. *Nature (London)* **1985**, 318, 162–163.
2. Osawa, E. Kagaku (Kyoto) **1970**, 25, 854–863 (in Japanese); *Chem. Abstr.* **1971**, 74, 75698v.
3. Yoshida, Z.; Osawa, E. *Aromaticity; Kagakudojin:* Kyoto, 1971, 174–178, (in Japanese).
4. Bochvar, D. A.; Gal'pern, E. G. *Dokl. Akad. Nauk SSSR* **1973**, 209, 610–612 (English translation Proc. Acad. Sci. USSR **1973**, 209, 239–241).
5. Stankevich, I. V.; Nikerov, M. V.; Bochvar, D. A. *Russian Chem. Rev.* **1984**, 53(7), 640–655.

6. Jones, D. E. H. *New. Sci.,* 32 (3 November 1966), p. 245.
7. Jones, D. E. H. *The Inventions of Daedalus; Freeman:* Oxford, 1982; pp. 118–119.
8. Dietz, T. G.; Duncan, M. A.; Powers, D. E.; Smalley, R. E. J. *Chem. Phys.* **1981**, 74, 6511–6512.
9. Kroto, H. W.; Allaf, W.; Balm, S. P. *Chem. Revs.* 91 1213–35 **1991.**
10. Weltner, W., Jr.; Van Zee, R. J. *Chem. Rev.* **1989**, 89, 1713–1747.
11. Heath, J. R.; O'Brien, S. C.; Curl, R. F.; Kroto, H. W.; Smalley, R.E. *Comments. Condens. Matter Phys.* **1987**, 13, 119–141.
12. Curl, R. F. and Smalley, R. E. *Science* **1988**, 242, 1017–1022.
13. Kroto, H. W. *Science* **1988**, 242, 1139–1145.
14. Meijer, G.; Bethune, D. S., *Chem. Phys. Letts,* **1990**, 175, 1–2.
15. Krätschmer, W.; Lamb, L. D.; Fostiropoulos, K.; Huffman, D. R. *Nature (London).,* **1990**, 347, 354–358.
16. Krätschmer, W.; Fostiropoulos, K.; Huffman, D. R. *Dusty Objects in the Universe,* Bussoletti, E.; Vittone, A. A. eds.; Kluwer : Dordrecht, 1990 (Conference in 1989).
17. Krätschmer, W.; Fostiropoulos, K.; Huffman, D. R. *Chem. Phys. Lett.* **1990**, 170, 167–170.
18. Kurosawa, A.; Richie, D. (ed), *Rashomon,* **1987**, Rutgers Univ. Press, New Brunswick.
19. Smalley, R. E., *The Sciences,* **1991**, Mar/Apr pp 22–28.
20. Kroto, H. W., *Angewandte Chem.,* **1992**, 31, 111–129.
21. Curl, R. F.; Smalley, R. E., *Scientific American.*
22. Krätschmer, W.; Huffman, D. R., in p5–9 *The Fullerenes, eds.* Kroto, H. W.; Walton, D. R. M. **1993**, Cambridge Univ. Press.
23. Baggott, J. *Perfect Symmetry, the Accidental Discovery of Buckminsterfullerene,* **1994**, Oxford Univ. Press.
24. Aldersey-Williams H, *The Most Beautiful Molecule in Chemistry, An Adventure in Chemistry,* **1995**, Aurum Press, London.
25. Krätshmer, W.; Schuster, H., *Von Fuller bis zu Fullerenen,* **1996**, Facetten Vieweg, Braunschweig.
26. Dettman, J., *Fullerene–Die Buckyballs erobern die Chemie,* **1994**, Birkhäuser Verlag, Basel.
27. Ball, P., *Designing the Molecular World,* **1994**, pp 13–53, Princeton.
28. Hirsch, A., *The Chemistry of the Fullerenes,* **1994**, Thieme, NY.
29. Kroto, H. W.; Walton, D. R. M., eds. *The Fullerenes,* **1993**, Cambridge Univ. Press.
30. Kroto, H. W.; Cox, D. E.; Fischer, J. E., eds, *The Fullerenes,* **1993**, Pergamon, Oxford.
31. Taylor, R., ed. *The Chemistry of the Fullerenes,* **1995**, World Science, Singapore.
32. Iijima, S., Nature **1991**, 58, 354.
33. Nickon, A.; Silversmith, E. F. *Organic Chemistry–The Name Game: Modern Coined Terms and Their Origins; Pergamon:* New York, 1987.
34. Fuller, R. B. *Inventions–The Patented Works of Buckminster Fuller; St. Martin's Press:* New York, **1983**.
35. Marx, R. W., *The Dymaxion World of Buckminsterfuller,* **1960**, Reinhold, New York.
36. Baldwin, J., *BuckyWorks,* **1996**, Wiley, New York.
37. Kroto, H. W., *Materials Res. Soc. Bull,* Nov **1994**, 19, 21–22.
38. Lawlor R.; *Sacred Geometry; Crossroad:* New York 1982.
39. Critchlow, K., *Time Stands Still,* **1979** London, Gordon Fraser, Forest Grove (Ore.), Int. Scholarly Book Service 1980.
40. Lee, H. D. P. (Transl.), *Plato's Timaeus and Critias,* **1971**, p73, Penguin, Harmondsworth, UK.
41. Van Vleck, J. H., *Rev. Mod. Phys.,* **1951**, 23 213.
42. Condon, E. U.; Shortley, G. H., *The Theory of Atomic Spectra,* **1967**, Cambridge Univ. Press.
43. Kroto, H. W., *Molecular Rotation Spectra,* **1992** Dover, New York (1st ed **1975** Wiley, Chichester).
44. Dixon, R. N.; Kroto, H. W., *Trans Faraday Soc,* **1963**, 59, 1484.
45. Douglas, A. E., *Ap. J.,* **1951**, 114, 466.

46. Kroto, H. W. *Chem. Soc. Rev.* **1982**, 11, 435–491.
47. Gier T E., J. *Am. Chem. Soc 83* (1961) 1769.
48. Hopkinson, M. J.; Kroto, H. W.; Nixon, J. F.; Simmons, N. P. C., J. *Chem. Soc Chem Commun.,* (1976) 513.
49. Becker G., Z. *Anorg. Allg. Chem.,* **1976**, 423, 242.
50. Hopkinson, M. J.; Kroto, H. W.; Nixon, J. F.; Simmons, N. P. C., *Chem. Phys. Letts,* **1976**, 42, 460.
51. Eastmond, R.; Walton, D. R. M., *Chem. Comm.* **1968**, 204.
52. Eastmond, R.; Johnson, T. R.; Walton, D. R. M., *Tetrahedron,* **1972**, 28 4601.
53. Johnson, T. R.; Walton, D. R. M., *Tetrahedron,* **1972**, 28, 5221.
54. Eaborn, C., *Chem in Britain,* **1970**, 6, 330.
55. Alexander, A. J.; Kroto, H. W.; Walton, D. R. M. J. *Mol. Spec.* **1976**, 62, 175–180.
56. A C Cheung, D M Rank, C H Townes, D D Thornton and W J Welch, *Phys Rev Lett. 21* (1968) 1701.
57. Kroto, H. W. *Int. Rev. Phys. Chem.* **1981**, 1, 309–376.
58. B E Turner, *Ap. J., 163* (1971) L35.
59. Avery, L. W.; Broten, N. W.; Macleod, J. M.; Oka, T.; Kroto, H. W. *Astrophys. J.* **1976**, 205, L173–L175.
60. Kirby, C.; Kroto, H. W.; Walton, D. R. M. J. *Mol. Spec.* **1980**, 261–265.
61. Kroto, H. W., Kirby, C., Walton, D. R. M.; Avery, L. W.; Broten, N. W.; Macleod, J. M., Oka, T. *Astrophys. J.* **1978**, **219**, L133–L137.
62. Kroto, H. W. *Proc. R. Inst.* **1986**, 58, 45–72.
63. Oka, T. J. *Mol. Spec.* **1978**, 72, 172–174.
64. Broten, N. W.; Oka. T.; Avery, L. W.; Macleod, J. M.; Kroto, H. W. *Astrophys. J.* **1978**, 223, L105–L107.
65. Herbst, E.; Klemperer, W. *Astrophys. J.* **1973**, 185, 505–533.
66. Dalgarno, A.; Black, J. H. *Rep. Prog. Phys.* **1976**, 39, 573–612.
67. E E Becklin, J A Frogel, A R Hyland, J Kristian and G Neugebauer, *Ap J.* 158 (1969) L133.
68. Michalopoulos, D. L.; Geusic, M. E.; Langridge-Smith, P. R. R.; Smalley, R. E J. *Chem. Phys.* **1984**, 80, 3556–3560.
69. Hintenberger, H.; Franzen, J.; Schuy, K. D. Z. *Naturforsch. Teil A* **1963**, 18A, 1236–1237.
70. Douglas, A. E. *Nature (London)* **1977**, 269, 130–132.
71. Herbig, G. H. *Astrophys. J.* **1975**, 196, 129–160.
72. Kroto, H. W.; Heath, J. R.; O'Brien, S. C.; Curl, R. F.; Smalley, R. E., *Ap. J.,* **1987**, 314, 352–355.
73. Rohlfing, E. A.; Cox, D. M.; Kaldor, A. J. *Chem. Phys.* **1984**, 81, 3322–3330.
74. Bloomfield, L. A.; Geusic, M. E.; Freeman, R. R.; Brown, W. L. *Chem. Phys. Lett.* **1985**, 121, 33–37.
75. Poulx M *Graphis 132 Garphis Preaa* Zurich 1967 p379.
76. Heath, J. R.; O'Brien, S. C.; Curl, R. F.; Kroto, H. W.; Smalley, R.E. *Comments. Condens. Matter Phys.* **1987**, 13, 119–141.
77. D W Thompson, *On Growth and Form,* Cambridge Univ Press (1942).
78. Schmalz, T. G.; Seitz, W. A.; Klein, D. J.; Hite, G. E. *Chem. Phys. Lett.* **1986**, 130, 203–207.
79. Heath, J. R.; O'Brien, S. C.; Zhang, Q.; Liu, Y.; Curl, R. F.; Kroto, H. W.; Smalley, R. E. *J. Am. Chem. Soc.* **1985**, 107, 7779–7780.
80. Kroto, H. W. *Nature (London)* **1987**, 329, 529–531.
81. Schmalz, T. G.; Seitz, W. A.; Klein, D. J.; Hite, G. E. J. *Am. Chem. Soc.* **1988**, 110, 1113–1127.
82. O'Brien Heath, J. R.; O'Brien, S. C.; Curl, R. F.; Kroto, H. W.; Smalley.
83. Cox, D. M.; Reichmann, K. C.; Kaldor, A. J. *Chem. Phys.* **1988**, 88, 1588–1597.
84. Fowler, P. W.; Steer, J. I. J. *Chem. Soc. Chem. Commun.* **1987**, 1403–1405.
85. Kroto, H. W. *Chem. Brit.* **1990**, 26, 40–45.
86. M Goldberg, *Tohoku Math. J.* 43 (1937) 104.

87. H S M Coxeter *Regular Polytopes* (Macmillan, New York 1963).
88. Kroto, H. W.; M[c]Kay, K. G. *Nature (London)*, **1988**, 331, 328–331.
89. Iijima, S. *J. Cryst. Growth*, **1980**, 5, 675–683.
90. Zhang, Q. L.; O'Brien, S. C.; Heath, J. R.; Liu, Y.; Curl, R. F.; Kroto, H. W.; Smalley, R. E. *J. Phys. Chem.* **1986**, 90, 525–528.
91. Newton, M. D.; Stanton, R. E. *J. Am. Chem. Soc.* **1986**, 108, 2469–2470.
92. Wu, Z. C; Jelski, D. A.; George, T. F. *Chem. Phys. Lett.* **1987**, 137, 291–294.
93. Disch, R. L.; Schulman, J. M. *Chem. Phys. Lett.* **1986**, 125, 465–466.
94. Larsson, S.; Volosov, A.; Rosén, A. *Chem. Phys. Lett.* **1987**, 137, 501–504.
95. Braga, M.; Larsson, S., Rosén, A.; Volosov, A. *Astron. Astrophys.;* **1991**, 245, 232–238.
96. Taylor, R.; Hare, J. P.; Abdul-Sada, A. K.; Kroto, H. W. *J. Chem. Soc. Chem. Commun.* **1990**, 1423-1425.
97. Johnson, R. D.; Meijer, G.; Bethune, D. S. *J. Am. Chem. Soc.* **1990**, 112, 8983-8984.
98. Avent, A. G.; Birkett, P. R.; Christides, C.; Crane, J. D.; Darwish, A. D.; Hitchcock, P. B.; Kroto, H. W.; Meidine, M,; Prassides, K.; Taylor, R.; Walton, D. R. M., *J. Mol. Struct.* **1994**, 325 1-11).
99. Prassides, K., Christides, C.; Thomas, I. M.; Mizuki, J.; Tanigaki, K.; Hirosawa, I.; Ebbesen, T. W., *Science,* **1994**, 263, 950-954.
100. Prassides, K.; Vavekis, K.; Kordatos, K.; Tanigaki, K.; Bendele, G. M.; Stephens, P. W., J. *Am. Chem. Soc.*, **1997,** 119, 834–835.
101. Prassides, K.; Keshavarz-K., M.; Hummelen, J. C.; Andreoni, W.; Giannozzi, P.; Beer, E.; Bellavia, L.; Christofolini, L.; Gonzalez, R.; Lappas, A.; Murata, Y.; Malecki, M.; Srdanov, V.; Wudl, F., *Science* **1994**, 271, 1833–1835.
102. Kroto, H. W.; J. P. Hare.; Sarkar, A.; Hsu, W-K.; Terrones, M.; Abeysenghe, J. R., *Materials Res. Soc. Bull,* Nov **1994**, 19, 51–55.

R. E. Smalley

RICHARD ERRETT SMALLEY

I was born in Akron, Ohio on June 6, 1943, one year to the day before D-Day, the allied invasion at Normandy. The youngest of four children, I was brought up in a wonderfully stable, loving family of strong Midwestern values. When I was three my family moved to Kansas City, Missouri where we lived in a beautiful large home in a lovely upper-middle class neighborhood. I grew up there (at least to the extent one can be considered to be grown up on leaving for college at age 18) and was convinced that Kansas City, Missouri was the exact center of the known universe.

My mother, Esther Virginia Rhoads, was the third of six children of Charlotte Kraft and Errett Stanley Rhoads, a wealthy manufacturer of furniture in the Kansas City area. She liked the unusual name Errett so much that she gave it to me as my middle name. She picked the name Richard after the crusading English king (the Lion-Hearted), but being a good American and suitably suspicious of royalty, she was fond of calling me "Mr. President" instead. She had big plans for me, and loved me beyond all reason.

My father, Frank Dudley Smalley, Jr., was the second of four children born to Mary Rice Burkholder and Frank Dudley Smalley (Sr.), a railroad mail clerk in Kansas City. Although my father went by the name of June (short for Junior), he never quite forgave his father for not having given him a name of his own, and for not having aspired to more in life. My father started work as a carpenter, and then as a printer's devil, working for the local newspaper, *The Kansas City Star,* and later for a farm implement trade journal, *Implement and Tractor.* By the time he retired in 1963 he had long since risen to be CEO of this company, and a group of several others that published trade journals in the booming agriculture industry throughout the Western Hemisphere. He was incredibly industrious, talented, and fascinated with both business and technology. He had a wonderfully analytic mind, and loved argument, open discussion, and homespun philosophy. During the depression in the early 1930's he married my mother (who fell in love with his blue eyes) and was promptly laid off from work. The story of his career is one of total dedication to both his work and his family, a dedication that held steady through a series of tribulatons, many of which I am only now beginning to appreciate. He loved me too, but he could see himself in me, and knew my failings through and through. Until late in life I was never quite good enough for my father, and I suppose that is part of what drives me even now, well after his death in 1992.

My interest in Science had many roots. Some came from my mother as she finished her B.A. Degree studies in college while I was in my early teens. She fell in love with science, particularly as a result of classes on the Foundations of Physical Science taught by a magnificent mathematics professor at the University of Kansas City, Dr. Norman N. Royall, Jr. I was infected by this professor second hand, through hundreds of hours of conversations at my mother's knees. It was from my mother that I first learned of Archimedes, Leonardo da Vinci, Galileo, Kepler, Newton, and Darwin. We spent hours together collecting single-celled organisms from a local pond and watching them with a microscope she had received as a gift from my father. Mostly we talked and read together. From her I learned the wonder of ideas and the beauty of Nature (and music, painting, sculpture, and architecture). From my father I learned to build things, to take them apart, and to fix mechanical and electrical equipment in general. I spent vast hours in a woodworking shop he maintained in the basement of our house, building gadgets, working both with my father and alone, often late into the night. My mother taught me mechanical drawing so that I could be more systematic in my design work, and I continued in drafting classes throughout my 4 years in high school. This play with building, fixing, and designing was my favorite activity throughout my childhood, and was a wonderful preparation for my later career as an experimentalist working on the frontiers of chemistry and physics.

The principal impetus for my entering a career in science, however, was the successful launching of Sputnik in 1957, and the then-current belief that science and technology was going to be where the action was in the coming decades. While I had been a rather erratic student for many years, I suddenly became very serious with my education at the beginning of my junior year in the fall of 1959. I set up a private study in the partly furnished, unheated attic of our home, and began to spend long hours in solitude studying and reading (and smoking cigarettes). This happened to be the year when I began to study chemistry for the first time. Luckily, these years were some of the best ever for the public school system in Kansas City, and my local high school, Southwest High, was one of the most effective anywhere in the US as measured by scores on standard achievement tests, and the fraction of students going on to college. My teacher, Victor E. Gustafson, was a great inspiration. He had just begun to teach the preceding year, and was full of love for his subject and for teaching, and had an as yet unblunted ambition to reach even the slowest of students. In addition, this was the first class I had ever taken with my sister, Linda, who was a year older than I, and was a far better student than I had ever been. The result was that by the end of the year, my sister and I finished with the top two grades in the class. We hardly ever missed a question on an exam. It was an exhilarating experience for me, and still ranks as the single most important turning point in my life, even from my current perspective of nearly four decades later. It was the proof of an existence theorem. After my junior year, I knew I could be successful at science. The next year I did equally well in physics with a wonderful professor, J. C. Edwards, but my soul had already been imprinted by my exposure to chemistry the year before.

My mother's youngest sibling, Dr. Sara Jane Rhoads, was one of the first women in the United States to ever reach the rank of full Professor of Chemistry. After earning her Ph.D. in 1949 with William von Eggers Doering, who was then at Columbia University, she devoted her life to teaching and research in the Department of Chemistry of the University of Wyoming. She received the Garvan Medal of the American Chemical Society in 1982 for her contributions to physical organic chemistry, particularly in the study of the Cope and Claisen rearrangements. She was the only scientist in our extended family and was one of the brightest and, in general, one of the most impressive human beings I have ever met. She was my hero. I used to call her, lovingly, "The Colossus of Rhoads". Her example was a major factor that led me to go into chemistry, rather than physics or engineering. One of the most enjoyable memories of my early life was the summer (1961) I spent working in her organic chemistry laboratory at the University of Wyoming. It was at her suggestion that I decided to attend Hope College that fall in Holland, Michigan. Hope had then (and still has now) one of the finest undergraduate programs in chemistry in the United States.

At Hope College I spent two years in fruitful study, but decided to transfer to the University of Michigan in Ann Arbor after my favorite professor, Dr. J. Harvey Kleinheksel, died of a heart attack, and the organic chemistry professor with whom I had hoped to do research, Dr. Gerrit Van Zyl, announced his retirement. While the next two years in Ann Arbor were successful, I had become so entangled in a stormy love affair with a lovely girl back at Hope College, that I was not able to concentrate as much on science as I should have. I did, however, learn a lot. Most of all I learned from my fellow students, and particularly from John Seely Brown, a graduate student in mathematics who lived in an apartment down the hall in a small house off campus (he is currently Director of Xerox's Palo Alto Research Center, PARC). John displayed an audacity of thought and intellectual ambition that I have rarely seen in any individual. My fellow housemates and I were infected with the notion that we could master any subject, and at times we did manage to at least feel that we got close.

By the time of my graduation in 1965, the job market for scientists in the United States was at an all-time high, and even chemistry graduates with just a BS degree were in great demand. Rather than proceeding directly to graduate school, I decided to take a job in the chemical industry in order to buy a bit of time to see what I really wanted to do in science, and to live a little in the "real" world. It turned out to be a terrific decision.

In the fall of 1965 I began work full time in Woodbury, New Jersey at a large polypropylene manufacturing plant owned by the Shell Chemical Company. I began as a chemist working in the quality control laboratory for the plant, a 24 hour a day operation that in the mid 60's was quite a wonderland of high technology. My first boss was a chemist named Donald S. Brath. He taught his young professionals that "chemists can do anything", and the time I worked under him was a wonderfully broadening experience. I was teamed up with chemical engineers at the plant to study problems with the quality of the polymer product. The Ziegler-Natta catalyst system then in use by Shell to produce isotactic polypropylene was no where near as efficient as those currently in use, and the level of inorganics remaining in the polymer was high. Much of what we were concerned with in those days revolved around this problem of high "ash" content and how it affected the downstream applications. These were fascinating days, involving huge volumes of material, serious real-world problems, with large financial consequences. I loved it.

After two years I moved up to the Plastics Technical Center at the same site in Woodbury, and devoted myself to developing analytical methods for various aspects of polyolefins, and of the materials involved in their manufacture, modification, and processing. Although I found my work at Shell highly enjoyable, I realized it was time to get on to graduate school, so I began to study seriously and to send out applications. At the time I was most interested in quantum chemistry, and received several offers for graduate assistantships in excellent schools. I was close to accepting an offer from the Theoretical Chemistry Institute at the University of Wisconsin when the automatic graduate student deferments from the Draft into the US military were eliminated. This was in early 1968, during a major buildup phase in the Vietnam War, and I decided it would be more prudent to remain at Shell for a while since my industrial deferment was still in effect.

In my off hours over the past few years I had met Judith Grace Sampieri, who was a wonderful young secretary at Shell. We were married on May 4 of 1968. Soon thereafter, even the industrial deferment was lost, and we decided that I might as well reapply for graduate school. Since Judy's family lived in New Jersey, I decided to apply to Princeton University, and was accepted. In the late fall of 1968 I was reclassified 1A for the draft and reported to the processing center in Newark for my physical. At the end of the day I ended up in the group who had passed. We were told to put our affairs in order since we would soon be called up. However, in a great stroke of luck, within a week, my wife told me she was pregnant, and within just a few more weeks my draft board reclassified me to some status I do not remember, save that it meant I would not be drafted. On June 9, 1969 Judy and I were blessed with the birth of a beautiful child, Chad Richard. Later that summer, I held him in my lap as Neil Armstrong first stepped out onto the Moon.

In the fall of 1969 I moved my new family up to Princeton to begin studies and research for the Ph.D. in the Department of Chemistry. I was lucky enough to be in the first group of graduate students to work with Elliot R. Bernstein who was just starting as an Assistant Professor at Princeton, after having spent a few years postdoctoral work at the University of Chicago with Clyde A. Hutchison III, following doctoral training with G. Wilse Robinson at CalTech. Elliot's research at the time involved detailed optical and microwave spectral probes of pure and mixed molecular single crystals cooled in liquid helium. I knew nothing about it at the time I joined the group. I was certain that it was going to be both experimentally and theoretically complex and challenging, but it seemed likely to be worth the effort. My research project was the detailed study of 1,3,5-triazine, a heterocyclic benzene analog that we expected would provide a poignant testing ground for theories of the Jahn-

Teller effect. In the end we found that the crystal field surrounding each molecule was insufficiently symmetrical to provide the tests we originally sought, but much was learned. Most importantly from my standpoint, I learned from Elliot Bernstein a penetrating, intense style of research that I had never known before, and I learned a great deal about the chemical physics of condensed phase and molecular systems.

In the summer of 1973 we moved to the south side of Chicago so I could begin a postdoctoral period with Donald H. Levy at the University of Chicago. Levy had studied gas-phase magnetic resonance with Alan Carrington, and had been doing some of the most impressive research anywhere in the world with microwave/optical double resonance and the Hanle effect on NO_2 and other open-shell small molecules. These were the earliest days when tunable dye lasers were beginning to transform molecular spectroscopy, and Levy's group was in the lead. The optical spectrum of NO_2 was the most troublesome problem for molecular spectroscopists. Even though it had only three atoms, the visible spectrum had far more structure than anyone could understand. But since NO_2 was readily available and it displayed an extensive absorption spectrum just where the new lasers could readily operate (500–640 nm), it was a favorite object for study. Don Levy and one of his students, Richard Solarz, had made some major advances with NO_2 earlier that summer, so after I arrived in Chicago I began to consider what I could do next. My biggest problem was that my training at Princeton had been in condensed matter spectroscopy, and the ultrahigh resolution gas-phase spectral techniques being used by the Levy group were going to take months to understand. The detailed physics of rotating polyatomic molecules with spin is extremely complex. I was familiar only with the physics of molecules frozen still in a crystal lattice near absolute zero.

When we first arrived in Chicago, Don Levy was in Germany for a several month-long visit, so I had an opportunity to do some extended reading and to prepare for the final oral exam for the Ph.D. degree back in Princeton. At that time in the Chemistry Department at Princeton, the final oral exam consisted of a defense of three original research proposals. I spent many hours in the Univ. Chicago chemistry department library reading recent journal articles, searching for possible topics for these research proposals. On one day I read a new paper by Yuan Lee and Stuart Rice on the crossed beam reaction of fluorine with benzene [*J. Chem. Phys.* **59**, 1427 (1973)] in one of Yuan's "universal" molecular beam apparatuses. It was the sort of experiment that was to lead to Yuan Lee sharing the Nobel Prize in 1986 with John Polanyi and Dudley Herschbach. I was deeply struck by a passage in the paper which said that the supersonic expansion used to make the benzene molecular beam was strong enough to cool out essentially all rotational degrees of freedom. That was just what I needed. Since I didn't understand rotating molecules yet, perhaps I could just stop them from rotating in the first place!

As a result of this exciting day in the Chicago library, one of the proposals I presented to the Princeton Ph.D. committee later that fall was to use a supersonic expansion to cool NO_2 to the point that only a single rotational state was populated, and then to use a tunable dye laser to study the now greatly simplified spectrum. I had found in further reading that the current supersonic expansion techniques actually would not get cold enough, so I added the further use of an electric resonance "state-selector" to do a final sorting out of just a single rotational state for study. I recommended, in fact, that the 10 meter state-selector beam machine of Lennard Wharton at Chicago could be used.

When Levy returned from Germany, I told him of this proposal, and we discussed it in some depth. He was intrigued, but was concerned that too much of the NO_2 would dimerize to N_2O_4 before sufficient cooling was obtained. A few weeks later we discussed it again, and became sufficiently excited to walk down the hall and ask Lennard Wharton what he thought. Len lit up like a light bulb.

Wharton argued that we should first do the experiment on NO_2 expanded in a supersonic free jet, and leave the much more elaborate state-selected experiment for later. I told him that wouldn't be cold enough–the lowest rotational temperature reported for a polyatomic molecule in a supersonic beam that I was aware of at that time was 30 K, still way too hot to achieve the simplification we needed. Wharton smiled wryly and swiveled in his chair to reach a research notebook from the shelf behind him. After reading a few pages he looked up and asked "would 3 K be cool enough?". He had already built a liquid hydrogen cryopumped supersonic beam source with argon, and in the research notebook had measured data for the velocity distribution showing the translational temperature was cooled to 3 K. That, I knew from my Ph.D. proposal, would be quite cool enough in the case of NO_2 to collapse the rotational population to just a few levels. We would simply mix in a percent or so of NO_2 into the argon and make a "seeded" supersonic beam. This would avoid the N_2O_4 formation that concerned Don Levy, and may just possibly cool the rotational degrees of freedom to near the translational temperature of the argon carrier gas. Thus began the collaboration that led to supersonic beam laser spectroscopy.

On the night of August 8, 1974 (the night Nixon resigned from the US Presidency) we recorded the first jet cooled spectrum of NO_2. The next morning Don Levy saw the spectrum for the first time, and immediately recognized its significance. Molecular physics had changed. Now we could study at least small polyatomic molecules with at the same penetrating level of detail previously attained only for atoms and diatomics.

A year later, Lennard Wharton came back from a trip to France where he had visited with Roger Campargue and learned of the concept of the "zone of silence" that exists in an expanding gas at sufficiently high densities. While this zone is surrounded by shock waves where the gas is heated to very high temperatures, within the zone the expanding gas is exactly as cold and unperturbed as it would be if the gas expanded into a perfect vacuum, forming no shock waves at all. Campargue had learned to fabricate a ultrasharp edged "skimmer" that could penetrate the "Mach disc" at end of the zone and transmit the gas streaming along the center line of the zone of silence to form the most intense, coldest supersonic beams ever produced. Wharton told Don Levy and me that using helium in such an apparatus we could easily get down to 1 K and perhaps even lower. I was stunned. I knew that 1 K was low enough to freeze out the rotational motion of even medium-sized molecules such as benzene and naphthalene, and all such molecules could now be studied without rotational congestion.

Later that same day in a hallway conversation Len Wharton and I realized we didn't need the skimmer. The probe laser beam could easily penetrate the shock waves without perturbation, and we could image just the fluorescence from the laser-excited ultracold molecules in the zone of silence. We quickly built a new apparatus that incorporated these ideas. With the spectroscopic insight of Don Levy and with a series of graduate students we published the pioneering papers on not only jet cooled spectra of ordinary molecules such as NO_2, and tetrazine, but also on the first van der Waals complexes with helium (e.g. HeI_2), and with the vital collaboration of Daniel Auerbach the first supersonic beam study of a metal atom–rare gas complex, NaAr.

In the summer of 1976 my family and I moved to Houston, Texas where I had accepted a position as assistant professor in the chemistry department at Rice University. I knew of Rice principally because of the beautiful laser spectroscopy that was being done there by Robert F. Curl, and I wanted to collaborate with him much the same as I had with Don Levy. The first supersonic beam apparatus I set up was a free jet machine similar to that I had used in Chicago, but adapted to use pulsed dye lasers in the ultraviolet so that we could study more ordinary molecules such as benzene. My first proposal to the National Science Foundation was for a much larger, more ambitious - apparatus that would for the first time use pulsed supersonic nozzles. With these pulsed devices mounted in a large chamber I expected we could attain a 10–100 fold increase in beam intensity and cooling, and by synchronizing with the pulsed lasers in both the visible and ultraviolet be able to study a vast array of large molecules, radicals, and clusters. Being the second apparatus we constructed, it was called "AP2".

With AP2 we quickly succeeded in setting the world's record for rotational cooling of a polyatomic molecule (0.17 K). We invented resonant two-photon ionization (R2PI) with time of flight mass spectrometric detection as a means of probing the spectrum of molecules in the supersonic beam. We used this to probe the structure and molecular dynamics of large aromatic molecules, particularly focussing on the question of intramolecular vibrational redistribution. We also developed a means of producing fragments of polyatomic molecules (free radicals such as benzyl and methoxy) by directing a pulsed laser into a specially designed pulsed supersonic nozzle, and studying these cooled in the supersonic beam.

In the late 1970s in collaboration with Andrew Kaldor and his group at Exxon we had extended the capabilities of AP2 so that we could study a large uranium containing molecule (a hexafluoroacetylacetonate-, tetahydrofuran-complexed form of UO_2). These were the days of the oil crisis, when there was widespread belief that nuclear fission using uranium was going to be the only long-term alternative. Exxon was working intensely on laser-based isotope separation schemes, and Kaldor was heading up a group to pursue the molecular route. Our experiment on AP2 ultimately revealed a beautiful sharpening of the infrared multiphoton dissociation spectrum of this volatile UO_2 complex cooled in the supersonic beam, just what Exxon was looking for. Unfortunately, we began to succeed with these experiments only after the nuclear release "event" at Three Mile Island on March 28, 1979. Within a year, Exxon made a corporate level decision to get out of the isotope separation business. But Kaldor had become so impressed with the capabilities of AP2 that he wanted his own at the corporate laboratories in Linden in any event. Under contract to Exxon, we developed a smaller version of the apparatus, and built two versions. One was kept at Rice and lived on for many years with a very productive science history. Logically, it was called AP3. The clone of AP3 was shipped to Exxon in late 1982.

After a few years of intensive research we found a way to use a pulsed laser directed into a nozzle to vaporize any material, allowing for the first time the atoms of any element in the periodic table to be produced cold in a supersonic beam. Most importantly, we developed a way to control the clustering of these atoms to small aggregates, which then were cooled in the supersonic expansion. Now for the first time it was possible to roam the periodic table and make detailed study of the properties of nanometer-scale particles consisting of a precise number of atoms. The field of metal and semiconductor cluster beams was born. We shipped Exxon this new accessory to their AP3 clone, and both groups then rapidly began to develop the new field.

As is now well known, the Kaldor group was the first to put carbon in a laser vaporization cluster beam apparatus, and see the amazing even-numbered distribution of carbon clusters that we now know to be the fullerenes. Within a year we repeated the same experiment, but now on an improved version of AP2 that had been modified for the study of semiconductor clusters. The story of what we discovered on this apparatus in September of 1985 has been told many times.

The subsequent development of my research in metal and semiconductor clusters, and the fullerenes is too involved to recount here. Increasingly, the tubular variant of the fullerenes has dominated our activities. Now our motto is "if it ain't tubes, we don't do it". We are convinced that major new technologies will be developed over the coming decades from fullerene tubes, fibers, and cables, and we are moving as fast as possible to bring this all to life.

Several years ago AP2 was dismantled and sold off in pieces to other research groups, and the main chamber where the first pulsed nozzle experiments were performed was sold off to a scrap metal dealer along the Houston Ship Channel. Now there are no supersonic beam machines of any type in the laboratory. Times change.

But life and science go on.

DISCOVERING THE FULLERENES

Nobel Lecture, December 7, 1996

by

RICHARD E. SMALLEY

Center for Nanoscale Science and Technology, Rice Quantum Institute, and Departments of Chemistry and Physics, Rice University, Houston, Texas 77005, USA

It is a thrill for me to be here today and to be the first of three speakers discussing the wonders of the fullerenes, an infinite new class of carbon molecules. My colleagues in this famous photograph (Figure 1) are also thrilled to be here in Stockholm this week to see "Bucky get the Prize". This picture was taken on September 11, 1985, the day before we sent off the manuscript describing the discovery of C_{60} to the editorial offices of *Nature* (1) (and only a few days after the discovery itself). Every one of the people in that photograph was critically involved in the discovery (with the exception of the one woman walking in the back–we still don't know who that mystery woman was), so you can understand that there is also some sadness in our hearts today. While the chemistry prize this year is for the discovery of the fullerenes, it is given to individuals, and this individual honor can be shared by no more than three. The Nobel Committee has done as well as they possibly can with this problem. We understand. But the sadness remains.

On the other hand, there are positive aspects to the limit of three. For example, I have asked what happens in those years when there is only one person receiving an award in physics or chemistry. I was told that you just get one lecture for that prize. Now that I am beginning to appreciate the full impact of having a long lecture from each of the three winners this year in each of the two fields, physics and chemistry, all on the same day and in the same room, I can see that one must set limits somewhere.

This discovery was one of the most spiritual experiences that any of us in the original team of five have ever experienced. The main message of my talk today is that this spiritual experience, this discovery of what Nature has in store for us with carbon, is still ongoing. So the title of my talk is not "The Discovery of the Fullerenes" but rather "Discovering the Fullerenes". Fullerene researchers worldwide are still engaged in this process of discovery.

The sense in which we are still in this process has to do with what the true essence of the 1985 fullerene discovery actually turned out to be. After all, the five people in that happy photograph (Figure 1), brilliant as they all are, were not the ones who first conceived of the truncated icosahedron. That was done several thousand years ago. Archimedes gets the credit for it, although one may reasonably suspect that icosahedra had been truncated long before Archimedes. Nor were we the first people to conceive that if you replaced the

Figure 1. Photograph of the research group that discovered the fullerenes at Rice University in September of 1985. Standing: Curl. Kneeling in front, left to right: O'Brien, Smalley, Kroto, and Heath.

vertices of that pattern with carbon atoms, and let the carbon do what it wanted to do, that that would be an interesting chemical object. That honor had already gone more than a decade before our discovery to E. G. Osawa (2, 3), the Japanese physical organic chemist who had perceived that carbon in that structure would be aromatic and would therefore probably be stable. And a large part of this honor had been earned even before that by David Jones (4), who in a wonderfully imaginative piece, had conceived of closed spheroidal cages made of graphene sheets somehow folded around. A little later, Jones realized that a pentagon would serve nicely as the required defect in an otherwise hexagonal lattice to produce a complex curvature (5). The notion that C_{60} would be a closed shell molecule with a very large HOMO-LUMO gap, which is a well-appreciated signature of chemical stability, fell to Bochvar, Gal'pern (6) and Stankevich (7) who actually did the relevant Hückel calculations in Russia well over a decade before we ever got into the game.

The conception of carbon being stable in the form of a truncated icosahedron really wasn't the discovery that is being honored this week. If that were, then Archimedes, Osawa, Jones, and/or one or more of these insightful Soviet scientists should have gotten the prize.

Instead, the discovery that garnered the Nobel Prize was the realization that carbon makes the truncated icosahedral molecule, and larger geodesic cages, all by itself. Carbon has wired within it, as part of its birthright ever since the beginning of this universe, the genius for spontaneously assembling into fullerenes. We now realize that all you need to do to generate billions of billions of these objects of such wonderful symmetry is just to make a vapor of carbon atoms and to let them condense in helium. Now we are still in the process of discovering all of the other consequences of the genius that is wired into carbon atoms. It isn't just a talent to make balls. It can also make tubes such as the short section shown in Figure 2.

Nearly all of us have long been familiar with the earlier known forms of pure carbon: diamond and graphite. Diamond, for all its great beauty, is not nearly as interesting as the hexagonal plane of graphite. It is not nearly as interesting because we live in a three-dimensional space, and in diamond each atom is surrounded in all three directions in space by a full coordination. Consequently, it is very difficult for an atom inside the diamond lattice to be confronted with anything else in this 3D world because all directions are already taken up. In contrast, the carbon atoms in a single hexagonal sheet of graphite (a "graphene" sheet) are completely naked above and below. In a 3D world this is not easy. I do not think we ever really thought enough about how special this is. Here you have one atom in the periodic table, which can be so satisfied with just three nearest neighbors in two dimensions, that it is largely immune to further bonding. Even if you offer it another atom to bond with from above the sheet–even a single bare carbon atom, for that matter–the only result is a mild chemisorption that with a little heat is easily undone, leaving the graphene sheet intact. Carbon has this genius of making a chemically stable two-dimensional, one-atom-thick membrane in a three-dimensional world. And that, I believe, is going to be very important in the future of chemistry and technology in general.

What we have discovered is that if you just form a vapor of carbon atoms

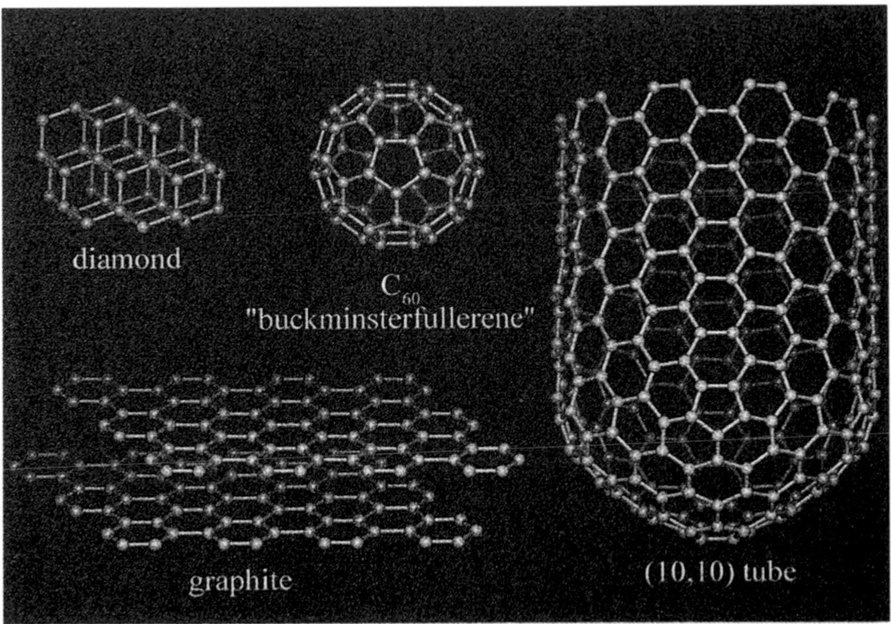

Figure 2. Four perfect crystalline forms of carbon: diamond, graphite, C_{60}, and a short length of a (10,10) fullerene nanotube showing the hemi-C_{240} end cap.

and let them condense slowly while keeping the temperature high enough so that as the intermediate species grow they can do what it is in their nature to do, there is a path where the bulk of all the reactive kinetics follow that goes to make spheroidal fullerenes. Now it turns out that in addition to this most symmetric of all possible molecules, C_{60}, and the other fullerene balls, it is possible by adding a few percent of other atoms (nickel and cobalt) to trick the carbon into making tubes. Of all possible tubes there is one tube that is special (8). It is the tube shown in Figure 2, the (10,10) tube. We are beginning to understand that what causes this tube to be the most favorite of all tubes is also wired within the instruction set of what it means to be a carbon atom. The propensity for bonding that causes C_{60} to be the end point of 30–40% of all the reactive kinetics, leads as well to this (10,10) tube. This detour on the road that otherwise leads to spheroidal fullerenes is taken if you somehow (with cobalt or nickel atoms) frustrate the ability of the open edge to curve in and close. The metal atoms prevent by local annealing the addition of the seventh, eighth, ninth pentagons, and insure by judicious choice of temperature and reaction rate that the growing tublet can anneal to its most energetically favored form.

The object shown in Figure 3 depicts one of the more fascinating new opportunities in the future of the fullerenes. It is a short piece of the (10,10) tube with chemically derivitized ends. One end is closed with a hemifullerene dome (actually one half of a very special fullerene, icosahedral C_{240}). The

Figure 3. Derivitized section of a (10,10) fullerene nanotube with one end open.

other end is intentionally left open. Since the closed end contains pentagons and is accordingly more reactive than the smooth, all-hexagon side of the nanotubes, there are techniques such as boiling in nitric acid (9) for heating the closed ends off of these tubes. We also know that if you take such a tube and put it in an oven and heat it to 1200 °C, it will spontaneously close back again. These ends, regardless of whether they are closed or open, are directly amenable to the formation of excellent C-O, C-N, or C-C covalent bonds (10) to attach nearly any molecule, enzyme, membrane, or surface to the end of the tube. You could attach one or several such objects (let's call them A) to the upper end and some other objects on the bottom (B). What is so stunning about this molecule, unlike any other molecule we have ever had before in chemistry, is that with this object "A" and "B" will communicate with each other by true metallic transport along the tube. The (10,10) tube is a quantum waveguide for electrons.

The band structure of graphene, the individual flat sheet of graphite is that of a zero gap semi-conductor. The valence band and conduction band meet at a point at the end of the Brillouin zone. There is a node in the density of states at Fermi energy, and accordingly it is not a very good conductor. In my youth on first hearing that graphite was a poor conductor–more like lead than gold–I had thought that the problem must be due to some sort of original sin that the carbon atom had made. The valence electrons of carbon tended to be localized, and were not freely able to move from one carbon atom to another through an extended sheet. In fact that is not the problem. The π electrons are perfectly itinerant in the graphene sheet, just as they are perfectly itinerant in the aromatic ring of benzene. In fact, it is the freedom of the electrons to move around the ring that gives the special chemical stability to aromatic molecules. The trouble with the electrical conductivity is that when you calculate the band structure of the hexagonal graphene sheet, by symmetry, there is a node in the density of states at the Fermi energy. Even if you were somehow able to replace every one of the carbon atoms in the hexagonal lattice with a gold atom, the band structure would still look the same. It is the symmetry of the hexagonal lattice that is the problem, not the itinerancy of the π electrons of carbon.

But now we realize that there is one (but only one) answer to this problem of making a metal out of pure carbon. If you take the graphene sheet and cut out a thin strip, curl it along it's length to form a long cylinder, and seal up the dangling bonds together to form the (10,10) tube as shown in Figure 2, the very symmetry that had been your enemy in preventing metallic behavior from the flat lattice, now becomes your friend. The symmetry of this tubular hexagonal lattice now insists that there will be two bands that cross at the Fermi energy, approximately two thirds of the way across the Brillouin Zone, as shown in the band structure of Figure 4. In addition, the cohesive rigidity of the σ-bond framework of the graphene sheet prevents the metallic π electrons from engendering a Peierls instability that normally plagues all such one-dimensional conductors (11). The (10,10) fullerene tube, and all (n,n) tubes in general, will be a molecular wire that is simultaneously a good me-

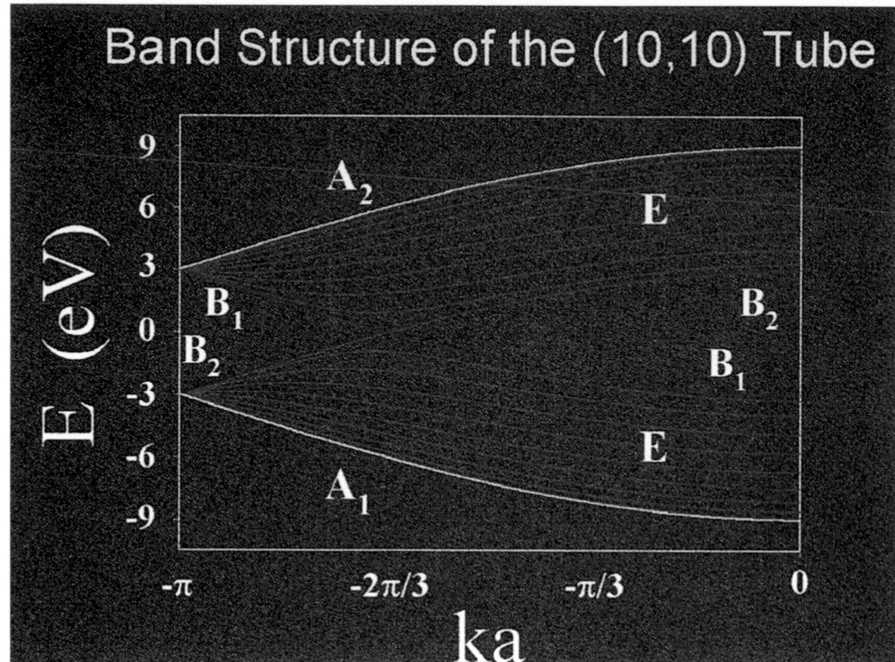

Figure 4. Electronic band structure of a (10,10) fullerene nanotube calculated using with tight binding methods, using zone-folding from the band structure of an infinite 2D graphene sheet. The two bands that cross the Fermi energy at ka = 2/3 have different symmetry, and guarantee that the tube will be a metallic conductor.

tallic conductor, and a good molecule, maintaining its structure and conductivity even when exposed to air and water in the real world.

I believe that in the future of chemistry, we are likely to see a vast new set of metallic fullerene molecules such as that shown in Figure 3 readily available from chemical supply houses. Imagine what the impact could be. Essentially, every technology you have ever heard of where electrons move from here to there, has the potential to be revolutionized by the availability of molecular wires made up of carbon. Organic chemists will start building devices. Molecular electronics could become reality.

This is where the fullerenes appear to be leading at the moment. One does not get a feeling that we are over with this discovery process yet. There may be many more wondrous properties of this one little atom in the periodic table than we have yet to appreciate.

Still there was a particular discovery that we celebrate this week. There was something about September of 1985. What was that? How did that happen? Since that original discovery we have been involved in a little bit of what I like to call the archeology of the buckyball: looking back in the written and oral history trying to decide what the roots of the fullerene discovery were. This discovery is principally about the way that carbon condenses, it's genius for forming clusters.

It has long been known that the carbon has a special ability to cluster in the gas phase at high temperatures. Unlike every other refractory element in the periodic table, the vapor of carbon in equilibrium with its solid at temperatures in the 3000–4000 K range is dominated by clusters, C_n, with substantial abundance of species as high as C_{15}. The first evidence of this extends back to early research on nuclear fission products by Hahn and Strassman in Germany (12). They noticed that carbon cluster ions up to C^+_{15} were produced in a high frequency arc with a graphite electrode, the arc being used for elemental analysis by mass spectrometry (13); Similar observations were made in the US about this same time in research associated with the Manhattan Project during the Second World War (14). By the early 1950's it was clear that there were sufficient numbers of small carbon clusters at equilibrium in the vapor to have a major affect on the measurement (15,16) of the heat of formation of C(gas), one of the most important constants in chemical thermodynamics. In 1959 Pitzer and Clementi (17,18) made the first serious quantum calculations of the structures responsible for this behavior in the vapor up to about 20 atoms, and concluded that they had the form of linear chains for C_2 up to about C_{10}, and above that they took the form of monocyclic rings–little 'Hoola Hoops' of pure carbon.

Although it was not commented on at the time, this is quite a remarkable result. Here carbon is able to make clusters that are so stable that they are the dominant species–substantially more abundant than C_1–in the gas phase even at a temperature of 3000–4000 K, and they do this with only a coordination number of two! All other refractory elements such as platinum, tungsten, or tantalum achieve their high cohesive energy by a close packing arrangement within the bulk crystal or liquid, with coordination number of 8 to 12. Even though the clusters of these metals in the gas phase also adopt compact structures (19), arranging as many atoms around each other in three dimensions as possible, they still do not have a sufficiently high cohesive energy to be abundant in the equilibrium vapor. Instead the vapor of these metals is almost completely monatomic. Above 1000 K the vapor in equilibrium with pure condensed phase of every element in the periodic table–except carbon–is dominantly either monatomic or diatomic. But here is carbon making so many of these large clusters that it throws off the measures of heat of formation, and in a show of chemical bonding chutzpah, doing this with two of it's three available dimensions for bonding "tied behind its back".

Looking back now at the data available in the literature on gas-phase clusters of pure carbon up to mid-1984, it is clear that there was no suggestion in the experiments of anything more interesting going on than these one-dimensional clusters (20,21). All the data appeared to be well-explained by the model of linear chains and monocyclic rings, and the cluster abundance dropped off so severely by the time the clusters were in the mid-twenty-atom size that no one was led to speculate what would happen as the clusters grew larger. In light of what we now know about the fullerenes, this would have been a very fruitful line of speculation. After all, at some point as the clusters grew larger they would certainly have to start trying structures that were two- or three-dimensional. What would these look like? Consideration of the dangling bond energies could reasonably have led to the speculation that graphene sheets would form but curl up to closed cages.

But, as far as we can determine, no such speculation ever occurred. While in the mid 1960's Jones (4) had the notion that graphene sheets could curl up to make "hollow molecules," Osawa (2) had already conjured up the notion of carbon in a soccer ball structure in early 1970, and Gal'pern (6) had completed the first of many Hückel calculations showing that it would be a closed shell molecule with a large HOMO-LUMO gap in 1973, but no one ever suggested these objects could form spontaneously in a condensing carbon vapor. The mystery of the buckyball was never so much that it would be a stable molecule once formed. After all, it violates no rules of organic chemistry. The secret laying there to be discovered is that part of carbon's "birth right" is the genius to form a chemically passive two-dimensional surface, to self-assemble fullerenes in general, and C_{60} in particular and in sensationally high yield, out of the chaos of a carbon vapor at thousands of degrees.

To trigger this realization, new data turned out to be necessary. Data on what happened when you allowed a carbon vapor to become supersaturated, allowed it to begin to condense, and the small clusters that were in equilibrium with the solid began to grow larger. That data had to wait for the invention of a new technique, something that would enable one to study the properties of carbon clusters in detail as they grow through the size range of 40 to 100 atoms where the dimensionality of the bonding does, in fact, increase from 1 to 2. It had to wait for the laser-vaporization cluster beam methods of the 1980's.

The laser-vaporization supersonic cluster beam technique was originally developed at Rice University in 1980–1981 as a means of studying clusters of virtually any element in the periodic table, including highly refractory metals (22–26) and semiconductors such as silicon (27) and gallium arsenide (28). The objective of this line of research was to explore the behavior of matter intermediate in size between atoms and bulk crystals. It grew out of decades of development of atomic and molecular beams, and in particular the development of seeded supersonic molecular beams as a means of "freezing out" the vast number of rotational and vibrational excitations which otherwise preclude detailed study of polyatomic molecules (29–32). In addition to enabling the study of common chemically stable polyatomic molecules, it was possible to generate supercold van der Waals clusters of these molecules with each other, and with other species, including at these ultralow temperatures even helium (33). By the use of intense pulsed laser irradiation within the supersonic nozzle it was possible to study highly reactive fragments of molecules, free radicals (34). Extension of the supersonic beam technique to seeded beams of refractory atoms and clusters was a direct outgrowth of this early free radical work.

Once beams of refractory clusters were available, a vast new area of research was opened, for each cluster may be thought of as a nanoscale crystalline particle that has a surface (in fact, most of it is surface). Supersonic metal cluster beams thereby provided a route to a new sort of surface science

(35). In my research group at Rice we were very heavily engaged in developing this new science. We developed and applied new methods for studying the electronic structure of the clusters by one- and two-photon laser photoionization with time-of-flight mass spectral detection (25, 36, 37), photodissociation, and photodepletion spectroscopy (38–40), and ultraviolet photoelectron spectroscopy (41). In order to study the surface chemistry of the nanoscale clusters, we developed techniques using a fast flow reactor attached to the end of the supersonic nozzle (42–44). We developed a variety of methods involving charged ions of the clusters levitated in a magnetic field and probed by ion cyclotron resonance spectroscopy (45–47), etc. The early 1980's were a very busy, very fruitful time in this research group at Rice University, and much was learned.

The laser vaporization supersonic beam source and the associated probe techniques that we developed to study the 2–200 atom metal and semiconductor clusters was effectively a new sort of microscope. It allowed one to "see" something of the nature of nanoscopic aggregates of atoms in a way that was entirely new, and very poignant. Whatever we measured for the cluster in the supersonic beam, we knew that it was the true property of that cluster traveling free in space. We developed the technique in order to bring a sort of intellectual tension to surface science: to make measurements so fundamental that theorists stayed awake at night trying to understand them. Although we had no notion of it at the time we were engaged in this enterprise in the early 1980's, we were building the instrument and the line of research that would discover the fullerenes. All one had to do is put carbon in this new "microscope", adjust the focus a little, and "see" the fullerenes revealed plainly for the first time.

As it happened, we at Rice were not the first to put carbon into the new microscope. In 1984 a group headed by Andrew Kaldor at Exxon used such an apparatus (actually one that had been designed and built at Rice) in a study of carbon clusters that was motivated by the desire to study coke buildup on reforming catalyst. Now, in a famous mass spectrum reproduced here as Figure 5, carbon clusters were evident extending out beyond 100 atoms (48), and it was immediately clear that a whole new world of interesting carbon clusters could exist that had never been seen before.

Three distinct regions characterized the mass spectrum: first the small clusters, containing fewer than 25 atoms, consisting of the chains and monocyclic rings (49) so well known from the earlier studies; second a new region between about 25 and 35 atoms in which few species of any sort were observed–a region that the Rice group came to call the "forbidden zone"; and, third, an even-numbered clusters distribution extending from the high 30's to well over 150 atoms. By early 1984 the laser-vaporization supersonic cluster beam technique had been used to study clusters broadly throughout the periodic table, and many "magic number" cluster distributions had been found and studied (35), but nothing remotely like this even-numbered distribution of carbon had ever been seen for any other element. Today this is still true: carbon is unique.

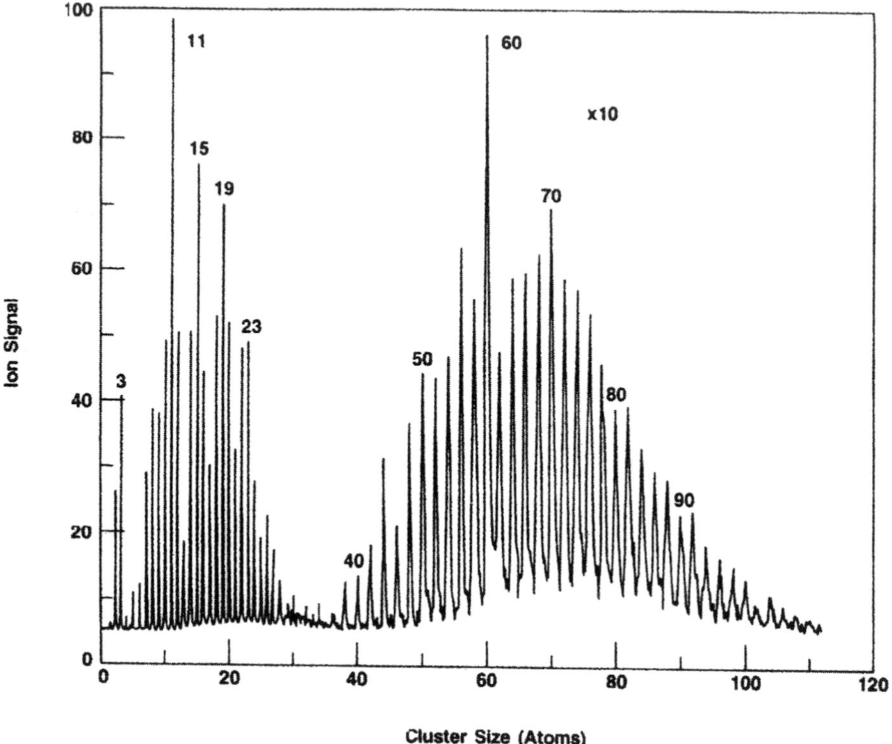

Figure 5. Mass spectrum of carbon clusters in a supersonic beam produced by laser vaporization of a carbon target in a pulsed supersonic nozzle operating with a helium carrier gas. The long distribution of even-numbered carbon clusters starting near C_{40} and extending to over C_{100} is due the fullerenes. This was the first published experiment that revealed the fullerene cluster distribution, although it was not appreciated as such at the time. Reproduced with permission of the authors, reference (48).

The even-numbered distribution seen by the Exxon group was due to the fullerenes. I suspect the members of that original group at Exxon still regret that they did not consider in more depth why the peak for C_{60} appeared to be about 20% more intense than its neighbors. But to be fair, at the time neither did we. This result of the Exxon group was widely known in the burgeoning metal and semiconductor cluster research world by the summer of 1984. I saw the result at a meeting, and discussed it with Andy Kaldor at some length. Bob Curl saw it. Harry Kroto saw it. Wolfgang Krätschmer and Donald Huffman saw it. None of us stopped to think just what might be the reason C_{60} was a little more prominent than the other even-numbered clusters. It did not stand out sufficiently above its neighbors to attract notice. As it turned out, the Exxon group simply had not "focussed" the new cluster beam microscope carefully enough.

At nearly the same time another research group, including one of my former graduate students, Michael Geusic, had built a supersonic cluster beam apparatus as well, and was engaged in early experiments with mass-selected semiconductor cluster ions (50). Having heard about the Exxon work, they put carbon in the apparatus, also observed the mysterious even-numbered large cluster distribution, and even selected out C^{+}_{60} for a photofragmentation experiment. But they, too, failed to experiment sufficiently with the nozzle conditions to appreciate the potential preeminence of C_{60}.

The necessary "focussing" of the new supersonic cluster beam "microscope" was finally performed in my laboratory at Rice in the September of 1985 (1, 51). Now the supremacy of C_{60}, was made clear. In thinking about what this all meant, we finally saw that C_{60} must be a closed spheroidal cage. No other explanation was consistent with the observed facts. The realization that all the even-numbered carbon cluster distribution was due to carbon in the form of hollow geodesic domes (fullerenes) came within a month, as a result of reactivity studies of these clusters (52) using the fast flow reactor on the end of the supersonic nozzle. Soccer ball C_{60} quickly became a sort of "Rosetta Stone" leading to the discovery of a new world of geodesic structures of pure carbon built on the nanometer scale.

This discovery episode has by now been so extensively covered in articles (53, 54), monographs (55, 56), and television documentaries, that there is little reason to repeat the details here. We succeeded where two other groups had failed for at least two reasons. First, we had evolved a better version of the "microscope". We had been the group to develop the supersonic cluster beam technique in the first place, and we were still leading the subsequent development and elaboration of its capabilities. The original apparatus, known affectionately by the students as "AP2" and shown in Figure 6, was able to handle much larger gas flows, and had a more advanced supersonic nozzle design than the machines of either of the other groups. Particularly important was the development of the rotating disc design for the cluster beam source that we had just recently completed for our semiconductor cluster

Figure 6. Photograph of the author climbing around on the top of a section of AP2, the supersonic laser vaporization cluster beam apparatus that discovered C_{60} and the fullerenes in the fall of 1985.

work with silicon, germanium (27), and gallium arsenide (28). As part of this project, it was necessary to anneal the semiconductor clusters in the supersonic nozzle source as much as possible. To simplify interpretation we need conditions so that the most energetically stable geometrical form would dominate the cluster distribution. For this reason the nozzle was fitted with a variety of down-stream flow restrictors to form what we called the "integrating cup", as shown in Figure 7.

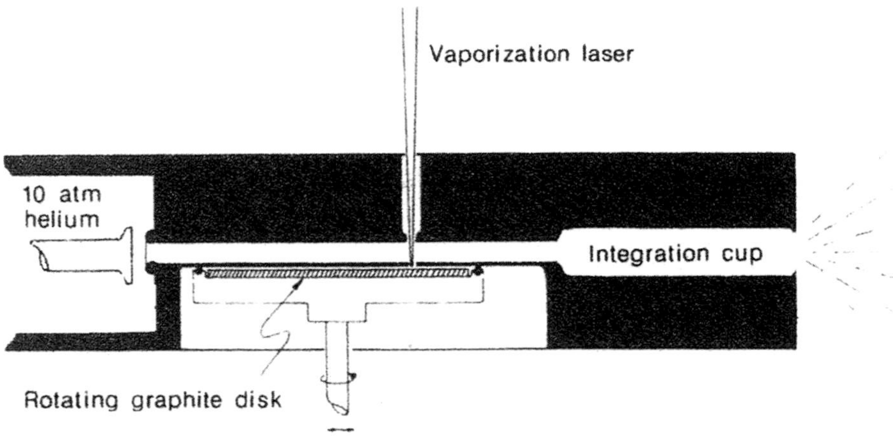

Figure 7. Schematic cross-sectional drawing of the supersonic laser-vaporization nozzle source used in the discovery of the fullerenes. Note the section labeled "integrating cup". Cluster "cooking" reactions in this zone were responsible for the C_{60} cluster becoming over 50 times more intense than any other cluster in the nearby size range. These up-clustering reactions with small carbon chains and rings reacted away nearly all clusters except for C_{60}, which because of its perfect symmetry survived.

It was our experience with the extra "chemical cooking" achieved by this device that made us appreciate that it was cluster chemistry that differentiated C_{60} from the others. The relative absence of further up-clustering reactions is, we realized, what ultimately makes the C_{60} peak stand out over 50 times more intensely than any other. And it is in thinking about how 60 carbon atoms can possibly avoid reactive edges, that one is led to the conclusion that C_{60} is a truncated icosahedron.

The second reason I believe we succeeded was, in a word, karma. With Bob Curl in our collaboration on semiconductor clusters, we had evolved one of the most intellectually demanding and penetrating styles of research I have ever witnessed in any research group. Sean O'Brien had evolved just the right version of the cluster nozzle to handle the difficulties of dealing with semiconductor discs, and Jim Heath had developed an amazing talent for making "science happen" on the machine. When Harry Kroto came, his intensity and scientific background blended in perfectly. Working with the students mostly in late afternoons and at night, and in daily marathon conversations with me, he kept the focus of our minds on the results coming out of AP2, under the hot hands of Heath and O'Brien. One way or another, over the years, each member of the team had paid the dues required to deserve to be there for the discovery of the fullerenes during those wonderful days in September 1985.

Contrary to most written accounts, I do not believe the discovery of the fullerenes had much to do with questions of astrophysics such as the mechanism of formation of interstellar carbon molecules. While this was certainly what brought Kroto to Texas, and this was without contest the reason we first put carbon into AP2 in preparation for his visit, in the end the connection of this line of research (57, 58) to the fullerene discovery was casual, not causal. The discovery of C_{60} and the fullerenes would have been made by AP2 or some other such instrument within a year or two in any event. Two other groups had already put carbon in a supersonic cluster beam machine for reasons that were much more mundane. At Exxon, as mentioned earlier, their principal concern was to understand carbon buildup on catalysts, while at AT&T Bell Laboratories the motivation derived from their long-term interest in semiconductors and the nanometer scale.

The notion that the discovery of the fullerenes came out of research into the nature of certain molecules in space is highly appealing to scientists. It is hard to think of any line of research that is less likely than interstellar chemistry to have some practical, technological impact back here on Earth. So if fullerenes turn out to lead to the technological wonders that some people (like me) believe are in our future, then perhaps one can argue that any research project could get lucky too, no matter how irrelevant to worldly problems it may currently seem. I have argued this way in the past, and I still believe there is some sense to it–but only a little. In fact, the fullerenes were discovered as a result of decades of research and development of methods to study first atoms, then polyatomic molecules, and ultimately nanometer-scale aggregates. It was well-funded research that at nearly every stage was justified by its perceived relevance to real world technological problems. To a great extent, many of these earlier bets as to the worldly significance of fundamental research actually paid off.

While it is fun to think about the wonderful role of serendipity in the story, one should also spend a bit of time comprehending the inevitability of the discovery as well. The only character of true genius in the story is carbon. Fullerenes are made wherever carbon condenses. It just took us a little while to find out.

REFERENCES

1. Kroto, H.W., Heath, J.R., O'Brien, S.C., Curl, R.F. & Smalley, R.E. *Nature* 318, 162–163 (1985).
2. Osawa, E. *Kagaku* 25, 854–863 (1970).
3. Osawa, E. *Philosophical Transactions of the Royal Society of London Series A – Physical Sciences and Engineering* 343, 1–8 (1993).
4. Jones, D.E.H. *New Scientist* 32, 245 (1966).
5. Jones, D.E.H. *The Inventions of Daedalus* (W. H. Freeman, Oxford, 1982).
6. Bochvar, D.A. & Gal'pern, E.G. *Proc. Acad. Sci. USSR* 209, 610–612 (1973).
7. Stankevich, I.V., Nikerov, M.V. & Bochvar, D.A. *Russian Chemical Reviews* 53, 640–655 (1984).
8. Thess, A., et al. *Science* 273, 483–487 (1996).
9. Tsang, S.C., Chen, Y.K., Harris, P.J.F. & Green, M.L.H. *Nature* 372, 159–162 (1994).

10. Hirsch, A. *Chemistry of the Fullerenes* (1995).
11. Mintmire, J.W., Dunlap, B.I. & White, C.T. *Physical Review Letters* 68, 631–634 (1992).
12. Hahn, O., Strassman, F., Mattauch, J. & Ewald, H. *Naturwiss* 30, 541–542 (1942).
13. Mattauch, J., Ewald, H., Hahn, O. & Strassman, F. *Zeitshrift für Physik* 120, 598–617 (1943).
14. Brewer, L., Gilles, P.W. & Jenkins, F.A. J. *Chem. Phys.* 16, 797 (1948).
15. Chupka, W.A. & Inghram, M.G. *Journal of Chemical Physics* 21, 1313 (1953).
16. Chupka, W.A. & Inghram, M.G. *Journal of Physical Chemistry* 59, 100–104 (1955).
17. Pitzer, K.S. & Clemente, E. *Journal of the American Chemical Society* 81, 4477 (1959).
18. Strickler, S.J. & Pitzer, K.S. in *Molecular Orbitals in Chemistry, Physics, and Biology* (eds. Lowdin, P.O. & Pullman, B.) 281 (Academic Press, New York, 1964).
19. Jarrold, M.F. *Journal of Physical Chemistry* 99, 11–21 (1995).
20. Dornenburg, E. & Hintenburger, H. Z. *Naturforsch.* 14A, 765–767 (1959).
21. Dornenburg, E. & Hintenberger, H. Z. *Naturforsch.* 16A, 532–534 (1961).
22. Diez, T.G., Duncan, M.A., Powers, D.E. & Smalley, R.E. *Journal of Chemical Physics* 74, 6511–6512 (1981).
23. Michalopoulos, D.L., Geusic, M.E., Hansen, S.G., Powers, D.E. & Smalley, R.E. *The Journal of Physical Chemistry* 86, 3914–3916 (1982).
24. Powers, D.E., et al. *The Journal of Physical Chemistry* 86, 2556–2560 (1982).
25. Powers, D.E., Hansen, S.G., Geusic, M.E., Michalopoulos, D.L. & Smalley, R.E. *Journal of Chemical Physics* 78, 2866–2881 (1983).
26. Hopkins, J.B., Langridge-Smith, P.R.R., Morse, M.D. & Smalley, R.E. *Journal of Chemical Physics* 78, 1627–1637 (1983).
27. Heath, J.R., et al. *Journal of Chemical Physics* 83, 5520–5526 (1985).
28. O'Brien, S.C., et al. *Journal of Chemical Physics* 84, 4074–4079 (1986).
29. Smalley, R.E., B, L.R., Levy, D.H. & Wharton, L. *The Journal of Chemical Physics* 61, 4363–4364 (1974).
30. Smalley, R.E., Wharton, L. & Levy, D.H. *The Journal of Chemical Physics* 63, 4977–4989 (1975).
31. Smalley, R.E., Wharton, L. & Levy, D.H. in *Chemical and Biochemical Applications of Lasers* (ed. Moore, C.B.) 1–43 (Academic Press, 1977).
32. Smalley, R.E., Wharton, L. & Levy, D.H. *Accounts of Chemical Research* 10, 139–145 (1977).
33. Smalley, R.E., Levy, D.H. & Wharton, L. *The Journal of Chemical Physics* 64, 3266–3276 (1976).
34. Powers, D.E., Hopkins, J.B. & Smalley, R.E. *The Journal of Physical Chemistry* 85, 2711–2713 (1981).
35. Smalley, R.E. in *Comparison of Ab Initio Quantum Chemistry with Experiment for Small Molecules* (ed. Bartlett, R.J.) 53–65 (D. Reidel Publishing Company, Boston, 1985).
36. Dietz, T.G., Duncan, M.A., Liverman, M.G. & Smalley, R.E. *Journal of Chemical Physics* 73, 4816–4821 (1980).
37. Michalopoulos, D.L., Geusic, M.E., Langridge–Smith, P.R.R. & Smalley, R.E. *Journal of Chemical Physics* 80, 3556–3560 (1984).
38. Morse, M.D., Hopkins, J.B., Langridge–Smith, P.R.R. & Smalley, R.E. *Journal of Chemical Physics* 79, 5316–5328 (1983).
39. Brucat, P.J., Zheng, L.–S., Pettiette, C.L., Yang, S. & Smalley, R.E. *Journal of Chemical Physics* 84, 3078–3088 (1986).
40. Brucat, P.J., et al. *Journal of Chemical Physics* 85, 4747–4748 (1986).
41. Cheshnovsky, O., et al. in *International Symposium on the Physics and Chemistry of Small Clusters* (eds. Jena, P., Kanna, S. & Rao, B.) 1–14 (Plenum Press, Richmond, Virginia, 1986).
42. Morse, M.D., Geusic, M.E., Heath, J.R. & Smalley, R.E. *Journal of Chemical Physics* 83, 2293–2304 (1985).
43. Geusic, M.E., Morse, M.D. & Smalley, R.E. *Journal of Chemical Physics* 82, 590–591 (1985).

44. Geusic, M.E., Morse, M.D., O'Brien, S.C. & Smalley, R.E. *Review of Scientific Instruments* 56, 2123–2130 (1985).
45. Alford, J.M., Williams, P.E., Trevor, D.J. & Smalley, R.E. *International Journal of Mass Spectrometry and Ion Processes* 72, 33–51 (1986).
46. Alford, J.M., Weiss, F.D., Laaksonen, R.T. & Smalley, R.E. *Journal of Physical Chemistry* 90, 4480–4482 (1986).
47. Elkind, J.L., Weiss, F.D., Alford, J.M., Laaksonen, R.T. & Smalley, R.E. *Journal of Chemical Physics* 88, 5215–5224 (1988).
48. Rohlfing, E.A., Cox, D.M. & Kaldor, A. *Journal of Chemical Physics* 81, 3322–3330 (1984).
49. Yang, S., et al. *Chemical Physics Letters* 144, 431–436 (1988).
50. Bloomfield, L.A., Geusic, M.E., Freeman, R.R. & Brown, W.L. *Chemical Physics Letters* 121, 33–37 (1985).
51. Curl, R.F. & Smalley, R.E. *Science* 242, 1017–1022 (1988).
52. Zhang, Q.L., et al. *The Journal of Physical Chemistry* 90, 525–528 (1986).
53. Hargittai, I. in *The Chemical Intelligencer* 6–54 (1995).
54. Smalley, R.E. *The Sciences* 22–28 (1991).
55. Baggott, J. *Perfect Symmetry, The Accidental Discovery of Buckminsterfullerene* 1–315 (Oxford University Press, Oxford, 1994).
56. Aldersey-Williams, H. *The Most Beautiful Molecule, The Discovery of the Buckyball* 1–340 (John Wiley & Sons, Inc., New York, 1995).
57. Heath, J.R., et al. *Journal of the American Chemical Society* 109, 359–363 (1987).
58. Kroto, H.W., Heath, J.R., O'Brien, S., Curl, R.F. & Smalley, R.E. *Astrophysics J* 314, 352– (1987).
59. Dresselhaus, M.S., Dresselhaus, G. & Eklund, P.C. *Science of Fullerenes and Carbon Nanotubes* 1–985 (Academic Press, San Diego, 1996).

Erratum

Page 15, line 8 should read as:

Cox and Kaldor[26] in an essentially the same apparatus. In the Rohlfing, Cox, and

Chemistry 1997

PAUL D. BOYER and JOHN E. WALKER

"for their elucidation of the enzymatic mechanism underlying the synthesis of adenosine triphosphate (ATP)"

and

JENS C. SKOU

"for the first discovery of an ion-transporting enzyme, Na+, K+-ATPase"

THE NOBEL PRIZE IN CHEMISTRY

Speech by Professor Bertil Andersson of the Royal Swedish Academy of Sciences.
Translation of the Swedish text.

Your Majesties, Your Royal Highnesses, Ladies and Gentlemen,

Life requires energy. Our muscles require energy when we move. We need energy to think. Energy input is required for the production of new biological molecules. This year's three Nobel Laureates in Chemistry have contributed in different ways to our knowledge of how living organisms can obtain and utilize energy. Common to their discoveries is the unique adenosine triphosphate (ATP) molecule, which can store and transport energy in all organisms, whether it be a simple bacterium, a dandelion, a finch or a human being. Large quantities of ATP must be formed and consumed. Each day an adult converts a quantity of ATP roughly equivalent to his or her own body weight, and in case of physical exertion, many times more.

All energy on earth originates from the sun. Green plants can absorb sunlight and convert it into chemical energy through the process of photosynthesis, in which carbon dioxide and water form sugar, starch and other complex carbon compounds. Other organisms, such as humans and animals, are in turn dependent on these carbon compounds as sources of energy, and they burn them with the help of oxygen. That is why we breathe. Nature can thus be said to have chosen a combination of solar and coal-fired power plants for its energy supply. Although these two energy conversion systems may seem different in purely technical terms, in many respects they operate in the same way in living cells. The most important similarity is that the energy released is utilized with the help of the ATP molecule.

According to Peter Mitchell, the 1978 Nobel Laureate in Chemistry, the energy released in photosynthesis and cell respiration initiates a stream of positively charged hydrogen ions. These hydrogen ions, in turn, drive the production of ATP with the aid of a membrane-bound enzyme called ATP synthase. Two of this year's Laureates, Paul Boyer and John Walker, have studied this important enzyme and have shown that it functions in a unique way. Among other things, they have demonstrated that ATP synthase can be compared to a molecular machine, whose rotating bent axle is driven in a stepwise process by "biological electricity"–that is, the flow of hydrogen ions. Because of the asymmetry of the rotating axle, three subunits of the enzyme assume different forms and functions: a first form that binds adenosine diphosphate (or ADP) and phosphate building blocks, a second form where these two molecules are chemically combined into a new ATP molecule, and a third form where the ATP that has been formed is released. In the next twist of the axle, the three subunits switch form and thus also function with each other, and another ATP molecule can be formed, and so on. This "binding change mechanism" was put forward by Boyer in the late 1970s, but only in 1994 did his ingenious model gain general acceptance among researchers. In August of that year, Walker and his colleagues published three-dimensional images of ATP synthase that had been obtained by X-ray analysis of enzyme crystals. These X-ray images, magnified several million times, showed how an asymmetrically elongated protein molecule interacted with three other protein units that all showed mutually different forms. Walker had finally revealed the detailed blueprint of the molecular machine and shown that Boyer's theory of ATP formation was correct.

Let me now leave ATP production and instead turn to the use of ATP. In 1957, Jens Christian Skou discovered an enzyme called sodium, potassium ATPase (or Na^+,K^+-ATPase), which maintains the right ion balance in living cells. This enzyme, too, can be described with a technical analogy: It functions as a biological "pump" that transports potassium ions into a cell, while transporting sodium ions in the opposite direction out of the cell. This is a process that requires a lot of energy, and up to one third of the ATP formed in the body may be used to drive the Na^+,K^+ pump. Today, we are also aware of a number of other ion pumps, which have been discovered as a consequence of Skou's pioneering work. All these ion pumps are prerequisites for various important life functions such as transmission of nerve impulses, muscle contraction and digestion. The effects of many pharmaceuticals, such as heart and ulcer medicines, are related to the action of cellular ion pumps.

Skou's discovery clearly illustrates the unpredictability of basic research–in 1957, no one could have imagined that his somewhat odd experiments, which consisted of studying the effects of various salts on tissue from a shore crab, would be important 40 years later to industrial operations and the production of new pharmaceuticals. Research breakthroughs and their applications cannot be custom-ordered. Instead they emerge from a combination of curiosity, scientific excellence and far-sightedness.

Dr. Boyer, Dr. Skou and Dr. Walker,

I have tried to describe how your pioneering studies on the enzymology of ATP metabolism have contributed to our understanding of how living cells can store and make use of energy. Your work has revealed new principles for enzyme function, opened up new areas of chemical research as well as providing the basis for biomedical applications for the benefit of mankind. In recognition of your services to chemistry the Royal Swedish Academy of Sciences has decided to confer upon you this year's Nobel Prize for Chemistry.

On behalf of the Academy, I wish to convey to you our warmest congratulations and I now ask you to receive the Prize from the hands of His Majesty the King.

Paul D. Boyer

PAUL D. BOYER

The first 21 years of my life were spent in Provo, Utah, then a city of about 15,000 people, beautifully situated at the foot of the Wasatch Mountains. Hardy Mormon pioneers had settled the area only 70 years before my birth in 1918. Provo was a well-designed city with stable neighbourhoods, a pride in its past and a spirit of unbounded opportunity. The geographical isolation and lack of television made world happenings and problems seem remote.

My father, Dell Delos Boyer, born in 1879 in Springville, Utah, came from the Pennsylvania Boyers, who in turn came from an earlier Bayer ancestry in what is now Holland and Germany. A small portion of my Boyer DNA has been traced to John Alden, famous as a Mayflower pilgrim who wooed for another and won for himself. Dad's education, at what was then the Brigham Young Academy, was delayed by the ill health he had endured in much of his youth. Through his ambition, and the sacrifices of his family, he acquired training in Los Angeles to become an osteopathic physician. He served humanity well. More by example than by word, my father taught me logical reasoning, compassion, love of others, honesty, and discipline applied with understanding. He also taught me such skills such as pitching horseshoes and growing vegetables. Dad loved to travel. Family trips to Yellowstone and to what are now national parks in Southern Utah, driving the primitive roads and cars of that day, were real adventures. Father became a widower when the youngest of my five siblings was only eight. Fifteen years later he married another fine woman. They shared many happy times, and she cared for him during a long illness as he died from prostate cancer at the age of 82. Prostate cancer also took the life of my only brother when he was 76. If our society continues to support basic research on how living organisms function, it is likely that my great grandchildren will be spared the agony of losing family members to most types of cancer.

Recently I scanned notes from a diary that my mother, Grace Guymon, wrote in her late teens, when living near Mancos, Colorado. The Guymons were among the Huguenots who fled religious persecution in France. My French heritage has been mixed with English and other nationalities as the Guymons descended. Mother's diary revealed to me more about her vitality and charm than I remembered from her later years, which were clouded by Addison's disease. She died in 1933, at the age of 45, just weeks after my fifteenth birthday. Discoveries about the adrenal hormones, that could have saved her life, came too late. Her death contributed to my later interest in studying biochemistry, an interest that has not been fulfilled in the sense that my accomplishments remain more at the basic than the applied level. Mother made a glorious home environment for my early years. During her long illness and after her death, all of the children helped with family chores. One of my less pleasant memories is of getting up in the middle of the night to use our allotted irrigation time to water the garden.

The large, gracious home provided by Mother and Dad at 346 North University Avenue has been replaced by a pizza parlor, although an inspection a few months ago revealed that the irrigation ditch for our garden area (now a parking lot) can still be found. Mother had a talent for home decorating. I often read from a set of the *Book of Knowledge* or *Harvard Classics* while lying in front of the fireplace, with a mantel designed and decorated by her. Staring into the glowing coals as a fire dims provided a wonderful milieu for a youthful imagination. I also remember such things as picnics in Provo Canyon, and the anticipation that I might get to lick the dasher after cranking the ice-cream freezer. My older brother, Roy, and I had a play-fight relationship. I still carry a scar on my nose from when I plunged (he pushed me!) through the mirror of the dining room closet. I am told that I had a bad temper, and remember being banished to the back hall until civility returned. Perhaps this temper was later sublimated into drive and tenacity, traits that may have come in part from my mother.

The great depression of the 1930s left lasting impressions on all our family. Father's patients became non-paying or often exchanged farm produce or some labor for medical care. Mother saved pennies to pay the taxes. The burden of paper routes and odd jobs to provide my spending money made it painful when my new Iver Johnson bicycle was stolen. We were encouraged to be creative. I recall mother's tolerance when she allowed me, at an early age, to take off the hinges and doors of cupboards if I would put them back on. My first exposure to chemistry came when I was given a chemistry set for Christmas. It competed for space in our basement with a model electric trains and an "Erector" set. After school the neighborhood yards were filled with shouts of play; games of "kick-the-can," "run-sheepy-run," "steal-the-sticks," as well as marbles, baseball and other activities. In our back yard we built tree houses, dug underground tunnels and secret passages, and made a small club house. The mountains above our house offered other outlets for adventuresome teenage boys. Days were spent in an abandoned cabin or sleeping under the sky in the shadow of Provo peak. We even took cultures of sour dough bread to the mountains and baked delicious biscuits in an a rusty stove. Mountain hikes instilled in me a life-long urge to get to the top of any inviting summit or peak.

Provo public schools were excellent. At Parker Elementary School, a few blocks from my home, I fell in love with my 3rd grade teacher, Miss McKay. Students who learned more easily were allowed to skip a grade, and I entered the new Farrer Junior High school at a younger age than my classmates. This handicapped me in two types of sporting events, athletics and courting girls. Girls did not want to dance with little Paul Boyer; boys were quite unimpressed with my physique. As I grew my status among fellows improved. Once I got into a scuffle in gym class, the instructor had the "combatants" put on boxing gloves, and I gave more than I received. It wasn't until late high school and early college that I gained enough size and skill to make me welcome on intramural basketball teams.

I was one of about 500 students of Provo High School, where the atmosphere was friendly, and scholarship and activities were encouraged by both students and faculty. I participated on debating teams and in student government, and served as senior class president. I still have a particularly high regard for my chemistry teacher, Rees Bench. I was pleased when he wrote in my Yearbook for graduation, "You have proven yourself as a most outstanding student." I graduated while still 16, and thought myself quite mature. I wish I had saved a copy of my valedictorian address. I suspect it may have sparkled with naivete.

It was always assumed that I would go to college. The Brigham Young University (BYU) campus was just a few blocks from my home and tuition was minimal. It was a small college of about 3,500 students, less than a tenth of its present size. As in high school, I enjoyed social and student government activities. Friendships abounded. New vistas were opened in a variety of fields of learning. Chemistry and mathematics seemed logical studies to emphasize, although I had little concept as to where they might lead. A painstaking course in qualitative and quantitative analysis by John Wing gave me an appreciation of the need for, and beauty of, accurate measurement. However, the lingering odor of hydrogen sulfide, used for metal identification and separation, called unwanted attention to me in later classes. "Prof" Joe Nichol's enthusiasm for general chemistry was superbly conveyed to his students. Professor Charles Maw excelled in transferring a knowledge of organic chemistry to his students. Biochemistry was not included in the curriculum.

Summers I worked as a waiter and managerial assistant at Pinecrest Inn, in a canyon near Salt Lake City. One summer a college friend and I lived there in a sheep camp trailer while managing a string of saddle horses for the guests to use. A different type of education came when as a member of a medical corps in the National Guard I spent several weeks in a military camp in California.

As my senior year progressed several career paths were considered; employment as a chemist in the mining industry, a training program in hotel management, the study of osteopathic or conventional medicine, or some type of graduate training. Little information was available about the latter possibility; but a few chemistry majors from BYU had gone on to graduate school. I have a tendency to be lucky and make the right choices based on limited information. A notice was posted of a Wisconsin Alumni Research Foundation (WARF) Scholarship for graduate studies. My application was approved, and the stage was set for a later phase of my career.

Before leaving Provo, a most important and fortunate event occurred. A beautiful and talented brunette coed, with one year of college to finish, indicated a willingness to marry me. She came from a large and loving family, impoverished financially by her father's death when she was 2 years old. She had worked and charmed her way nearly through college. My savings were limited and hers were negative. But it was clear that my choice was to have her join with me in the Wisconsin adventure or take my chances when I returned a year later. It was an easy decision. Paul, who had just turned 21, and Lyda Whicker, 20, were married in my father's home on August 31, 1939. Five days later we left by train to Wisconsin for my graduate study.

A few months after our arrival our new marriage almost ended. I was admitted to the student infirmary with diagnosed appendicitis. Through medical mismanagement my appendix ruptured and I became deathly ill. Sulfanilamides, discovered a few years earlier by Domagk, saved my life. Last summer I read an outstanding book, *The Forgotten Plague: How the Battle Against Tuberculosis Was Won and Lost,* by Frank Ryan. The book gives a stirring account, the first I have read, of Domagk's research and how he was not allowed to leave Hitler's Germany to receive the 1939 Nobel Prize.

Fortunately, the Biochemistry Department at the University of Wisconsin in Madison was outstanding and far ahead of most others in the country. A new wing on the biochemistry building had recently been opened. The excitement of vitamins, nutrition and metabolism permeated the environment. Steenbock had recently patented the irradiation of milk for enrichment with vitamin D. Elvehjem's group had discovered that nicotinic acid would cure pellagra. Petersen's group was identifying and separating bacterial growth factors. Link's group was isolating and identifying a vitamin K antagonist from sweet clover. Patents for the use of dicoumarol as a rat poison and as an anticoagulant sweetened the coffers of the WARF, the Foundation that supported my scholarship. Among younger faculty an interest in enzymology and metabolism was blossoming.

Married graduate students were rare, and the continuing economic depression made jobs hard to find. But my remarkable wife soon found a good job, and I settled into graduate studies. During our Wisconsin years she gained a perspective of art while employed in Madison's leading art retail outlet. It was years later before Lyda finished a college degree, became a professional editor at UCLA, and worked with me on the eighteen-volume series of *The Enzymes.* Our contacts in graduate school and through Lyda's employment gave us life-long friends; one was Henry Lardy, from South Dakota farm country. He and I were assigned to work under Professor Paul Phillips. Henry was highly talented, and it was my good fortune to work along side him. Phillips' main interests were in reproductive and nutritional problems of farm animals. Henry developed an egg yolk medium for sperm storage that revolutionized animal breeding.

We were encouraged by Phillips to explore metabolic and enzyme interests. I did not realize that it was unusual to be able step across the hall and attend a symposium on respiratory enzymes in which such biochemical giants as Otto Meyerhof, Fritz Lipmann, and Carl Cori spoke. Evening research discussion groups with keen young faculty such as Marvin Johnson and Van Potter, centered on enzymes and metabolism, broadened and sharpened our perspectives. One evening I presented my and Henry's evidence for the first known K^+ activation of an enzyme, pyruvate kinase. Henry kept score on the interruptions for questions or discussions–some 35 as I recall. This superb training environment set the base for my career.

My Ph.D. degree was granted in the spring of 1943, the nation was at war, and I headed for a war project at Stanford University. A few weeks after my arrival in California, on my birthday, July 31, our daughter Gail was born. I became somewhat more involved in home duties and more deeply in love with Lyda.

The wartime Committee on Medical Research sponsored a project at Stanford University on blood plasma proteins, under the direction of J. Murray Luck, founder of the nonprofit *Annual Review of Biochemistry* and other Reviews. Concentrated serum albumin fractionated from blood plasma was effective in battlefield treatment of shock. When heated to kill microorganisms and viruses, the solutions of albumin developed cloudiness from protein denaturation. The principal goal of our research project was to find some way to stabilize the solutions so that they would not show this behavior. Our small group found that acetate gave some stabilization and butyrate was better. This led to the discovery that long chain fatty acids would remarkably stabilize serum albumin to heat denaturation, and would even reverse the denaturation by heat or concentrated urea solutions. Other compounds with hydrophobic portions and a negative charge, such as acetyl tryptophan, were also effective. Our stabilization method was quickly adopted and is still in use. From the Stanford studies I gained experience with proteins and a growing respect for the beauty of their structures.

In marked contrast to the University of Wisconsin, Biochemistry was hardly visible at Stanford in 1945, consisting of only two professors in the chemistry department. The war project at Stanford was essentially completed, and I accepted an offer of an Assistant Professorship at the University of Minnesota, which had a good biochemistry department. But my local War Draft Board in Provo, Utah, had other plans and I became a member of the U.S. Navy. The Navy did not know what to do with me, the war with Japan was nearly over, and I became what is likely the only seaman second-class that has had a nearly private laboratory at the Navy Medical Research Institute in Bethesda, Maryland. In less than a year I returned to civilian life. In the spring of 1946 I, my wife, and now two daughters, Gail and Hali, became Minnesotans. But I had unknowingly acquired a latent California virus to be expressed years later.

Minnesota has generally competent and honest public officials, good support of the schools and cultural amenities, and an excellent state university. It was a fine place to rear a family, and soon our third child, Douglas, was born. A golden era for biochemistry was just starting. The NIH and NSF research grants were expanding at a rate equal to, or even ahead, of the growing number of meritorious applications. The G.I. bill provided financial support that brought excellent and mature graduate students to campus. New insights into metabolism, enzyme action, and protein structure and function were being rapidly acquired.

Housing was almost unavailable in the post war years. Initially we coped with an isolated, rat-infested farm house. In 1950, after my academic competence seemed satisfactorily established, we built a home not far from the St. Paul campus where the Department of Biochemistry was located. I served as contractor, plumber, electrician, finish carpenter etc. My warm memories of this home include looking at a sparkling, snow-covered landscape, while seated at the desk in the bedroom corner that served as my study, and struggling with the interpretation of some puzzling isotope exchanges accompanying an enzyme catalysis. The understanding that developed was rewarding and perhaps one of my best intellectual efforts. However, it did not seem that the approach would give answers to major problems.

During my early years at Minnesota I conducted an evening enzyme seminar. One participant in our lively discussions was a promising graduate student from another department, Bo Malmstrom, who became a renowned scientist in his field, and is now a retired professor from the University of Göteborg. In 1952 my family spent a memorable summer at the Woods Hole Marine Biological Laboratories on Cape Cod. A sabbatical period on a Guggenheim Fellowship in Sweden in 1955 was especially rewarding. There I did research at both the Wenner-Gren Institute of the University of Stockholm with Olov Lindberg and Lars Ernster, and at the Nobel Medical Institute, working with Hugo Theorell's group. Professor Theorell received a Nobel Prize that year, exposing us to the splendor and formality of the Nobel festivities.

Along the way, I was gratified to receive the Award in Enzyme Chemistry of the American Chemical Society in 1955. In 1959–60 I served as Chairman of the Biochemistry Section of the American Chemical Society. In 1956 I accepted a Hill Foundation Professorship and moved to the medical school campus of the University of Minnesota in Minneapolis. Much of my group's research was on enzymes other than the ATP synthase. But solving how oxidative phosphorylation occurred remained one the most challenging problems of biochemistry, and I could not resist its siren call. Mildred Cohn reported that mitochondria doing oxidative phosphorylation catalyzed an exchange of the phosphate and water oxygens, an intriguing capacity. An able physicist and a pioneer in mass spectrometry, Alfred Nier, made gaseous ^{18}O and facilities available to me, and some experiments were run using this heavy isotope of oxygen. However, much of our effort over several years was directed toward attempting to detect a possible phosphorylated intermediate in ATP (adenosine triphosphate) synthesis using ^{32}P as a probe. The combined efforts of some excellent graduate students and postdocs, most of whom went on to rewarding academic careers, culminated in the discovery of a new type of phosphorylated protein, a catalytic intermediate in ATP formation with a phosphoryl group attached to a histidine residue.

By then, time and queries had stimulated the latent California virus. Change was underway. In the summer of 1963, I and a group of graduate students and postdocs who came with me, activated laboratories in the new wing of the chemistry building at the University of California in Los Angeles (UCLA), located on a beautiful campus at the foot of the Santa Monica mountains. We soon found that the enzyme-bound phosphohistidine we had discovered was an intermediate in the substrate level phosphorylation of the citric acid cycle. It was not a key to oxidative phosphorylation. The experience reminds me of a favorite saying: Most of the yield from research efforts comes from the coal that is mined while looking for diamonds.

In 1965 I accepted the Directorship of a newly created Molecular Biology Institute (MBI) at UCLA, in part because of my disappointment that oxidative phosphorylation had resisted our efforts. A building that was promised failed to materialize, but through luck and persistence adequate funds were obtained, partly from private resources, and promising faculty were recruited. The objective was to promote basic research on how living cells function at the molecular level. I believe the best research is accomplished by a faculty member with a small group of graduate students and postdocs, who freely design, competently conduct and intensely evaluate experiments. To spend time with such a group I soon found ways to reduce my administrative chores. Probes of oxidative phosphorylation continued, and, as 1971 approached, we hit pay dirt. We recognized the first main postulate of what was to become the binding change mechanism for ATP synthesis, namely that energy input was not used primarily to form the ATP molecule, but to promote the release of an already formed and tightly bound ATP

In the following decade, the other two main concepts of the mechanism were revealed, namely that the three catalytic sites participate sequentially and cooperatively, and that our, and other, data could be best explained by what was termed a rotational catalysis. These previously unrecognized concepts in enzymology provided motivation and excitement within my research group. Richard Cross, a postdoctoral fellow trained with Jui Wang at Yale, capably probed tightly bound ATP. Jan Rosing, a gifted experimentalist from Bill Slater's group in Amsterdam, and Celik Kayalar, an intelligent, innovative graduate student from Turkey, formed a productive pair that unveiled essential facets of cooperative catalysis. David Hackney, a postdoc from Dan Koshland's stable of budding scientists at Berkeley, was an intellectual leader in our ^{18}O experimentation that led to rotational catalysis. Dan Smith, Michael Gresser, Linda Smith, and Chana Vinkler (from Israel) as postdocs, and Lee Hutton, Gary Rosen and Glenda Choate as graduate students, established the participation of bound intermediates in rapid mixing and quenching experiments, and conducted ^{18}O exchange experiments that clarified and supported our mechanistic postulates.

In ensuing years, other aspects of the complex ATP synthase were explored that solidified our feeling that the binding change mechanism was likely valid and general, and promoted its acceptance in the field. I will resist telling you here about the number, properties, and function of the six nucleotide binding sites, of the probes that agreed with rotational catalysis, of the unraveling of the complex Mg^{2+} and ADP inhibition, of the generality of the mechanism and other synthase properties revealed by studies with chloroplasts, *E. coli,* and Kagawa's thermophilic bacterium. It was a pleasure to work on such problems with Teri Melese, a postdoc who excelled in enthusiasm as well as capability, and Zhixiong Xue, an exceptional graduate student that I first met while leading a biochemical delegation to China, with Raj Kandpal a scholarly postdoc from India, with the productive postdocs John Wise (from Alan Senior's lab) and Rick Feldman (from David Sigman's lab), with Janet Wood during her sabbatical, and with June-Mei Zhou and Ziyun Du (on leave from Academia Sinica laboratories in China) as well as Dan Wu, Steven Stroop, and Karen Guerrero as graduate students. Special mention should be made of three excellent Russian researchers, Vladimir Kasho, Yakov Milgrom and Marat Murataliev, from the laboratory of Vladimir Skulachev, a respected leader in bioenergetics. With the latter two I am now writing what will likely be my last paper reporting research results. Other welcome postdocs, visitors, and graduate students at UCLA worked with other problems, including the Na^+,K^+-ATPase that Skou first isolated, and the related Ca^{++} transporting ATPase of the sarcoplasmic reticulum. During these active years it was a pleasure to receive peer recognition in the form of the Rose Award of the American Society for Biochemistry and Molecular Biology, the preeminent society in my field (I served as its President many years earlier).

An unexpected benefit of my career in biochemistry has been travel. The information exchanged and gained at scientific conferences and visits has been tremendously important for progress in my laboratory. My travelophilic wife and I thoroughly enjoyed being guests of the Australian and South African biochemical societies while visiting their countries. Meetings or laboratory visits in Japan, Sweden, France, Germany, Russia, Italy, Wales, Argentina, Iran, and elsewhere gave us a world perspective. Manuscripts that have to be produced, sometimes a bit unwillingly, offer the challenge to present speculation and perspective often not welcome by editors of prestigious journals. It was in a volume from a conference dedicated to one of the giants of the bioenergetics field, Efraim Racker, that the designation "the binding change mechanism" was introduced. Conferences at the University of Wisconsin provided opportunity to publish thoughts about rotational catalysis that had not been enthusiastically endorsed at Gordon Conferences, where information is exchanged without publication. These travels have strong scientific justification. They provided the opportunity for exchange of information, to test new ideas, to gain new perspective, and to avoid unnecessary experiments. The milieu encourages innovation and planning, as well as providing a stimulus and vitality that fosters research progress.

Other events that make up a lifetime continued. Through fortunate circumstances, Lyda and I obtained a building lot at a price that a professor could afford, in the hills north of UCLA, overlooking the city and ocean. The home we built (I was again contractor and miscellaneous laborer) has served as a focal point for family activities, and a temporary residence for grandchildren attending UCLA. The home meant much for my research, as I could readily move between home and lab, and the ambiance created was supportive for study and writing.

The study of life processes has given me a deep appreciation for the marvel of the living cell. The beauty, the design, and the controls honed by years of evolution, and the ability humans have to gain more and more understanding of life, the earth and the universe, are wonderful to contemplate. I firmly believe that our present and future knowledge of all that we are and what surrounds us depends on the tools and approaches of science. I was struck by how well Harold Kroto, one of last year's Nobelists, presented what are some of my views in his biographical sketch. As he stated, "I am a devout atheist–nothing else makes sense to me and I must admit to being bewildered by those, who in the face of what appears to be so obvious, still believe in a mystical creator." I wonder if in the United States we will ever reach the day when the man-made concept of a God will not appear on our money, and for political survival must be invoked by those who seek to represent us in our democracy.

It is disappointing how little the understanding that science provides seems to have permeated into society as a whole. All too common attitudes and approaches seem to have progressed little since the days of Galileo. Religious fundamentalists successfully oppose the teaching of evolution, and by this decry the teaching of critical thinking. We humans have a remarkable ability to blind ourselves to unpleasant facts. This applies not only to mystical and religious beliefs, but also to long-term environmental consequences of our actions. If we fail to teach our children the skills they need to think clearly, they will march behind whatever guru wears the shiniest cloak. Our political processes and a host of human interactions are undermined because many have not learned how to gain a sound understanding of what they encounter.

The major problem facing humanity is that of the survival of our selves and our progeny. In my less optimistic moments, I feel that we will continue to decimate the environment that surrounds us, even though we know of our folly and of what has happened to others. Humans could become quite transient occupants of planet earth. The most important cause of our problem is over population, which nature, as with other species, will deal with severely. I hear the cry from capable environmental leaders and organizations for movement toward sustainable societies. They are calling for sensible approaches to steer us away from impending disaster. But their voices remain largely unheard as those with power, and those misled by religious or nationality concerns, become immersed in unimportant, self-centered and short-range pursuits.

ENERGY, LIFE, AND ATP

Nobel Lecture, December 8, 1997

by

PAUL D. BOYER

University of California at Los Angeles, Department of Chemistry and Biochemistry, California 90095-1469, USA

OVERVIEW

I have a deep appreciation for the unusual and unexpected chain of events that has brought me the Nobel Award. It is my good fortune to be a spokesman for a considerable number of outstanding researchers in the field of bioenergetics whose efforts have revealed an unusual and novel mechanism for one of nature's most important enzymes. Over 50 years ago a vital cellular process called oxidative phosphorylation was demonstrated. The process was recognized as the major way that our bodies capture energy from foods to be used for a myriad of essential cellular functions, but how it occurred was largely unknown. The intervening years have seen much progress. Today I will tell you how contributions of my research group in the 1970s led to new hypotheses that helped overcome the limitations of old paradigms, which were no longer applicable. We gained further support of the hypotheses and clarified other aspects of the process in the 1980s and early 1990s. Then as John Walker, my co-recipient will relate, the X-ray structural data from his group became available. The structural information, about the catalytic portion of the enzyme for the phosphorylation, supported the most novel and least accepted aspect of our hypotheses. Now on this occasion, John and I can tell you how a truly remarkable molecular machine accomplishes the oxidative phosphorylation that was left unexplained for over half a century.

A key player in the process is called ATP, the abbreviation for adenosine triphosphate. At the time I was a graduate student, Fritz Lipmann (1) recognized the broad role ATP played in biological energy capture and use. The adenosine portion for our purposes can be regarded as a convenient handle to bind the ATP to enzymes. It is the three phosphate groups attached in a row, particularly the last two, that participate in energy capture. When the energy stored in ATP is used, the terminal anhydride bond is split, forming adenosine diphosphate (ADP) and inorganic phosphate (P_i). The resynthesis of ATP, coupled to energy input, is catalyzed by an enzyme called ATP synthase, present in abundance in intracellular membranes of animal mitochondria, plant chloroplasts, bacteria and other organisms. The ATP made by your ATP synthase is transported out of the mitochondria and used for the function of muscle, brain, nerve, kidney, liver and other tissues, and for transport and for making a host of compounds that the cell needs. The ADP and phosphate formed when ATP is used return to the mitochondria and ATP is made again using the energy from the oxidations. I estimate that the net synthesis of ATP is the most prevalent chemical reaction that occurs in your body. Indeed, because plants and microorganisms capture and use energy by the same reaction, and the amount of biomass is large, the formation and use of ATP is the principal net chemical reaction occurring in the whole world. This is obviously a very important reaction. How does it occur?

All living cells contain hundreds of large, specialized protein molecules called enzymes. These catalyze the hundreds of chemical reactions that are necessary for the cell to function. Among these are the reactions by which energy is captured by the mitochondria, which are packed into muscle, brain and other cells. Inside the mitochondria and imbeded in its membranes are enzymes that catalyze oxidation of the food you eat. They essentially burn it, using oxygen and producing carbon dioxide and water, in a series of small steps, each catalyzed by a special enzyme. The oxygen you are breathing now is carried by the hemoglobin of your red blood cells, then it reaches the mitochondria where it oxidizes iron atoms that are part of a specialized enzyme, which in turn oxidizes other enzymes in a respiratory chain. The blood stream carries the carbon dioxide produced to the lungs for exhaling. The sequence of oxidations liberates protons and promotes a charge that tends to force protons across the membrane. Similarly, in chloroplasts light energy is coupled to the formation of protonmotive force. This protonmotive force, as shown by the 1978 Nobelist Peter Mitchell (2), causes protons (hydrogen ions) to be translocated through the ATP synthase accompanied by formation of ATP. The important and very difficult question that remained unanswered for many years was how the ATP synthase uses the protonmotive force to make ATP.

ATP SYNTHASE

First I will summarize what is now known about the ATP synthase, then convey aspects of how this knowledge was attained. The enzyme uses a novel mechanism that has catalytic steps different from any that had been seen before with other enzymes. A sketch that depicts the enzyme function is available on the Nobel Foundation internet site. A similar sketch was provided in a recent paper from Richard Cross's laboratory (3). The ATP synthase has three copies each of large α and β subunits, with three catalytic sites located mostly on the β subunit at the interface of the α and β subunits. A γ subunit core and smaller δ and ε subunits complete a portion known as F_1, with a subunit composition in order of decreasing size designated as $\alpha_3\beta_3\gamma\delta\varepsilon$. This portion of the enzyme was first isolated in the laboratory of a splendid investigator, Efraim Racker, and shown to act as an ATPase (4). Several leading investigators in the bioenergetic field were trained in Racker's laboratory.

[1] Abbreviations used for the F_1-ATPase from various sources are: From heart mitochondria MF_1, from chloroplasts CF_1, from *E. coli* EcF_1, from Kagawa's thermophilic bacterium TF_1.

The F_1-ATPase[1] catalyzes ATP hydrolysis but not ATP synthesis. The rest of the enzyme, imbedded in the membrane, is known as F_0; in *E. coli* the F_0 contains a large subunit *a*, two copies of a subunit *b* and probably 12 copies of a much smaller *c* subunit. The F_0 of the mitochondrial enzyme is much more complex. The designation F_1F_0-ATPase is sometimes used in the literature for the complete ATP synthase.

During net ATP synthesis the three catalytic sites on the enzyme, acting in sequence, first bind ADP and phosphate, then undergo a conformational change so as to make a tightly bound ATP, and then change conformation again to release this ATP. These changes are accomplished by a striking rotational catalysis driven by a rotating inner core of the enzyme, which in turn is driven by the protons crossing the mitochondrial membrane. I share the view that revealing the mechanism of the ATP synthase is a fine achievement of modern biochemistry. I am also keenly aware that this achievement comes from the sum of the research of many members of the bioenergetics community, who deserve a major share in the recognition of the accomplishment. But the Nobel awards tend to make heroes of only one or a few of those responsible. It is my good fortune to be addressing you today because my research group, strongly dependent on the information provided by others, gained the first insights into three unusual features of the ATP synthase catalysis. These unusual features are energy-linked binding changes that include release of a tightly bound ATP, sequential conformational changes of three catalytic sites to accomplish these binding changes, and a rotary mechanism that drives the conformational changes. These features had not been recognized previously in enzymology.

EARLY PROBES

In the mid-1950s, some 12 years after receiving my Ph.D., some experiments on how ATP is made were conducted in my laboratory. One concerned the capture of energy in glycolysis. We found that the oxidation of glyceraldehyde 3-phosphate could occur without the participation of inorganic phosphate (5), suggesting participation of an acyl enzyme intermediate. Extension of these experiments, and salient findings in Racker's group (6), demonstrated that a sulfhydryl group on the enzyme was acylated and the acyl enzyme was cleaved by inorganic phosphate to form 1,3-diphosphoglycerate, which in turn transferred a phosphoryl group to ADP to make ATP. The demonstration that two covalent intermediates, the acyl enzyme and the phosphorylated substrate, preceded ATP formation made it seem logical to seek for similar intermediates in oxidative phosphorylation. As we and others learned years later, this was not a useful approach.

Of more relevance to ATP synthase were experiments with ^{18}O and ^{32}P, initiated because of the demonstration by Mildred Cohn that mitochondria would catalyze a rapid exchange of phosphate oxygens with those of water (7). We found from the ^{32}P experiments that the overall reaction of oxidative phosphorylation was dynamically reversible (8). The ^{18}O experiments revealed the striking finding that the exchange of inorganic phosphate oxygens with water was occurring even more rapidly. As illustrated in Fig. 1, we attributed this to the formation of a covalent intermediate, which was then cleaved by inorganic phosphate. We tried unsuccessfully to separate out fractions from mitochondria that would catalyze the first step leading to the formation of an intermediate in oxidative phosphorylation. It was some sixteen years later that we found the simple explanation that no intermediate was formed, and that the rapid ^{18}O exchange resulted from the rapid and reversible formation of a tightly bound ATP.

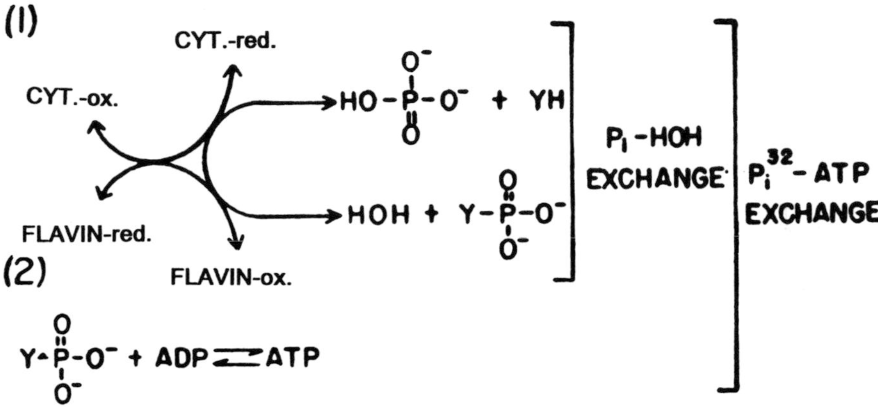

Figure 1. Phosphate oxygen exchange and dynamic reversal of oxidative phosphorylation. Adapted from (7). In this early study covalent intermediates were proposed to explain an oxygen exchange more rapid than the overall reaction reversal.

In the 1960s we embarked on another, only partially successful, series of experiments. By using ^{32}P as a sensitive tracer, we found in mitochondria a ^{32}P-labeled protein that was an intermediate between inorganic phosphate and ATP. We identified this as a previously unrecognized phosphorylated protein, with a phosphoryl group attached to a histidine residue. We mistakenly thought we had identified an intermediate in oxidative phosphorylation, but subsequently found it to be an intermediate in GTP or ATP formation by the succincyl CoA synthetase of the citric acid cycle (9). We were reaching for a gold but got a bronze instead.

^{18}O EXCHANGES AND A NEW CONCEPT

For several years we mostly studied other problems, including taking a look at active transport in *E. coli*. This study gave evidence for an intermediate and unidentified energized state (10), but we did not characterize this state or pay enough attention to the rumblings coming from Peter Mitchell's laboratory. It was difficult for me to accept protonmotive force as a driving agent for ATP formation when I could not visualize a logical way this could occur. But the lure of the ATP synthase continued, and we tried to get leads with photophosphorylation by spinach thylakoid membranes as well as oxidative phosphorylation by heart mitochondria. The use of the ^{18}O exchange measurements to study the process provided a crucial insight. The types of exchange that can be measured are readily understood with the aid of the diagram in Fig. 2. The box in Fig. 2 represents a catalytic site. ADP and P_i can bind and be converted to a tightly bound ATP. The water formed freely interchanges with medium water. Reversal of this reaction results in the incorporation of one water oxygen into the bound P_i. If the P_i can tumble freely at the catalytic site, when bound ATP is again formed there are three chances out of four that it will contain a water oxygen. Various exchanges of phosphate oxygens with water oxygens are measurable, as shown in Table 1. The oxygen exchanges thus provide sensitive probes of reaction steps that otherwise might be hidden.

The ^{18}O probes revealed a puzzling aspect, namely that the intermediate

Table 1. Exchanges of phosphate oxygens with water oxygens catalyzed by ATP synthase.

Exchange	Measurement
Intermediate $P_i \rightleftarrows$ HOH	Hydrolysis of γ-^{18}O-ATP and determination of ^{18}O in P_i formed
Intermediate ATP $\rightleftarrows$ HOH	Synthesis of ATP from ^{18}O-P_i and determination of ^{18}O in ATP formed
Medium $P_i \rightleftarrows$ HOH	Determination of loss of ^{18}O from ^{18}O-P_i when P_i binds, undergoes exchange, and returns to the reaction medium
Medium ATP $\rightleftarrows$ HOH	Determination of loss of ^{18}O from γ-^{18}O-ATP when ATP binds, undergoes exchange, and returns to the reaction medium

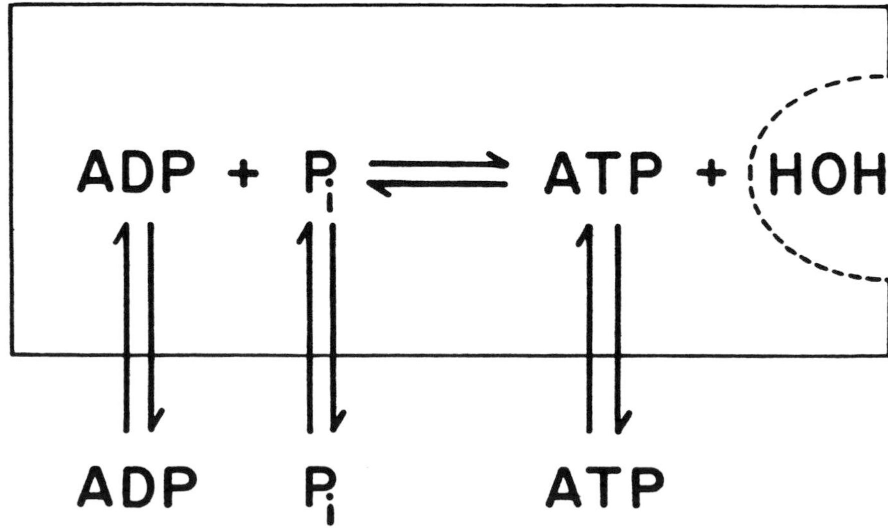

Figure 2. Binding, interconversion, and release steps for oxygen exchanges.

$P_i \rightleftarrows$ HOH was unusually insensitive to uncouplers of oxidative phosphorylation. As shown in Fig. 3, even though the potent uncoupler called S-13 allowed oxidation to proceed without net ATP synthesis, the rapid exchange of phosphate and water oxygens continued. The significance of this was not grasped for some time. But one day, while listening to a seminar that I did not understand, the oxygen exchange data churned in my mind. It became clear to me that the results could be explained if the energy from oxidations was not used to *make* the ATP molecule, but instead was used to bring about a *release* of a tightly bound ATP. The reversible formation of the tightly bound ATP molecule could continue at the catalytic site without involving protonmotive force, and give rise to the uncoupler-insensitive oxygen exchange. We now had a new concept for oxidative phosphorylation and were anxious to call it to the attention of the field. The editors of the *Journal of Biological Chemistry* declined the opportunity to publish this new concept. I used the privilege of my recent membership in the National Academy of Sciences (11, Fig.4) to publish this first feature of what was to become the binding change mechanism of ATP synthesis. Independently, Slater's group, based on the presence of tightly bound nucleotides on the isolated F_1-ATPases, also suggested that energy input might be involved in their release (12).

Our feeling that the new concept was valid was strengthened by companion studies with the ATPase activity of muscle myosin. Data from my and from

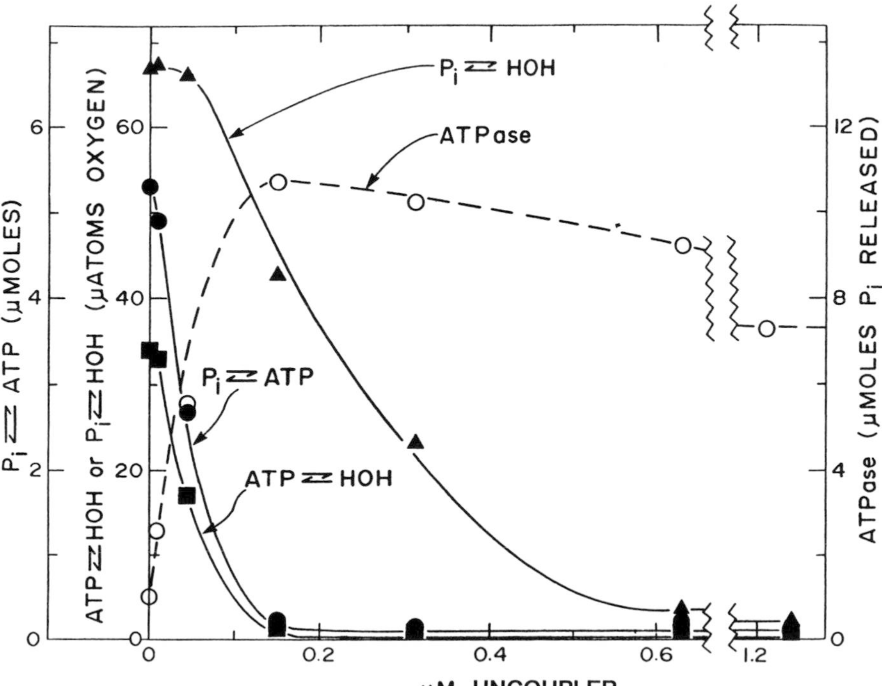

Figure 3. The insensitivity of phosphate oxygen exchange to an uncoupler of oxidative phosphorylation as compared to other measured reactions. Various uncouplers gave similar results; in this experiment an uncoupler known as "S-13" was used.

Koshland's laboratories (13, 14) had shown that myosin could catalyze both a medium $P_i \rightleftarrows HOH$ and an intermediate $P_i \rightleftarrows HOH$ exchange. It seemed possible that myosin might be able to spontaneously form a tightly bound ATP from medium ADP and P_i. Experiments showed this to be the case (15). This and other salient properties of myosin had also been revealed in contemporary studies by Trentham and associates (16, 17). Importantly, the oxygen exchange could be quantitatively accounted for by the rate of formation and cleavage of the bound ATP.

Not all bioenergeticists readily accepted the concept that a prime function of energy input was to bring about the release of a tightly bound ATP. For example, Mitchell preferred a mechanism in which the protons migrated to the catalytic site and induced the formation of ATP from ADP and P_i. It seemed logical to me that proton translocation was linked to ATP release indirectly through protein conformational changes (18). Without my being informed, my publication was accompanied by a rebuttal from Mitchell (19), and I thus presented a more complete model for conformational coupling (20). With time this indirect manner in which proton translocation drives ATP formation has become generally accepted, but this does not detract from Mitchell's salient recognition of protonmotive force as a means of capturing energy for ATP synthesis and active transport.

Reprinted from
Proc. Nat. Acad. Sci. USA
Vol. 70, No. 10, pp. **2837–2839, October 1973**

A New Concept for Energy Coupling in Oxidative Phosphorylation Based on a Molecular Explanation of the Oxygen Exchange Reactions

(protein conformational change/uncouplers/mitochondria)

PAUL D. BOYER, RICHARD L. CROSS, AND WILLIAM MOMSEN

The Molecular Biology Institute and The Department of Chemistry, University of California, Los Angeles, Calif. 90024

Contributed by Paul D. Boyer, June 25, 1973

ABSTRACT **The $P_i \rightleftarrows HOH$ exchange reaction of oxidative phosphorylation is considerably less sensitive to uncouplers than the $P_i \rightleftarrows ATP$ and $ATP \rightleftarrows HOH$ exchanges. The uncoupler-insensitive $P_i \rightleftarrows HOH$ exchange is inhibited by oligomycin. These results and other considerations suggest that the relatively rapid and uncoupler-insensitive $P_i \rightleftarrows HOH$ exchange results from a rapid, reversible hydrolysis of a tightly but noncovalently bound ATP at a catalytic site for oxidative phosphorylation, concomitant with interchange of medium and bound P_i. Such tightly bound ATP has been demonstrated in submitochondrial particles in the presence of uncouplers, P_i, and ADP, by rapid labeling from $^{32}P_i$ under essentially steady-state phosphorylation conditions. These results lead to the working hypothesis that in oxidative phosphorylation energy from electron transport causes release of preformed ATP from the catalytic site. This release could logically involve energy-requiring protein conformational change.**

The beginning of the binding change mechanism.

Figure 4. From the publication presenting a new concept for oxidative phosphorylation (11).

CATALYTIC COOPERATIVITY

We were now launched on an exciting period of research. As we probed mitochondrial oxidative phosphorylation further by ^{32}P and ^{18}O isotope exchanges, some puzzling aspects emerged. For example, when submitochondrial particles capable of oxidative phosphorylation were hydrolyzing ATP, a lively medium ATP ⇄ HOH exchange occurred. Removal of product ADP stopped this exchange (Fig. 5), although the reversal of ATP hydrolysis was still occurring on the enzyme. Somehow the lack of medium ADP to bind to the enzyme was stopping the release of ATP. It was not apparent how this could occur if the simple scheme of Fig. 2 was used to explain the oxygen exchanges. Similarly, during net synthesis of ATP, removal of the medium ATP stopped the medium P_i ⇄ HOH exchange. An explanation for why these oxygen exchanges were blocked, and for other related observations, was suggested by one of my graduate students, Celik Kayalar from Turkey. Celik said he could account for these results if the catalytic sites had to work cooperatively, so that ATP could not be released from one site unless ADP and P_i were available to bind at another site, or that P_i could not be released from one catalytic site unless ATP were available to bind at another catalytic site. Celik, together with Jan Rosing from Holland, demonstrated and characterized sequential and cooperative participation of catalytic sites with the synthase in submitochondrial particles capable of or during oxidative phosphorylation (21, 22). In addition, their results gave evidence that the binding changes accompanying proton translocation also promoted the tight binding of P_i.

Adolfsen and Moudrianakis suggested that site-site cooperativity might occur with the separated F_1-ATPase, based on the observation that a tightly

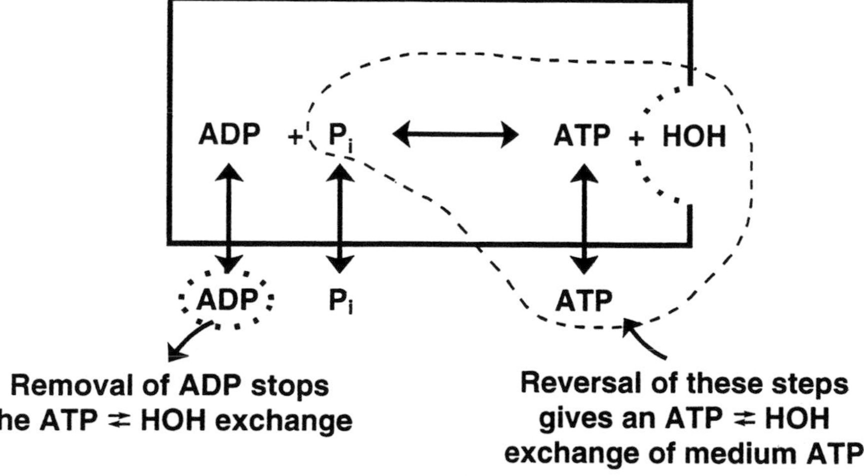

Figure 5. Removal of medium ADP stops the medium ATP ⇄ HOH exchange.

bound ADP was released when ATP was cleaved by a bacterial F_1-ATPase (23). Experiments in my laboratory revealed that, as we had found with the ATP synthase, a strong cooperativity of catalytic sites occurs with the isolated ATPase. When the MF_1 hydrolyzes relatively high concentrations of ATP, the P_i formed contains only slightly more than the one water oxygen required for the hydrolysis (Fig. 6). But as the ATP concentration is lowered, an instructive change occurs. The hydrolysis velocity is of course lowered, but the number of water oxygens appearing in each P_i formed increases to almost four. It can be calculated that nearly 400 reversals of bound ATP hydrolysis occur before the P_i formed is released (10). The bound ADP and P_i formed can not be released until ATP is available to bind at another catalytic site[2].

Experiments had now made it seem likely that an unexpected catalytic cooperativity was a prominent feature of the ATP synthase. At that time the prevailing view was that the enzyme had only two catalytic sites, and a diagram depicting a bi-site mechanism appeared in my 1977 review article in the

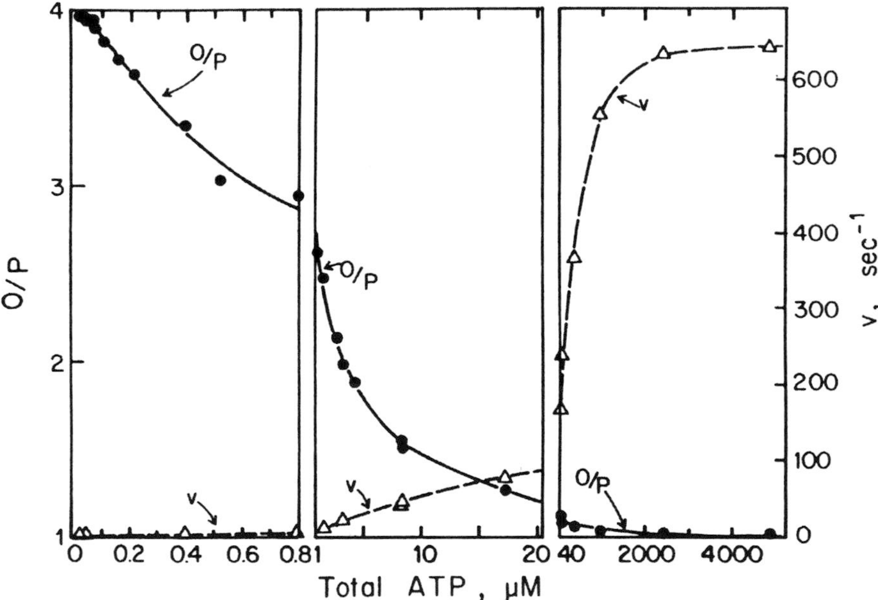

Figure 6. Effect of decrease in ATP concentration on the extent of water oxygen incorporation each phosphate released and on net hydrolysis velocity as catalyzed by the MF1-ATPase. Adapted from (10).

[2] We missed obtaining this striking result about a decade earlier. At that time Efraim Racker came to my laboratory with some F_1 ATPase so we could find if it catalyzed an oxygen exchange when ATP was hydrolyzed. We conducted the reaction at a relatively high ATP concentration, and found the incorporation of only one water oxygen, as required for the cleavage reaction. Harvey Penefksy made a similar observation (24). If we had measured what happened as the ATP concentration was lowered, we would have revealed the catalytic site cooperativity then. But we had no reason to suspect that the enzyme catalysis for each substrate cleaved would change so dramatically with substrate concentration. The strong catalytic cooperativity that we later demonstrated had not been described previously for any other enzyme.

Annual Review of Biochemistry (25). We recognized, however, that if the enzyme were found to have three catalytic sites, a tri-site mechanism, as currently known to occur, would be likely (22, 26). The crucial point was that a tightly bound ATP could not be released until ADP and P_i bound at a second or a second and third catalytic site and the binding changes driven by proton translocation occur. This positive cooperativity meant that at low substrate concentrations, during either net synthesis or hydrolysis of ATP, a tightly bound ATP should still be present at a catalytic site. We undertook experiments to find if this was so. These tests were made with submitochondrial particles (27) or chloroplast thylakoids (28) so that net ATP formation was occurring with ADP concentrations far below the apparent K_m of ADP for maximal phosphorylation rates. About one tightly bound ATP committed to net ATP formation was found on each synthase. Such data give good evidence that strong positive catalytic cooperativity takes place under conditions where ATP synthesis is actually occurring. Additional findings from our and from other laboratories consistent with or favoring the catalytic cooperativity are summarized elsewhere (29).

In 1982, Feldman and Sigman demonstrated that the CF_1-ATPase or the ATP synthase on chloroplast thylakoids, which have a tightly bound catalytic-site ADP, would slowly form an equilibrium concentration of bound ATP from relatively high concentrations of medium P_i (30, 31). Their characterization of this single site catalysis supported our concept of tight ATP formation without coupling to protonmotive force. Factors that promote formation of ATP at catalytic sites of myosin and F_1-ATPases likely include the very tight preferential binding of ATP and, as suggested by deMeis (32), low water activity.

Also in 1982, the acceptance of catalytic cooperativity by the field was considerably enhanced by the determination in Penefsky's laboratory of the rate constants for the interconversion, binding, and release steps for MF_1 exposed to ATP at molar concentrations less than the molarity of the enzyme, conditions that gave what was termed uni-site catalysis (33, 34). An additional important contribution from the same laboratory was the demonstration that the hydrolysis of a trinitrophenyl ATP bound at a single catalytic site was markedly increased by the binding of a second trinitrophenyl-ATP (35).

A number researchers subsequently found a slow uni-site catalysis with different F_1-ATPases. However, an inability to see a definite uni-site catalysis with TF_1 raised the question as to whether the cooperativity we had observed was a general phenomenon of F_1-ATPases (36). That slow uni-site catalysis was indeed occurring was demonstrated by the increase in intermediate $P_i \rightleftarrows$ HOH exchange as ATP concentration was lowered (37). For a number of years there appeared to be a general acceptance that a slow uni-site rate occurs and that the catalytic rate is markedly accelerated when ATP binds to additional sites. It was thus somewhat surprising when a quite recent claim appeared that MF_1 depleted of bound nucleotides did not show a slow uni-site catalysis (38). However, this claim is not experimentally sound; slow uni-site catalysis occurs with either native or nucleotide depleted MF_1 (39, 40).

RELATED EXPERIMENTS

My laboratory group at this time also had an experimental interest in the Na^+K^+-ATPase that Prof. Skou has presented. There was uncertainty whether the phosphoryl group that became attached to the enzyme as an intermediate in the catalysis was on a glutamyl or an aspartyl residue. We developed a borohydride reduction method that established that the group was attached to an aspartyl residue (41, 42). We also discovered that the enzyme in the presence of K^+ and Mg^{2+} catalyzed a rapid exchange of oxygens of P_i with water oxygens, attributable to a dynamic reversal of enzyme phosphorylation (43). This was and remains (44) a useful way to probe this step of the reaction sequence. The related Ca^{++}-activated sarcoplasmic reticulum ATPase was likewise found to catalyze a rapid $P_i \rightleftarrows$ HOH exchange (45).

Our attention was also directed toward the capacity of yeast pyrophosphatase to catalyze a $P_i \rightleftarrows$ HOH exchange (46). We revealed that this exchange was due to a reversible formation of an enzyme-bound pyrophosphate (47) and the details of the exchange process were elucidated (48, 47). It was important to us that rapid mixing experiments showed that the rate of formation and cleavage of the bound pyrophosphate accounted for the oxygen exchange (47). As mentioned above, this was shown previously for the bound ATP and oxygen exchange catalyzed by myosin. For the pyrophosphatase exchange, Hackney developed a theoretical analysis of the distribution of ^{18}O-labeled species of P_i (49) that was to serve us well in studies we had underway on oxidative phosphorylation. The $P_i \rightleftarrows$ HOH exchange catalyzed by the sarcoplasmic reticulum ATPase was also shown by rapid mixing and quenching experiments to result from the dynamic reversal of the formation of the phosphorylated enzyme intermediate (50). Such results, and the demonstration by Wimmer and Rose that the ATP $\rightleftarrows$ HOH exchange catalyzed by mitochondria resulted from the reversible cleavage of the terminal P-O-P bond (51), gave us confidence that in oxidative phosphorylation and photophosphorylation the oxygen exchanges we observed were due to the reversible hydrolysis of tightly bound ATP.

THE NUMBER OF CATALYTIC SITES

Meanwhile studies in other laboratories were revealing the subunit stoichiometry of the F_1-ATPase. As noted in a review by Penefsky covering literature up through 1978 (24), considerable controversy remained. The difficulty of obtaining satisfactory molecular weights and subunit quantitation made it hard to get a clear choice between the presence of two or three copies of the major α and β subunits. Reports that measurements with EcF_1 and TF_1 isolated from bacteria grown on [^{14}C]-amino acids (52, 53) favored a stoichiometry of $\alpha_3\beta_3\gamma_1$ seemed convincing to us. Reports on the composition of CF_1 strongly supported presence of three each of the large subunits (54). On the basis of these and other developments, the field soon widely accepted the composition of F_1-ATPases as $\alpha_3\beta_3\gamma\delta\epsilon$. All of our further experiments have been based on such a stoichiometry of subunits.

The number of nucleotide binding sites on the enzyme remained controversial until about a decade ago. Both α and β subunits were shown to have nucleotide binding sites. Reports in 1982 for MF_1 (55) and in 1983 for EcF_1 (56) gave good evidence for the presently accepted values of six potential nucleotide binding sites per enzyme. However, as late as 1987 claims were still made for only three nucleotide binding sites on CF_1 (57) and four for the liver F_1 (58). Subsequent data for CF_1 (59, 60) and the liver enzyme (61), as well as the highly conserved sequence of the β subunits, support the present view that all F_1-ATPases have six nucleotide binding sites, although differing considerably in affinity.

Chemical derivatization studies, such as those in Bragg's laboratory (62) and summarized in reviews (63, 29) showed that all three β subunits, although with identical amino acid sequence, had distinctly different chemical properties. Such heterogeneity was a prominent reason why we considered it likely that all three β subunits passed through different conformations during catalysis. The participation of all three β subunits in a cooperative, sequential manner was supported, but not proven, by observations (over twenty are given in an earlier review (29)) that derivatization of only one site per enzyme would nearly or completely block catalysis. We were also impressed by studies in Futai's laboratory showing that one defective mutant β subunit stopped catalysis (64), and by related mutational studies in Senior's laboratory (65) that favored the participation of three equivalent β subunits for catalysis.

There has, however, been considerable delay in reaching a general acceptance that three catalytic sites participate in an equivalent manner. A single catalytic and two regulatory sites have been proposed (66, 67). Various models with only two catalytic sites have been suggested (68, 69, 70, 71, 72, 73), as well as a 1991 model with four functioning catalytic sites arranged in two alternate pairs (74, 75). A 1989 review by Tiedge and Schafer (76) stresses symmetrical considerations and favors equivalent β subunit participation. Various models, and a 1991 review favoring a two-site model (77), were appraised in a review prepared in 1992, in which I attempted to consider any experiments not in harmony with the binding change mechanism (29). The conclusion I reached is that very likely three sites participate in an equivalent manner. Subsequent events (see 78) have strengthened this conclusion, although some doubts of which I am not aware may remain. The probability that three sites participate equivalently has guided experiments in my laboratory since the presence of three β subunits first seemed likely.

ROTATIONAL CATALYSIS

Toward the end of the 1970s, we initiated experiments that led to the postulation of the third feature of the binding change mechanism. The presence of three copies of the major α and β subunits and single copies of the γ, δ, and ε made it unlikely that all three β subunits could have identical interactions with single copy subunits. In particular, interactions with the larger γ subunits seemed likely to be crucial. McCarty's laboratory had reported that, with chloroplasts, light increased the reactivity of -SH groups on the subunit and that modifications in the γ subunit increased the leakage of protons across the coupling membrane (79, 80). This and other evidence suggested that the γ subunit interacted strongly with the catalytic β subunit. The growing information about the synthase gave a base for the interpretation of additional experiments with ^{18}O that were underway in my laboratory.

Water highly labeled with ^{18}O had became more available, and by nuclear magnetic resonance, as demonstrated by Cohn (81), or mass spectrometry we could measure what we designated as the ^{18}O isotopomers of P_i, containing 0, 1, 2, 3, or 4 ^{18}O atoms. Then when ATP synthesis or hydrolysis occurs with highly ^{18}O-labeled substrates, under conditions where appreciable oxygen exchange occurs, the distribution of isotopomers formed can be measured. If all the catalytic sites involved behave identically, the distributions of ^{18}O isotopomers would conform to a statistically predicted pattern. The results observed in a typical experiment for hydrolysis of ^{18}O-ATP by F_1 ATPase are given in Fig. 7 (82). They show that the distribution of isotopomers conforms very closely to that expected for identical behavior of all catalytic sites. The data rule out the possible participation of two types of catalytic sites. As shown by one example in Fig. 7, this would give a markedly different distribution of

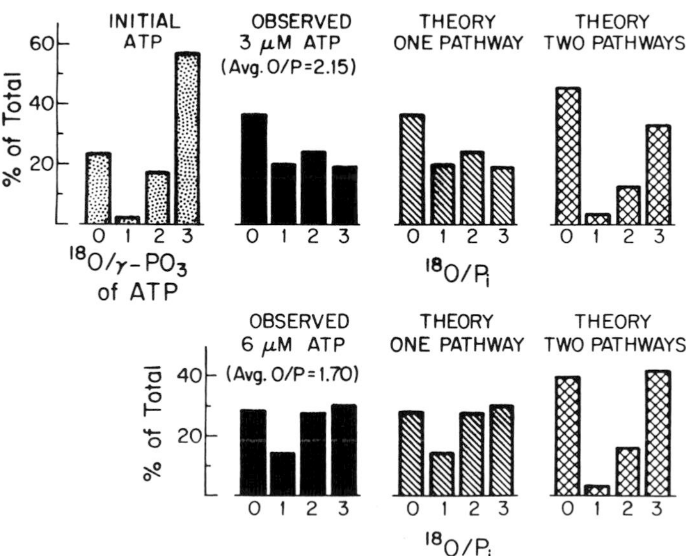

Figure 7. Distribution of ^{18}O-isotopomers of P_i formed from γ-^{18}O-ATP by MF1-ATPase hydrolysis at two relatively low ATP concentrations. The observed average number of water oxygens incorporated (O/P ratio) and distribution of species with 0 to 4 ^{18}O atoms are shown. Also shown is the theoretical distribution for one pathway as expected if the probability for exchange instead of release of bound P_i was 0.73 with 3 μM ATP and 0.55 with 6 μM ATP. This is compared to the expected distributions if two pathways were operative, one with a high an one with a low probability of exchange, that would give the observed total amount of oxygen exchange. Adapted from (82).

isotopomers. Importantly, experiments with the net ATP synthesis by chloroplast and mitochondrial ATP synthases also showed that all catalytic sites behave identically (83, 84, 85). The tests were sensitive and revealing; if steps of substrate binding, interconversion or release, or their concentration dependencies differed among catalytic sites, this should have been revealed in the ^{18}O experiments.

I was again confronted with unexplained results. Although it might be possible to bring a similar residue or residues on minor subunits into contact with each of the three β subunits, the interactions would not be expected to be identical. The situation might be analogous to the family of serine proteases, where markedly different sequences can appropriately position a serine residue. But the resulting proteases do not conduct their catalyses identically. To me, there seemed only one way that all catalytic sites could proceed sequentially and identically, with modulation by one or more single-copy, minor subunits. This was by a rotational catalysis, in which large catalytic subunits moved rotationally around a smaller asymmetric core. Such consideration, together with what was known about the structure of the enzyme, resulted in the postulate of rotational catalysis, presented at Gordon Research Conferences and elsewhere (86, 87, 88). A sketch of our view as presented at that time is shown in Fig. 8 (88). The internal core was likened to a cam shaft that modulated the conformation of the β subunits. The probability that the core was asymmetric was strengthened when amino acid sequence data became

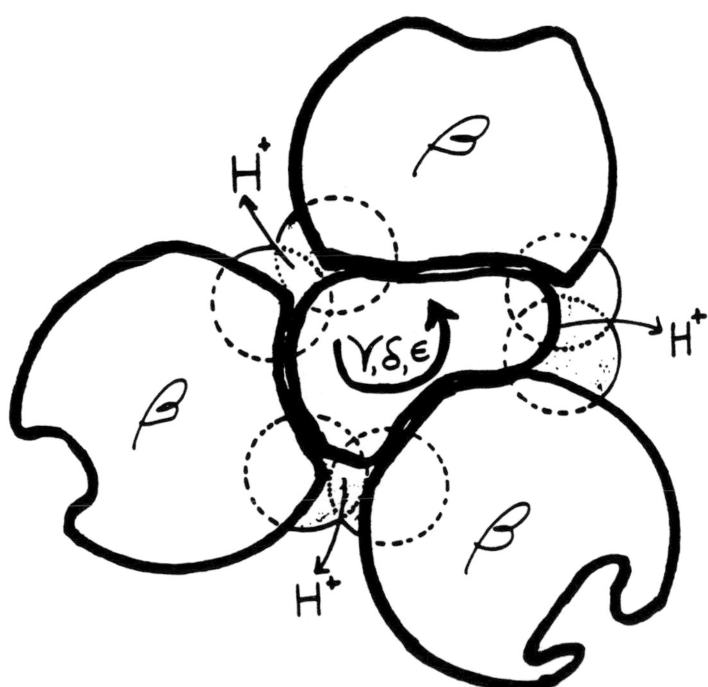

Figure 8. A sketch of possible rotational catalysis as used for 1980 presentations and discussions.

available (89); this gave no indications of possible tripartite symmetry of the minor subunits.

Later other suggestions were made of possible rotational features in the ATP synthase catalysis. Increased information about the structure of the F_0 portion of the synthase made some type of circular motion in the F_0 attractive. Cox *et al.* suggested rotational movement of circularly arranged *c* subunits (90, 91). Hoppe and Sebald visualized an oligomeric core of *c* subunits rotating against subunit *a* or *b* (92), a suggestion that still seems pertinent. Mitchell proposed a rotational model that exposed catalytic sites to a proton channel through the γ subunit (93).

The homogeneity of catalysis demonstrated by the ^{18}O technique also, to my mind, ruled out postulates, as mentioned earlier, that only two β subunits were involved in catalysis, with the other serving a regulatory function. Considerations of the need for symmetry in subunit interactions made it unlikely that two sites could alternate in catalysis identically. Their interactions on one side could not be identical with those on another side at the same stages of catalysis.

We attempted some assessments of subunit positional interchange as required by a rotational catalysis. The MF_1-ATPase after labeling one β subunit with radioactive DCCD (dicyclohexylcarbodiimide) still retained some activity. A different β subunit reacted with 2-azido-ATP. After catalytic turnover, the reactivity toward DCCD and 2-azido-ATP was randomized, as expected if a change in relative position and conformation had occurred (94). In another approach, we observed that a mild cross-linking of subunits stopped catalysis, and that cleavage of the cross-linker restored activity (95). A report from another laboratory that cross linking of the β and γ subunits did not stop catalysis (96), I regarded as inconclusive (29). None of these experiments were as edifying as those that came later from other laboratories (see below). It seemed apparent that an adequate evaluation of the possibility of rotational catalysis would need to await the knowledge of the 3-dimensional structure of the F_1-ATPase. In a review I prepared in the spring of 1992, I summarized the case for rotational catalysis at that stage (29). This included the need for a second attachment between F_0 and F_1 to act as a stator, and the suggestion that present evidence indicated that the δ subunit of the *E. coli* enzyme, or the analogous OSCP of mitochondria, could help serve this function, a prediction that has found support in recent experiments (97, 98). Attachment of a stator to the exterior of an α subunit might be partly responsible for the asymmetry of the α subunits, an asymmetry that is retained during catalysis (99). This may be analogous to the symmetry of the internal rotation of a motor not being disrupted by bolting the motor to a bench.

The occurrence of a rotational catalysis was dramatically supported by the X-ray structure for the major portion of MF_1, attained by Abrahams, Leslie, Lutter and Walker (100). This structure served as the base for innovative demonstrations of rotation in the laboratories of Cross (2, 101, 102), Capaldi (103, 104, 105), and Kagawa (97). Sabbert et al. demonstrated rotation by sophisticated fluorescent techniques (106, 107), and Noji et al. demonstrated rotation visually (108). Such developments allowed me to title a recent review as "The ATP Synthase–A Splendid Molecular Machine" (78). These more recent aspects of the ATP synthase story are more appropriately the subject of my able co-recipient John Walker's lecture. But before you have the opportunity to hear from him, I want to discuss some additional important and unsettled facets of the ATP synthase catalysis.

SOME ADDITIONAL ASPECTS

Acceptance of the binding change mechanism over the past two decades has been fostered by clarification of a number of unusual aspects of the synthase action, some of which are mentioned here. The number and properties of nucleotide binding sites needed clarification. With the use of the 2-azido-ATP, introduced for studies with F_1-ATPases by Abbot et al. (109), we established where catalytic and noncatalytic sites resided with the F_1-ATPase from different sources (59, 110, 111, 112). The characteristics of the Mg^{2+} and tightly bound ADP inhibition of the F_1-ATPase, that had harassed our, and many other, earlier studies, were established (113, 114). A role for the noncatalytic nucleotides in enabling the inhibition to be overcome was uncovered (115, 116).

A direct estimation of how many catalytic sites were filled during photophosphorylation was accomplished (117). The results gave evidence that near maximal rates of ATP synthesis were attained when a second, and not a second and a third, site were loaded with substrates. The consideration of these results, other earlier data, and recent experiments on site filling in $MF_{1,}$ have led to refinements in how I consider the binding change mechanism to operate. Salient points from earlier data are that the rate of ATP formation during uni-site catalysis is much slower than the rate of ATP formation when rapid photophosphorylation is occurring, and that during photophosphorylation about one tightly bound ATP per synthase is present. In previous depictions of the mechanism (Fig. 9), after a binding change a site is depicted as having a tightly bound ADP and P_i that is being reversibly converted to tightly bound ATP, while waiting for the next binding change. We now propose that during active net ATP synthesis the interconversion of sites is as depicted in Fig. 10. As a site to which ADP and P_i have added is converted to a tight site, the capacity for the rapid formation of the terminal covalent bond in ATP is also acquired, such that essentially all the bound ADP and P_i are converted to bound ATP. A site with tightly bound ADP and $P_{i,}$, as in Fig. 9, may not be a compulsory intermediate. The next rapid binding change brings about the release of the ATP to the medium.

All ATP made in oxidative phosphorylation (118) or photophosphorylation (119, 84) contains about 0.4-1.1 water oxygens. This means that some rapid reversal of ATP formation has occurred. Indeed, during net oxidative phosphorylation by mitochondria, rapid reversal of the overall process is demonstrated by ^{32}P measurements (9, 118). Thus it seems likely that the rapid incorporation of some water oxygen results from the reversal of a bind-

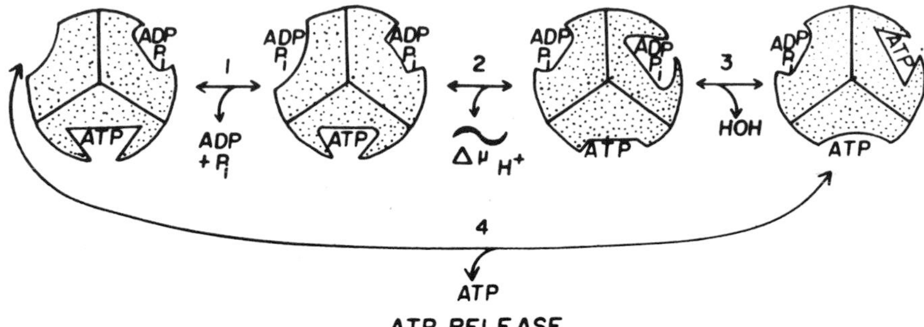

Figure 9. A typical tri-site model for cooperativity including tightly bound ADP and P_i as an intermediate.

ing change step of Fig. 10. When chloroplasts doing net ATP synthesis are separated from medium nucleotides by centrifugation and washing, bound P_i drops off and the catalytic site is left with tightly bound ADP. This is the ADP that in presence of Mg^{2+} results in a strong inhibition of ATPase activity.

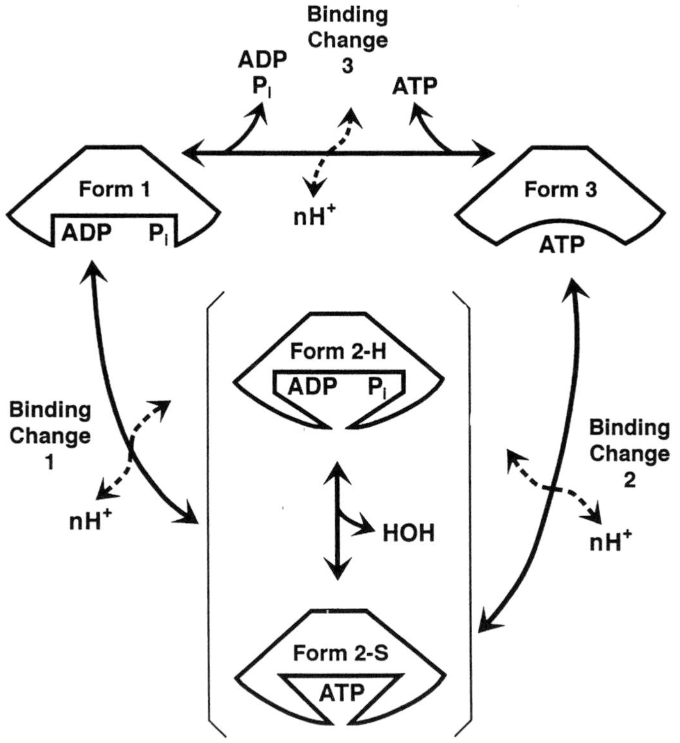

Figure 10. A proposal of how a catalytic site on the ATP synthase is modified by successive binding changes. When ADP + P_i add to Form 1, and adequate protonmotive force is present, both rapid formation and tight binding of ATP arise during Binding Change 1. Most of the site assumes the conformation of Form 2-S, and the ATP becomes loosely bound in Binding Change 2. When ATP adds to Form 3, and no protonmotive force is present, both rapid formation and tight binding of ADP + P_i arise during Binding Change 2. Most of the site assumes the conformation of Form 2-H, and the ADP + P_i become loosely bound during Binding Change 1. Both site occupancy on Forms 1 and 3 and protonmotive force modulate the quasi-equilibrium of Form 2.

However, when protonmotive force is applied, such tightly bound ADP is released to the medium without delay in the first binding change (120).

Other recent experiments pertinent to site occupancy during ATP hydrolysis by MF_1 were based on competition between ATP and trinitrophenyl-ATP (TNP-ATP). They revealed that TNP-ATP could bind strongly to a third catalytic site for which ATP which had a K_d in the millimolar concentration range. The near maximal ATPase rate was attained at considerably less than 1 mM ATP (121). This result, and further characterization of the transition from uni-site to multi-site catalysis and initial velocity measurements, are best explained by the filling of only two catalytic sites being necessary for near maximal rates of ATP hydrolysis (40). Interestingly, ADP had a considerably higher affinity than ATP for the third empty site of the MF_1. Our present hypothesis about catalytic site occupancy during rotational catalysis is depicted in Fig. 11. During rapid synthesis one site has a bound ATP and a site to its left (as viewed from above the F_1 portion of the synthase) can preferentially bind ADP and P_i. When adequate protonmotive force is present, rapid ATP synthesis ensues. The filling of a third site with ADP and P_i at higher substrate concentrations results in little rate acceleration. During net ATP hydrolysis, when protonmotive force is weak or absent, the preferential binding of ATP to a site to the right of the tight ATP site can result in a near maximal hydrolysis rate. Filling of the third site at millimolar concentrations of ATP gives little rate acceleration. Nature appears to have designed a way that ATP synthesis occurs with ADP addition to a site that has low affinity for ATP, helping to obviate ATP inhibition of its own synthesis.

The recognition of the principal features of the ATP synthase catalysis

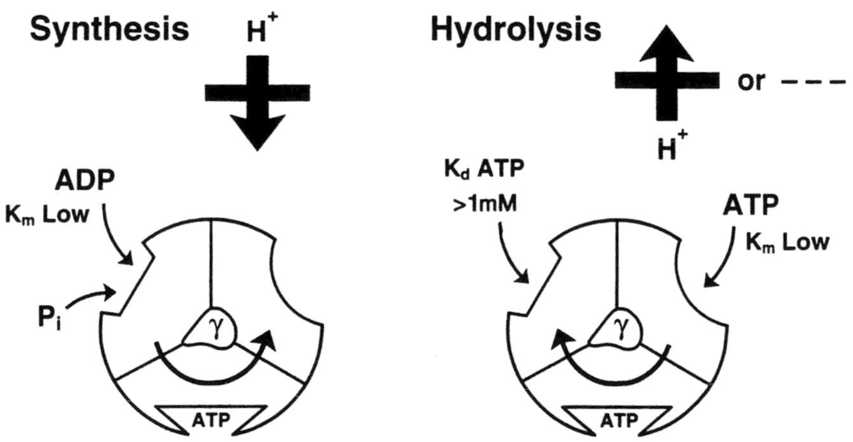

Figure 11. A proposal that near maximum rates of hydrolysis by F_1-ATPase or synthesis by ATP synthase occurs with the filling of only two sites.

creates many opportunities for gaining a better understanding of this remarkable enzyme. I will be an interested spectator in these developments. I believe that societies will, and should continue to, devote some of their resources to basic scientific research, even if the only return is the satisfaction that comes from the knowledge of how living processes occur. An additional justification is that such knowledge underlies past and future gains for attaining a healthy life. As summarized by Ernster (122), the oxygen we use to make ATP is also a toxic substance, resulting in production of harmful free radicals. The mitochondrion is particularly susceptible to such damage, and knowledge of the enzymes involved in energy capture and use may give insight into, and help find how to prevent. unwanted damage.

A final acknowledgment–I am exceptionally fortunate to have been a biochemist over the past decades when so much has been accomplished in my field. Participation in a series of researches that has revealed an unusual rotational catalysis by a vital enzyme has been warmly gratifying. I am indebted to the society that has made this possible, to my wife, Lyda, for her devotion and guidance given freely to help me and our children find our way, and to the universities and government agencies that provided the environment and the financial support for my researches.

REFERENCES

1. Lipmann, F. (1941) *Adv. Enzymol.* **1,** 99–152
2. Mitchell, P. (1979) *Science* **206,** 1148–1159
3. Duncan, T. M., Bulygin, V. V., Zhou, Y., Hutcheon, M. L., and Cross, R. L. (1995) *Proc. Natl. Acad. Sci. U. S. A.* **92,** 10964–10968
4. Penefsky, H. S., Pullman, M. E., Datta, A., and Racker, E. (1960) *J. Biol. Chem.* **235,** 3330–3336
5. Segal, H. L., and Boyer, P. D. (1953) *J. Biol. Chem.* **204,** 265–280
6. Krimsky, I., and Racker, E. (1952) *J. Biol. Chem.* **198,** 721–730
7. Cohn, M. (1953) *J. Biol. Chem.* **201,** 739–744
8. Boyer, P. D., Falcone, A. S., and Harrison, W. H. (1954) *Nature* **174,** 401–404
9. O'Neal, C. C., and Boyer, P. D. (1984) *J. Biol. Chem.* **259,** 5761–5767
10. Klein, W. L., and Boyer, P. D. (1972) *J. Biol. Chem.* **247,** 7257–7265
11. Boyer, P. D., Cross, R. L., and Momsen, W. (1973) *Proc. Natl. Acad. Sci.* **70,** 2837–2839
12 Harris, D. A., Rosing, J., van deStadt, R. J., and Slater, E. C. (1973) *Biochim. Biophys. Acta.* **314,** 149–153
13. Dempsey, M. E., Boyer, P. D., and Benson, E. S. (1963) *J. Biol. Chem.* **238,** 2708–2715
14. Levy, H. M., Sharon, N., Lindemann, E., and Koshland, D. E. (1960) *J. Biol. Chem.* **235,** 2628–2633
15. Wolcott, R. G., and Boyer, P. D. (1975) *J. Supramol. Structure* **3,** 154–161
16. Bagshaw, C. R., Eccleston, J. F., Eckstein, F., Goody, R. S., Gutfreund, H., and Trentham, D. R. (1974) *Biochem. J.* **141,** 351–364
17. Bagshaw, C. R., Trentham, D. R., Wolcott, R. G., and Boyer, P. D. (1975) *Proc. Natl. Acad. Sci. U. S. A.* **72,** 2592–2596
18. Boyer, P. D. (1975) *FEBS Lett.* **50,** 91–94
19. Mitchell, P. (1975) *FEBS Lett.* **50,** 95–97
20. Boyer, P. D. (1975) *FEBS Lett.* **58,** 1–6
21. Rosing, J., Kayalar, C., and Boyer, P. D. (1977) *J. Biol. Chem.* **252,** 2478–2485
22. Kayalar, C., Rosing, J., and Boyer, P. D. (1977) *J. Biol. Chem.* **252,** 2486–2491
23. Adolfsen, R., and Moudrianakis, E. N. (1976) *Arch. Biochem. Biophys.* **172,** 425–433

24. Penefsky, H. S. (1979) *Advances in Enzymol. and Related Areas Mol. Biol.* **49,** 223–280
25. Boyer, P. D. (1977) *Annu. Rev. Biochem.* **46,** 955–966
26. Kayalar, C. (1977) Ph. D. Thesis, University of California, Los Angeles
27. Gresser, M., Cardon, J., Rosen, G., and Boyer, P. D. (1979) *J. Biol. Chem.* **254,** 10649–10653
28. G., Gresser, M., Vinkler, C., and Boyer, P. D. (1979) *J. Biol. Chem.* **254,** 10654–10661
29. Boyer, P. D. (1993) *Biochim. Biophys. Acta* **1140,** 215–250
30. Feldman, R. I., and Sigman, D. S. (1982) *J. Biol. Chem.* **257,** 1676–1683
31. Feldman, R., and Sigman, D. S. (1983) *J. Biol. Chem.* **258,** 12178–12183
32. De Meis, L. (1989) *Biochim. Biophys. Acta* **973,** 333–349
33. Grubmeyer, C., Cross, R., and Penefsky, H. S. (1982) *J. Biol. Chem.* **257,** 12092–12100
34. Cross, R. L., Grubmeyer, C., and Penefsky, H. S. (1982) *J. Biol. Chem.* **257,** 12101–12105
35. Hisabori, T., Muneyuki, E., Odaka, M., Yokoyama, K., Mochizuki, K., and Yoshida, M. (1992) *J. Biol. Chem.* **267,** 4551–4556
36. Yohda, M. K., and Yoshida, M. (1987) *J. Biochem.* **102,** 875–883
37. Kasho, V. N., Yoshida, M., and Boyer, P. D. (1989) *Biochemistry* **28,** 6949–6954
38. Reynafarje, D. B., and Pedersen, P. L. (1996) *J. Biol. Chem.* **271,** 32546–32550
39. Milgrom, Y., and Cross, R. L. (1997) *J. Biol. Chem.***272,** 32211–32214
40. Milgrom, Y., Murataliev, M. B., and Boyer, P. D. (1998) *Biochem. J.* **330,** 1307–1043
41. Degani, C., and Boyer, P. D. (1973) *J. Biol. Chem.* **248,** 8222–8226
42. Degani, C., Dahms, A. S., and Boyer, P. D. (1974) *Ann. N. Y. Acad. Sci.* **242,** 77–79
43. Boyer, P.D., de Meis, L., Carvalho, M.G.C., and Hackney, D.D. (1977) *Biochemistry* **16,** 136–40
44. Kasho, V. N., Stengelin, M., Smirnova, I. N., and Faller, L. D. (1997) *Biochemistry* **36,** 8045–8052
45. Kanazawa, T., and Boyer, P. D. (1973) *J. Biol. Chem.* **248,** 3163–3172.
46. Cohn, M. (1958) *J. Biol. Chem.* **230,** 369–379
47. Janson, C. A., Degani, C., and Boyer, P. D. (1979) *J. Biol. Chem.* **254,** 3743–3749
48. Hackney, D. D., and Boyer, P. D. (1978) *Proc. Natl. Acad. Sci. U. S. A.* **75,** 3133–3137
49. Hackney, D. D. (1980) *J. Biol. Chem.* **255,** 5320–5328
50. Boyer, P. D., de Meis, L., Carvalho, M. G. C., and Hackney, D. D. (1977) *Biochemistry* **16,** 136–140
51. Wimmer, M. J., and Rose, I. A. (1977) *J. Biol. Chem.* **252,** 6769–6775
52. Bragg, P. D., and Hou, C. (1975) *Arch. Biochem. Biophys.* **167,** 311–321
53. Kagawa, Y., Sone, N., Yoshida, M., Hirata, H., and Okamoto, H. (1976) *J. Biochem.* **80,** 141–151
54. Merchant, S., Shaner, S. L., and Selman, B. R. (1983) *J. Biol. Chem.* **258,** 1026–1031
55. Cross, R. L., and Nalin, C. M. (1982) *J. Biol. Chem.* **257,** 2874–2881
56. Wise, J. G., Duncan, T. M., Latchney, L. R., Cox, D. N., and Senior, A. E. (1983) *Biochem. J.* **215,** 343–350
57. McCarty, R. E., and Hammes, G. G. (1987) *Trends Biochem. Sci.* **12,** 234–237
58. Williams, N., Hullihen, J., and Pedersen, P. L. (1987) *Biochemistry* **26,** 162–169
59. Xue, Z., Zhou, J. M., Melese, T., Cross, R. L., and Boyer, P. D. (1987) *Biochemistry* **26,** 3749–3753
60. Girault, G., Berger, G., Galmiche, J. M., and Andre, F. (1988) *J. Biochem. Chem.* **263,** 14690–14695
61. Guerrero, K. J., and Boyer, P. D. (1988) *Biochem. Biophys. Res. Comm.* **154,** 854–860
62. Bragg, P. D., and Hou, C. (1990) *Biochim. Biophys. Acta* **1015,** 216–222
63. Vignais, P. V., and Lunardi, J. (1985) *Annu. Rev. Biochem.* **54,** 977–1014
64. Noumi, T., Taniai, M., Kanazawa, H., and Futai, M. (1986) *J. Biol. Chem.* **261,** 9196–9201
65. Rao, R., and Senior, A. E. (1987) *J. Biol. Chem.* **262,** 17450–17454
66. Wang, J. H., Joshi, V., and Wu, J. C. (1986) *Biochemistry* **25,** 7996–8001
67. Wang, J. H., Cesana, J., and Wu, J. C. (1987) *Biochemistry* **26,** 5527–5533

68. Di Pietro, A., Penin, F., Godinot, C., Gautheron, D. C. (1980) *Biochemistry* **19,** 5671–5678
69. Bullough, D. A., Verburg, J. G., Yoshida, A., and Allison, W. A. (1987) *J. Biol. Chem.* **262,** 11675–11683
70. Leckband, D., and Hammes, G. G. (1987) *Biochemistry* **26,** 2306–2311
71. Issartel, J. P., Dupuis, A., Junardi, J., and Vignais, P. V. (1991) *Biochemistry* **30,** 4726–4730
72. Ysern, X., Amzel, L. M., and Pedersen, P. L. (1988) *J. Bioenerg. Biomemb.* **29,** 423–450
73. Fromme, P., and Gräber, P. (1989) *FEBS Lett.* **259,** 33–36
74. Shapiro, A. B., and McCarty, R. E. (1991) *J. Biol. Chem.* **266,** 4194–4200
75. Shapiro, A. B., Gibson, K. D., Scheraga, H. A., and McCarty, R. E. (1991) *J. Biol. Chem.* **266,** 17277–17285
76. Tiedge, H., and Schafer, G. (1989) *Biochim. Biophys. Acta* **977,** 1–9
77. Berden, J. A., Hartog, A. F., and Edel, C. M. (1991) *Biochim. Biophys. Acta* **1057,** 151–156
78. Boyer, P. D. (1997) *Annu. Rev. Biochem.* **66,** 717–749
79. McCarty, R. E., and Fagan, J. (1973) *Biochemistry* **12,** 1503–1507
80. Moroney, J. V., and McCarty, R. E. (1979) *J. Biol. Chem.* **254,** 8951–8955
81. Cohn, M., and Hu, A. (1978) *Proc. Natl. Acad. Sci. U. S. A.* **75,** 200–205
82. Hutton, R. L., and Boyer, P. D. (1979) *J. Biol. Chem.* **254,** 9990–9993
83. Hackney, D. D., and Boyer, P. D. (1978) *J. Biol. Chem.* **253,** 3164–3170
84. Hackney, D. D., Rosen, G., and Boyer, P. D. (1979) *Proc. Natl. Acad. Sci. U. S. A.* **76,** 3646–3650
85. Kohlbrenner, W. E., and Boyer, P. D. (1983) *J. Biol. Chem.* **258,** 10881–10886
86. Boyer, P. D., and Kohlbrenner, W. E. (1981) in *Energy Coupling in Photosynthesis* (Selman, B., and Selman-Reiner, S., eds) pp. 231–240, Elsevier/North Holland, New York
87. Gresser, M. J., Myers, J. A., and Boyer, P. D. (1982) *J. Biol. Chem.* **257,** 12030–12038
88. Boyer, P. D. (1983) in *Biochemistry of Metabolic Processes* (Lennon, D. L. F., Stratman, F. W., and Zahlten, R. N., eds) pp. 465–477, Elsevier-Biomed., New York
89. Kanazawa, H., Kayano, T., Mabuchi, K., and Futai, M. (1981) *Biochem. Biophys. Res. Comm.* **103,** 604–612
90. Cox, G. B., Jans, D. A., Fimmel, A. L. A., Gibson, F., and Hatch, L. (1984) *Biochim. Biophys. Acta* **768,** 201–208
91. Cox, G. B., Fimmel, A. L., Gibson, F., and Hatch, L. (1986) *Biochim. Biophys. Acta* **849,** 62–69
92. Hoppe, J., and Sebald, W. (1984) *Biochim. Biophys. Acta* **768,** 1–27
93. Mitchell, P. (1985) *FEBS Lett.* **181,** 1–7
94. Melese, T., and Boyer, P. D. (1985) *J. Biol. Chem.* **260,** 15398–15401
95. Kandpal, R. P., and Boyer, P. D. (1987) *Biochim. Biophys. Acta* **890,** 97–105
96. Musier, K. M., and Hammes, G. G. (1987) *Biochemistry* **26,** 5982–5988
97. Kagawa, W., and Hamamoto, T. (1996) *J. Bioenerg. Biomembr.* **28,** 421–431
98. Ogilvie, I., Aggeler, R., and Capaldi, R. A. (1997) *J. Biol. Chem.* **272,** 19621–19624
99. Kironde, F.A.S., and Cross, R. L. (1977) *J. Biol. Chem.* **262,** 3488–3495
100. Abrahams, J. P., Leslie, A. G. W., Lutter, R., and Walker, J. E. (1994) *Nature* **370,** 621–628
101. Zhou, Y., Duncan, T. M., Bulygin, V. V., Hutcheon, M. L., and Cross, R. L. (1996) *Biochim. Biophys. Acta* **1275,** 96–100
102. Cross, R. L., and Duncan, T. M. (1996) *J. Bioenerg. Biomembr.* **28,** 403–408
103. Aggeler, R., and Capaldi, R. A. (1996) *J. Biol. Chem.* **271,** 13888–13891
104. Tang, C., and Capaldi, R. A. (1996) *J. Biol. Chem.* **271,** 3018–3024
105. Feng, Z., Aggeler, R., Haughton, M., and Capaldi, R. A. (1996) *J. Biol. Chem.* **271,** 17986–17989
106. Sabbert, D., Engelbrecht, S., Junge, W. (1996) *Nature* **381,** 623–625
107. Sabbert, D., Engelbrecht, S., Junge, W. (1997) *Proc. Natl. Acad. Sci. U. S. A.* **94,** 2312–2317

108. Noji, H., Yasuda, R., Yoshida, M., Kinosita, K., Jr. (1997) *Nature* 299–312
109. Abbott, M. S., Czarnecki, J. J., and Selman, B. R. (1984) *J. Biol. Chem.* **259,** 12271–12278
110. Cross, R. L., Cunningham, D., Miller, C. G., Xue, Z., Zhou, J. M., and Boyer, P. D. (1987) *Proc. Natl. Acad. Sci. U. S. A.* **84,** 5715–5719
111. Xue, Z., Miller, C. G., Zhou, J. M., and Boyer, P. D. (1987) *FEBS Lett.* **223,** 391–394
112. Wise, J. G., Hicke, B. J., and Boyer, P. D. (1987) *FEBS Lett.* **223,** 395–401
113. Guerrero, K. J., Xue, Z., and Boyer, P. D. (1990) *J. Biol. Chem.* **265,** 16280–16287
114. Murataliev, M. B., Milgrom, Y. M., and Boyer, P. D (1991) *Biochemistry* **30,** 8305–8310
115. Milgrom, Y. M., Ehler, L. L., and Boyer, P. D. (1991) *J. Biol. Chem.* **266,** 11551–11558
116. Murataliev, M. B., and Boyer, P. D. (1992) *Eur. J. Biochem.* **209,** 681–687
117. Zhou, J. M., and Boyer, P. D. (1993) *J. Biol. Chem.* **268,** 1531–1538
118. Berkich, D. A., Williams, G. D., Masiakos, P. T., Smith, M. B., Boyer, P. D., and LaNoue, K. F. (1991) *J. Biol. Chem.* **266,** 123–129
119. Avron, M., and Sharon, N. (1960) *Biochem. Biophys. Res. Comm.* **2,** 336–339
120. Rosing, J., Smith, D. J., Kayalar, C., and Boyer, P. D. (1976) *Biochem. Biophys. Res. Comm.* **72,** 1–8
121. Murataliev, M. B., and Boyer, P. D. (1994) *J. Biol. Chem.* **269,** 15431–15439
122. Ernster, L. (1986) *Chemica Scripta* **26,** 525–534

John E. Walker.

JOHN E. WALKER

I was born in Halifax, Yorkshire on January 7th, 1941 to Thomas Ernest Walker and Elsie Walker (née Lawton). My father was a stone mason, and a talented amateur pianist and vocalist. I was brought up with my two younger sisters, Judith and Jennifer, in a rural environment overlooking the Calder valley near Elland, and then in Rastrick. I received an academic education at Rastrick Grammar School, specializing in Physical Sciences and Mathematics in the last three years. I was a keen sportsman, and became school captain in soccer and cricket. In 1960, I went to St. Catherine's College, Oxford, and received the B.A. degree in Chemistry in 1964.

In 1965, I began research on peptide antibiotics with E. P. Abraham in the Sir Willian Dunn School of Pathology, Oxford, and was awarded the D. Phil. degree in 1969. During this period, I became aware of the spectacular developments made in Cambridge in the 1950s and early 1960s in Molecular Biology through a series of programmes on BBC television given by John Kendrew, and published in 1966 under the title "The Thread of Life". These programmes made a lasting impression on me, and made me want to know more about the subject. Two books, "Molecular Biology of the Gene" by J. D. Watson, first published in 1965, and William Hayes' "Bacterial Genetics" helped to assuage my appetite for more information. My knowledge of this new field was extended by a series of exciting lectures for graduate students on protein structure given in 1966 by David Phillips, the new Professor of Molecular Biophysics at Oxford. Another series of lectures given by Henry Harris, the Professor of Pathology and published in book form under the title "Nucleus and Cytoplasm", provided more food for thought.

Then followed a period of five years working abroad, from 1969–1971, first at The School of Pharmacy at the University of Wisconsin, and then from 1971–1974 in France, supported by Fellowships from NATO and EMBO, first at the CNRS at Gif-sur-Yvette and then at the Institut Pasteur.

Just before Easter in 1974, I attended a research workshop in Cambridge entitled "Sequence Analysis of Proteins". It was sponsored by EMBO (The European Molecular Biology Organization), and organised by Ieuan Harris from the Medical Research Council's Laboratory of Molecular Biology (LMB) and by Richard Perham from the Cambridge University Department of Biochemistry. At the associated banquet, I found myself sitting next to someone that I had not met previously, who turned out to be Fred Sanger. In the course of our conversation, he asked if I had thought about coming back to work in England. I jumped at the suggestion, and with some trepidation, approached Ieuan Harris about the possibility of my joining his group. After discussions with Fred Sanger, it was agreed that I could come to the Protein and Nucleic Acid Chemistry (PNAC) Division at the LMB for three months from June 1974. More than 23 years later, I am still there.

It goes without saying that this encounter with Fred Sanger and Ieuan Harris transformed my scientific career. In 1974, the LMB was infused throughout its three Divisions with a spirit of enthusiasm and excitement for research in molecular biology led by Max Perutz (the Chairman of the Laboratory), Fred Sanger, Aaron Klug, Francis Crick, Sidney Brenner, Hugh Huxley, John Smith and César Milstein, which was coupled with extraordinary success. For example, along the corridor from my laboratory Fred was inventing his methods for sequencing DNA, immediately across the corridor César Milstein and Georges Köhler were inventing monoclonal antibodies, and elsewhere in the building, Francis Crick and Aaron Klug and their colleagues were revealing the structures of chromatin and transfer RNA. Fred's new DNA sequencing methods were applied first to the related bacteriophages fX174 and G4, and then to DNA from human and bovine mitochondria. I analyzed the sequences of the proteins from G4 and from mitochondria using direct methods. These efforts led to the discovery of triple overlapping genes in G4 where all three DNA phases encode proteins, and to the discovery that subunits I and II of cytochrome c oxidase were encoded in the DNA in mitochondria. Later on, I helped to uncover details of the modified genetic code in mitochondria.

In 1978, I decided to apply protein chemical methods to membrane proteins, since this seemed to be both a challenging and important area. Therefore, in search of a suitable topic, I read the literature extensively. The enzymes of oxidative phosphorylation from the inner membranes of mitochondria were known to be large membrane bound multi-subunit complexes, but despite their importance, they had been studied hardly at all from a structural point of view. Therefore, the same year, I began a structural study of the ATP synthase from bovine heart mitochondria and from eubacteria. These studies resulted eventually in a complete sequence analysis of the complex from several species, and in the atomic resolution structure of the F_1 catalytic domain of the enzyme from bovine mitochondria, giving new insights into how ATP is made in the biological world. Michael Runswick has worked closely with me throughout this period, and has made contributions to all aspects of our studies.

In 1959, I received the A. T. Clay Gold Medal. I was awarded the Johnson Foundation Prize by the University of Pennsylvania in 1994, in 1996, the CIBA Medal and Prize of the Biochemical Society, and The Peter Mitchell Medal of the European Bioenergetics Congress, and in 1997 The Gaetano Quagliariello Prize for Research in Mitochondria by the University of Bari, Italy. In 1995, I was elected a Fellow of the Royal Society. In 1997, I was made a Fellow of Sidney Sussex College, Cambridge and became an Honorary Fellow of St. Catherine's College, Oxford.

I married Christina Westcott in 1963. We have two daughter, Esther, aged 21 and Miriam, aged 19. At present, both of them are university students, studying Geography and English, respectively, at Nottingham-Trent and Leeds Universities.

ATP SYNTHESIS BY ROTARY CATALYSIS

Nobel Lecture, December 8, 1997

by

JOHN E. WALKER

The Medical Research Council Laboratory of Molecular Biology, Hills Road, Cambridge, CB2 2QH, U. K.

Biological energy comes from the sun. Light energy harvested by photosynthesis in chloroplasts and phototropic bacteria, becomes stored in carbohydrates and fats. This stored energy can be released by oxidative metabolism in the form of adenosine triphosphate (ATP), and used as fuel for other biological processes. ATP is a high energy product of both photosynthesis and oxidative metabolism, and in textbooks, it is often referred to as the chemical currency of biological energy.

In the 1960s and early 1970s, the field of oxidative metabolism was dominated by a debate about the nature of the intermediate between NADH (a key product from carbohydrate and fat metabolism) and ATP itself. Many bioenergeticists believed in and sought evidence for a high energy covalent chemical intermediate. The issue was resolved by Peter Mitchell, the Nobel Laureate in Chemistry in 1978. He established that in mitochondria, energy is released from NADH via the electron transport chain and used to generate a chemical potential gradient for protons across the inner membrane of the organelle. He referred to this gradient as the proton motive force (pmf; also designated as $\Delta\mu_{H+}$). It was demonstrated that the pmf is harnessed by the ATP synthesizing enzyme (ATP synthase) to drive the synthesis of ATP from ADP and inorganic phosphate, not only in mitochondria, but also in eubacteria and chloroplasts [1].

Because of Peter Mitchell's efforts, the pmf became established as a key intermediate in biological energy conversion. In addition to being employed in ATP synthesis, it is also used by various membrane bound proteins to drive the transport of sugars, amino acids and other substrates and metabolites across biological membranes. The pmf powers the rotation of flagellae in motile bacteria. In newly born children and in hibernating animals, it is converted directly into heat by uncoupling the mitochondria in brown adipose tissue. Some bacteria that live in saline conditions generate a sodium motive force to act as an equivalent intermediate in their energy conversion processes [2]. The general notion of creating a proton (or sodium) motive force and then using it as a source of energy for other biochemical functions is known as chemiosmosis.

Today, the general outlines of chemiosmosis are well established. It is accepted that during electron transport in mitochondria, redox energy derived from NADH is used by three proton pumping enzymes called complex I

(NADH:ubiquinone oxidoreductase), complex III (ubiquinone:cytochrome c oxidoreductase) and complex IV (cytochrome c oxidase). They act consecutively and produce the pmf by ejecting protons from the matrix (the inside) of the organelle (see Figure 1). Until recently, the workings of these chemiosmotic proton pumps was obscure, but currently our understanding is being transformed by the application of modern methods of molecular biology for the analysis of their structures and functions. For example, two independent atomic resolution structures of cytochrome c oxidase isolated from bovine mitochondria [3, 4] and from the bacterium, *Paracoccus denitrificans* [5, 6], are guiding mutational and spectroscopic experiments that are providing new insights into its mechanism. Partial structures of mitochondrial complex III from two different species [7] will have a similar impact soon on our understanding of that enzmye. Complex I, the third proton pump in mitochondria, is an assembly of at least 43 different polypeptides in mammals, with a combined molecular mass in excess of 900,000 [8]. In addition, it has a non-covalently bound flavin mononucletide, and at least five iron-sulphur clusters that act as redox centres. A consequence of this extreme complexity is that the structural analysis of complex I is less advanced than those of complexes III and IV, although the general outline of the complex has been established by electron microscopy [9, 10].

THE ATP SYNTHASE

Since 1978, my colleagues and I have concentrated on analyzing the structure of the ATP synthase, another multisubunit complex from mitochondria, where it is found in the inner membrane alongside the three proton pumping enzymes (see Figure 1). Similar complexes are found in chloroplast and eubacterial membranes. Throughout our endeavours, we have been motivated by the expectation that detailed knowledge of its structure would lead to a deeper understanding of how ATP is made. A substantial part of our efforts has been directed at establishing the subunit compositions of the ATP synthesizing enzymes from various sources, and with determining the primary sequences of the subunits [11–31]. These rather extensive analyses helped to show that the overall structure of the ATP synthase, and hence the general principles governing its operation, are very similar in mitochondria, chloroplasts and eubacteria, although the enzymes from the various sources differ in the details of both their sequences and subunit compositions (see Table 1). It is also known that the ATPases from various sources differ in the mechanisms that regulate their catalytic activities [32].

In 1962, the inside surface of the inner membranes of bovine heart mitochondria was found by electron microscopic examination to be lined with mushroom shaped knobs about 100 Å in diameter (Figure 2A) [33]. Later on, similar structures were found to be associated with the thylakoid membranes of chloroplasts (Figure 2B) [34], and with the inner membranes of eubacteria (Figure 2C). At the time of their discovery, the function of these membrane bound knobs was not known, but they were thought to be proba-

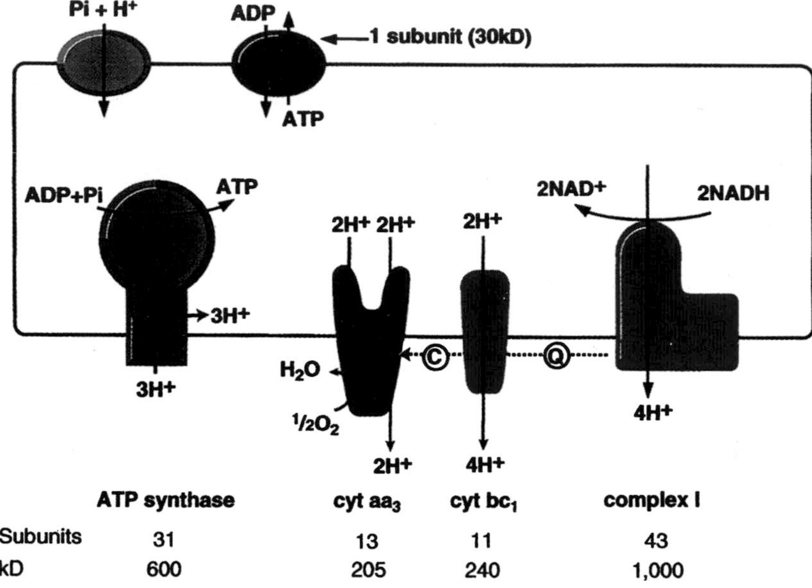

Figure 1. The enzyme complexes of oxidative phophorylation in the inner membranes of mammalian mitochondria. Three proton pumps, complex I, complex III (cytochrome bc_1) and complex IV (cytochrome aa_3) convert redox energy in NADH into the proton motive force (pmf) by ejecting protons from the matrix of the mitochondrion. The ATP synthase uses the energy of the pmf to produce ATP from ADP and phosphate. ADP and phosphate are brought into the mitochondrion by related proetin carriers. External ADP is exchanged for internal ATP, making the newly synthesized ATP available for many biological functions. Dotted lines indicate electron pathways. Q and C are the mobile electron carriers ubiquinone and cytochrome c, respectively.

Table 1. Equivalent subunits in ATP synthases in bacteria, chloroplasts, and bovine mitochondria

Type	Bacteria	Chloroplasts	Mitochondria
F_1	α	α	α
	β	β	β
	γ	γ	γ
	δ	δ	OSCP
	ε	ε	δ
	–	–	ε
F_0	a	a (or x)	a (or ATPase 6)
	b[a]	b and b′ (or I and II)	b
	c	c (or III)	c
Supernumerary	–	–	F_6
	–	–	inhibitor
	–	–	A6L
	–	–	d
	–	–	e
	–	–	f
	–	–	g

[a] ATP synthases in *E. coli* and bacterium PS 3 (both eight-subunit enzymes) have two identical copies of subunit b per complex. Purple non-sulphur bacteria and cyanobacteria appear to have nine different subunits, the extra subunits (known as b′) being a homologue of b. Similarly, chloroplast enzymes are made of nine non-identical subunits, and the chloroplast subunits known as I and II are the homologues of b and b′.

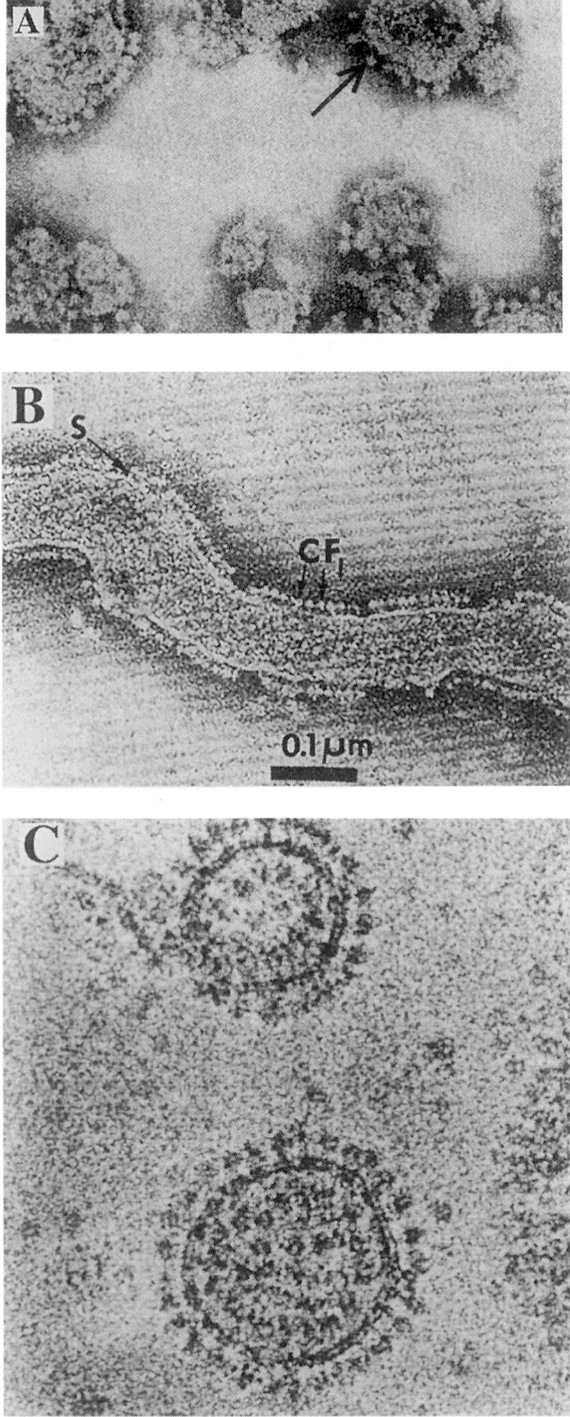

Figure 2. Knobs associated with biological membranes detected by electron microscopy in negative stain. (A), Inside-out vesicles from bovine heart mitochondria; (B), thylakoid membranes form pea chloroplasts; (C), inside-out vesicles from *E. coli.* Different magnifications have been used in parts (A)–(C), and the knobs are all about 100 Å in diameter. Reproduced with permission from references [88, 89].

bly important in biological energy conversion, and hence they were named "the fundamental particles of biology" [35]. In a brilliant series of biochemical reconstitution experiments conducted in the 1960s, Efraim Racker established that they were the ATP synthesizing enzyme complex (for example see references [36–38], and hence that these early micrographs were the first glimpses of its structure.

We now know that the head of the mushroom is a globular protein complex (known as the F_1 domain), where the catalytic sites of the ATP synthase lie. The F_1 part is attached to the membrane sector by a slender stalk about 45 Å long. The hydrophobic membrane domain (known as F_o) transports protons back through the energized membrane into the matrix, somehow releasing energy in this process and making it available to drive ATP synthesis in the catalytic F_1 domain. For some years, it has been accepted that three protons are transported back through the F_o membrane sector for each ATP molecule that is formed in F_1 [1]. However, recent experiments suggest that the chloroplast enzyme transports four protons through the membrane for each ATP that is made [39]. Therefore, it appears that either one or, less likely, both of these values of the H^+:ATP ratio are incorrect, or that the mechanisms and structures of the mitochondrial and chloroplast enzymes differ significantly in their F_o domains. This point will be elaborated later in a consideration of the possible structure and mechanism of F_o.

The general model of the mechanism of the ATP synthase that will be developed below is that the F_o membrane domain contains a rotating molecular motor fuelled by the proton motive force. It is proposed that this motor is mechanically coupled to the stalk region of the enzyme, and that the rotation of the stalk affects the catalytic domain and makes the three catalytic sites pass through a cycle of conformational states in which first, substrates are bound and sequestered, then second, ATP is formed from the sequestered substrates, and finally the newly synthesized ATP is released from the enzyme. This cycle of interconversion of the three catalytic sites is part of a binding change mechanism of ATP synthesis developed by Paul Boyer (see Figure 3)

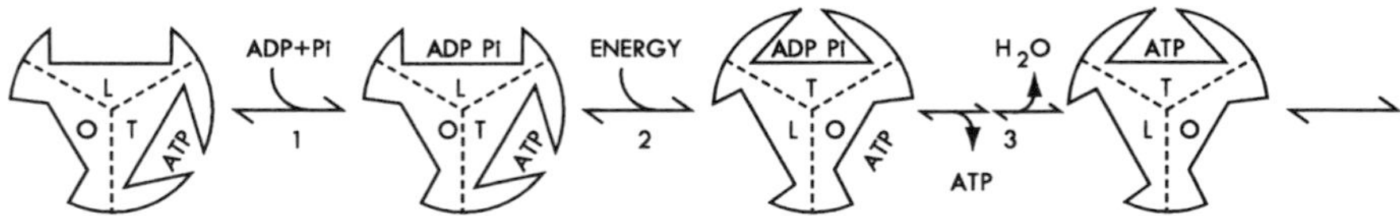

Figure 3. An energy dependent binding change mechanism of ATP synthesis. The catalytic sites in the β-subunits interact and interconvert between three forms: O, open, site with very low affinity for substrates and catalytically inactive; L, loose site, loosely binding substrates and catalytically inactive; T, tight site, tightly binding substrates and catalytically active. The proton induced conformational changes convert a T-site with bound ATP into an open site, releasing the bound nucleotide. Concomittantly, an L-site with loosely bound ADP and phosphate is converted to a T-site where substrates are bound and ATP froms. Fresh substrates bind to an O-site converting it to an L-site, and so on. One third of the catalytic cycle is illustrated. Reproduced with permission from reference [40].

[40]. One of its basic tenets is that the energy requiring steps in this cycle of ATP synthesis are in the binding of substrates and in the release of products. This mechanism also implies that the ATP synthase must be an asymmetrical structure, and structural asymmetry is also implicit in the molar ratios of the subunits of F_1-ATPase. Indeed, it is now clear from the asymmetrical features that are inherent in the enzyme's structure that there can be no structurally symmetrical states in the catalytic cycle.

THE STRUCTURE OF F_1-ATPase

An important practical point about the ATP synthase complex that has influenced the strategy for analyzing its structure is that, as Racker had shown in the 1960s, the globular catalytic domain F_1-ATPase can be detached from the membrane domain and studied separately in aqueous solution. Subsequently, Alan Senior and Harvey Penefsky purified the bovine F_1 complex [41, 42] and Penefsky demonstrated that it is an assembly of five different kinds of polypeptides, which he called α, β, γ, δ and ε [42]. Eventually, it was accepted that they were assembled in the complex in the molar ratios 3α:3β:1γ:1δ:1ε. Hence, each F_1 particle is an assembly of nine polypeptides, and in the bovine heart enzyme, their combined molecular mass is about 371,000 (see Table 2).

Table 2. The subunits of bovine F_1-ATPase

Subunit	MWt	Function
α	55,247	Nucleotide binding
β	51,705	Nucleotides, catalysis
γ	30,141	Link to F_o
δ	15,065	Stalk
ε	5,632	Stalk
α3β3γ1δ1ε1	371,694	

Both α- and β-subunits bind nucleotides, and we now know that the catalytic nucleotide binding sites lie almost entirely within the β-subunits (see Figure 4). The nucleotides bound to α-subunits remain associated during the catalytic cycle and do not participate directly in ATP synthesis. What they are doing remains mysterious, although both structural and regulatory functions have been suggested.

In 1981, we found that the sequences of the α- and β-subunits were related weakly through most of their length [12, 13], and Matti Saraste and I wondered which regions of the sequences were contributing to the nucleotide binding sites. By examination of the known primary and atomic structures of adenylate kinase [43], and of the sequence of myosin from *Caenorhabditis elegans* (at the time unpublished information, made available to me by Jonathan Karn), we were able to propose that two short degenerate sequence motifs common to adenylate kinase, myosin and the α- and β-subunits, of F_1-ATPase were involved in helping to form their nucleotide binding pockets [44].

This proposal has had far greater consequences than we ever imagined at the time of its publication in 1982. Over the years, one of the two motifs (see

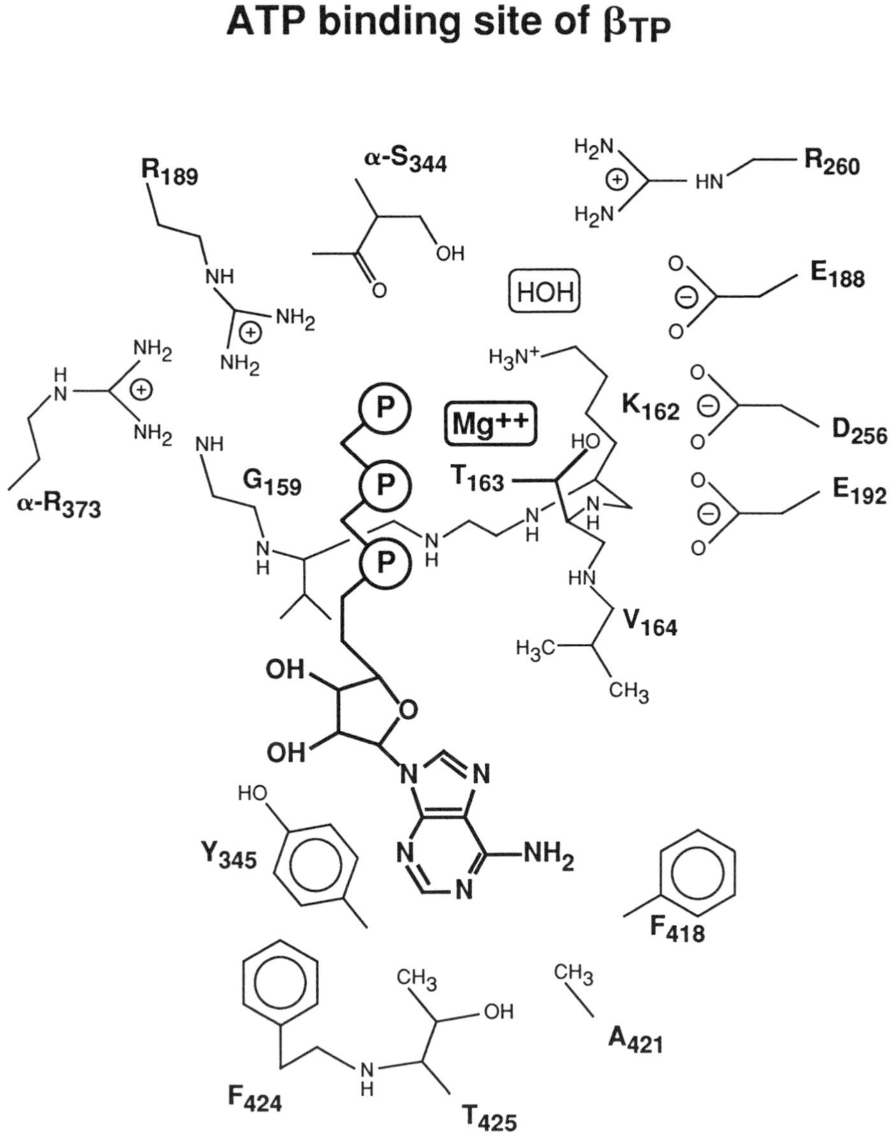

ATP binding site of β_{TP}

Figure 4. The nucleotide binding site in the bTP-subunit of bovine F_1-ATPase. Except for αTP-Arg373 and the main chain carbonyl and the side-chain of αTP-Ser344, all of the amino acids are in bTP. Residues 159–164 are part of the P-loop sequence. The ordered water molecule is poised for nucleophilic attack on the terminal phosphate and is activated by Glu188. As an incipient negative charge develops on the terminal phosphate in a penta-coordinate transition state of ATP hydrolysis, it will be stabilized by the guanidinium of α-Arg 373. Reproduced with permission from reference [50].

Figure 5) has become established as a reliable indicator of the presence of purine nucleotide binding sites in proteins of known sequence, but of unknown biochemical function. One spectacular demonstration of its predictive value has been in the identification of members of the widely dispersed family of ABC (adenosine binding cassette) transport proteins [45], which includes the cystic fibrosis protein and multi-drug resistance proteins. Another early success was its help in the identification of the oncogene protein p21 as a GTPase [46]. The atomic structure of F_1-ATPase described below, has shown that, as in other proteins, the two sequence motifs describe amino acids that are involved in forming the phosphate binding region of its nucleotide binding sites. For this reason, one of the sequences is often referred to as the P-loop (phosphate binding loop) sequence [47] (see Figure 5).

In the determination of the atomic structure of F_1-ATPase from bovine heart mitochondria, the key problem that had to be solved was how to grow crystals of the protein complex that would diffract X-rays to appropriately high resolution. Crystals were obtained at the beginning of our efforts, but, as often happens, they diffracted X-rays rather poorly. Therefore, over a period

P-LOOP SEQUENCES IN F1-ATPase

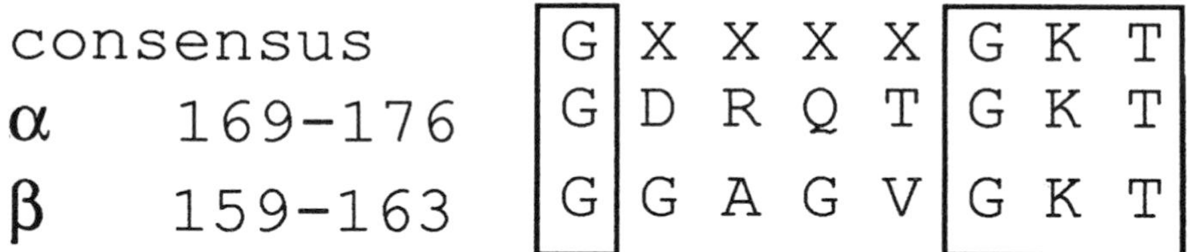

Figure 5. The phosphate binding loop sequence motif in the α- and β-subunits of bovine F_1-ATPase. In some purine binding sites, the final threonine is replaced by a serine residue.

of seven years, factors influencing crystal formation were studied systematically, and the diffraction properties of the crystals were improved gradually. Eventually, suitable cystals were obtained in 1990. In retrospect, the important factors for obtaining these crystals were to use highly pure preparations of enzyme from which trace impurities had been eliminated, to remove endogenous bound nucleotides and to replace them with a non-hydrolyzable chemical analogue of ATP, namely 5'-adenylylimidodiphosphate (AMP-PNP) which has the effect of locking the complex in a unique conformation, and to grow the crystals in the presence of deuterium oxide instead of water. By exposing the crystals to X-rays at a synchrotron source, it was demonstrated that crystals of bovine F_1-ATPase grown under these conditions diffracted to at least 2.8 Å resolution [48, 49]. Drs. René Lutter and Rose Todd were key collaborators during this critical period. At this point, for the first time the structural analysis of F_1-ATPase appeared to be a realistic possibility. Our chances of success with such a large protein complex were increased significantly by collaborating with Dr. Andrew Leslie, who is a professional protein crystallographer. Together with him and a post-doctoral visitor, Dr. J. P. Abrahams, we were able to arrive at an atomic resolution structure [50] surprisingly rapidly, given the size of the complex, and the associated problems of collecting and processing the X-ray diffraction data.

The structural model of bovine F_1-ATPase contains 2,983 amino acids. Except for short disordered stretches at their N-terminals, the sequences of the α- and β-subunits were traced in their entirety. Three α-helical segments of the γ-subunit corresponding to residues 1–45, 73–90, and 209–272 were also built into the model. The segments linking these three segments are also disordered, as are the entire δ- and ε-subunits (150 and 50 amino acids respectively). These disordered regions of the γ-, δ- and ε-subunits, comprising in total about 300 amino acids, probably lie beneath the $\alpha_3\beta_3$ sub-complex, where the γ-subunit protrudes from the structure (the γ-subunit is blue in Figures 6 and 7). The protrusion is probably a vestige of the 45 Å stalk that links the F_1 and F_o domains in the complete ATP synthase complex.

The structural model shows that the three α-subunits and the three β-subunits are arranged in alternation around a sixfold axis of pseudo-symmetry provided by an α-helical structure in the single γ-subunit (see Figures 6 and 7). Despite the large excesses of AMP-PNP and ADP in the mother liquor surrounding the crystals, five, and not six, nucleotides are bound to each enzyme complex. An AMP-PNP molecule is found in each α-subunit and a fourth one in one β-subunit. An ADP molecule is bound to the second β-subunit, and the third β-subunit has no bound nucleotide at all. A comparison of the structures of the three chemically identical β-subunits in the F_1 complex provides an explanation. The two catalytic β-subunits that have AMP-PNP and ADP bound to them (known as β_{TP} and β_{DP}, respectively; see Figures 7b and 7c) have different but rather similar conformations, whereas the structure of the third β-subunit, to which no nucleotide has bound (known as $\beta_{E;}$ see Figure 7d), differs substantially from the other two, particularly in the central domain where the nucleotides are bound in β_{TP} and β_{DP}. In β_E, the C-terminal half of this central domain, together with the bundle of six α-helices that form the C-terminal domain, have rotated away from the sixfold axis of pseudo-symmetry. This disruption of the central domain removes the capacity of the β_E-subunit to bind nucleotides. Therefore, the crystal structure of F_1-ATPase contains the asymmetry required by the binding change mechanism, and the three different conformations of the subunits β_E, β_{TP} and β_{DP} could be interpreted as representing the "open", "loose" and "tight" states, respectively.

More recent extensive crystallographic analyses of F_1-ATPase with bound antibiotic inhibitors [51, 52], of enzyme occupied with ADP and ATP (K. Braig, M. Montgomery, A. G. W. Leslie and J. E. Walker, unpublished work), of enzyme inhibited by ADP and aluminium fluoride (K. Braig, I. Menz, M. Montgomery, A. G. W. Leslie and J. E. Walker, unpublished work), and of en-

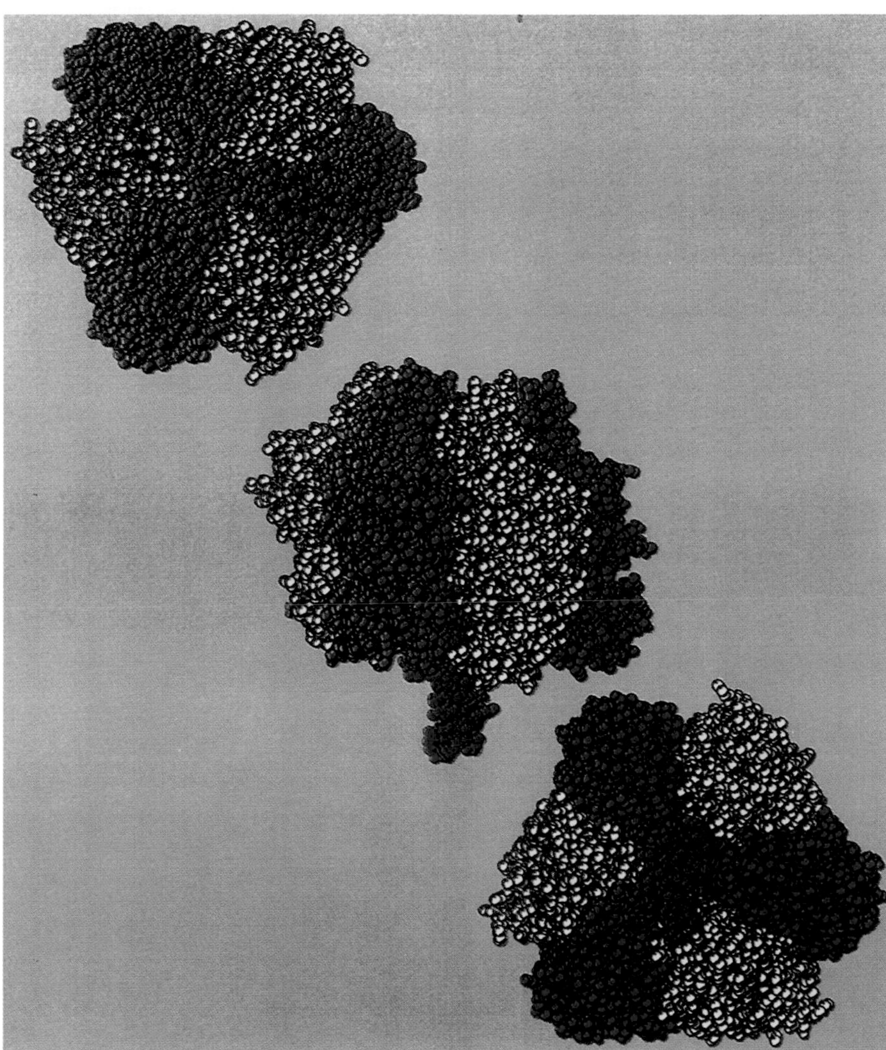

Figure 6. The three-dimensional structure of bovine F_1-ATPase shown in solid representation. The red, yellow and blue parts correspond to α-, β- and γ-subunits respecively. From top to bottom, respectively, the complex is viewed from above (towards the membrane in the intact ATP synthase), from the side, and from beneath.

zyme covalently inhibited with 4-chloro-7-nitrobenzofluorazan (G. Orriss, A. G. W. Leslie and J. E. Walker, unpublished work), leave little room for any doubt that the high resolution structure described above represents a state in the active cycle of the enzyme.

The most exciting aspect of the structure is that it suggests a mechanism for interconverting the three catalytic subunits through the cycle of conformations required by the binding change mechanism. It appears from the structure (Figure 7) that the nucleotide binding properties of the three catalytic β-subunits are modulated by the central α-helical structure in the γ-sub-

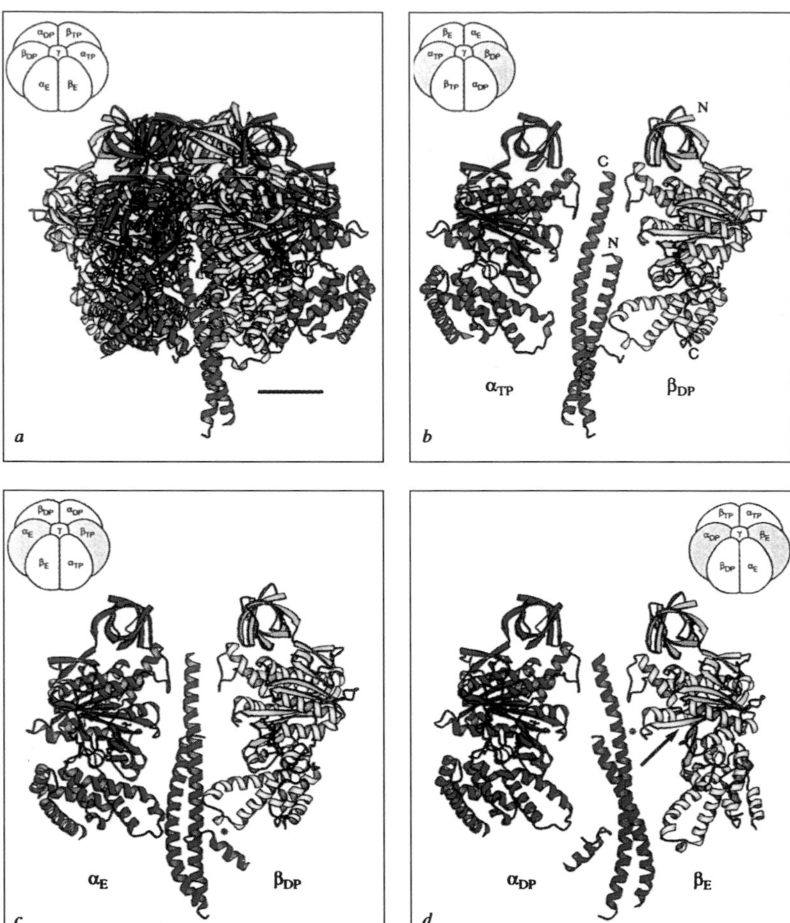

Figure 7. The three-dimensional structure of bovine F_1-ATPase shown in ribbon representation. The colour code for subunits is the same as used in Figure 6, and nucleotides are black, in a "ball-and-stick" representation. The axis of pseudo-symmetry is vertical. AMP-PNP is bound to the three α-subunits, and to the β-subunit defined as β_{TP}. Subunit β_{DP} has bound ADP and subunit BE has no associated nucleotide. Subunits αTP, αE and αDP occupy the same relative positions as the corresponding β-subunits, but are rotated by -60°. The relationships of the various α- and β-subunits to each other is summarized in the icon in the top left or right corner of parts (A)–(D). In parts (C)–(D), the shaded part of the icon shows which subunits are depicted. Subunit α_{TP} contributes to the nucleotide binding site of β_{TP}, and similarly for α_{DP} and α_E. Subunits α and γ are numbered from 1–510 and 1–272, respectively. By convention, the fifth amino acid (serine) in subunit β is residue 1 and the first four amino acids (Ala.Ala.Gln.Ala) are referred to as residues –1 to –4. The C-terminal amino acid is residue 478. (A) A view of the entire F_1 particle in which subunits α_E and β_E point towards the viewer, revealing the anti-parallel coiled-coil of the N- and C-terminal helices of the γ-subunit through the open interface between them. The bar is 20 Å long. (B) Subunits α_{TP}, γ and β_{DP} from a similar viewpoint to (A), but rotated 180° about the axis of pseudo-symmetry. The N- and C-termini of the β- and γ-subunit are shown. (C) Subunits α_E, g and β_{TP} from a similar viewpoint to (A), but rotated by -60°. The asterisk indicates an interaction of the loop containing the DELSEED sequence and the γ-subunit. (D) Subunits α_{DP}, γ, and β_E from a similar viewpoint to (A), but rotated by 60°. The arrow indicates the disruption of the β-sheet in the nucleotide binding domain. β_E-Asp 316, β_E-Thr 318 and β_E-Asp 323 in a loop of the nucleotide binding domain make H-bonds with residues γ-Arg 254 and γ-Gln 255 from the C-terminal helix, 6 Å below the hydrophobic sleeve. The asterisk indicates a loop that makes an interaction with the C-terminal part of the γ-subunit. Reproduced with permission from reference [50].

unit. This structure is curved, and its coiled-coil region is likely to have rigidity. In the crystal structure, its curvature appears to be imposing the "open state" on the β_E-subunit by pushing against the C-terminal domain of the protein, forcing the nucleotide binding domain to split and hinge outwards, thereby removing its nucleotide binding capacity. Therefore, by inspection of the model, the simplest way of interconverting the three conformations of β-subunits would be to rotate this central α-helical structure. As it is rotated, the curvature of the γ-subunit moves away from the β_E conformation, allowing that subunit to close, progessively entrapping substrates and allowing ATP to form spontaneously. At the same time, the next β-subunit will be progressively opened by the effect of the curvature of the γ-subunit, allowing the ATP that has already formed in its nucleotide binding site to be released.

EVIDENCE FOR A ROTARY MECHANISM IN ATP SYNTHASE

The suggestion that ATP synthesis involves the cyclic modulation of nucleotide binding properties of the three catalytic β-subunits by rotation of the γ-subunit was attractive because it provided a reasonable structural basis for the binding change mechanism. However, the proposal was based upon the interpretation of a static atomic model, and there was no proof that the enzyme operated in this way in reality. Therefore, in the laboratories of Richard Cross, Rod Capaldi and Wolfgang Junge, various experiments were carried out to test the rotary hypothesis [53–55]. These experiments were all consistent with the rotary model, and they provided convincing evidence of the movement of the γ-subunit through 90–240°, but conclusive proof of repeated net rotations through 360° was lacking.

Early in 1997, clear evidence of continuous rotation of the γ-subunit relative to the surrounding $\alpha_3\beta_3$ subcomplex was provided by a spectacular experiment conducted at the Tokyo Institute of Technology by Masasuke Yoshida and colleagues [56]. The essence of this experiment (summarized in Figure 8) was to bind the $\alpha_3\beta_3\gamma_1$ sub-complex to a nickel coated glass surface in a unique orientation, by introducing the nickel binding sequence (histidine)$_{10}$ at the N-terminals of the β-subunits. A cysteine residue, introduced by mutagenesis into the exposed tip of the γ-subunit, distal from the nickel surface, was biotinylated, thereby allowing a fluorescently labelled biotinylated actin filament to be attached via an intermediate streptavidin molecule, which has four biotin binding sites. The actin filaments were 1–3 μm long, and their ATP dependent anticlockwise rotation could be seen in a fluorescence microscope. The rate of rotation was approximately once per second, about one fiftieth of the rate anticipated from the turnover number of the fully active enzyme. The reduced activity can be attributed to the load of the actin filament and to the absence of the δ- and ε-subunits.

In addition to its impact in providing direct visual proof of the rotation of the γ-subunit, one crucially important aspect of this experiment is that it establishes the order of interconversion of the three conformations of β-sub-

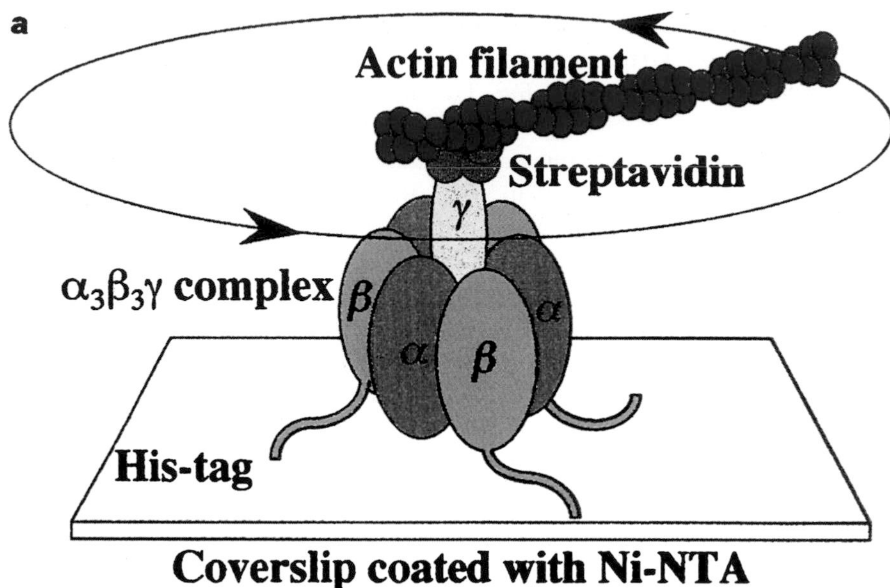

Figure 8. Experimental observation of the ATP-depenedent rotation of the γ-subunit of the $\alpha_3\beta_3\gamma_1$ sub-complex. For explanation, see the text. Reproduced with permission from reference [56].

units observed in the crystal structure. During ATP hydrolysis, the β_{DP} subunit converts into the β_E state, β_E changes to β_{TP} and β_{TP} changes to β_{DP}. During ATP synthesis, it is reasonable to assume that the γ-subunit rotates in the opposite direction, and that the order of conformational changes in the β-subunits is the reverse of those occurring during ATP hydrolysis.

GENERATION OF ROTATION BY PROTON TRANSPORT THROUGH THE F_0 MEMBRANE DOMAIN

Once convincing evidence of rotation of the γ-subunit had been obtained, among the next questions for consideration were: how is rotation generated by proton transport through F_o, and what is the nature of the connections between the F_1 and the F_o domains? High resolution structure is likely to be crucial in providing clear answers to both questions, but as yet there is no such structure for either the F_o domain or the central stalk between F_1 and F_o, except for the ordered protrusion of the γ-subunit. Structures of the isolated ε-subunit [57, 58] and of a fragment of subunit d [59] of the *E. coli* enzyme have been established. The bacterial ε-subunit (equivalent to the bovine δ-subunit [60]) probably interacts with the γ-protrusion, and the bacterial δ-subunit (equivalent of the bovine OSCP subunit [60]) appears to interact with the N-terminal region of the α-subunits (*vide infra*), but the precise locations and functions of these proteins in ATP synthase are unclear.

Despite the lack of a detailed structure of the F_o domain, evidence about the arrangement of the membrane subunits is accumulating. The simplest F_o

domain characterized so far is the one found in eubacterial enzymes exemplified by *E. coli.* It has three constituent subunits named a, b and c assembled in the molar ratios $a_1b_2c_{9-12}$ [61, 62]. The uncertainty in the number of c subunits per complex is a consequence of the experimental difficulties associated with making appropriate measurements. Secondary structural models for all three subunits have been advanced by interpretation of the sequences of the bacterial F_o subunits [63]. That of subunit a has been interpreted variously as indicating the presence of five, six or seven hydrophobic membrane spanning subunits in the protein, but recent experimental evidence indicates that the correct value is five [64]. In its C-terminal and in penultimate α-helices are found positively charged amino acids that are essential for a functional F_o. Subunit b is anchored in the membrane by a single N-terminal α-helix [65]. The remainder of this protein is highly charged and may form a homo-dimer by making a parallel α-helical coiled-coil. This polar extramembrane region interacts with subunits in F_1 [66]. The model of subunit c has two antiparallel transmembrane α-helices linked by an extra-membranous loop, and this has been shown by nmr studies to be the structure of the protein in a chloroform-methanol solvent mixture [64]. Cysteine residues introduced by mutation into the loop region and into the tip of the γ-subunit form a disulphide link under oxidizing conditions, showing that in the intact enzyme, these regions are either in contact (possibly transiently) or close to each other [67]. Similar kinds of experiment provide evidence of interactions between subunits e and c [68]. The most important feature of subunit c is the side chain carboxyl of an aspartate residue, which is buried in the membrane in the C-terminal α-helix. This carboxyl group is conserved throughout all known sequences of c-subunits, and it is required for transmembrane proton transport. [61, 63] Depending on the exact number of subunits c per complex, there are 9–12 such buried carboxyls in each ATP synthase complex.

The mitochondrial ATP synthase complex contains subunits that are equivalent to subunits a and c (see Table 1). Mitochondrial subunit b is also probably the equivalent of its bacterial homonym, although it appears to have two anti-parallel transmembrane a-helices at its N-terminus, rather than one, and there is only one b-subunit per enzyme complex. Presumably the role of the second b-subunit in the bacterial complex is performed by other subunits that are unique to the mitochondrial complex (for example subunits d and F_6), but such matters will only become clear when more detailed structural evidence becomes available for both bacterial and mitochondrial enzymes. Nonetheless, the bacterial and mitochondrial F_o domains have many features in common, and it is highly probable that they operate by very similar mechanisms.

Models have also been advanced about the association of the F_o subunits in the membrane domain. One such model suggests that the c-subunits form a ring by interactions though their C-terminal α-helices, with the N-terminal α-helices outside the ring [69], and annular structures of c-subunits have been visualized by atomic force microscopy [70, 71]. Some preliminary experimental evidence of this arrangement has also been obtained by formation of disulphide cross links after introduction of cysteine residues at appropriate sites (N. J. Glavas & J. E. Walker, unpublished work). In addition, the atomic force microscopy images indicate that the b-subunits are placed peripheral to a ring of c-subunits [71]. Currently, there is no experimental evidence that shows whether the a subunit lies outside or within the c-annulus.

Wolfgang Junge has proposed a model of how proton transport through F_o might generate a rotary motion [72] (see Figure 9). The essence of this hypothetical rotary motor is that the essential carboxyls in the c subunits are arranged around the external circumference of the c-annulus. Part of the external surface of the annulus interacts with subunit a, and in this region the carboxyls are negatively charged. It is envisaged that subunit a has an inlet port on the external surface of the membrane which allows a proton to neutralize one of the negatively charged carboxylates. The resulting un-ionized carboxyl will find its way by thermal vibrations to its preferred environment in contact with the phospholipid bilayer. The neutralization of this carboxylate at one point of the circumference is accompanied by re-ionization of another one further around the circumference of the c-annulus, by release of the proton on the opposite side of the membrane via an exit port in subunit a, and regeneration of a negative charge in the c-annulus:subunit a interface. These protonation-deprotonation events result in a rotary movement of the c-annulus. The rotation brings the next negatively charged carboxyl to the inlet port

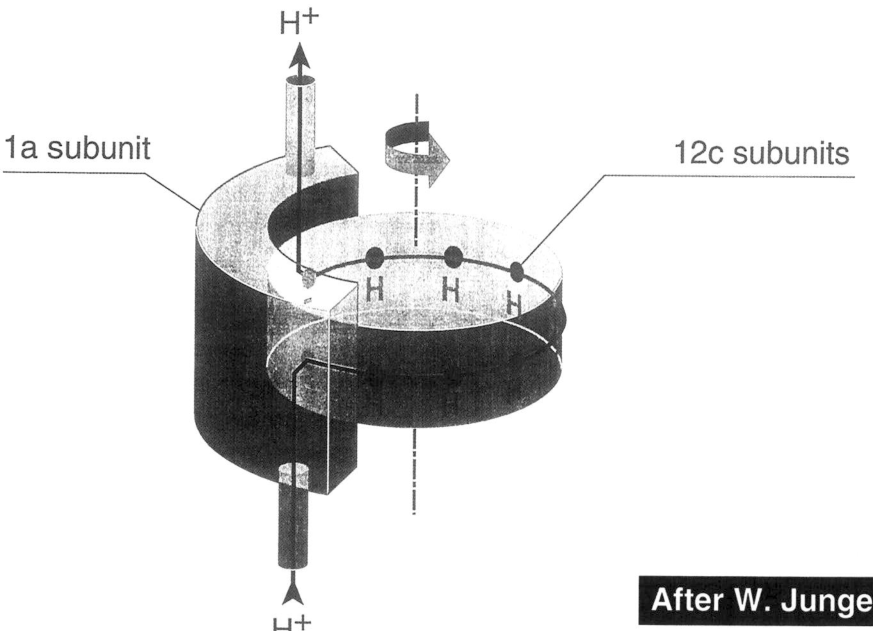

Figure 9. A hypothetical model for generation of rotation by proton transport through the F_o domain of ATP synthase. The central cylindrical part is made of c-subunits, the external curved structure is the single a subunit. The heavy black line indicates the path of the protons. For further explanation, see the text.

where in turn it is neutralized by another proton. The accompanying release of another proton via the exit port generates further rotation. It is envisaged that this rotary device is directly coupled to the γ-subunit.

The synthesis of each ATP molecule requires a rotation of the γ-subunit by 120°, and so each complete rotation of the γ-subunit in F_1 produces three ATP molecules. In a hypothetical proton motor with a c-annulus consisting of twelve c-subunits, this corresponds to the sequential neutralization of four carboxylates by proton binding (and accompanying sequential proton release from four others). In other words, a H^+: ATP ratio of four is compatible with a molar ratio of c-subunits in F_o of twelve. Likewise, a H^+:ATP ratio of three requires nine c-subunits in each F_o.

CONNECTIONS BETWEEN THE F_1 AND F_o DOMAINS OF ATP SYNTHASE

The evidence for the interaction of the central γ-subunit and the associated ε-subunit in bacterial F_1 with the loop region of a c-subunit in F_o has been described above. It is likely that these proteins are the principal components making the central interaction between F_1 and F_o, visualized as the central 45 Å stalk. There is accumulating evidence for a second link between F_1 and F_o. It is known that the bacterial δ-subunit and the equivalent bovine OSCP subunit interact with the N-terminal part of the α-subunits [73, 74], which are now known to be on top of F_1 distal from the membrane domain. Since the 1960s, it has been known that bovine OSCP is required for binding F_1 to F_o, implying that it interacts with subunits in the membrane domain. Therefore, the inescapable conclusion is that OSCP (and possibly also subunit δ in the bacterial and chloroplast enzymes) extends from the top of F_1, down its external surface, to a region associated with the membrane domain. In the bovine enzyme, the N-terminal part of OSCP interacts with the N-terminal part of the a-subunits [75, 76]. Additionally, the interactions of bovine OSCP with various F_o components have been studied by *in vitro* reconstitution experiments [77]. They indicate that OSCP binds mainly to the polar extramembranous part of subunit b (referred to as b′), and not with subunits d and F_6. However, subunits d and F_6 bind to the b′-OSCP complex to forming a stoichiometric complex containing one copy of each of the four proteins. This complex (unfortunately named as "stalk") makes no strong interactions with a complex of bovine subunits δ and ε [78], consistent with the bovine "stalk" complex being separate from the central 45 Å stalk.

By single particle analysis of negatively stained samples of monodisperse-bovine ATP synthase [79–82], Simone Karrasch has obtained electron microscopic evidence for a second peripheral stalk (see Figure 10). A similar feature has been observed independently in the *E. coli* enzyme (S. Wilkens and R. A. Capaldi, personal communication), and in a V-type ATPase (a relative of ATP synthase) from *Clostridium fervidus* [83]. The image of the bovine enzyme also contains another novel feature, seen as a disc shaped structure or collar

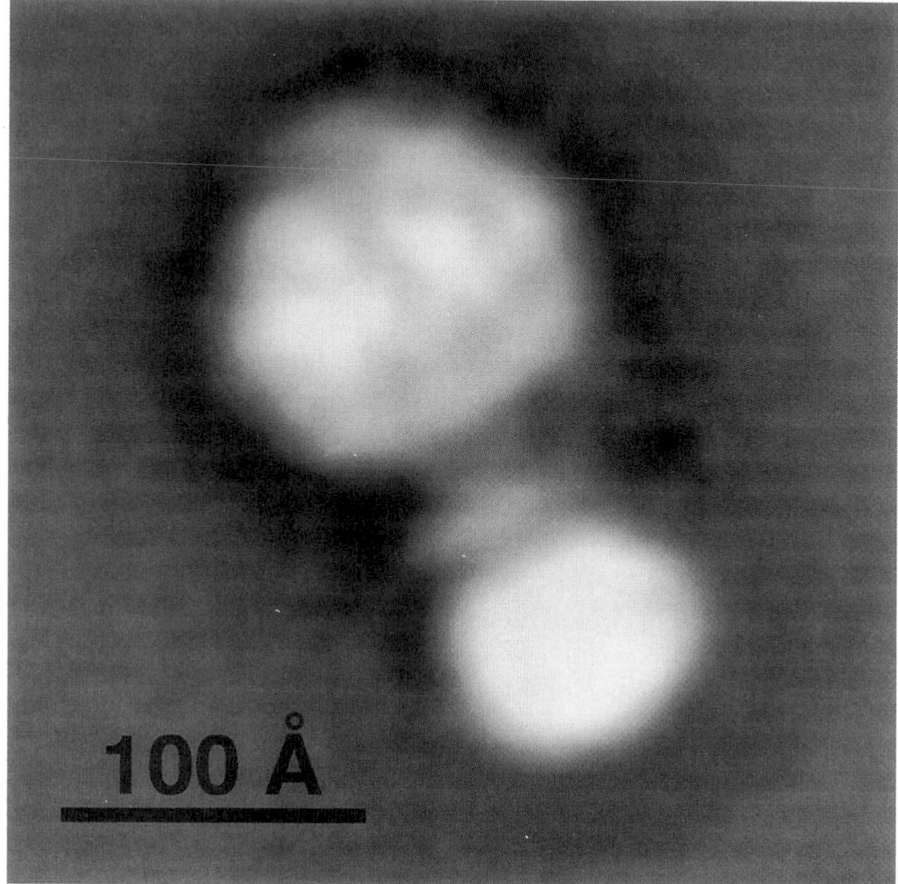

Figure 10. Image of bovine ATP synthase obtained by averaging 4940 single negatively stained particles of the enzyme. The plane of the membrane with which the assembly is associated in mitochondria, runs upwards, left to right, at an angle of 45°. The larger sphere is the F_1 catalytic domain, and the smaller sphere beneath it is F_o (or part of F_o). The new features are the collar apparently surrounding the central stalk, and, on the right, the faint peripheral connection between the collar and F_1 which may serve as a "stator". There may also be extra material on top of the F_1 domain that is not present in isolated F_1-ATPase (for more detailed discussion, see the text).

evidently sitting between the central stalk and F_o domains, and a similar collar was observed in the Clostridial enzyme. It remains to be established which subunits of the bovine enzyme are responsible for the peripheral stalk and the collar. The OSCP and subunit b are likely candidates for being components of the former, and the collar may be composed of parts of the c subunits.

At present, we can only speculate about the role of the peripheral stalk, but it may be acting as a stator to counter the tendency of the $\alpha_3\beta_3$ sub-complex to follow the rotation of the central γ-subunit. However, it is unlikely that this peripheral stalk remains bound in a unique position because the nature of the surface of the $\alpha_3\beta_3$ sub-complex must change cyclically in response to the rotating γ-subunit, and hence the preferred binding site of the peripheral structure must also move in the same way. The role of the collar is an even bigger mystery.

EVOLUTION OF ATP SYNTHASE AND ITS RELATIONSHIP TO OTHER ENZYMES

In bacterial operons coding for the subunits of ATP synthase, the genes for F_1 and F_o subunits are often clustered separately. This arrangement suggested the possibility that the two domains have evolved separately as structural modules [84], similar to the "modular evolution" of heads and tails of bacteriophages [85, 86]. If this suggestion were correct, it might be expected that the F_1 and F_o modules would be found to fulfill other biochemical functions elsewhere in biology.

It is too early to know whether there is any significant structural and mechanistic relationship between F_o and the motors that drive the flagellae of bacteria, although this remains a distinct possibility. However, there is emerging evidence of significant similarities, both functional and structural, between the F_1 domain of ATP synthase and DNA and RNA helicases that employ energy released by hydrolysis of ATP or GTP to separate the strands of nucleic acid duplexes (for example, see reference [87]). Therefore, the rotary mechanism of ATP synthase may turn out to be the first example of a more general principle in enzyme catalysis.

CONCLUDING REMARKS

The high resolution structure of the catalytic domain of the ATP synthesizing enzyme has provided new insights into our understanding of how ATP is made in biology. Nevertheless, challenging structural experiments lie ahead in the quest to understand the generation of rotation by transmembrane proton transport through its membrane sector. A short poem written by Robert Frost provides an appropriate summary of the current state of affairs.

We dance round in a ring and suppose
The secret sits in the middle and knows

I am glad to have played a role in arriving at our present level of understanding of ATP synthesis, and in the future I hope to contribute to the revelation of the secret sitting in the middle.

ACKNOWLEDGEMENTS

From what I have written, it is obvious that the work summarized above is the outcome of the joint efforts of many extremely able and highly valued colleagues and collaborators, including Ph. D. students and post-doctoral visitors, whom I have been privileged to lead. To all of them, I offer my thanks for their contributions. I am particularly grateful to The Medical Research Council, which has supported my work unstintingly for more that 20 years.

Many post-doctoral visitors were supported by Fellowships from The European Molecular Biology Organization, and others by Fellowships from the European Community and the Human Frontiers of Science Project. I thank these organizations also for their support.

REFERENCES

1. Nicholls, D. G. & Ferguson, S. J. (1992). In *Bioenergetics 2* Academic Press, London, San Diego.
2. Dimroth, P. (1991). Na^+-coupled alternative to H^+ coupled primary tarnsport system in bacteria. *Bioessays* **13**, 463–468.
3. Tsukihara, T., Aoyama, H., Yamashita, E., Tomizaki, T., Yamaguchi, H., Shinzawa-Itoh, K., Nakashima, R., Yaono, D. & Yoshikawa, S. (1995). Structures of the metal sites of oxidized cytochrome c oxidase at 2.8 Å resolution. *Science* **269**, 1069–1074.
4. Tsukihara, T., Aoyama, H., Yamashita, E., Tomizaki, T., Yamaguchi, H., Shinzawa-Itoh, K., Nakashima, R., Yaono, D. & Yoshikawa, S. (1996). The whole structure of the 13-subunit oxidized cytochrome c oxidase at 2.8 Å. *Science* **272**, 1136–1144.
5. Iwata, S., Ostermeier, C., Ludwig, B. & Michel, H. (1995). Structure at 2.8 Å resolution of cytochrome c oxidase from *Paracoccus denitrificans. Nature* **376**, 660–669.
6. Ostermeier, C., Harrenga, A., Ermler, U. & Michel, H. (1997). Structure at 2.7 Å of the *Paracoccus denitrificans* two-subunit cytochrome c oxidase with an antibody F_v fragment. *Proc. Natl. Acad. Sci. U.S.A.* **94**, 10547–10553.
7. Xia, D., Yu, C. A., Kih, H., Xia, J. Z., Kachurin, A. M., Zhang, L., Yu, L. & Deisenhofer, J. (1997). Crystal structure of the cytochrome bc_1 complex from bovine heart mitochondria. *Science* **277**, 60–66.
8. Walker, J. E. (1992). The NADH:ubiquinone oxidoreductase (complex I) of respiratory chains. *Qu. Rev. Biophys.* **25**, 253–324.
9. Guénbaut, V., Vincentelli, R., Mills, D., Weiss, H. & Leonard, K. (1997). Three-dimensional structure of NADH-dehydrogenase from *Neurospora crassa* by electron microscopy and conical tilt reconstruction. *J. Mol. Biol.* **265**, 409–418.
10. Grigorieff, N. (1997). Three-dimensional structure of bovine NADH:ubiquinone oxidoreductase (complexI) at 22 Å in ice. *J. Mol. Biol.* In press.
11. Gay, N. J. & Walker, J. E. (1981). The *atp* operon: nucleotide sequence of the promoter and the genes for the membrane proteins and the δ subunit of *Escherichia coli* ATP synthase. *Nucleic Acids Res.* **9**, 3919–3926.
12. Gay, N. J. & Walker, J. E. (1981). The *atp* operon: nucleotide sequence of the region encoding the α subunit of *Escherichia coli* ATP synthase. *Nucleic Acids Res.* **9**, 2187–2194.
13. Saraste, M., Gay, N. J., Eberle, A., Runswick, M. J. & Walker, J. E. (1981). The *atp* operon: nucleotide sequence of the genes for the γ, β and ε subunits of *Escherichia coli* ATP synthase. *Nucleic Acids Res.* **9**, 5287–5296.
14. Runswick, M. J. & Walker, J. E. (1983). The amino acid sequence of the b subunit of ATP synthase from bovine heart mitochondria. *J. Biol. Chem.* **258**, 3081–3089.
15. Tybulewicz, V. L. J., Falk, G. & Walker, J. E. (1984). *Rhodopseudomonas blastica atp* operon: Nucleotide sequence and transcription. *J. Mol. Biol.* **179**, 185–214.
16. Walker, J. E., Saraste, M. & Gay, N. J. (1984). The *unc* operon: nucleotide sequence, regulation and structure of ATP synthase. *Biochim. Biophys. Acta* **768**, 164–200.
17. Falk, G., Hampe, A. & Walker, J. E. (1985). Nucleotide sequence of the *Rhodospirillum rubrum atp* operon. *Biochem. J.* **228**, 391–407.
18. Walker, J. E., Fearnley, I. M., Gay, N. J., Gibson, B. W., Northrop, F. D., Powell, S. J., Runswick, M. J., Saraste, M. & Tybulewicz, V. L. J. (1985). Primary structure and subunit stoichiometry of F_1-ATPase from bovine mitochondria. *J. Mol. Biol.* **184**, 677–701.
19. Cozens, A. L., Walker, J. E., Phillips, A. L., Huttly, A. K. & Gray, J. C. (1986). A sixth subunit of ATP synthase, an F_o component, is encoded in the pea chloroplast genome. *EMBO J.* **5**, 217–222.

20. Fearnley, I. M. & Walker, J. E. (1986). Two overlapping genes in bovine mitochondrial DNA encode membrane components of ATP synthase. *EMBO J.* **5**, 2003–2008.
21. Cozens, A. L. & Walker, J. E. (1987). The organization and sequence of the genes for ATP synthase subunits in the cyanobacterium *Synechococcus* 6301: support for an endosymbiotic origin of chloroplasts. *J. Mol. Biol.* **194**, 359–383.
22. Walker, J. E., Gay, N. J., Powell, S. J., Kostina, M. & Dyer, M. R. (1987). ATP synthase from bovine mitochondria: sequences of imported precursors of oligomycin sensitivity conferral protein, factor 6 and adenosine triphosphatase inhibitor protein. *Biochemistry* **26**, 8613–8619.
23. Walker, J. E., Runswick, M. J. & Poulter, L. (1987). ATP synthase from bovine mitochondria: characterization and sequence analysis of two membrane associated subunits and of their corresponding c-DNAs. *J. Mol. Biol.* **197**, 89–100.
24. Falk, G. & Walker, J. E. (1988). DNA sequence of a gene cluster coding for subunits of the F_o membrane sector of ATP synthase in *Rhodospirillum rubrum*. *Biochem. J.* **254**, 109–122.
25. Walker, J. E., Powell, S. J., Vinas, O. & Runswick, M. J. (1989). ATP synthase from bovine mitochondria:complementary DNA sequence of the import precursor of a heart isoform of the alpha subunit. *Biochemistry* **28**, 4702–4708.
26. Runswick, M. J., Medd, S. M. & Walker, J. E. (1990). The δ subunit of ATP synthase from bovine heart mitochondria. Complementary DNA sequence of its import precursor cloned with the aid of the polymerase chain reaction. *Biochem. J.* **266**, 421–426.
27. Viñas, O., Powell, S. J., Runswick, M. J., Iacobazzi, V. & Walker, J. E. (1990). The epsilon subunit of ATP synthase from bovine heart mitochondria: complementary DNA, expression in bovine tissues and evidence of homologous sequences in man and rat. *Biochem. J.* **265**, 321–326.
28. Walker, J. E., Lutter, R., Dupuis, A. & Runswick, M. J. (1991). Identification of the subunits of F_1F_o-ATPase from bovine heart mitochondria. *Biochemistry* **30**, 5369–5378.
29. Van Walraven, H. S., Lutter, R. & Walker, J. E. (1993). Organization and sequence of genes for subunits of ATP synthase in the thermophilic cyanobacterium *Synechococcus* 6716. *Biochem. J.* **294**, 239–251.
30 Collinson, I. R., Runswick, M. J., Buchanan, S. K., Fearnley, I. M., Skehel, J. M., van Raaij, M. J., Griffiths, D. E. & Walker, J. E. (1994). The F_o membrane domain of ATP synthase from bovine heart mitochondria: purification, subunit composition and reconstitution with F_1-ATPase. *Biochemistry* **33**, 7971–7978.
31. Collinson, I. R., Skehel, J. M., Fearnley, I. M., Runswick, M. J. & Walker, J. E. (1996). The F_1F_o-ATPase complex from bovine heart mitochondria: the molar ratio of the subunits in the stalk region linking the F_1 and F_o domains. *Biochemistry* **35**, 12640–12646.
32. Walker, J. E. (1994). The regulation of catalysis in ATP synthase. *Curr. Opinion Struct. Biol.* **4**, 912–918.
33. Fernández-Morán, H. (1962). Cell-membrane ultrastructure. Low-temperature electron microscopy and x-ray diffraction studies of lipoprotein components in lamellar systems. *Circulation* **26**, 1039–1065.
34. Vambutas, V. K. & Racker, E. (1965). Partial resolution of the enzymes catalysing photophosphorylation. Stimulation of photophossphorylation by a preparation of a latent, Ca^{++}-dependent adenosine triphosphatase from chloroplasts. *J. Biol. Chem.* **240**, 2660–2667.
35. Green, D. E. (1964). The Mitochondrion. *Scientific American* **210**, 63–74.
36. Kagawa, Y. & Racker, E. (1966). Partial resolution of the enzymes catalyzing oxidative phosphorylation. Correlation of morphology and function in submitochondrial particles. *J. Biol. Chem.* **241**, 2475–2482.
37. Kagawa, Y. & Racker, E. (1966). Partial resolution of the enzymes catalyzing oxidative phosphorylation. Properties of a factor conferring oligmycin sensitivity on mitochondrial adenosine triphosphatase. *J. Biol. Chem.* **241**, 2461–2466.

38. Kagawa, Y. & Racker, E. (1966). Partial resolution of the enzymes catalyzing oxidative phosphorylation. Reconstruction of oligomycin-sensitive adenosine triphosphatase. *J. Biol. Chem.* **241**, 2467–2474.
39. Van Walraven, H., Strotmann, H., Schwartz, O. & Rumberg, B. (1996). The H^+/ATP ratio of the ATP synthase from the thiol modulated chloroplasts and two cyanobacterial strains is four. *FEBS Lett.* **379**, 309–313.
40. Boyer, P. D. (1993). The binding change mechanism for ATP synthase–some probabilities and possibilities. *Biochim. Biophys. Acta* **1140**, 215–250.
41. Brooks, J. C. & Senior, A. E. (1972). Methods for the purification of each subunit of the mitochondrial oligomycin-insensitive adenosine triphosphatase. *Biochemistry* **11**, 4675–4678.
42. Knowles, A. F. & Penefsky, H. S. (1972). The subunit structure of beef heart mitochondrial adenosine triphosphatase. *J. Biol. Chem.* **247**, 6616–6623.
43. Pai, E. F., Sachsenheimer, W., Schirmer, R. H. & Schulz, G. E. (1977). Substrate positions and induced-fit in crystalline adenylate kinase. *J. Mol. Biol.* **114**, 37–45.
44. Walker, J. E., Saraste, M., Runswick, M. J. & Gay, N. J. (1982). Distantly related sequences in the α and β subunits of ATP synthase, myosin, kinases and other ATP requiring enzymes and a common nucleotide binding fold. *EMBO J.* **1**, 945–951.
45. Higgins, C. F. (1992). ABC transporters: from microorganisms to man. *Ann. Rev. Cell Biol.* **8**, 67–113.
46. Gay, N. J. & Walker, J. E. (1983). Homology between human bladder carcinoma oncogene product and mitochondrial ATP synthase. *Nature* **301**, 262–264.
47. Koonin, E. V. (1993). A superfamily of ATPases with diverse functions containing either classical or deviant ATP-binding motif. *J. Mol. Biol.* **229**, 1165–1174.
48. Walker, J. E., Fearnley, I. M., Lutter, R., Todd, R. J. & Runswick, M. J. (1990). Structural aspects of proton pumping ATPases. *Phil. Trans. Royal Soc.* **326**, 367–378.
49. Lutter, R., Abrahams, J. P., van Raaij, M. J., Todd, R. J., Lundqvist, T., Buchanan, S. K., Leslie, A. G. W. & Walker, J. E. (1993). Crystallization of F_1–ATPase from bovine heart mitochondria. *J. Mol. Biol.* **229**, 787–790.
50. Abrahams, J. P., Leslie, A. G. W., Lutter, R. & Walker, J. E. (1994). Structure at 2.8 Å resolution of F_1-ATPase from bovine heart mitochondria. *Nature* **370**, 621–628.
51. van Raaij, M. J., Abrahams, J. P., Leslie, A. G. W. & Walker, J. E. (1996). The structure of bovine F_1-ATPase complexed with the antibiotic inhibitor aurovertin. *Proc. Natl. Acad. Sci. U.S.A.* **93**, 6913–6917.
52. Abrahams, J. P., Buchanan, S. K., van Raaij, M. J., Fearnley, I. M., Leslie, A. G. W. & Walker, J. E. (1996). The structure of bovine F_1-ATPase complexed with the peptide antibiotic efrapeptin. *Proc. Natl. Acad. Sci. U.S.A.* **93**, 9420–9424.
53. Duncan, T. M., Bulygin, V. V., Zhou, Y., Hutcheon, M. L. & Cross, R. L. (1995). Rotation of subunits during catalysis by *Escherichia coli* F_1-ATPase. *Proc. Natl. Acad. Sci. U. S. A.* **92**, 10964–10968.
54. Capaldi, R. A., Aggeler, R., Wilkens, S. & Grüber, G. (1996). Structural changes in the γ and ε subunits of the *Escherichia coli* F_1F_0-type ATPase during energy coupling. *J. Bioenerget. Biomemb.* **28**, 397–401.
55. Sabbert, D., Engelbrecht, S. & Junge, W. (1995). Intersubunit rotation in active F-ATPase. *Nature* **381**, 623–625.
56. Noji, H., Yasuda, R., Yoshida, M. & Kinosita Jr, K. (1997). Direct observation of the rotation of F_1-ATPase. *Nature* **386**, 299–302.
57. Wilkens, S., Dahlquist, F. W., McIntosh, L. P., Donaldson, L. W. & Capaldi, R. A. (1995). Structural features of the e subunit of the *Escherichia coli* ATP synthase determined by NMR spectroscopy. *Nature Struct. Biol.* **2**, 961–967.
58. Uhlin, U., Cox, G. B. & Guss, J. M. (1997). Crystal structure of the ε-subunit of the proton-translocating ATP synthase from *Escherichia coli. Structure* **5**, 1219–1230.
59. Wilkens, S., Dunn, S. D., Chandler, J., Dahlquist, F. W. & Capaldi, R. A. (1997). Solution structure of the N-terminal domain of the δ-subunit of the *E. coli* ATP synthase. *Nature Struct. Biol.* **4**, 198–201.

60. Walker, J. E., Runswick, M. J. & Saraste, M. (1982). Subunit equivalence in *Escherichia coli* and bovine heart mitochondrial F_1F_o ATPases. *FEBS Lett.* **146**, 393–396.
61. Fillingame, R. H. (1996). Membrane sectors of F- and V-type H^+-transporting ATPases. *Current Opinion Struct. Biol.* **6**, 491–498.
62. Foster, D. L. & Fillingame, R. H. (1982). Stoichiometry of subunits in the H^+ -ATPase complex of *Escherichia coli. J. Biol. Chem.* **257**, 2009–2015.
63. Fillingame, R. H. (1990). Molecular mechanics of ATP synthesis by F_1F_o-type H^+-transporting ATPases. *The Bacteria* **12**, 345–391.
64. Fillingame, R. H., Girvin, M. E., Jiang, W., Valiyaveetil, F. & Hermolin, J. (1998). Subunit interactions coupling H^+ transport and ATP synthesis in F_1F_o ATP synthase. *Acta Physiol. Scand.* In the press.
65. Walker, J. E., Saraste, M. & Gay, N. J. (1982). *E. coli* F_1-ATPase interacts with a membrane protein component of a proton channel. *Nature* **298**, 867–869.
66. Dunn, S. D. (1992). The polar domain of the b subunit of *Escherichia coli* F_1F_o-ATPase forms an elongated dimer that interacts with the F_1 sector. *J. Biol. Chem.* **267**, 7630–7636.
67. Watts, S. D., Zhang, Y., Fillingame, R. H. & Capaldi, R. A. (1995). The gamma subunit in the *Escherichia coli* ATP synthase complex (ECF_1F_o) extends through the stalk and contacts the c subunits of the F_o part. *FEBS Lett.* **368**, 235–238.
68. Zhang, Y. & Fillingame, R. H. (1995). Subunits coupling H^+ transport and ATP synthesis in the *Escherichia coli* ATP synthase: Cys-Cys crosslinking of F_1 subunit ε to the polar loop of F_o subunit c. *J. Biol. Chem.* **270**, 24609–24614.
69. Groth, G. & Walker, J. E. (1997). Model of the c-subunit oligomer in the membrane domain of F-ATPases. *FEBS Lett.* **410**, 117–123.
70. Singh, S., Turina, P., Bustamente, C. J., Keller, D. J. & Capaldi, R. A. (1996). Topographical structure of membrane-bound *Escherichia coli* F_1F_o ATP synthase in aqueous buffer. *FEBS Lett.* **397**, 30–34.
71. Takeyasu, K., Omote, H., Nettikadan, S., Tokumasu, F., Iwamoto-Kihara, A. & Futai, M. (1996). Molecular imaging of *Escherichia coli* F_oF_1-ATPase in reconstituted membranes by atomic force microscopy. *FEBS Lett.* **392**, 110–113.
72. Junge, W., Sabbert, D. & Engelbrecht, S. (1996). Rotary catalysis by F-ATPase: real time recording of intersubunit rotation. *Ber. Bunsenges. Phys. Chem.* **100**, 2014–2019.
73. Dunn, S. D., Heppel, L. A. & Fullmer, C. S. (1980). The NH_2-terminal portion of the α-subunit of the *Escherichia co*li F_1 -ATPase is required for binding the δ-subunit. *J. Biol. Chem.* **255**, 6891–6896.
74. Hundal, T., Norling, B. & Ernster, L. (1983). Lack of ability of trypsin-treated mitochondrial F_1-ATPase to bind to the oligomycin sensitivity conferring protein (OSCP). *FEBS Lett.* **162**, 5–10.
75. Joshi, S., Javed, A. A. & Gibbs, L. C. (1992). Oligomycin sensitivity-conferring protein (OSCP) of mitochondrial ATP synthase. The carboxy terminal region of OSCP is essential for the reconstitution of oligomycin-sensitive H^+-ATPase. *J. Biol. Chem.* **267**, 12860–12867.
76. Joshi, S., Pringle, M. J. & Siber, R. (1986). Topology and function of "stalk" proteins in the bovine mitochondrial H^+-ATPase. *J. Biol. Chem.* **261**, 10653–10658.
77. Collinson, I. R., van Raaij, M. J., Runswick, M. J., Fearnley, I. M., Skehel, J. M., Orriss, G., Miroux, B. & Walker, J. E. (1994). ATP synthase from bovine heart mitochondria: *in vitro* assembly of a stalk complex in the presence of F_1-ATPase and in its absence. *J. Mol. Biol.* **242**, 408–421.
78. Orriss, G. L., Runswick, M. J., Collinson, I. R., Miroux, B., Fearnley, I. M., Skehel, J. M. & Walker, J. E. (1996). The δ- and ε-subunits of bovine F_1-ATPase interact to form a heterodimeric subcomplex. *Biochem. J.* **314**, 695–700.
79. Lutter, R., Saraste, M., van Walraven, H. S., Runswick, M. J., Finel, M., Deatherage, J. F. & Walker, J. E. (1993). F_1F_o-ATPase from bovine heart mitochondria: development of the purification of a monodisperse oligomycin sensitive ATPase. *Biochem. J.* **295**, 799–806.

80. Walker, J. E., Collinson, I. R., Van Raaij, M. J. & Runswick, M. J. (1995). Structural analysis of ATP synthase (F_1F_o-ATPase) from bovine heart mitochondria. *Methods in Enzymol.* 163–190.
81. Buchanan, S. K. & Walker, J. E. (1996). Large scale chromatographic purification of F_1F_o-ATPase and complex I from bovine heart mitochondria. *Biochem. J.* **318**, 343–349.
82. Groth, G. & Walker, J. E. (1996). ATP synthase from bovine heart mitochondria: reconstitution into unilamellar phospholipid vesicles of the pure enzyme in a functional state. *Biochem. J.* **318**, 351–357.
83. Boekema, E., Ubbink-Kok, T., Lolkema, J. S., Brisson, A. & Konings, W. N. (1997). Visualization of a peripheral stalk in V-type ATPase: evidence for the stator structure essential to rotational catalysis. *Proc. Natl. Acad. Sci. U.S.A.* **94**, In the press.
84. Walker, J. E. & Cozens, A. L. (1986). Evolution of ATP synthase. *Chemica Scripta* **26B,** 263–272.
85. Botstein, D. (1980). A theory for molecular evolution of bacteriophages. *Ann. N. Y. Acad. Sci.* **354**, 484–491.
86. Casjens, S. & Hendrix, R. (1974). Comments on the arrangement of the morphogenetic genes of bacteriophage lambda. *J. Mol. Biol.* **90**, 20–23.
87. Yu, X. & Egelman, E. H. (1997). The RecA hexamer is a structural homologue of ring helicases. *Nature Struct. Biol.* **4**, 101–104.
88. Gogol, E. P., Aggeler, R., Sagerman, M. & Capaldi, R. A. (1989). Cryoelectron microscopy of *Escherichia coli* F_1 adenosine triphosphatase decorated with monoclonal antibodies to individual subunits of the complex. *Biochemistry* **28**, 4717–4724.
89. Weissman, G. & Claibourne, R. (1975). Cell Membranes: Biochemistry, Cell Biology and Pathology.

JENS C. SKOU

I was born on the 8th of October 1918 into a wealthy family in Lemvig, a town in the western part of Denmark. The town is nicely situated on a fjord, which runs across the country from the Kattegat in East to the North Sea in West. It is surrounded by hills, and is only 10 km, i.e. bicycling distance, from the North Sea, with its beautiful beaches and dunes. My father Magnus Martinus Skou together with his brother Peter Skou were timber and coal merchants.

We lived in a big beautiful house, had a nice summer house on the North Sea coast. We were four children, I was the oldest with a one year younger brother, a sister 4 years younger and another brother 7 years younger. The timber-yard was an excellent playground, so the elder of my brothers and I never missed friends to play with. School was a minor part of life.

When I was 12 years old my father died from pneumonia. His brother continued the business with my mother Ane-Margrethe Skou as passive partner, and gave her such conditions that there was no change in our economical situation. My mother, who was a tall handsome woman, never married again. She took care of us four children and besides this she was very active in the social life in town.

When I was 15, I went to a boarding school, a gymnasium (high school) in Haslev a small town on Zealand, for the last three years in school (student exam). There was no gymnasium in Lemvig.

Besides the 50–60 boys from the boarding section of the school there were about 400 day pupils. The school was situated in a big park, with two football fields, facilities for athletics, tennis courts and a hall for gymnastics and handball. There was a scout troop connected to the boarding section of the school. I had to spend a little more time preparing for school than I was used to. My favourites were the science subjects, especially mathematics. But there was plenty of time for sports activities and scouting, which I enjoyed. All the holidays, Christmas, Easter, summer and autumn I spent at home with my family.

After three years I got my exam, it was in 1937. I returned to Lemvig for the summer vacation, considering what to do next. I could not make up my mind, which worried my mother. I played tennis with a young man who studied medicine, and he convinced me that this would be a good choice. So, to my mother's great relief, I told her at the end of August that I would study medicine, and started two days later at the University of Copenhagen.

The medical course was planned to take 7 years, 3 years for physics, chemistry, anatomy, biochemistry and physiology, and 4 years for the clinical subjects, and for pathology, forensic medicine, pharmacology and public health. I followed the plan and got my medical degree in the summer of 1944.

I was not especially interested in living in a big town. On the other hand it was a good experience for a limited number of years to live in, and get acquainted with the capital of the country, and to exploit its cultural offers. Art galleries, classical music and opera were my favourites.

For the first three years I spent the month between the semesters at home studying the different subjects. For the last 4 years the months between the semesters were used for practical courses in different hospital wards in Copenhagen.

It was with increasing anxiety that we witnessed to how the maniac dictator in Germany, just south of our border, changed Germany into a madhouse. Our anxiety did not become less after the outbreak of the war. In 1914 Denmark managed to stay out of the war, but this time, in April 1940 the Germans occupied the country. Many were ashamed that the Danish army were ordered by the government to surrender after only a short resistance. Considering what later happened in Holland, Belgium and France, it was clear that the Danish army had no possibility of stopping the German army.

The occupation naturally had a deep impact on life in Denmark in the following years, both from a material point of view, but also, what was much worse, we lost our freedom of speech. For the first years the situation was very peculiar. The Germans did not remove the Danish government, and the Danish government did not resign, but tried as far as it was possible to minimize the consequences of the occupation. The army was not disarmed, nor was the fleet. The Germans wanted to use Denmark as a food supplier, and therefore wanted as few problems as possible.

The majority of the population turned against the Germans, but with no access to weapons, and with a flat homogeneous country with no mountains or big woods to hide in, the possibility of active resistance was poor. So for the first years the resistance only manifested itself in a negative attitude to the Germans in the country, in complicating matters dealing with the Germans as far as possible, and in a number of illegal journals, keeping people informed about the situation, giving the information which was suppressed by the German censorship. There was no interference with the teaching of medicine.

The Germans armed the North Sea coast against an invasion from the allied forces. Access was forbidden and our summer house was occupied. My grandmother had died in 1939, and we four children inherited what would have been my father's share. For some of the money my brother and I bought a yacht, and took up sailing, and this has since been an important part of my leisure time life. After the occupation the Germans had forbidden sailing in the Danish seas except on the fjord where Lemvig was situated, and another fjord in Zealand.

The resistance against the Germans increased as time went on, and sabotage slowly started. Weapons and ammunition for the resistance movement began to be dropped by English planes, and in August 1943 there were general strikes all over the country against the Germans with the demand that the Government stopped giving way to the Germans. The Government consequently resigned, the Germans took over, the Danish marine sank the fleet and the army was disarmed. An illegal Frihedsråd (the Danish Liberation Council) revealed itself, which from then on was what people listened to and took advice from.

Following this, the sabotage against railways and factories working for the Germans increased, and with this arrests and executions. One of our medical classmates was a German informer. We knew who he was, so we could take care. He was eventually liquidated by unknowns. We feared a reaction from Gestapo against the class, and stayed away from the teaching.

The Germans planned to arrest the Jews, but the date, the night between the first and second October 1943 was revealed by a high placed German. By help from many, many people the Jews were hidden. Of about 7000, the Germans caught 472, who were sent to Theresienstadt where 52 died. In the following weeks illegal routes were established across the sea, Øresund, to Sweden, and the Jews were during the nights brought to safety. From all sides of the Danish society there were strong protests against the Germans for this encroachment on fellow-countrymen.

In May and June 1944, we managed to get our exams. A number of our teachers had gone underground, but their job was taken over by others. We could not assemble to sign the Hippocratic oath, but had to come one by one at a place away from the University not known by others.

I returned to my home for the summer vacation. The Germans had taken over part of my mother's house, and had used it for housing Danes working for the Germans. This was extremely unpleasant for my mother, but she would not leave her house and stayed. I addressed the local German commander, and managed to get him to move the "foreigners" from the house at least as long as we four children were home on holiday.

The Germans had forbidden sailing, but not rowing, so we bought a canoe and spent the holidays rowing on the fjord.

After the summer holidays I started my internship in a hospital in Hjørring in the northern part of the country. I first spent 6 months in the medical ward, and then 6 months in the surgical ward. I became very interested in surgery, not least because the assistant physician, next in charge after the senior surgeon, was very eager to teach me how to make smaller operations, like removing a diseased appendix. I soon discovered why. When we were on call together and we during the night got a patient with appendicitis, it happened–after we had started the operation–that he asked me to take over and left. He was then on his way to receive weapons and explosives which were dropped by English planes on a dropping field outside Hjørring. I found that this was more important than operating patients for appendicitis, but we had of course to take care of the patients in spite of a war going on. He was finally caught by the Gestapo, and sent to a concentration camp, fortunately not in Germany, but in the southern part of Denmark, where he survived and was released on the 5th of May 1945, when the Germans in Denmark surrendered.

I continued for another year in the surgical ward. It was here I became interested in the effect of local anaesthetics, and decided to use this as a subject for a thesis. Thereafter I got a position at the Orthopaedic Hospital in Aarhus as part of the education in surgery.

In 1947 I stopped clinical training, and got a position at the Institute for Medical Physiology at Aarhus University in order to write the planned doctoral thesis on the anaesthetic and toxic mechanism of action of local anaesthetics.

During my time in Hjørring I met a very beautiful probationer, Ellen Margrethe Nielsen, with whom I fell in love. I had become ill while I was on the medical ward, and spent some time in bed in the ward. I had a single room and a radio, so I invited her to come in the evening to listen to the English radio, which was strictly forbidden by the Germans – but was what everybody did.

After she had finished her education as a nurse in 1948, she came to Aarhus and we married. In 1950 we had a daughter, but unfortunately she had an inborn disease and died after 1 1/2 year. Even though this was very hard, it brought my wife and I closer together. In 1952, and in 1954 respectively we got two healthy daughters, Hanne and Karen.

The salary at the University was very low, so partly because of this but also because I was interested in using my education as a medical doctor, I took in 1949 an extra job as doctor on call one night a week. It furthermore had the advantage that I could get a permission to buy a car and to get a telephone. There were still after war restrictions on these items.

I was born in a milieu which politically was conservative. The job as a doctor on call changed my political attitude and I became a social democrat. I realized how important it is to have free medical care, free education with equal opportunities, and a welfare system which takes care of the weak, the handicapped, the old, and the unemployed, even if this means high taxes. Or as phrased by one of our philosophers, N.F.S. Grundtvig, "a society where few have too much, and fewer too little".

We lived in a flat, so the car gave us new possibilities. We wanted to have a house, and my mother would give us the payment, but I was stubborn, and wanted to earn the money myself. In 1957 we bought a house with a nice garden in Risskov, a suburb to Aarhus not far away from the University.

I am a family man, I restricted my work at the Institute to 8 hours a day, from 8 to 4 or 9 to 5, worked concentratedly while I was there, went home and spent the rest of the day and the evening with my wife and children. All weekends and holidays, and 4 weeks summer holidays were spent with the family. In 1960 we bought an acre of land on a cliff facing the beach 45 minutes by car from Aarhus, and built a small summer house. From then on this became the centre for our leisure time life. We bought a dinghy and a rowing boat with outboard motor and I started to teach the children how to sail, and to fish with fishing rod and with net.

Later, when the girls grew older, we bought a yacht, the girls and I sailed in the Danish seas, and up along the west coast of Sweden. My wife easily gets seasick, but joined us on day tours. Later the girls took their friends on sail tours.

In wintertime the family skied as soon as there was snow. A friend of mine, Karl Ove Nielsen, a professor of physics, took me in the beginning of the 1960s at Easter time on an 8-day cross country ski tour through the high mountain area in Norway, Jotunheimen. We stayed overnight in the Norwegian Tourist Association's huts on the trail, which were open during the Easter week. It was a wonderful experience, but also a tour where you had to take all safety precautions. It became for many years a tradition. Later the girls joined us, and they also took some of their friends. When the weather situation did not allow this tour, we spent a week in more peaceful surroundings either in Norway with cross-country skiing or in the Alps with slalom. We still do, now with the girls, their husbands and the grandchildren. Outside the sporting activities, I spend much time listening to classical music, and reading, first of all biographies.

When the children left home, one for studying medicine and the other architecture, my wife, worked for several years as a nurse in a psychiatric hospital for children, then engaged herself in politics. She was elected for the County Council for the social democrats, and spent 12 years on the council, first of all working with health care problems. She was also elected to the county scientific ethical committee, which evaluates all research which involves human beings. Later she was elected co-chairman to the Danish Central Scientific Ethical Committee, which lay down the guidelines for the work on the local committees, and which is an appeal committee for the local committees as well as for the doctors. She has worked 17 years on the committees and has been lecturing nurses and doctors about ethical problems.

I had no scientific training when I started at the Institute of Physiology in 1947. It took me a good deal of time before I knew how to attack the problem I was interested in and get acquainted with this new type of work. The chairman, Professor Søren L. Ørskov was a very considerate person, extremely helpful, patient, and gave me the time necessary to find my feet. During the work I got so interested in doing scientific work that I decided to continue and give up surgery. The thesis was published as a book in Danish in 1954, and written up in 6 papers published in English. The work on the local anaesthetics, brought me as described in the following paper to the identification of the sodium-potassium pump, which is responsible for the active transport of sodium and potassium across the cell membrane. The paper was published in 1957. From then on my scientific interest shifted from the effect of local anaesthetics to active transport of cations.

In the 1940s and the first part of the 1950s, the amount of money allocated for research was small. Professor Ørskov, fell chronically ill. His illness developed slowly so he continued in his position, but I, as the oldest in the department after him, had partly to take over his job. This meant that besides teaching in the semesters I had to spend two months per year examining orally the students in physiology.

The identification of the sodium-potassium pump gave us contact to the outside scientific world. In 1961, I met R.W. Berliner at an international Pharmacology meeting in Stockholm. He mentioned the possibility of obtaining a grant from National Institutes of Health (NIH). I applied and got a grant for two years. The importance of this was not only the money, but that it showed interest in the work we were doing.

In 1963, Professor Ørskov resigned and I was appointed professor and chairman. In the late 1950s and especially in the 1960s, more money was allocated to the Universities, and also more positions. Due to the work with the sodium-potassium pump, it became possible to attract clever young people, and the institute staff in a few years increased from 4 to 20–25 scientists. This had also an effect on the teaching. I got a young doctor, Noe Næraa, who had expressed ideas about medical teaching, to accept a position at the Institute. He started to reorganise our old fashioned laboratory course, we got new modern equipment, and thereafter we also reorganised the teaching, made it problem-oriented with teaching in small classes. My scientific interest was membrane physiology, but I wanted also to find people who could cover other aspects of physiology, so we ended up with 5–6 groups who worked scientifically with different physiological subjects.

In 1972 we got a new statute for the Universities, which involved a democratization of the whole system. The chairman was no longer the professor (elected by the board of chairmen which made up the faculty), but he/she was now elected by all scientists and technicians in the Institute and could be anybody, scientist or technician. This was of course a great relief for me because I could get rid of all my administrative duties. A problem was, however, that I got elected as chairman, but later others took over. In the beginning it was very tedious to work with the system, not least because everybody thought that they should be asked and take part in every decision. Later we learned to hand over the responsibility to an elected board at the Institute.

In these years the money to the Institutes came from the Faculty, which got it from the University (which got it from the State). The money was then divided inside the institute by the chairman, and later by the elected board. It was usually sufficient to cover the daily expenses of the research. External funds were only for bigger equipment. Besides research-money we had a staff of very well trained laboratory assistants, whose positions–as well as the positions of the scientific staff–was paid by the University. The institute every year sent a budget for the coming year, to the faculty, who then sent a budget for the faculty to the University, and the University to the State.

This way of funding had the great advantage that there was not a steady pressure on the scientists for publication and for sending applications for external funds. It was a system that allowed everybody to start on his/her own project, independently, and test their ideas. Nobody was forced from lack of money to join a group which had money and work on their ideas. It was also a system which could be misused, by people who were not active scientifically. With an elected board it proved difficult to handle such a situation. Not least because the very active scientists tried to avoid being elected–i.e. it could be the least active who actually decided. In practice, however, the not very active scientists usually accepted to do an extra job with the teaching, thus relieving the very active scientists from part of the teaching burden.

In the 1980s this was changed, the money for science was transferred to centralized (state) funds, and had to be applied for by the individual scientist.

Not an advantage from my point of view. Applications took a lot of time, it tempted a too fast publication, and to publish too short papers, and the evaluation process used a lot of manpower. It does not give time to become absorbed in a problem as the previous system.

My research interest was concentrated around the structure and function of the active transport system, the Na^+,K^+-ATPase. A number of very excellent clever young scientists worked on different sides of the subject, either their own choice or suggested by me. Each worked independent on his/her subject. Scientists who took part in the work on the Na^+,K^+-ATPase and who made important contributions to field were, P.L. Jørgensen (purification and structure), I. Klodos (phosphorylation), O. Hansen (effect of cardiac glycosides and vanadate), P. Ottolenghi (effect of lipids), J. Jensen (ligand binding), J. G. Nørby (phosphorylation, ligand binding, kinetics), L. Plesner (kinetics), M. Esmann (solubilization of the enzyme, molecular weight, ESR studies), T. Clausen (hormonal control), A.B. Maunsbach and E. Skriver from the Institute of Anatomy in collaboration with P.L. Jørgensen (electron microscopy and crystallization), and I. Plesner from the Department of Chemistry (enzyme kinetics and evaluation of models). We also had many visitors.

We got many contacts to scientists in different parts of the world, and I spent a good deal of time travelling giving lectures. In 1973 the first international meeting on the Na^+,K^+-ATPase was held in New York. The next was 5 years later in Århus, and thereafter every third year. The proceedings from these meetings have been a very valuable source of information about the development of the field.

My wife joined me on many of the tours and we got friends abroad. Apart from the scientific inspiration the travelling also gave many cultural experiences, symphony concerts, opera and ballet, visits to Cuzco and Machu Picchu in Peru, to Uxmal and Chichén Itzá on the Yucatan Peninsula, and to museums in many different countries. Not to speak of the architectural experiences from seeing many different parts of the world. And not least it gave us good friends.

It is not always easy to keep your papers in order when travelling. Sitting in the airport in Moscow in the 1960s waiting for departure to Khabarovsk in the eastern part of Siberia, we–three Danes on our way to a meeting in Tokyo–realized that we had forgotten our passports at the hotel in town. There were twenty minutes to departure and no way to get the passports in time. We asked Intourist what to do. There was only one boat connection a week from Nakhodka, where we should embark to Yokohama, so they suggested that we should go on, they would send the passports after us. I had once had a nightmare, that I should end my days in Siberia. When we after an overnight flight arrived in Khabarovsk we were met by a lady who asked if we were the gentlemen without passports. We could not deny, and she told us that they would not arrive until after we had left Khabarovsk by train to Nakhodka. But they would send them by plane to Vladivostok and from there by car to Nakhodka. To our question if we could leave Siberia without our passports the answer was no. When the train the following morning stopped in Nakhodka, a man came into the sleeping car and asked if we were the gentlemen without passports. To our "yes" he said "here you are", and handed over the passports. Amazing. We had an uncomplicated boat trip to Yokohama.

It was not as easy some years later in Argentina. I had been at a meeting in Mendoza, had stopped in Cordoba on the way, had showed passport in and out of the airports without problems. Returning to Buenos Aires to leave for New York, the man at the counter told me that my passport had expired three months earlier, and according to rules I had to return direct to my home country. I argued that I was sure I could get into the U.S., but he would not give way. We discussed for half an hour. Finally shortly before departure he would let me go to New York if he could reserve a plane out of New York to Denmark immediately after my arrival. He did the reservation, put a label on my ticket with the time of departure, and by the second call for departure I rushed off, hearing him saying "You can always remove the label". In New York, I stepped to the rear end of the line, hoping the man at the counter would be tired when it was my turn. He was not. I asked if I had to return to Denmark. "There is always a way out" was his answer, "No, go to the other counter, sign some papers, pay 5 dollars, and I let you in".

In 1977, I was offered the chair of Biophysics at the medical faculty. It was a smaller department, with 7 positions for scientists, of which 5 were empty, which meant that we could get positions for I. Klodos and M. Esmann, who had fellowships. Besides J. G. Nørby and L. Plesner moved with us. The two members in the Institute, M. J. Mulvany and F. Cornelius became interested in the connection between pump activity and vasoconstriction, and reconstitution of the enzyme into liposomes, respectively, i.e. all in the institute worked on different sides of the same problem, the structure and function of the Na^+,K^+-ATPase. We got more space, less administration, and I were free of teaching obligations.

We all got along very well, lived in a relaxed atmosphere, inspiring and helping each other, cooperating, also with the Na^+,K^+-ATPase colleagues left in the Physiological Institute. And even if we all worked on different sides of the same problem, there were never problems of interfering in each others subjects, or about priority.

In 1988, I retired, kept my office, gave up systematic experimental work and started to work on kinetic models for the overall reaction of the pump on computer. For this I had to learn how to programme, quite interesting, and amazing what you can do with a computer from the point of view of handling even complicated models. And even if my working hours are fewer, being free of all obligations, the time I spent on scientific problems are about the same as before my retirement.

I enjoy no longer having a meeting calendar, I enjoy to go fly-fishing when the weather is right, and enjoy spending a lot of time with my grandchildren.

THE IDENTIFICATION OF THE SODIUM-POTASSIUM PUMP

Nobel Lecture, December 8, 1997

by

JENS C. SKOU

Department of Biophysics, University of Aarhus, Ole Worms Allé 185, DK-8000 Aarhus C, Denmark.

Looking for the answer.

You hunt it,
you catch it,
You fool yourself;
the answer,
is always,
a step ahead.
J. C. S.

INTRODUCTION

The cell membrane separates the cell from the surrounding medium. In 1925 Gorter and Grendel[1] extracted the lipids from red blood cells, spread them in a monomolecular layer on a water phase, and measured the compressed area. It was about the double of the surface area of the extracted red blood cells. They suggested that the cell membrane is a bilayer of lipids with the charged head groups of the phospholipids facing the water phase on the two sides of the membrane, and with the hydrocarbon chains meeting in the middle of the membrane. The thickness of the membrane is about 40 Å, much too small to be seen under a microscope.

The surface tension of invertebrate eggs and other cells is however, much lower than for a water-lipid interphase. To explain this, Danielli and Davson in 1935[2] suggested that there is an adsorbed layer of proteins on each side of the lipid bilayer. They also introduced a layer of non-oriented lipids in the middle of the bilayer.

In the cytoplasm there are proteins for which the cell membrane is impermeable. At the cell pH they carry negative charges, neutralized by potassium, K^+, which in the cell is at a concentration of about 150 meq/l, while outside the cell the K^+ concentration is 4 meq/l. The difference in K^+ concentration was explained as due to a Donnan effect of the proteins, and that the membrane is permeable to K^+. The Donnan effect gives an osmotic pressure, which is higher inside the cell than outside, and since the membrane is permeable to water, water will flow in, and as the lipid bilayer cannot resist a hydrostatic pressure the cell will swell and finally burst.

The higher osmotic pressure inside the cell is opposed by a high concentration of sodium, Na^+, outside, 140 meq/l, while the concentration inside is low, about 10–20 meq/l. A problem is to explain how this difference in the Na^+ concentration on the two sides of the membrane can be maintained. There are two possible explanations:

One is that the membrane is impermeable to Na^+, that it is an equilibrium situation. This view was advocated by Conway, an Irish biochemist from Dublin. Boyle and Conway published in 1941 a paper in which they showed that for muscle fibers soaked in solutions with varying K^+ concentrations, the calculated intracellullar K^+ concentrations, based on this assumption agreed with the measured[3]. There were, however, two problems. One was that this agreement did not hold at the normal physiological concentrations of K^+, only at higher concentrations. The other was the requirement of Na^+ impermeability. But how then to explain that there is sodium in the cell, even if it is at a low concentration. Conway gave no answers to these problems.

The other possibility is that the membrane is permeable to Na^+, and that there are secretory, energy dependent processes in the cell, which compensate for the steady influx of Na^+, a steady state distribution. This view was advocated by R.B. Dean in a paper also published in 1941 entitled: "Theories of Electrolyte Equilibrium in Muscle"[4]. Referring to investigations by L.A. Heppel (1939,1940), by L.A. Heppel and C.L.A Schmidt (1938), and by H.B. Steinbach (1940) (see ref. 4) on muscle fibers, which had shown that the muscle membrane, contrary to the view held by Conway, is permeable to Na^+, Dean concluded: "the muscle can actively move potassium and sodium against concentration gradients … this requires work. Therefore there must be some sort of a pump possibly located in the fiber membrane, which can pump out sodium or, what is equivalent, pump in potassium."

In the following decade, helped by the introduction of radioactive isotopes of Na^+, and K^+[5], it was shown not only from experiments on muscle fibers, but on red blood cells, on nerves, and on frog skin that the membrane is permeable to Na^+ as well as to K^+, (for references see an extensive review by Ussing[6]). An energy dependent efflux of Na^+ is therefore necessary.

However, Conway strongly defended his view about the impermeability to sodium, and only reluctantly gave way. He admitted that there may be a certain permeability for Na^+ and thereby a need for a pump, but that low permeability for Na^+ is the main explanation of the concentration gradient. Conway's concern was, that it is a waste of energy to have a membrane permeable to Na^+, and then spend energy to pump Na^+ out. Krogh[7] had in a Croonian lecture in 1946 entitled "The active and passive exchange of inorganic ions through the surfaces of living cells and through living membranes generally" criticised Conway's view about impermeability to sodium. Krogh concluded: "The power of active transport of ions is of a common occurrence both in the vegetable and the animal kingdom and is possibly a general characteristic of the protoplasmic surface membrane". Conway replied the same year in a paper in Nature[8]: "Krogh … considers the apparent impermeability to sodium as due to an active extrusion, sodium ions entering the (muscle) fibers as fast, if indeed not faster, than potassium. The following may then be considered: the minimal energy required for extrusion of sodium ions from the normal frog's sartorius if sodium enters as fast as potassium". A calculation showed that the energy requirement was about twice the resting metabolism of the muscle. Ussing[9] came to the same result, but he explained what the apparent problem was. Only a part of the measured Na^+ flux is due to active transport of Na^+, the other part is due to an exchange across the membrane of Na^+ from the one side for Na^+ from the other side, a Na:Na exchange, which is energetically neutral, and which gives no net flux of Na^+ across the membrane. Taking this into account, the energy available is more than sufficient. However, Conway's concern was relevant in the sense that with the knowledge at that time about membrane function, an active transport seemed energetically an expensive way to solve the osmotic problem of the cell.

In the 1940s and first half of the 1950s the concept of active transport developed[6]. It was defined as a transport against an electrochemical gradient[10]. It was shown that the active efflux of sodium was coupled to an influx of potassium, a pump, and that the substrate for the transport was energy rich phosphate esters (for references see[11]).

But what was the nature of the pump? With the information available in the beginning of the 1950s it was possible to foresee that the pump is a membrane bound protein with enzymatic activity, which has ATP as substrate and is activated by Na^+ on the cytoplasmic side, and by K^+ on the extracellular side. But nobody apparently thought that way. A reason may be that the membrane according to the model by Danielli and Davson is a bilayer of lipids with no room for proteins inside the bilayer spanning the bilayer. On the contrary it was assumed that protein in the bilayer would destabilize the cell membrane. It was assumed that the protein was in the interphase between the lipids and the water on the two sides of the membrane.

THE WAY TO THE SODIUM PUMP[12–13].

My scientific interest was the mechanism of action of local anaesthetics. I held a position at the Institute of Physiology at the University of Århus, and was using this problem as subject for a thesis. In 1953, I had finished a series of experiments on the problem, which was published in 1954 in book form in Danish, and accepted by the faculty to be defended for the medical doctor's degree[14]. It was also published as 6 papers in English[15].

I had received my medical degree in the summer 1944, and started my internship at a hospital in Hjørring in the northern part of the country, six months in the medical ward followed by six months in the surgical ward. I became interested in surgery, and after my internship I continued for another year in the surgical ward. We had no anaesthetists, and to avoid the unpleasant ether narcosis, we used whenever possible spinal and local anaesthesia. From the teaching of pharmacology I knew the Meyer-Overton theory[16–17], that there is a correlation between solubility of general anaesthetics in lipids, and the anaesthetic potency. General anaesthetics are non-polar substances, while local anaesthetics are weak bases, which at the physiological pH exist as a mixture of charged and uncharged molecules. I wondered which of the two components is the anaesthetic component, and whether a correlation, similar to that for the general anaesthetics, existed for the local anaesthetics. I decided to use this problem as a subject for a thesis. After the two years at the hospital in Hjørring, I took up a position for a year at the Orthopaedic Hospital in Aarhus, and after this, in 1947 I applied for and received a position at the Institute of Medical Physiology at Aarhus University.

Aarhus University was young, founded 19 years earlier. There were Institutes for Anatomy, Biochemistry and Physiology as a beginning of a Medical Faculty, which was not completed until 1957. These were the only biological Institutes at the Campus, and the scientific biology milieu was poor, with little or no contact with the outside scientific world. We were three young doctors besides the Professor, Søren L. Ørskov, in the department each working on our thesis. None of us had any scientific background, but Ørskov was very helpful, patient and let us take the necessary time. We had an intake of 140 medical students a year so the teaching load was heavy. After having passed physiology the students had to continue at the medical faculty in Copenhagen for their medical degree.

I used the intact sciatic nerve of frog legs as a test object for measuring the blocking potency of five different local anaesthetics, which are weak bases, and of butanol as a representative of a nonpolar blocking agent. After removing the sheath around the nerve to get easier access to the single nerve fibers, I measured the blocking concentration as a function of time, and from this the minimum blocking concentration at infinite time of exposure could be determined. The concentrations necessary varied from the weakest to the strongest of the local anaesthetics by a factor of 1:920 (with butanol included, 1:13.500).

The order of anaesthetic potency and solubility in lipids were the same, but the quantitative correlation was poor, i.e. local anaesthetics did not follow the Meyer-Overton rule for general anaesthetics.

I was looking for another test object. As the cell membrane is a bilayer of lipids, I decided to use a monomolecular layer of lipids on a Krebs-Ringer water phase as a model for a water cell membrane interphase. The inspiration came from reading about Langmuir's work on monomolecular layers of lipids on a water phase in "The Physics and Chemistry of Surfaces" by N.K. Adam, and that Schulmann had applied capillary active drugs in the waterphase beneath the monolayer, and observed that they penetrated up into the monolayer[18].

In a Langmuir trough the area of the monolayer can be measured as a function of the pressure that the monolayer exerts on a floating barrier, which separates the monolayer from the pure water phase without the monolayer. My first experiments were with a monolayer of stearic acid. At a given area per molecule, which also means at a given surface pressure, application of the local anaesthetics to the water phase gave an increase in pressure, indicating that the local anaesthetics penetrated up into the monolayer, and the pressure increased with the concentration. There was a certain correlation between anaesthetic potency and pressure increase, but quantitatively not clear enough. However, with a monolayer of lipids extracted from the sciatic frog nerves there was a reasonably good correlation. The order of the concentrations necessary to increase the pressure followed the order of anaesthetic potency. And the minimum blocking concentration of the five local anaesthetics, which as mentioned varied by a factor of 1:920, gave a pressure increase in the monolayer at a certain area which was of the same order, they varied by a factor of 1:3.2. Also the effect of a change in pH on the local anaesthetic potency correlated reasonable well to the effect of pH on the pressure increase at a given area.

The rising phase of the nerve impulse, the depolarization is due to a transient increase in permeability of the membrane for Na^{+}[19]. The molecular basis for this was unknown, but it seemed unlikely that it was connected to the lipids. I assumed that the permeability increase was on proteins in the membrane. The monolayer results suggested to me, that the effect of the penetration of local anaesthetics into the lipid part of the nerve membrane was a blocking of the conformational change in proteins in the membrane, which gave the increase in permeablity to Na^+.

To test this I wanted to see if pressure in a monolayer could influence the enzymatic activity of a protein in the monolayer, and take this as indication of an effect of pressure on conformation. And if there was an effect, then form a monolayer of a mixture of lipids and the enzyme, and test if penetration of local anaesthetics into the monolayer had an effect on the enzymatic activity. For this I needed an enzyme with high activity, which was related to membrane function. A candidate was acetylcholinesterase, which was then being prepared from electric eel by Professor Nachmansohn at Columbia University in New York. It had the further advantage that it involved a visit to New York.

The Professor of Physiology in Copenhagen Einar Lundsgaard was a close friend of David Nachmansohn and introduced me. I had planned to spend August in New York, take a break at the end of August and beginning of September to attend the 19th International Congress of Physiology in Montreal, and then return to New York in September*. This would fit with my teaching schedule. Nachmansohn would not be in New York until September, as he spent the summer at the Marine Biological Station in Woods Hole, he therefore suggested I should join him there in August. In September he would return to New York. I agreed, although I did not know what to do in Woods Hole; there was no access to electric eel.

Scientists interested in the function of the nervous system came from all over the world to Woods Hole during the summer, to use the giant axons from squids as test objects. Coming from a young University with a poor scientific milieu, this was like coming to another planet. The place was bubbling with scientific activity. I realized that science is a serious affair and not just a

* Not July and August as mentioned in references 12 and 13.

temporary hobby for young doctors writing a thesis in order to qualify for a clinical career. And also that it is competitive. I listened to lectures, met people whose names I knew from the textbooks, and from the literature, spent time in the laboratories looking on, and learning from the experiments.

In between I did some reading, and in a paper written by Nachmansohn, it was mentioned that B. Libet[20] in 1948 had shown that there is an ATP hydrolysing enzyme in the sheath part of the giant axon from squid: an ATPase. As ATP is the energy source in cells I wondered what the function could be of an ATPase in the membrane of a nerve. Situated in the membrane I assumed that it was a lipoprotein, and this was what I needed for the monolayer experiments. I decided to look for the enzyme when I came home.

I prepared acetylcholinesterase at Columbia University in September from the electric eel. Back in Aarhus I continued the monolayer experiments.

I had no access to giant axons in Aarhus, but decided to look for the putative nerve membrane ATPase in crab nerves, because the crab nerve, like the giant axon, has no myelin sheath. In 1954 I arranged with a fisherman south of Aarhus to send me some crabs, and started to isolate the sciatic nerve from the legs. The nerves were homogenized and the membrane pieces isolated by a differential centrifugation.

The experiments showed that the membrane fractions had a low magnesium (Mg^{2+}) activated ATP hydrolysing enzyme activity. Addition of Na^+ besides Mg^{2+} gave a slight increase in activity. K^+ had no effect in the presence of Mg^{2+}, Fig. 1. However the activity varied from experiment to experiment. Calcium (Ca^{2+}) was excluded as the reason for the variations. After having spent November–December trying to find a solution I gave up and went on my Christmas holiday. I resumed the experiments in June the following year, but still without being able to get reproducible results, went on summer holidays. Returning in August, I made a Na^+ salt of ATP and a K^+ salt and found to my surprise that the activity with the K^+ salt was higher than with the Na^+ salt. This could not be due to a difference in ATP but to an effect of K^+ which differed from that of Na^+. But why in this experiment and not in the previous experiments, where K^+ had no effects in the presence of Mg^{2+}. The answer was that in the experiments with the K^+ salt of ATP there was Na^+ in the medium. In other words the enzyme needed a combined effect of Na^+ and K^+ for activation. I then started a systematic investigation of the combined effect of the two cations.

As seen from Fig. 1, K^+ has two effects in the presence of Na^+. It activates, the higher the concentration is of Na^+. The K^+ affinity for the activating effect is high. At higher concentrations the activating effect of K^+ decreases, and the apparent affinity for K^+ for this effect decreases with an increase in the Na^+ concentration. With 3 mM Na^+ it is seen that K^+ not only inhibits its own activation, but also the small activation due to Na^+. The results suggest that there are two sites on the enzyme, one where Na^+ is necessary for activation, and another where K^+ activates when Na^+ is bound to the former. K^+ in higher concentrations competes for Na^+ at the Na^+ site, and by displacing Na^+ from the site decreases the activity.

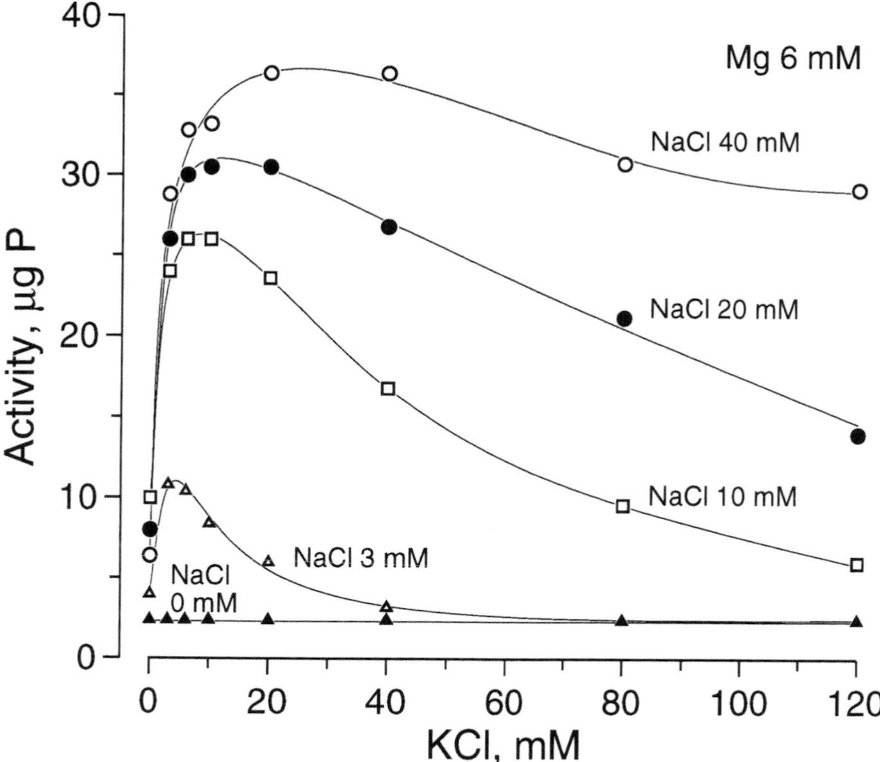

Figure 1. Effect of Na^+, and of K^+ in the presence of Mg^{2+} on the hydrolysis of ATP by membrane fractions isolated from nerves of the shore crab *Carcinus maenas*. Test solution contained 30 mM histidine HCL as buffer, pH 7.2, 3 mM ATP, 6 mM Mg^{2+}, Na^+ and K^+ in concentrations shown on the Figure. Activity is given as µg P (inorganic phosphate) hydrolysed from ATP in 30 min. at 36 °C (Reprinted by permission from.[27])

I now understood the reason for the varying results. With little effect of Na^+ on the activity, and of no effect of K^+, I had not bothered whether or not there was Na^+ or K^+ in the medium. I got ATP as an insoluble barium salt, which was converted to a soluble Na^+ or K^+ salt, sometimes the one sometimes the other, and in between instead of homogenising the nerves in glucose I used a 0.58 M KCl solution. It never occurred to me that there could be a combined effect of Na^+ and of K^+.

The problem then was, what was the physiological function of the enzyme? I was interested in the effect of local anaesthetics on nerve conduction, and my first reaction was that this was the Na^+ channel, which opens for the influx of Na^+ leading to the nerve impulse. I soon rejected the idea because the opening of the channel is voltage dependent, and not dependent on ATP. The other possibility was that it was part of or the sodium pump. I had little knowledge about the active transport of Na^+ and of K^+, I therefore started to look into the literature to see what the substrate was for the active transport in nerves. I had limited access to the literature, so there were few papers I read about active transport. The closest I could come was that A.L. Hodgkin and

R.D. Keynes[21] had shown that poisoning giant axons with dinitrophenol, cyanide or azide, decreased the active transport of sodium, suggesting that high energy phosphate esters are the substrate. And as ATP is a high energy phosphate ester, I thought it likely that it could be the substrate.

There were two papers I did not read, one was by G. Gardos[22] from Budapest published in 1954 in a Hungarian journal, which I did not have access to, and which was not cited in the paper by Hodgkin and Keynes. In this Gardos showed that ATP supported the active uptake of K^+ in red blood cells. The other was a paper published in 1956 by Hodgkin and Keynes[23], in which they reported that injecting ATP in a cyanide poisoned giant axon gave no dramatic recovery of the active extrusion of sodium. Fortunately, I did not see this last paper until I had sent my paper on the crab nerve enzyme for publication. In 1957, the experiment was repeated by Caldwell and Keynes[24], with the result that ATP was the substrate for the active transport.

In a discussion after a paper by R.D. Keynes on "Electrolytes and Nerve Activity"[25] at an international symposium on Neurochemistry in Aarhus in 1956, I showed the results with the crab nerve ATPase[26]. The same year I wrote the paper, and suggested from the characteristics of the effect of the cations, and that fact that ATP was a substrate, that the enzyme was involved in the active transport of Na^+ across the cell membrane. I considered putting the word pump in the title, but found it too provocative, so it became "The Influence of Some Cations on an Adenosine Triphosphatase from Peripheral Nerves". No wonder that few people noticed that this enzyme had to do with active transport of Na^+. It was published in 1957[27].

With my little knowledge about active transport, I was unaware of the importance of the observation on the crab nerves. Parallel with the crab nerve experiments I continued the experiments on the monolayer with the acetylcholinesterase. In 1958, I presented a paper at the 4th International Congress of Biochemistry in Vienna on "The Influence of the Degree of Unfolding and the Orientation of the Side Chains on the Activity of a Surface Spread Enzyme".

There was one important experiment I had not done. I realized this when I met Robert Post at the conference. I knew Robert from Woods Hole. We had spent time in the same laboratory, and I had driven with him and his wife Elisabeth in their car from Woods Hole to the International Congress for Physiology in Montreal. He told me that he had since worked with active transport of Na^+ and K^+ in red blood cells, and had shown that the stoichiometry between the Na^+ transport out of, and of K^+ into, the cell was 3 to 2[28].

I told him about the $Na^+ + K^+$ activated crab nerve enzyme, and that it seemed to be part of or the sodium pump. His reaction suggested to me that this was more important than surface spread enzymes. "Is it inhibited by ouabain"? he asked. "What is ouabain" was my reply. He then told me that Schatzmann in Switzerland in 1954 had shown that cardiac glycosides, of which ouabain is the most water soluble, specifically inhibits the active transport in red blood cells[29]. When Robert Post came to Aarhus after the conference I had the answer. The enzyme was inhibited by Ouabain, even if the sensitivity of the crab nerve enzyme is much lower than the sensitivity of the transport in red blood cells. It convinced Robert that the enzyme had to do with active transport. I had learned that red blood cells were a classical test object for experiments on active transport, and had started to look for the enzyme in these cells. Robert asked if he could go on with these experiments when he returned to U.S.. I had no experience with this test object, and as he had the experience I agreed, and continued the experiments with the crab nerve enzyme[30], and looked for the enzyme in other tissues[31].

In 1959, after I had given my first paper on the crab nerve enzyme at the 21th International Congress for Physiology in Buenos Aires, Professor Hodgkin, who was the great name in neurophysiology and came from the famous Cambridge University, invited me for lunch to hear more about the enzyme. His interest suggested to me that the observation was of a certain importance.

Looking back, it was a very simple experiment to identify the pump. Just break the membrane and by this gain access to the Na^+ site on the inside and the K^+ site on the outside, add some ATP and test for the combined effect of Na^+ and K^+. It ought to have been done by someone who worked in the transport field and knew about active transport. I felt like an intruder in a field that was not mine.

There was however, much luck involved. First, that Nachmansohn had invited me to Woods Hole, where I learned about the giant axons and read about the observation by Libet. Next, that from the monolayer experiments I became interested in membrane proteins, especially lipoproteins, and therefore took notice of Libet's observation. Finally, that I chose crab nerves as a test object. I learned later, that after homogenization of most other tissues in order to break the membrane, the membrane pieces form vesicles, which must be opened by treatment with detergent, in order to get access to both sides of the membrane, and thereby to see the combined effect of Na^+ and K^+. This is not the case with the crab nerve membranes.

From then on the Na^+,K^+-ATPase took me away from the monolayer experiments. I never did the planned experiments on the effect of the local anaesthetics on a monolayer of a mixture of lipids and proteins; and I never used the Na^+,K^+-ATPase for monolayer experiments, first of all because it was not until 1980 that it became possible to extract the enzyme from a membrane in a pure stable water-soluble form, and secondly that was only possible with the use of detergent[32]. A problem would have been the detergent, but I had lost interest in monolayers.

In 1960, Robert Post published the paper on the red blood cell experiments, "Membrane adenosine triphosphatase as a participant in the active transport of sodium and potassium in the human erythrocytes" [33]. In this he convincingly showed that in the red blood cells there is a Na^++K^+ activated ATPase also, and that the effect of the cations on the activity correlated with the effect on transport. Robert Post was known in the transport field, and his paper had a better title than mine, so it attracted more attention.

In the following few years many papers were published, which showed that the enzyme could be found in many different tissues, and evidence was given for its involvement in active transport (for references see[34]).

In 1965, so much evidence was at hand, that in a review paper I could conclude that the enzyme system fulfilled the following requirements for a system responsible for the active transport across the cell membrane: 1) it is located in the cell membrane; 2) on the cytoplasmic side, it has a higher affinity for Na^+ than for K^+; 3) it has an affinity for K^+ on the extracellular side, which is higher than for Na^+; 4) it has enzymatic activity and catalyzes ATP hydrolysis; 5) the rate of ATP hydrolysis depends on cytoplasmic Na^+ as well as on extracellular K^+; 6) it is found in all cells that have coupled active transport of Na^+ and K^+; 7) the effect of Na^+ and of K^+ on transport in intact cells, and on the activity of the isolated enzyme, correlates quantitatively; and 8) the enzyme is inhibited by cardiac glycosides, and the inhibitory effect on the active fluxes of the cations correlates with the inhibitory effect on the isolated enzyme system[34].

The enzyme was named the Na^+- and K^+- activated ATPase, or Na^+, K^+-ATPase.

THE Na^+,K^+-ATPase AS AN ENERGY TRANSDUCER

Returning to Conway's concern about the waste of energy. Conway was right in the sense that even if the cell membrane is permeable to sodium, and therefore an energy requiring pump is needed, the ground permeability to sodium of the cell membrane is low, which is necessary in order that the maintenance of the gradient for sodium does not become energetically too costly for the cell. But what is the meaning of having a membrane permeable to sodium and then spend energy, 10–60% of the cell metabolism, to keep Na^+ out of the cell and K^+ within the cell?

1) As mentioned above it solves the osmotic problem due to the presence of impermeable protein anions in the cytoplasm.

But besides this, the gradients for Na^+ into and K^+ out of the cell sustained by the pump represent an energy source, which is used

2) for the creation of a membrane potential. The cell membrane is more permeable to K^+ than to Na^+, which means that K^+ flows out of the cell faster than Na^+ into the cell. This leads to a diffusion potential across the membrane, negative on the inside, which slows the K^+ outflux and increases the Na^+ influx until the potential reaches a value of about –70 mV, at which point the rates of the two fluxes are equal; a steady state situation. The membrane potential is the basis for the function of all excitable tissue. The nerve impulse is a depolarization of the membrane potential due to a transient increase in permeability to Na^+, with influx of Na^+ followed by a repolarization due to an outflux of K^+[35]. This leads to an increase in intracellular Na^+ and a decrease in K^+, which subsequently must be compensated for by the pump.

3) for transport of other substances in and out of the cell, see Fig. 2. In the cell membrane there are a number of protein molecules, which act as co- or countertransporters, or sym- and antiporters[36]. There are cotransporters which use the gradient for Na^+ into the cell to transport glucose[37] or amino acids[38] into cells to a higher concentration inside than outside, a Na^+/Cl^-, a

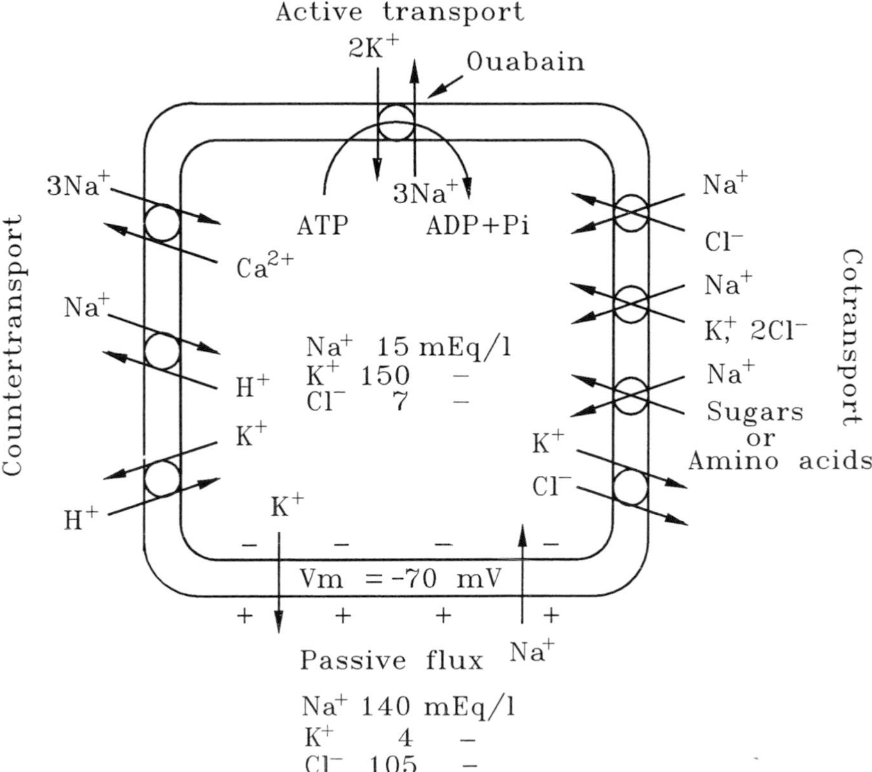

Figure 2. Transport across the cell membrane. For explanation see text. (Reproduced by permission from[73]).

$Na^+/K^+/2Cl^-$ and a K^+/Cl^- cotransporter, which may be involved in volume regulation (for references see[39]). A $3Na^+/Ca^{2+}$ countertransporter of importance for regulation of the intracellular Ca^{2+} concentration[40], and a Na^+/H^+ countertransport for regulation of internal pH. In addition, there are other examples.

4) for transepithelial transport in intestine, kidney and secretory glands.

The Na^+, K^+-pump thus acts as an energy transducer, that converts the chemical energy from the hydrolysis of ATP into another form of energy, a concentration gradient which is used for exchange of substances across the cell membrane. This is named a secondary active transport, while the transport, which is directly dependent on the energy from a chemical reaction, the active Na^+, K^+ transport, is named a primary active transport. The Na^+, K^+-ATPase has thus a key function in the exchange of substances across the cell membrane.

A problem was where to place the Na,K-ATPase in the membrane. The transport system, which is a protein, must have access to both sides of the membrane. As mentioned previously there is no room for proteins in the bilayer of the Danielli-Davson membrane model. Neither is there in Robertson's unit membrane model from 1959[41], which replaced it. The unit membrane was a bilayer of lipids with proteins arranged asymmetrically on the two sides of the membrane, but still with no proteins spanning the membrane. It was not until 1972 that a suitable model was introduced by Singer and Nicolson[42]. In their fluid mosaic membrane model there are globular proteins embedded with their non-polar parts in the bilayer, and with the polar parts facing the two sides of the bilayer. They can move laterally in the membrane, but do not flip-flop. They form pathways for the transport of hydrophilic substances across the membrane.

Other mammalian ion transporting pumps were identified in the following years. A Ca^{2+}-ATPase in sarcoplasmic reticulum in muscle responsible for the transport of Ca^{2+} out of the muscle fiber[43], a sarcolemma Ca^{2+}-ATPase isolated from red blood cells which transports Ca^{2+} out of the cell[44], a H^+,K^+-ATPase in the stomach, which transports H^+ out of the cells in exchange for K^+, producing the stomach acid[45]. They all have in common with the Na^+,K^+-ATPase that the reaction with ATP involves a phosphorylation, and they are therefore named P-type ATPases. A number of other P-type ATPases have been identified in bacteria and fungi (see[46]).

In the years that followed the identification of the sodium-potassium pump, many scientists from many countries took part in the elucidation of the structure of the system, and of the reaction steps in the transport process. This cannot be covered in the present lecture, but for those interested I shall refer to an extensive review by I. Glynn 1985[11], to the Proceedings from the International Conferences on the Na^+,K^+-ATPase held every third year, of which the latest is from 1996[47], to recent reviews[48–50], and to the recently published book by J. D. Robinson: "Moving Questions. A History of Membrane Transport and Bioenergetics"[46].

There is, however, one question I would like shortly to touch, without going into details, namely this: what is our present view on the way the system transports the cations?

A MODEL FOR THE TRANSPORT REACTION (see Figure 3).

The model in Fig. 3 is based on the so called Albers-Post scheme, a reaction scheme in which the system reacts consecutively with the cations, and in which the reaction with ATP in the presence of Na^+ leads to a phosphorylation, and the following reaction with K^+ to a dephosphorylation[51–56].

The transport system consists of a carrier part located in the innermost part of the membrane and with the ATP binding part on the cytoplasmic side of the membrane (see Fig. 3). The carrier is in series with a narrow channel which spans half to two thirds of the membrane and opens to the extracellular side of the membrane[57–58] (in Fig. 3 the channel part is shown only as the opening to the outside). The carrier exists in two major different conformations[51–56, 59–65], E_1 which is the high affinity sodium form, and E_2, which is the high affinity potassium form. Each of the two conformations can exist in a phosphorylated[51–56,] and a non-phosphorylated form[59–65], and each of the two major conformations has subconformations marked with primes in the

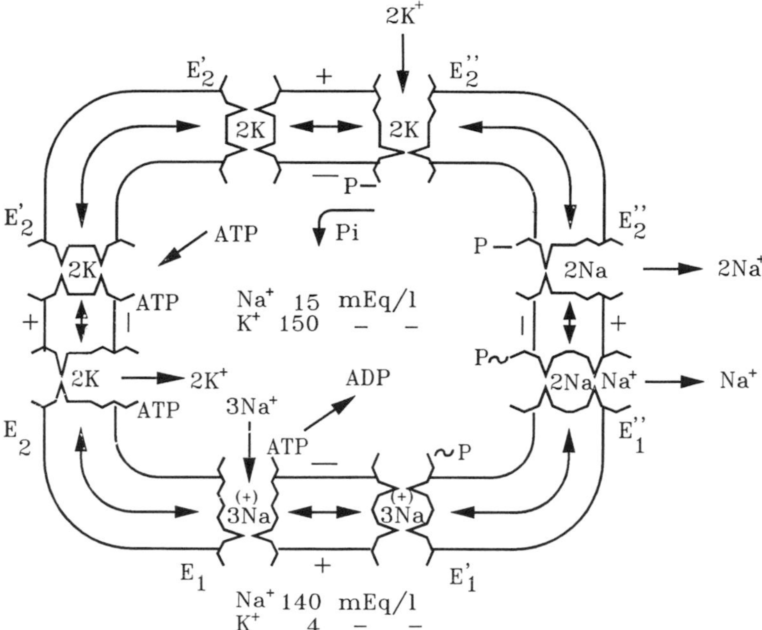

Figure 3. The active transport reaction. For explanation see text. (Reproduced by permission from[73]).

model. The carrier part has two gates, energy barriers, which can open and close access to the channel side and to the cytoplasmic side, respectively.

Referring to the model, with the system in the E_1 conformation with the gate to the cytoplasmic side open but closed to the channel side, 3 Na^+ are bound, two of them as a neutral complex and the third one as an ion. The system is catalytically active. The affinity for ATP is high (apparent K_D is about 0.1 μM[59–60]), and with ATP bound the enzyme is phosphorylated[51–56, 66–67], to an ADP sensitive form, which means that a high energy phosphate bond is formed, $E'_1 \sim P$, wich closes the gate to the cytoplasmic side. The 3 Na^+ are occluded inside the carrier part[68]. The following two steps, involve conformational changes and rearrangements of the phosphate bond from a high energy to a low energy bond, $E_1' \sim P$ to E_2''-P[51–56, 66–67], and with deocclusion and release of one Na^+ from $E_1'' \sim P$[67] to the extracellular side of the membrane. The transition to the E_2''-P form deoccludes, opens the gate for the remaining 2 Na^+, and the affinity is shifted to a low Na^+ high K^+ affinity. By this, the 2 Na^+ are exchanged for 2 K^+ from the extracellular medium, E_2''-PNa_2 to E_2''-PK_2. With 2 K^+ bound the system is dephosphorylated, the extracellular gate closes and the 2 K^+ are occluded, $E_2'(K_2)$[69–70]. ATP bound to $E_2'(K_2)$ on the cytoplasmic side, with an affinity which is low (apparent K_D is about 0.1 mM), increases the rate by which the cytoplasmic gate opens and decreases the affinity for K^+, $E_2'(K_2)$ATP to E_2K_2ATP[69]. The 2 K^+ are exchanged for 3 Na^+ from the cytoplasmic medium, and this closes the cycle.

The transport system is a very efficient pump. It pumps 3 Na^+ out of and 2

K^+ into the cell for each ATP hydrolysed to ADP and Pi, and for this it uses 70%–85% of the free energy of the hydrolysis of ATP (see[50]). With the normal intra- and extracellular concentrations of Na^+ and of K^+ , the activity of the pump is 10–15% of maximum, i.e. the pump has a considerable reserve power. At 37°C, pH 7.4 the enzyme turns over at a rate of about 160 per sec.

In the transport reaction one net positive charge is carried to the outside. The transport is therefore electrogenic, but due to a high permeability of the cell membrane to Cl^-, it only adds a few mV to the membrane potential. Each of the two conformations E_1 and E_2 seem to have 2 negative charged binding sites on which the positive charge of the cations are neutralized, in addition the E_1 conformation mentioned above binds a third Na^+ in ionic form[71–72].

In order to obtain transport of the cations against an electrochemical gradient coupled to a chemical reaction, the hydrolysis of ATP, it is necessary that there is a tight coupling between each of the steps in the catalytic reaction, and each of the corresponding steps in the translocation reaction, as suggested in the model. The shift in affinities, and the opening and closing of the gates, are coupled to the stepwise degradation of ATP from phosphorylation with a high energy bond, to a low energy bond, and finally to dephosphorylation. This divides the translocation into steps in which local gradients are created along which the ions can flow, and thereby they are transported against their overall gradient.

The model is a working hypothesis, which explain a good deal of the experimental observations. There are however observations which do not fit with the model. And even though we have much information about the structure, there are many unanswered questions, especially at the molecular level. What is the nature of the gates, where are the binding sites for the cations, and of what nature, and how does the system discriminate between Na^+ and K^+, etc.

CONCLUSION

It may seem disappointing that 30 years of work, since the conclusion that the membrane bound Na^+,K^+-ATPase is identical with the Na^+,K^+-pump, has not given us an understanding of the basic molecular events behind the transport. However, considering that the problem is to reveal how 1320 amino acids inside a volume of $60 \times 60 \times 100$ Å^3 can be assembled to a very efficient machine, which can convert the chemical energy from the hydrolysis of ATP into work, namely the transport of cations against their electrochemical gradient, and which can distinguish between so closely related cations as Na^+ and K^+, it can be of no surprise that progress is slow.

Thirty years ago it seemed impossible to find a way to purify the enzyme not to speak of getting it into solution and reconstitute it into liposomes. Nobody dared to dream about knowing the sequence, and yet this and much much more have been accomplished.

REFERENCES

1. Gorter, E., and F. Grendel, *J. Exp. Med.*, **41**, 439 (1925).
2. Danielli, J. F., and H. Davson, *J. Cell. Comp. Physiol.*, **5**, 495 (1935).
3. Boyle, P. J., and E. F. Conway, *J. Physiol.*, **100**, 1 (1941).
4. Dean, R. B., *Biological Symposia*, **3**, 331 (1941)
5. Hevesy, G., *Enzymologia*, **5**, 138 (1938)
6. Ussing, H. H., *Physiol. Rev.*, **29**, 129 (1949)
7. Krogh, A., *Proc. Roy. Soc. London B.*, **133**, 140 (1946).
8. Conway, E. J., *Nature*, **157**, 715 (1946).
9. Ussing, H. H., *Nature*, **160**, 262 (1947).
10. Rosenberg, T., *Acta Chem. Scand.*, **2**, 14 (1948).
11. Glynn, I. M., in: *The Enzymes of Biological Membranes*, edited by A. N. Martonosi, New York: Plenum, vol. **3**, 28 (1985).
12. Skou, J. C., *Biochim. Biophys. Acta*, **1000**, 435 (1989)
13. Skou, J. C., in: *Membrane Transport: People and Ideas* (ed. D. Tosteson), American Physiol. Soc. p. 155 (1989).
14. Skou, J. C., *Lokalanestetika.* Thesis, Universitetsforlaget Aarhus, (1954).
15. Skou, J. C., *J. Pharm. Pharmacol.*, **13**, 204 (1961).
16. Meyer, H. H., *Arch. Exp. Path. Pharmak.*, **42**, 109 (1899).
17. Overton, E., *Studien über die Narkose zugleich ein Beitrag zur allgemeinen Pharmakologie*, Jena: Fisher (1901).
18. Adam, N. K., *The Physics and Chemistry of Surfaces*, London, Oxford Univ.Press (1941).
19. Hodgkin, A. L., and B. Katz, *J. Physiol.*, **108**, 37 (1949).
20. Libet, B., *Federation Proc.*, **7**, 72 (1948).
21. Hodgkin, A. L., and R. D. Keynes, *J. Physiol. Lond.*, **128**, 28 (1955).
22. Gardos, G., *Acta Physiol. Scient. Hung.*, **6**, 191 (1954).
23. Hodgkin, A. L., and R. D. Keynes, *J. Physiol. Lond.*, **131**, 592 (1956).
24. Caldwell, P. C. and R. D. Keynes, *J. Physiol. Lond.*, **137**, 12P (1957).
25. Keynes, R. D., in: *Metabolism of the Nervous System*, (ed. D. Richter). Pergamon, London, p. 159 (1957).
26. Skou, J. C., in: *Metabolism of the Nervous System*, (ed. D. Richter). Pergamon, London, p. 173 (1957).
27. Skou, J. C., *Biochim. Biophys. Acta*, **23**, 394 (1957).
28. Post, R. L., and P. Jolly, *Biochim. Biophys. Acta*, **25**, 118 (1957).
29. Schatzmann, H. J., *Helv. Physiol. Pharmacol. Acta*, **11**, 346 (1953).
30. Skou, J. C., *Biochim. Biophys. Acta*, **42**, 6 (1960).
31. Skou, J. C., *Biochim. Biophys. Acta*, **58**, 314 (1962).
32. Esmann, M., J. C. Skou, and C. Christiansen, *Biochim. Biophys. Acta*, 567, 410 (1979)
33. Post, R. L., C. R. Merritt, C. R. Kinsolving, and C. D. Albright, *J. Biol. Chem.*, **235**, 1796 (1960).
34. Skou, J. C., *Physiol. Rev.*, **45**, 596 (1965).
35. Hodgkin, A. L., and A. F. Huxley. *J. Physiol.*, **116**, 449 (1952)
36. Mitchell, P., *Biochem. Soc. Symp.*, **22**, 142 (1963).
37. Crane, R. K., *Comp. Biochem.*, **35**, 43 (1983).
38. Kromphardt, H., H. Grobecker, K. Ring, and E. Heinz, *Biochim. Biophys. Acta*, **74**, 549 (1963).
39. Hoffmann, E. K., and L. O. Simonsen, *Physiol. Rev.*, 69, 315 (1989)
40. Pitts, B. J. R., *J. Biol. Chem.*, **254**, 6232 (1979).
41. Robertson, J. D., *Biochem. Soc. Symp.*, **16**, 3 (1959).
42. Singer, S. J., and G. L. Nicolson, *Science*, **175**, 720 (1972).
43. Hasselbach, W., and M. Makinose, *Biochem. Z.*, **333**, 518 (1961).
44. Schatzmann, H. J., *Experientia*, **22**, 364 (1966).
45. Stewart, B., B. Wallmark, and G. Sachs, *J. Biol. Chem.*, **256**, 2682 (1981)

46. Robinson, J. D., *Moving Questions. A History of Membrane Transport and Bioenergetics*, American Physiological Soc., N.Y. (1997) .
47. *Na/K-ATPase and Related Transport ATPases: Structure, Mechanism and Regulation* (eds. Beaugè, L. A., D. C. Gadsby, and P. J. Garrahan), Annals New York Academy Sci. vol. 834, New York (1997).
48. Jørgensen, P. L., in: *Molecular Aspects of Transport Proteins* (ed. De Pont J. J. H. M. M.), Elsevier Service Publisher B. V., p.1 (1992).
49. Glynn, I. M., *J. Physiol.*, **462**, 1 (1993).
50. Cornelius, F., *Biomembranes*, Vol. **5**, 133 (1996).
51. Albers, R. W., *Annu. Rev. Biochem.*, **36**, 727 (1967).
52. Albers, R. W., S. Fahn, and G. J. Koval, *Proc. Natl. Acad. Sci. USA*, **50**, 474 (1963).
53. Fahn, S., G. J. Koval, and R. W. Albers, *J. Biol. Chem.*, **243**, 1993 (1968).
54. Post, R. L., and S. Kume, *J. Biol. Chem.*, **248**, 6993 (1973).
55. Post, R. L., S. Kume, T. Tobin, B. Orcutt, and A. K. Sen, *J. Gen. Physiol.*, **54**, 306S (1969).
56. Post, R. L., A. K. Sen, and A. S. Rosenthal, *J. Biol. Chem.*, **240**, 1437 (1965).
57. Läuger, P., *Biochim. Biophys. Acta*, **552**, 143 (1979).
58. Gadsby, D. C., R. F. Rakowsky, and P. De Weer, *Science*, **260**, 100 (1993).
59. Nørby, J. G., and J. Jensen, *Biochim. Biophys. Acta*, **233**, 104 (1971).
60. Hegyvary, C., and R. L. Post, *J. Biol. Chem.*, **246**, 5234 (1971).
61. Jørgensen, P. L., *Biochim. Biophys. Acta*, **401**, 399 (1975).
62. Karlish, S. J. D., and D. W. Yates, *Biochim. Biophys. Acta*, **527**, 115 (1978).
63. Karlish, S. J. D., D. W. Yates, and I. M. Glynn, *Biochim. Biophys. Acta*, **525**, 252 (1978).
64. Skou, J. C., and M. Esmann, *Biochim. Biophys. Acta*, **647**, 232 (1981).
65. Karlish, S. J. D., *J. Bioenerg. Biomembr,*. **12**, 111 (1980).
66. Nørby, J. G., I. Klodos, and N. O. Christiansen, *J. Gen. Physiol.*, **82**, 725 (1983).
67. Yoda, S., and A. Yoda, *J. Biol. Chem.*, **26**, 1147 (1986).
68. Glynn, I. M., Y. Hara, and D. E. Richards, *J. Physiol.*, **351**, 531 (1984).
69. Post, R. L., C. Hergyvary, and S. Kume, *J. Biol. Chem.*, **247**, 6530 (1972).
70. Beaugé, L. A., and I. M. Glynn, *Nature*, **280**, 510 (1979).
71. Nakao, M., and D. C. Gadsby, *Nature*, **323**, 628 (1986).
72. Goldschleger, R., S. J. D. Karlish, A. Raphaeli, and W. D. Stein, *J. Physiol.* **387**, 331 (1987).
73. Skou, J. C., *News in Physiological Sciences*, **7**, 95 (1992).

Erratum

Page 182, line 14 should read as:

little or no contact with the outside scientific world. We were three young medical

Chemistry 1998

WALTER KOHN

"for his development of the density-functional theory"

and

JOHN A. POPLE

"for his development of computational methods in quantum chemistry"

THE NOBEL PRIZE IN CHEMISTRY

Speech by Professor Björn Roos of the Royal Swedish Academy of Sciences.
Translation of the Swedish text.

Your Majesties, Your Royal Highness, Ladies and Gentlemen,

Man is fantastic. Through his studies of nature he has brought order to chaos. He has created a language, mathematics, which makes it possible to formulate his knowledge about nature in a small number of simple sentences. Not only do these sentences summarize in a concentrated manner our knowledge about nature and matter, they can also be used to make predictions. With the aid of computer simulations, we can make weather forecasts, calculate the structural integrity of bridges, the aerodynamical characteristics of airplanes, etc. Today, we celebrate the fact that mathematics has invaded chemistry, that by means of theoretical calculations we can predict a large variety of chemical phenomena. Professors Walter Kohn and John Pople have individually made fundamental contributions to this development.

An atom consists of a nucleus and electrons. The motion of the electrons is described by the laws of quantum mechanics. When these laws were formulated more than 70 years ago, researchers realized immediately that in them was contained the explanation of the chemical bond. It was realized that if the quantum mechanical equations could be solved, one would be able to explain how atoms are bound together to form molecules. It would be possible to explain why molecules look as they do, what their properties are, and how they react with each other to form new molecules. A theoretical description of all of chemistry was within reach.

To achieve all this was, however, not easy. The equations are complicated and in the beginning it was only possible to solve them for the simplest cases. The science of applying quantum mechanics to chemical phenomena therefore developed only slowly. It was only in the beginning of the 1960s, when scientists could start using computers, that this development started to make rapid progress. John Pople was one of the scientists who understood at an early stage the potential that computers provided. He realized that if quantum chemistry was going to become important in chemistry, one had to develop methods that were effective and which could be used to compute interesting properties like the structure of molecules and binding energies for the chemical bonds. They also had to be easy to use for the general chemist, who could not be expected to be knowledgeable about all the subtle details of quantum chemistry. Pople was able to fulfill these conditions through a series of crucial innovations and improvements. He designed a new tool, which could be used to study molecules and compute their properties. This tool is a computer program called GAUSSIAN. It contains a theoretical model chemistry where quantum mechanical equations are solved through a series of more and more refined approximations. Pople's methods are used today by thousands of scientists at universities and companies around the world to study a large variety of problems in chemistry and biochemistry.

The methods that John Pople has developed seek approximative solutions to quantum mechanical equations, where the fundamental quantity is the so-called wave function, which describes the motion of all electrons. In two landmark articles from 1964 and 1965, Walter Kohn showed an alternative way in which quantum mechanical equations can be approximated. He showed that there is a one-to-one correspondence between the energy of a quantum mechanical system and its electron density, which is a function of three positional coordinates only and is therefore much easier to handle than the complicated wave function, which depends on the positions of all electrons. He also developed a method, which made it possible to construct a set of equations, which could be used to determine the energy and electron density. This approach, called density functional theory, has developed during the last ten years into a versatile computational tool with many applications in chemistry. Due to its simplicity, it can be applied to larger molecules than the wave function based methods. Density functional theory has made it possible to study the mechanisms of chemical reactions in enzymes, for example when water is transformed into oxygen in photosynthesis.

Due to extraordinary circumstances, Professor Walter Kohn is not with us today. We hope to see him at the Prize Ceremony next year instead.

Professor John Pople,

I have tried to describe in a few words how Professor Kohn's and your work has led to a new revolution in chemistry. You have made fundamental contributions to the field of quantum chemistry, with the result that chemists and biochemists today have a new tool, which they can use to study chemical phenomena at a molecular level. This is an outstanding achievement.

On behalf of the Royal Swedish Academy of Sciences, I wish to convey to you our warmest congratulations and I now ask you to receive the Prize from the hands of His Majesty the King.

Walter Kohn

WALTER KOHN

I suppose I am not the first Nobelist who, on the occasion of receiving this Prize, wonders how on earth, by what strange alchemy of family background, teachers, friends, talents and especially accidents of history and of personal life he or she arrived at this point. I have browsed in previous volumes of "Les Prix Nobel" and I know that there are others whose eventual destinies were foreshadowed early in their lives – mathematical precocity, champion bird watching, insatiable reading, mechanical genius. Not in my case, at least not before my late teens. On the contrary: An early photo of my older sister and myself, taken at a children's costume party in Vienna – I look about 7 years old – shows me dressed up in a dark suit and a black top hat, toy glasses pushed down my nose, and carrying a large sign under my arm with the inscription "Professor Know-Nothing".

Here then is my attempt to convey to the reader how, at age 75, I see my life which brought me to the present point: a long-retired professor of theoretical physics at the University of California, still loving and doing physics, including chemical physics, mostly together with young people less than half my age; moderately involved in the life of my community of Santa Barbara and in broader political and social issues; with unremarkable hobbies such as listening to classical music, reading (including French literature), walking with my wife Mara or alone, a little cooking (unjustifiably proud of my ratatouille); and a weekly half hour of relaxed roller blading along the shore, a throwback to the ice-skating of my Viennese childhood. My three daughters and three grandchildren all live in California and so we get to see each other reasonably often.

I was naturalized as an American citizen in 1957 and this has been my primary self-identity ever since. But, like many other scientists, I also have a strong sense of global citizenship, including especially Canada, Denmark, England, France and Israel, where I have worked and lived with a family for considerable periods, and where I have some of my closest friends.

My feelings towards Austria, my native land, are – and will remain – very painful. They are dominated by my vivid recollections of 1 1/2 years as a Jewish boy under the Austrian Nazi regime, and by the subsequent murder of my parents, Salomon and Gittel Kohn, of other relatives and several teachers, during the holocaust. At the same time I have in recent years been glad to work with Austrians, one or two generations younger than I: Physicists, some teachers at my former High School and young people (Gedenkdiener) who face the dark years of Austria's past honestly and constructively.

On another level, I want to mention that I have a strong Jewish identity and – over the years – have been involved in several Jewish projects, such as the establishment of a strong program of Judaic Studies at the University of California in San Diego.

My father, who had lost a brother, fighting on the Austrian side in World War I, was a committed pacifist. However, while the Nazi barbarians and their collaborators threatened the entire world, I could not accept his philosophy and, after several earlier attempts, was finally accepted into the Canadian Infantry Corps during the last year of World War II. Many decades later I became active in attempts to bring an end to the US-Soviet nuclear arms race and became a leader of unsuccessful faculty initiatives to terminate the role of the University of California as manager of the nuclear weapons laboratories at Los Alamos and Livermore. I offered early support to Jeffrey Leiffer, the founder of the student Pugwash movement which concerns itself with global issues having a strong scientific component and in which scientists can play a useful role. Twenty years after its founding this organization continues strong and vibrant. My commitment to a humane and peaceful world continues to this day. I have just joined the Board of the Population Institute because I am convinced that early stabilization of the world's population is important for the attainment of this objective.

After these introductory general reflections from my present vantage point I would now like to give an idea of my childhood and adolescence. I was born in 1923 into a middle class Jewish family in Vienna, a few years after the end of World War I, which was disastrous from the Austrian point of view. Both my parents were born in parts of the former Austro-Hungarian Empire, my father in Hodonin, Moravia, my mother in Brody, then in Galicia, Poland, now in the Ukraine. Later they both moved to the capital of Vienna along with their parents. I have no recollection of my father's parents, who died relatively young. My maternal grandparents Rappaport were orthodox Jews who lived a simple life of retirement and, in the case of my grandfather, of prayer and the study of religious texts in a small nearby synagogue, a Schul as it was called. My father carried on a business, Postkartenverlag Brueder Kohn Wien I, whose main product was high quality art postcards, mostly based on paintings by contemporary artists which were commissioned by his firm. The business had flourished in the first two decades of the century but then, in part due to the death of his brother Adolf in World War I, to the dismantlement of the Austrian monarchy and to a worldwide economic depression, it gradually fell on hard times in the 1920s and 1930s. My father struggled from crisis to crisis to keep the business going and to support the family. Left over from the prosperous times was a wonderful summer property in Heringsdorf at the Baltic Sea, not far from Berlin, where my mother, sister and I spent our summer vacations until Hitler came to power in Germany in 1933. My father came for occasional visits (The firm had a branch in Berlin). My mother was a highly educated woman with a good knowledge of German, Latin, Polish and French and some acquaintance with Greek, Hebrew and English. I believe that she had completed an academically oriented High School in Galicia. Through her parents we maintained contact with traditional Judaism. At the same time my parents, especially my father, also were a part of the secular artistic and intellectual life of Vienna.

After I had completed a public elementary school, my mother enrolled me in the Akademische Gymnasium, a fine public high school in Vienna's inner city. There, for almost five years, I received an excellent education, strongly oriented toward Latin and Greek, until March 1938, when Hitler Germany annexed Austria. (This so-called Anschluss was, after a few weeks, supported by the great majority of the Austrian population). Until that time my favorite subject had been Latin, whose architecture and succinctness I loved. By contrast, I had no interest in, nor apparent talent for, mathematics which was routinely taught and gave me the only C in high school. During this time it was my tacit understanding that I would eventually be asked to take over the family business, a prospect which I faced with resignation and without the least enthusiasm.

The Anschluss changed everything: The family business was confiscated but my father was required to continue its management without any compensation; my sister managed to emigrate rather promptly to England; and I was expelled from my school.

In the following fall I was able to enter a Jewish school, the Chajes Gymnasium, where I had two extraordinary teachers: In physics, Dr. Emil Nohel, and in mathematics Dr. Victor Sabbath. While outside the school walls arbitrary acts of persecution and brutality took place, on the inside these two inspired teachers conveyed to us their own deep understanding and love of their subjects. I take this occasion to record my profound gratitude for their inspiration to which I owe my initial interest in science. (Alas, they both became victims of Nazi barbarism).

I note with deep gratitude that twice, during the Second World War, after having been separated from my parents who were unable to leave Austria, I was taken into the homes of two wonderful families who had never seen me before: Charles and Eva Hauff in Sussex, England, who also welcomed my older sister, Minna. Charles, like my father, was in art publishing and they had a business relationship. A few years later, Dr. Bruno Mendel and his wife Hertha of Toronto, Canada, took me and my friend Joseph Eisinger into their family. (They also supported three other young Nazi refugees). Both of these families strongly encouraged me in my studies, the Hauffs at the East Grinstead County School in Sussex and the Mendels at the University of Toronto. I cannot imagine how I might have become a scientist without their help.

My first wife, Lois Kohn, gave me invaluable support during the early phases of my scientific career; my present wife of over 20 years, Mara, has supported me in the latter phases of my scientific life. She also created a wonderful home for us, and gave me an entire new family, including her father Roman Vishniac, a biologist as well as a noted photographer of pre-war Jewish communities in Eastern Europe, and her mother Luta. (They both died rather recently, well into their nineties).

After these rather personal reminiscences I now turn to a brief description of my life as a scientist.

When I arrived in England in August 1939, three weeks before the outbreak of World War II, I had my mind set on becoming a farmer (I had seen too many unemployed intellectuals during the 1930s), and I started out on a training farm in Kent. However, I became seriously ill and physically weak with meningitis, and so in January 1940 my "acting parents", the Hauffs, arranged for me to attend the above-mentioned county school, where – after a period of uncertainty – I concentrated on mathematics, physics and chemistry.

However, in May 1940, shortly after I had turned 17, and while the German army swept through Western Europe and Britain girded for a possible German air-assault, Churchill ordered most male "enemy aliens" (i.e., holders of enemy passports, like myself) to be interned ("Collar the lot" was his crisp order). I spent about two months in various British camps, including the Isle of Man, where my school sent me the books I needed to study. There I also audited, with little comprehension, some lectures on mathematics and physics, offered by mature interned scientists.

In July 1940, I was shipped on, as part of a British convoy moving through U-boat-infested waters, to Quebec City in Canada; and from there, by train, to a camp in Trois Rivieres, which housed both German civilian internees and refugees like myself. Again various internee-taught courses were offered. The one which interested me most was a course on set-theory given by the mathematician Dr. Fritz Rothberger and attended by two students. Dr. Rothberger, from Vienna, a most kind and unassuming man, had been an advanced private scholar in Cambridge, England, when the internment order was issued. His love for the intrinsic depth and beauty of mathematics was gradually absorbed by his students.

Later I was moved around among various other camps in Quebec and New Brunswick. Another fellow internee, Dr. A. Heckscher, an art historian, organized a fine camp school for young people like myself, whose education had been interrupted and who prepared to take official Canadian High School exams. In this way I passed the McGill University Junior Matriculation exam and exams in mathematics, physics and chemistry on the senior matriculation level. At this point, at age 18, I was pretty firmly looking forward to a career in physics, with a strong secondary interest in mathematics.

I mention with gratitude that camp educational programs received support from the Canadian Red Cross and Jewish Canadian philantropic sources. I also mention that in most camps we had the opportunity to work as lumberjacks and earn 20 cents per day. With this princely sum, carefully saved up, I was able to buy Hardy's Pure Mathematics and Slater's Chemical Physics, books which are still on my shelves.

In January 1942, having been cleared by Scotland Yard of being a potential spy, I was released from internment and welcomed by the family of Professor Bruno Mendel in Toronto. At this point I planned to take up engineering rather than physics, in order to be able to support my parents after the war.

The Mendels introduced me to Professor Leopold Infeld who had come to Toronto after several years with Einstein. Infeld, after talking with me (in a kind of drawing room oral exam), concluded that my real love was physics and advised me to major in an excellent, very stiff program, then called mathematics and physics, at the University of Toronto. He argued that this program would enable me to earn a decent living at least as well as an engineering program.

However, because of my now German nationality, I was not allowed into the chemistry building, where war work was in progress, and hence I could not enroll in any chemistry courses. (In fact, the last time I attended a chemistry class was in my English school at the age of 17.) Since chemistry was required, this seemed to sink any hope of enrolling. Here I express my deep appreciation to Dean and head of mathematics, Samuel Beatty, who helped me, and several others, nevertheless to enter mathematics and physics as special students, whose status was regularized one or two years later.

I was fortunate to find an extraordinary mathematics and applied mathematics program in Toronto. Luminous members whom I recall with special vividness were the algebraist Richard Brauer, the non-Euclidean geometer, H.S.M. Coxeter, the aforementioned Leopold Infeld, and the classical applied mathematicians John Lighton Synge and Alexander Weinstein. This group had been largely assembled by Dean Beatty. In those years the University of Toronto team of mathematics students, competing with teams from the leading North-American Institutions, consistently won the annual Putman competition. (For the record I remark that I never participated). Physics too had many distinguished faculty members, largely recruited by John C. McLennan, one of the earliest low temperature physicists, who had died before I arrived. They included the Raman specialist H.L. Welsh, M.F. Crawford in optics and the low-temperature physicists H.G. Smith and A.D. Misener. Among my fellow students was Arthur Shawlow, who later was to share the Nobel Prize for the development of the laser.

During one or two summers, as well as part-time during the school year, I worked for a small Canadian company which developed electrical instruments for military planes. A little later I spent two summers, working for a geophysicist, looking for (and finding!) gold deposits in northern Ontario and Quebec.

After my junior year I joined the Canadian Army. An excellent upper division course in mechanics by A. Weinstein had introduced me to the dynamics of tops and gyroscopes. While in the army I used my spare time to develop new strict bounds on the precession of heavy, symmetrical tops. This paper, "Contour Integration in the Theory of the Spherical Pendulum and the Heavy Symmetrical Top" was published in the Transactions of the American Mathematical Society. At the end of one year's army service, having completed only 2 1/2 out of the 4-year undergraduate program, I received a war-time bachelor's degree "on – active – service" in applied mathematics.

In the year 1945–6, after my discharge from the army, I took an excellent crash master's program, including some of the senior courses which I had missed, graduate courses, a master's thesis consisting of my paper on tops and a paper on scaling of atomic wave-functions.

My teachers wisely insisted that I do not stay on in Toronto for a Ph.D, but financial support for further study was very hard to come by. Eventually I was thrilled to receive a fine Lehman fellowship at Harvard. Leopold Infeld recommended that I should try to be accepted by Julian Schwinger, whom he knew and who, still in his 20s, was already one of the most exciting theoretical physicists in the world.

Arriving from the relatively isolated University of Toronto and finding myself at the illustrious Harvard, where many faculty and graduate students had just come back from doing brilliant war-related work at Los Alamos, the MIT Radiation Laboratory, etc., I felt very insecure and set as my goal survival for at least one year. The Department Chair, J.H. Van Vleck, was very kind and referred to me as the Toronto-Kohn to distinguish me from another person who, I gathered, had caused some trouble. Once Van Vleck told me of an idea in the band-theory of solids, later known as the quantum defect method, and asked me if I would like to work on it. I asked for time to consider. When I returned a few days later, without in the least grasping his idea, I thanked him for the opportunity but explained that, while I did not yet know in what subfield of physics I wanted to do my thesis, I was sure it would not be in solid state physics. This problem then became the thesis of Thomas Kuhn, (later a renowned philosopher of science), and was further developed by myself and others. In spite of my original disconnect with Van Vleck, solid state physics soon became the center of my professional life and Van Vleck and I became lifelong friends.

After my encounter with Van Vleck I presented myself to Julian Swinger requesting to be accepted as one of his thesis students. His evident brilliance as a researcher and as a lecturer in advanced graduate courses (such as wave-guides and nuclear physics) attracted large numbers of students, including many who had returned to their studies after spending "time out" on various war-related projects.

I told Schwinger briefly of my very modest efforts using variational principles. He himself had developed brilliant new Green's function variational principles during the war for wave-guides, optics and nuclear physics (Soon afterwards Green's functions played an important role in his Nobel-Prize-winning work on quantum electrodynamics). He accepted me within minutes as one of his approximately 10 thesis students. He suggested that I should try to develop a Green's function variational method for *three*-body scattering problems, like low-energy neutron-deuteron scattering, while warning me ominously, that he himself had tried and failed. Some six months later, when I had obtained some partial, very unsatisfactory results, I looked for alternative approaches and soon found a rather elementary formulation, later known as Kohn's variational principle for scattering, and useful for nuclear, atomic and molecular problems. Since I had circumvented Schwinger's beloved Green's functions, I felt that he was very disappointed. Nevertheless he accepted this work as my thesis in 1948. (Much later L. Fadeev offered his celebrated solution of the three-body scattering problem).

My Harvard friends, close and not so close, included P.W. Anderson, N. Bloembergen, H. Broida (a little later), K. Case, F. De Hoffman, J. Eisenstein, R. Glauber, T. Kuhn, R. Landauer, B. Mottelson, G. Pake, F. Rohrlich, and C. Slichter. Schwinger's brilliant lectures on nuclear physics also attracted many students and postdocs from MIT, including J. Blatt, M. Goldberger, and J.M. Luttinger. Quite a number of this remarkable group would become lifelong friends, and one – J.M. "Quin" Luttinger – also my closest collaborator for 13 years, 1954–66. Almost all went on to outstanding careers of one sort or another.

I was totally surprised and thrilled when in the spring of 1948 Schwinger offered to keep me at Harvard for up to three years. I had the choice of being a regular post-doctoral fellow or dividing my time equally between research and teaching. Wisely – as it turned out – I chose the latter. For the next two years I shared an office with Sidney Borowitz, later Chancellor of New York University, who had a similar appointment. We were to assist Schwinger in his work on quantum electrodynamics and the emerging field theory of strong interactions between nucleons and mesons. In view of Schwinger's deep physical insights and celebrated mathematical power, I soon felt almost completely useless. Borowitz and I did make some very minor contributions, while the greats, especially Schwinger and Feynman, seemed to be on their way to unplumbed, perhaps ultimate depths.

For the summer of 1949, I got a job in the Polaroid laboratory in Cambridge, Mass., just before the Polaroid camera made its public appearance. My task was to bring some understanding to the mechanism by which charged particles falling on a photographic plate lead to a photographic image. (This technique had just been introduced to study cosmic rays). I therefore needed to learn something about solid state physics and occasionally, when I encountered things I didn't understand, I consulted Van Vleck.

It seems that these meetings gave him the erroneous impression that I knew something about the subject. For one day he explained to me that he was about to take a leave of absence and, "since you are familiar with solid state physics", he asked me if I could teach a course on this subject, which he had planned to offer. This time, frustrated with my work on quantum field theory, I agreed. I had a family, jobs were scarce, and I thought that broadening my competence into a new, more practical, area might give me more opportunities.

So, relying largely on the excellent, relatively recent monograph by F. Seitz, "Modern Theory of Solids", I taught one of the first broad courses on Solid State Physics in the United States. My "students" included several of my friends, N. Bloembergen, C. Slichter and G. Pake who conducted experiments (later considered as classics) in the brand-new area of nuclear magnetic resonance which had just been opened up by E. Purcell at Harvard and F. Bloch at Stanford. Some of my students often understood much more than I, but they were charitable towards their teacher.

At about the same time I did some calculations, suggested by Bloembergen, on the recently discovered, so-called Knight shift of nuclear magnetic resonance, and, in this connection, returning to my old love of variational methods, developed a new variational approach to the study of wavefunctions in periodic crystals.

Although my appointment was good for another year and a half, I began actively looking for a more long-term position. I was a naturalized Canadian citizen, with the warmest feelings towards Canada, and explored every Canadian University known to me. No opportunities presented themselves. Neither did the very meager US market for young theorists yield an academic offer. At this point a promising possibility appeared for a position in a new Westinghouse nuclear reactor laboratory outside of Pittsburgh. But during a visit it turned out that US citizenship was required and so this possibility too vanished. At that moment I was unbelievably lucky. While in Pittsburgh, I stayed with my Canadian friend Alfred Schild, who taught in the mathematics department at the Carnegie Institute of Technology (now Carnegie Mellon University). He remarked that F. Seitz and several of his colleagues had just left the physics department and moved to Illinois, so that – he thought – there might be an opening for me there. It turned out that the Department Chair, Ed Creutz was looking rather desperately for somebody who could teach a course in solid state physics and also keep an eye on the graduate students who had lost their "doctor-fathers". Within 48 hours I had a telegram offering me a job!

A few weeks later a happy complication arose. I had earlier applied for a National Research Council fellowship for 1950–51 and now it came through. A request for a short postponement was firmly denied. Fortunately, Ed Creutz agreed to give me a one-year leave of absence, provided I first taught a compressed course in solid state physics. So on December 31, 1950 (to satisfy the terms of my fellowship) I arrived in Copenhagen.

Originally I had planned to revert to nuclear physics there, in particular the structure of the deuteron. But in the meantime I had become a solid state physicist. Unfortunately no one in Copenhagen, including Niels Bohr, had even heard the expression "Solid State Physics". For a while I worked on old projects. Then, with an Indian visitor named Vachaspati (no initial), I published a criticism of Froehlich's pre-BCS theory of superconductivity, and also did some work on scattering theory.

In the spring of 1951, I was told that an expected visitor for the coming year had dropped out and that the Bohr Institute could provide me with an Oersted fellowship to remain there until the fall of 1952. Very exciting work was going on in Copenhagen, which eventually led to the great "Collective Model of the Nucleus" of A. Bohr and B. Mottelson, both of whom had become close friends. Furthermore my family and I had fallen in love with Denmark and the Danish people. A letter from Niels Bohr to my department chair at Carnegie quickly resulted in the extension of my leave of absence till the fall of 1952.

In the summer of 1951, I became a substitute teacher, replacing an ill lecturer at the first summer school at Les Houches, near Chamonix in France, conceived and organized by a dynamic young French woman, Cécile Morette De Witt. As an "expert" in solid state physics, I offered a few lectures on that subject. Wolfgang Pauli, who visited, when he learned of my meager knowledge of solids, mostly metallic sodium, asked me, true to form, if I was a professor of physics or of sodium. He was equally acerbic about himself. Some 50 years old at the time, he descibed himself as "a child-wonder in menopause" ("ein Wunderkind in den Wechseljahren"). But my most important encounter was with Res Jost, an assistant of Pauli at the ETH in Zurich, with whom I shared an interest in the so-called inverse scattering problem: given asymptotic information, (such as phase-shifts as function of energy), of a particle scattered by a potential V(r), what quantitative information can be inferred about this potential? Later that year, we both found ourselves in Copenhagen and addressed this problem in earnest. Jost, at the time a senior fellow at the Institute for Advanced Study in Princeton, had to return there before we had finished our work. A few months later, in the spring of 1952, I received an invitation from Robert Oppenheimer, to come to Princeton for a few weeks to finish our project. In an intensive and most enjoyable collaboration, we succeeded in obtaining a complete solution for S-wave scattering by a spherical potential. At about the same time I.M. Gel'fand in the Soviet Union published his celebrated work on the inverse problem. Jost and I remained close lifelong friends until his death in 1989.

After my return to Carnegie Tech in 1952, I began a major collaboration with N. Rostoker, then an assistant of an experimentalist, later a distinguished plasma theorist. We developed a theory for the energy band structure of electrons for periodic potentials, harking back to my earlier experience with scattering, Green's functions and variational methods. We showed how to determine the bandstructure from a knowledge of purely geometric structure constants and a small number (~ 3) of scattering phase-shifts of the potential in a single sphericalized cell. By a different approach this theory was also obtained by J. Korringa. It continues to be used under the acronym KKR. Other work during my Carnegie years, 1950–59, includes the image of the metallic Fermi Surface in the phonon spectrum (Kohn anomaly); exponential localization of Wannier functions; and the nature of the insulating state.

My most distinguished colleague and good friend at Carnegie was G.C. Wick, and my first PhD's were D. Schechter and V. Ambegaokar. I also greatly benefitted from my interaction with T. Holstein at Westinghouse.

In 1953, with support from Van Vleck, I obtained a summer job at Bell Labs as assistant of W. Schockley, the co-inventor of the transistor. My project was radiation damage of Si and Ge by energetic electrons, critical for the use of the recently developed semiconductor devices for applications in outer space. In particular, I established a reasonably accurate energy threshold for permanent displacement of a nucleus from its regular lattice position, substantially smaller than had been previously presumed. Bell Labs at that time was without question the world's outstanding center for research in solid state physics and, for the first time, gave me a perspective over this fascinating, rich field. Bardeen, Brattain and Schockley, after their invention of the transistor, were the great heroes. Other world class theorists were C. Herring, G. Wannier and my brilliant friend from Harvard, P.W. Anderson. With a few interruptions I was to return to Bell Labs every year until 1966. I owe this institution my growing up from amateur to professional.

In the summer of 1954 both Quin Luttinger and I were at Bell Labs and began our 13-year long collaborations, along with other work outside our professional "marriage". (Our close friendship lasted till his death in 1997). The all-important impurity states in the transistor materials Si and Ge, which govern their electrical and many of their optical properties, were under intense experimental study, which we complemented by theoretical work using so-called effective mass theory. In 1957, I wrote a comprehensive review on this subject. We (mostly Luttinger) also developed an effective Hamiltonian in the presence of magnetic fields, for the complex holes in these elements. A little later we obtained the first non-heuristic derivation of the Boltzman transport equation for *quantum mechanical* particles. There followed several years of studies of many-body theories, including Luttinger's famous one-dimensional "Luttinger liquid" and the "Luttinger's theorem" about the conservation of the volume enclosed by a metallic Fermi surface, in the presence of electron electron interaction. Finally, in 1966, we showed that superconductivity occurs even with purely repulsive interactions – contrary to conventional wisdom and possibly relevant to the much later discovery of high-T_c superconductors.

In 1960, when I moved to the University of California San Diego, California, my scientific interactions with Luttinger, then at Columbia University, and with Bell Labs gradually diminished. I did some consulting at the nearby General Atomic Laboratory, interacting primarily with J. Appel. My university colleagues included G. Feher, B. Maple, B. Matthias, S. Schultz, H. Suhl and J. Wheatley, – a wonderful environment. During my 19-year stay there I typically worked with two postdocs and four graduate students. A high water mark period were the late 1960s, early 1970s, including N. Lang, D. Mermin, M. Rice, L. J. Sham, D. Sherrington, and J. Smith.

I now come to the development of density functional theory (DFT). In the fall of 1963, I spent a sabbatical semester at the École Normale Supérieure in Paris, as guest and in the spacious office of my friend Philippe Nozières. Since my Carnegie days I had been interested in the electronic structure of alloys, a subject of intense experimental interest in both the physics and metallurgy departments. In Paris I read some of the metallugical literature, in which the concept of the effective charge e* of an atom in an alloy was prominent, which characterized in a rough way the transfer of charge between atomic cells. It was a *local* point of view in *coordinate space*, in contrast to the emphasis on *delocalized* waves in *momentum space*, such as Bloch-waves in an average periodic crystal, used for the rough description of substitutional alloys. At this point the question occurred to me whether, in general, an alloy is *completely* or only partially characterized by its electronic density distribution n(r): In the back of my mind I knew that this was the case in the Thomas-

Fermi approximation of interacting electron systems; also, from the "rigid band model" of substitutional alloys of neighboring elements, I knew that there was a 1-to-1 correspondence between a weak perturbing potential $\delta v(r)$ and the corresponding small change $\delta n(r)$ of the density distribution. Finally it occurred to me that for a single particle there is an explicit elementary relation between the potential $v(r)$ and the density, $n(r)$, of the groundstate. Taken together, these provided strong support for the conjective that the density $n(r)$ completely determines the external potential $v(r)$. This would imply that $n(r)$ which integrates to N, the total number of electrons, also determines the total Hamilton H and hence *all* properties derivable from H and N, e.g. the wavefunction of the 17th excited state, $\Psi^{17}(r_1,...,r_N)$! Could this be true? And how could it be decided? Could two *different* potentials, $v_1(r)$ and $v_2(r)$, with associated different groundstates $\Psi_1(r_1,...,r_N)$ and $\Psi_2(r_1,...,r_N)$ give rise to the *same* density distribution? It turned out that a simple 3-line argument, using my beloved Rayleigh Ritz variational principle, confirmed the conjecture. It seemed such a remarkable result that I did not trust myself.

By this time I had become friends with another inhabitant of Nozière's office, Pierre Hohenberg, a lively young American, recently arrived in Paris after a one-year fellowship in the Soviet Union. Having completed some work there he seemed to be "between" problems and I asked if he would be interested in joining me. He was. The first task was a literature search to see if this simple result was already known; apparently not. In short order we had recast the Rayleigh-Ritz variational theorem for the groundstate energy in terms of the density $n(r)$ instead of the many electron wave function Ψ, leading to what is now called the Hohenberg Kohn (HK) variational principle. We fleshed out this work with various approximations and published it.

Shortly afterwards I returned to San Diego where my new postdoctoral fellow, Lu J. Sham had already arrived. Together we derived from the HK variational principle what are now known as the Kohn-Sham (KS) equations, which have found extensive use by physicists and chemists, including members of my group.

Since the 1970s I have also been working on the theory of surfaces, mostly electronic structure. The work with Lang in the early 1970s, using DFT, picked up and carried forward where J. Bardeen's thesis had left off in the 1930s.

In 1979, I moved to the University of California, Santa Barbara to become the initial director of the National Science Foundation's Institute for Theoretical Physics (1979–84). I have continued to work with postdoctoral fellows and students on DFT and other problems that I had put aside in previous years. Since the middle 1980s, I have also had increasing, fruitful interactions with theoretical chemists. I mention especially Robert Parr, the first major theoretical chemist to believe in the potential promise of DFT for chemistry who, together with his young co-workers, has made major contributions, both conceptual and computational.

Since beginning this autobiographical sketch I have turned 76. I enormously enjoy the continuing progress by my younger DFT colleagues and my own collaboration with some of them. Looking back I feel very fortunate to have had a small part in the great drama of scientific progress, and most thankful to all those, including family, kindly "acting parents", teachers, colleagues, students, and collaborators of all ages, who made it possible.

ELECTRONIC STRUCTURE OF MATTER – WAVE FUNCTIONS AND DENSITY FUNCTIONALS

Nobel Lecture, January 28, 1999

by

WALTER KOHN

Department of Physics, University of California, Santa Barbara, CA 93106-9530, USA

I. INTRODUCTION

The citation for my share of the 1998 Nobel Prize in chemistry refers to the "development of the density functional theory". The initial work on Density Functional Theory (DFT) was reported in two publications: the first with Pierre Hohenberg in 1964[1] and the next with Lu J. Sham[2] in 1965. This was almost 40 years after E. Schroedinger[3] published his first epoch-making paper marking the beginning of wave-mechanics. The Thomas-Fermi theory, the most rudimentary form of DFT, was put forward shortly afterwards[4, 5] and received only modest attention.

There is an oral tradition that, shortly after Schroedinger's equation for the electronic wave-function Ψ had been put forward and spectacularly validated for simple small systems like *He* and H_2, P.M. Dirac declared that chemistry had come to an end – its content was entirely contained in that powerful equation. Too bad, he is said to have added, that in almost all cases, this equation was far too complex to allow solution.

In the intervening more than six decades enormous progress has been made in finding approximate solutions of Schroedinger's wave equation for systems with several electrons, decisively aided by modern electronic computers. The outstanding contributions of my Nobel Prize co-winner John Pople are in this area. The main objective of the present account is to explicate DFT, which is an alternative approach to the theory of electronic structure, in which the electron density distribution $n(r)$, rather than the many electron wavefunction plays a central role. I felt that it would be useful to do this in a comparative context; hence the wording "Wavefunctions and Density Functionals" in the title.

In my view DFT makes two kinds of contribution to the science of multiparticle quantum systems, including problems of electronic structure of molecules and of condensed matter:

The first is in the area of fundamental *understanding*. Theoretical chemists and physicists, following the path of the Schroedinger equation, have become accustomed to think in a truncated *Hilbert space of single particle orbitals*. The spectacular advances achieved in this way attest to the fruitfulness of this perspective. However, when high accuracy is required, so many Slater determinants are required (in some calculations up to ~ 10^9!) that *comprehension* becomes difficult. DFT provides a complementary perspective. It focuses on quantities in the real, 3-dimensional coordinate space, principally on the electron density $n(r)$ of the groundstate. Other quantities of great interest are: the exchange correlation hole density $n_{xc}(r,r')$ which describes how the presence of an electron at the point r depletes the total density of the other electrons at the point r'; and the linear response function, $\chi(r,r';\omega)$, which describes the change of total density at the point r due to a perturbing potential at the point r', with frequency ω. These quantities are *physical*, independent of representation, and easily *visualisable* even for very large systems. Their understanding provides transparent and complementary insight into the nature of multiparticle systems.

The second contribution is practical. Traditional multiparticle wavefunction methods when applied to systems of *many particles* encounter what I call an exponential wall when the number of atoms, N, exceeds a critical value which, for "chemical accuracy", currently is in the neighborhood of $N_0 \approx 10$ (to within a factor of about 2) for a system without symmetries. A major improvement along present lines in the analytical and/or computational aspects of these methods will lead to only modest increases in N_0. Consequently, problems requiring the simultaneous consideration of very many interacting atoms, $N/N_0 \gg 1$, such as large organic molecules, molecules in solution, drugs, DNA, etc. overtax these methods. On the other hand, in DFT, computing time T rises much more moderately with the number of atoms, currently as $T \sim N^{\alpha}$ with $\alpha \approx 2$–3, with ongoing progress in bringing α down towards $\alpha \approx 1$ (so-called linear scaling). The current state of the art of applied DFT can handle systems with up to $N_0' = O(10^2)$–$O(10^3)$ atoms.

The following figures and legends illustrate what can currently be achieved. In these examples the number of atoms is $O(10^2)$ and the number of electrons several times larger.

In Section 1, I shall talk about traditional wavefunction methods and contrast their great success for few-atom systems with their fundamental limitations in dealing with very-many-atom systems.

Section 2 deals with DFT against the backdrop of wavefunction methods. The basic theory is summarized: First the original Hohenberg-Kohn (HK) variational principle, where $n(r)$ is the variational variable, is described. This is followed by the Kohn-Sham (KS) self-consistent single-particle equations which involve the well-defined exchange – correlation functional, $E_{xc}[n(r)]$. In principle, when used with the exact E_{xc}, these single particle equations incorporate *all* many-body effects.[1]

Next the physics of $E_{xc}[n(r)]$ is discussed in terms of the concept of the exchange correlation hole $n_{xc}(r,r')$. I have found the concept of "nearsightedness" useful which, in the present context, says that the exchange correlation hole $n_{xc}(r,r')$ for an electron at the point r is largely determined by $\mu - v_{eff}(\tilde{r})$, where μ is the chemical potential and $v_{eff}(\tilde{r})$ is the effective single particle

[1] It is however known that for some density distributions $E_{xc}[n(r)]$ cannot be defined.

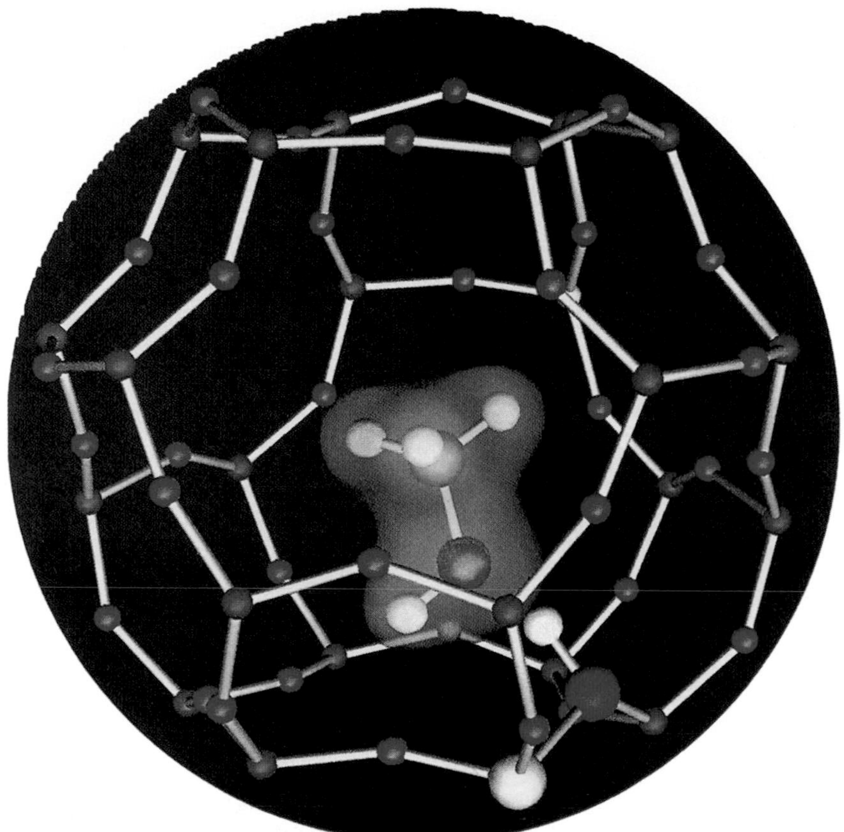

Figure 1. Methanol inside a cage of the zeolite sodalite. Zeolites are crystalline arrays of cages built of silicon (blue), aluminum (yellow), and oxygen (red) atoms. For each *Al* atom one must have a positive counter-ion (in this case H^+ (white)). A methanol molecule is inside the cage (carbon is green) where it can react with the proton. DFT calculations have assigned and clarified the IR spectra, have determined the binding sites of methanol, and have calculated the activation energy for the reaction. Acid catalysis in zeolites is widely used in the chemical industry. (After E. Nusterer, P. Bloechl and Karlheinz Schwarz, Angew. Chem. **35**, 175).

potential for $\tilde{r}$ *near* r. Although nearsightedness becomes a well defined concept only for metallic systems which are very large, it has been found to be useful also for systems as small as a single atom.

There follows a brief discussion of approximations for E_{xc}, which reflect nearsightedness, and other general principles.

Parts III-V discuss applications of DFT to electronic groundstates, as well as a host of generalizations to other electronic and non-electronic systems.

Finally a few concluding remarks and speculations are offered.

II. SCHROEDINGER WAVEFUNCTIONS – FEW VERSUS MANY ELECTRONS

The foundation of the theory of electronic structure of matter is the non-relativistic Schroedinger equation for the many-electron wavefunction Ψ,

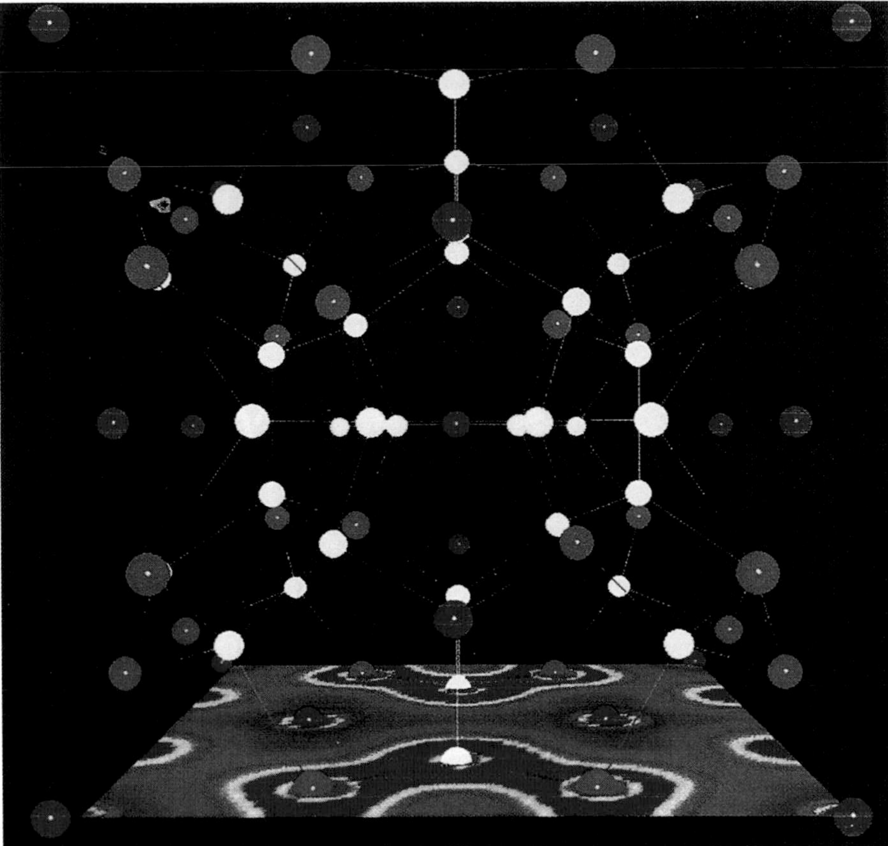

Figure 2. The geometric structure of the clathrate $Sr_8Ga_{16}Ge_{30}$ (Sr red, Ga blue, Ge white) and its charge density in a plane bisecting the centers of the cages. DFT calculations have shown that the Sr atoms are weakly bound and scatter phonons effectively, thereby, lowering thermal conductivity. However, contrary to intuitive expectations, the Sr atoms do not donate electrons to the frame and are practically neutral. Conductivity is due to electrons traveling through the frame, not through the one-dimensional Sr "wires" in the structure; there is thus little scattering of conduction electrons, by Sr vibrations. For these reasons, the compound is a metal with a large Seebeck coefficient (unlike ordinary metals). The calculation suggests that other compounds of this type may be even better thermoelectrics. (Theory by N.P. Blake and H. Metiu, submitted for publication).

$$\{-\frac{\hbar^2}{2m}\sum_j \nabla_j^2 - \sum_{j,\ell}\frac{Z_\ell e^2}{|\, r_j - R_\ell \,|} + \frac{1}{2}\sum_{j\neq j'}\frac{e^2}{|\, r_j - r'_j \,|} - E\}\Psi = 0\,, \quad (2.1)$$

where r_j are the positions of the electrons and R_l, Z_l the positions and atomic numbers of the nuclei; $\hbar$, m and e are the conventional fundamental constants; and E is the energy. This equation reflects the Born-Oppenheimer approximation, in which – for purposes of studying electron-dynamics – the much heavier nuclei are considered as fixed in space. This paper will deal largely with non-degenerate groundstates. The wavefunction Ψ depends on the positions and spins of the N electrons but in this paper spins will generally not be explicitly indicated. Thus

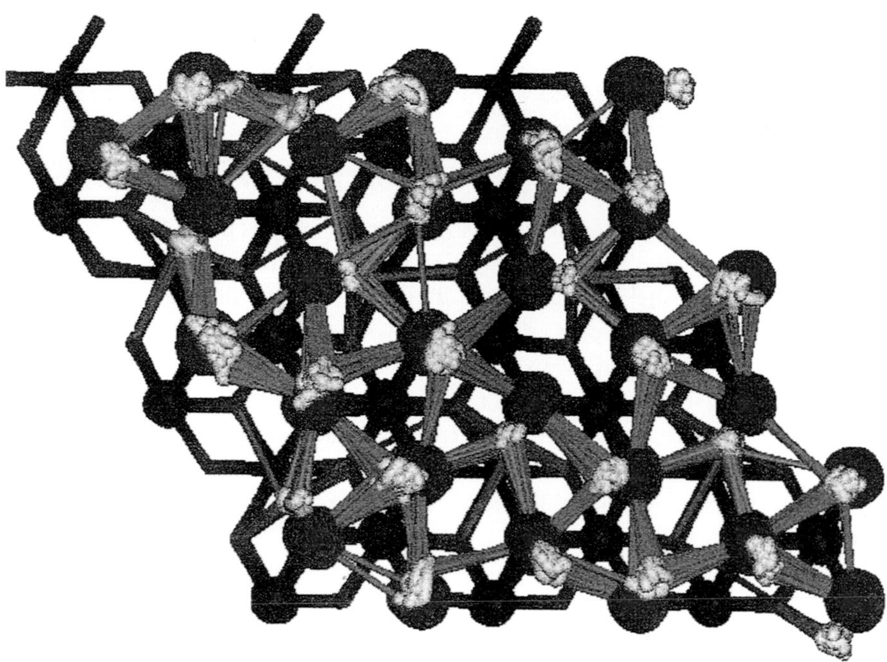

Figure 3. Fully Hydroxylated Aluminum (0001) Surface. (Red-O; blue-interior Al; grey-H-atoms; the green lines are H-bonds). Each surface Al-atom in Al_2O_3 has been replaced by 3 H-atoms. The figure represents a superposition of configurations in a molecular dynamics simulation at regular intervals of 1 ps. These calculations help to understand the complex dynamics of water adsorption on aluminum (K.C. Haas *et al.*, Science 282, 265 (1998)).

$$\Psi = \Psi(r_1, r_2, ...r_{\mathcal{N}}). \tag{2.2}$$

The Pauli principle requires that

$$P_{jj'}\Psi = -\Psi \ , \tag{2.3}$$

where $P_{jj'}$ permutes the space and spin coordinates of electrons j and j'. All physical properties of the electrons depend parametrically on the R_1 , in particular the density $n(r)$ and total energy E which play key roles in this paper:

$$n(r) = n(r; R_1, ...R_{\mathcal{N}}) \ , \tag{2.4}$$

$$E = E(R_1, ...R_{\mathcal{N}}) \ , \tag{2.5}$$

where $\mathcal{N}$ is the number of nuclei.

A. Few Electron Systems–the H_2 Molecule

The first demonstrations of the power of the Schroedinger equation in chemistry were calculations of the properties of the simplest multielectron molecule, H_2: Its experimental binding energy[2] and internuclear separation are

[2] This is the observed dissociation energy plus the zero point energy of 0.27 eV.

$$E_{exp} : D = 4.75eV, \quad R = 0.740 \text{ Å}. \tag{2.6}$$

The earliest quantum theoretical estimate was made by Heitler and London in 1927[6], who used the Ansatz

$$\Psi_{HL} = A\left[\varphi_H(r_1 - R_1)\varphi_H(r_2 - R_2) + \varphi_H(r_1 - R_2)\varphi_H(r_2 - R_1)\right]\chi_0 , \tag{2.7}$$

where $\varphi_H(r_1 - R_1)$ is the orbital wavefunction of electron 1 in its *atomic* groundstate around a proton located at R_1, etc; χ_0 denotes the spin singlet function; and A is the normalization. The components of this wavefunction describe two hydrogen atoms, at R_1 and R_2, with spins pointing in opposite directions. The combination satisfies the reflection symmetry of the molecule and the Pauli principle. The expectation value of the Hamiltonian as a function of $R = |R_1 - R_2|$ was calculated. Its minimum was found to occur at $R = 0.87\text{Å}$, and the calculated dissociation energy was $3.14eV$, in semiquantitative agreement with experiment. However the errors were far too great for the typical chemical requirements of $|\delta R| \leq 0.01$ Å and $|\delta D| \leq 0.1eV$.

An alternative Ansatz, analogous to that adopted by Bloch for crystal electrons, was made by Mullikan in 1928[7]:

$$\Psi_{BM} = \varphi_{mol}(r_1)\varphi_{mol}(r_2) \cdot \chi_0 , \tag{2.8}$$

where

$$\varphi_{mol}(r_1) = A'(\varphi_H(r_1 - R_1) + \varphi_H(r_1 - R_2)) , \tag{2.9}$$

and A' is the appropriate normalization constant. In this function both electrons occupy the same molecular orbital $\varphi_{mol}(r)$. The spin function χ_0 is again the antisymmetric singlet function. The results obtained with this function were $R = 0.76$ Å, and $D = 2.65eV$, again in semiquantitative agreement with experiment.

The Mullikan Ansatz can be regarded as the simplest version of a more general, so-called Hartree-Fock Ansatz, the Slater determinant

$$\Psi_{HF} = \frac{1}{2^{1/2}} Det \mid \varphi_m(r_1)\alpha(1)\varphi_m(r_2)\beta(2) \mid , \tag{2.10}$$

where $\varphi_m(r)$ is a general molecular orbital and α and β denote up and down spin functions. For given $R \equiv |R_1 - R_2|$, minimization with respect to $\varphi_m(r)$ of the expectation value of H leads to the non-local Hartree Fock equations[8] for the molecular orbital $\varphi_m(r)$, whose solution gives the following results: $R = 0.74$ Å, $D = 3.63eV$.

The most complete early study of H_2 was undertaken by James and Coolidge in 1933[9]. They made the very general variational Ansatz

$$\Psi_{JC} = \Psi(r_1, r_2)\chi_0 , \tag{2.11}$$

where $\Psi(r_1,r_2)$ is a general, normalized function of r_1 and r_2, symmetric under interchange of r_1 and r_2 and respecting the spatial symmetries of the molecule. The trial function Ψ was written as depending on a number of parameters, $p_1,p_2..p_M$, so that for given $|R_1 - R_2|$, the expectation value of the Schroedinger Hamiltonian in Ψ, an upper bound to the true groundstate energy, became a function of the parameters p_j, $E = E(p_1,...,p_M)$. The calculations were made with M up to 13. Minimization of $E(p_1,...,p_M)$ with respect to the p_j resulted in $R = 0.740$ Å and $D = 4.70$ *eV*, in very good agreement with experiment. More recent variational calculations of the same general character give theoretical results whose errors are estimated to be much smaller than experimental uncertainties, and other theoretical corrections.

Before leaving the variational calculation for H_2, we want to make a *very* rough "guesstimate" of the number of parameters M needed for a satisfactory result.

The number of continuous variables of $\Psi(r_1,r_2)$ is $6 - 1 = 5$, the reduction by 1 reflecting axial symmetry. Let us call the number of parameter per variable needed for the desired accuracy p. Since a fractional accuracy of $O(10^{-2})$ is needed for the energy, implying a fractional accuracy of $O(10^{-1})$ in Ψ, we guess that $3 \ll p \ll 10$. Hence $M = p^5 = 3^5 - 10^5 \approx 10^2 - 10^5$.

By using symmetries and chemical and mathematical insights, this number can be significantly reduced. Such relatively modest numbers are very managable on today's (and even yesterday's) computers.

It is thus not surprising that *for sufficiently small molecules, wavefunction methods give excellent results.*

B. Many Electrons – Encountering an Exponential Wall

In the same spirit as our last "guesstimates" for H_2, let us now consider a general molecule consisting of N atoms with a total of N interacting electrons, where $N \gg 10$ say. We ignore symmetries and spin, which will not affect our general conclusions. Reasoning as before, we see that the number M of parameter required is

$$M = p^{3N}\ , \quad 3 \leq p \leq 10. \tag{2.12}$$

The energy needs to be minimized in the space of these M parameters. Call $\bar{M}$ the maximum value feasible with the best available computer software and hardware; and $\bar{N}$ the corresponding maximum number of electrons. Then, from Eq. (2.12) we find

$$\bar{N} = \frac{1}{3}\frac{log\bar{M}}{logp}\ . \tag{2.13}$$

Let us optimistically take $\bar{M} \approx 10^9$ and $p = 3$. This gives the shocking result

$$\bar{N} = \frac{1}{3}\ \frac{9}{0.48} = 6(!)\ . \tag{2.14}$$

In practice, by being "clever", one can do better than this, perhaps by one half order of magnitude, up to say $\bar{N} \approx 20$. But the exponential in Eq. (2.12) represents a "wall" severely limiting $\bar{N}$.

Let us turn this question around and ask what is the needed M for $N = 100$. By Eq. (2.12) and taking $p = 3$ we find

$$M \approx 3^{300} \approx 10^{150}(!) \tag{2.15}$$

I cannot foresee an advance in computer science which can minimize a quantity in a space of 10^{150} dimensions. Of course, estimates like Eq. (2.15) are very rough and only their logarithm should be taken seriously. But the "exponential wall" is real and reflects the interconnectedness of $\Psi(r_1,...,r_N)$ in the $3N$ dimensional configuration space defined by the requirement that all r_j be inside the 3D region containing the molecule.

We conclude that traditional wavefunction methods, which provide the "required" chemical accuracy are generally limited to molecules with a small total number of chemically active electrons, $N \lesssim O(10)$.

C. Some Meta-physical-chemical Considerations

The following remarks are related to a very old paper by one of my teachers, J.H. Van Vleck,[11] in which he discusses a problem with many-body wavefunctions, later referred to as the Van Vleck catastrophy.

I begin with a provocative statement. *In general the many-electron wavefunction $\Psi(r_1,...,r_N)$ for a system of N electrons is not a legitimate scientific concept, when $N \geq N_0$, where $N_0 \approx 10^3$.*

I will use two criteria for defining "legitimacy": a) That Ψ can be calculated with sufficient accuracy and b) can be recorded with sufficient accuracy.

Construction of an accurate approximation to Ψ.

Without leaving the context of wavefunctions, I shall call the approximate wavefunction $\tilde{\Psi}$ sufficiently accurate if

$$|\,\tilde{\Psi}, \Psi)\,|^2 \gg 0.5 \;, \tag{2.16}$$

a rather liberal requirement. (One could equally well choose 0.9 or 0.1).

Consider now the example of N' non-overlapping identical n-electron molecules with exact wave-functions $\psi_l(r_1,...,r_n)$, and approximate wave-functions $\tilde{\psi}_l(r_1,...,r_n)$. Let us take $n = 10$ and posit that a very accurate $\tilde{\psi}_l$ can be calculated with

$$|\,(\tilde{\psi}_\ell, \psi)\,| = 1 - \epsilon \;\; where \;\; \epsilon = 10^{-2} \;, \tag{2.17}$$

again a liberal estimate.

Then, for an N'-molecule system with $N' \approx 10^2$, and thus $N = 10^3$ electrons $(\tilde{\Psi}, \Psi) = (1 - \varepsilon)^{N'} \approx e^{-N'\varepsilon} \approx e^{-1} \approx 0.37$, i.e., almost acceptable by the criterion (2.16). Note however, that for $N = 10^4$, $(\hat{\Psi}, \Psi) \approx e^{-10} \approx 5 \times 10^{-5}$ so that $|\tilde{\Psi}, \Psi|^2 \sim 3 \times 10^{-9}(!)$ – the exponential wall is again there, in another form. For fully interacting systems the situation is much worse and our estimate of $N_0 \approx 10^3$ is probably much too high.

Can this problem ever be overcome along present lines of thought? I think not. Even if there were *no* computational limits, other physical effects, such as relativistic or radiative corrections which may be minor for systems of small N exponentiate when N exceeds N_0.

(It is obvious that the estimates made above have only logarithmic validity).

Recording of $\Psi(r_1, r_2, \ldots, r_N)$.

Let us now assume that somehow we have obtained an accurate approximation to $\tilde{\Psi}$, in the sense of Eq (2.16), and wish to record it so it can be reproduced at a later time. How many bits are needed? Let us take q bits per variable. Then the total number of bits is

$$B = q^{3N} \tag{2.18}$$

For $q = 3$, a very rough fit, and $N = 10^3$, $B = 10^{1500}$, a quite unrealistic number. (The total number of baryons in the accessible universe is estimated as $10^{80\pm}$).

Having attempted to discredit the very-many-electron wavefunction $\Psi(r_1, \ldots, r_N)$, for *many* electrons I must, of course, recall two well-known facts: physically/chemically interesting quantities, like total energy E, density $n(r)$, pair correlation function $g(r, r')$, etc. depend on only very *few variables* and, formally, can be thought of as obtained by tracing over all other variables, e.g.

$$n(r) = N \int \Psi^*(r, r_2, \ldots, r_N)\Psi(r, r_2, \ldots, r_N) dr_2 \ldots dr_N \; ; \tag{2.19}$$

and that some $\tilde{\Psi}$'s which, by the criterion (1.16) are hopelessly "bad" for large N, give respectable and even very accurate results for these contracted quantities. Of course not *every* bad trial-$\tilde{\Psi}$ will give good results for these quantities, and the question of how one discriminates the useful "bad" $\tilde{\Psi}$, from the vast majority of useless "bad" $\tilde{\Psi}$'s requires much further thought. This issue is related, I believe, to the concept of "nearsightedness" which I have recently suggested[10].

In concluding this section I remark that DFT, while *derived* from the N-particle Schroedinger equation, is finally expressed entirely in terms of the density $n(r)$, in the Hohenberg-Kohn formulation,[1] and in terms of $n(r)$ and *single*-particle wavefunctions $\psi_j(r)$, in the Kohn-Sham formulation[2]. This is why it has been most useful for systems of very many electrons where wavefunction methods encounter and are stopped by the "exponential wall".

III. DENSITY FUNCTIONAL THEORY–BACKGROUND

In the fall of 1963, I was spending a sabbatical semester at the École Normale Supérieure in the spacious office of Philippe Nozières. A few weeks after my arrival Pierre Hohenberg, also a visitor from the US, joined forces with me. Ever since my period at the Carnegie Institute of Technology (1950-1959) I had been interested in disordered metallic alloys, partly because of the excellent metallurgy department and partly because of the interesting experimental program of Emerson Pugh, in Physics, on substitutional alloys of Cu with the adjacent elements in the periodic table, such as Cu_xZn_{1-x}. These alloys were viewed in two rather contradictory ways: As an *average* periodic crystal with non-integral atomic number $\bar{Z} = xZ_1 + (1-x) Z_2 (Z_1 = 29,\ Z_2 = 30)$. This model nicely explained the linear dependence of the electronic specific heat on *x*. On the other hand the low temperature resistance is roughly proportional to $x(1-x)$, reflecting the degree of disorder among the two constituents. While isolated *Cu* and *Zn* atoms are, of course, neutral, in a *Cu–Zn* alloy there is transfer of charge between *Cu* and *Zn* unit cells on account of their chemical differences. The electrostatic interaction energy of these charges is an important part of the total energy. Thus in considering the energetics of this system there was a natural emphasis on the electron density distribution $n(r)$.

Now a very crude theory of electronic energy in terms of the electron density distribution, $n(r)$, the Thomas-Fermi (TF) theory, had existed since the 1920s[4] [5]. It was quite useful for describing some qualitative trends, e.g. for *total* energies of atoms, but for questions of chemistry and materials science, which involve valence electrons, it was of almost no use; for example it did not lead to any chemical binding. However the theory had one feature which interested me: It considered interacting electrons moving in an external potential $v(r)$, and provided a highly over-simplified one-to-one implicit relation between $v(r)$ and the density distribution $n(r)$:

$$n(r) = \gamma(\mu - v_{eff}(r))^{3/2} \qquad \left(\gamma = \frac{1}{3\pi^2}\left(\frac{2m}{\hbar^2}\right)^{3/2}\right), \tag{3.1}$$

$$v_{eff}(r) \equiv v(r) + \int \frac{n(r')}{|\,r - r'\,|}dr', \tag{3.2}$$

where μ is the *r*-independent chemical potential; Eq. (3.1) is based on the expression

$$n = \gamma(\mu - v)^{3/2} \tag{3.3}$$

for the density of a *uniform*, non-interacting, degenerate electron gas in a constant external potential *v*; and the second term in (3.2) is just the classically computed electrostatic potential times (–1), generated by the electron density distribution $n(r')$. Since Eq. (3.1) ignores gradients of $v_{eff}(r)$ it was clear that the theory would apply best for systems of slowly varying density.

In subsequent years various refinements (gradient-, exchange-and correlation corrections) were introduced, but the theory did not become significantly more useful for applications to the electronic structure of matter. It was clear that TF theory was a rough representation of the exact solution of the many-electron Schroedinger equation, but since TF theory was expressed in terms of $n(r)$ and Schroedinger theory in term of $\Psi(r_1,...,r_N)$, it was not clear how to establish a strict connection between them.

This raised a general question in my mind: Is a *complete, exact* description of groundstate electronic structure in terms of $n(r)$ possible in principle. A key question was whether the density $n(r)$ *completely* characterized the system. It was true in TF theory, where $n(r)$, substituted in Eq. (3.1) yields, $(v_{eff}(r) - \mu$ and, by (3.2), $(v(r) - \mu)$. In addition, $n(r)$ also yields the total number of electrons by integration. Thus the physical system is completely specified by $n(r)$. It was also simple to check that the same was true for any 1-particle system, as well as for a weakly perturbed, interacting, uniform electron gas

$$v(r) = v_0 + \lambda v_1(r) \quad (\lambda \ll 1) \ , \tag{3.4}$$

$$n(r) = n_0 + \lambda n_1(r) + \ldots \ , \tag{3.5}$$

for which $v_1(r)$ can be explicitly calculated in term of $n_1(r)$ by means of the wave-number-dependent susceptibility of the uniform gas. This suggested the hypothesis that *a knowledge of the groundstate density of $n(r)$ for any electronic system, (with or without interactions) uniquely determines the system.* This hypothesis became the starting point of modern DFT.

IV. THE HOHENBERG-KOHN FORMULATION OF DENSITY FUNCTIONAL THEORY

A. The Density n(r) as the Basic Variable

The Basic Lemma of HK. The groundstate density $n(r)$ of a bound system of interacting electrons in some external potential $v(r)$ determines this potential uniquely[1].

Remarks:

(1) The term "uniquely" means here up to an uninteresting additive constant.

(2) In the case of a degenerate groundstate, the lemma refers to *any* groundstate density $n(r)$.

(3) This lemma is mathematically rigorous.

The proof is very simple. We present it for a non-degenerate groundstate.

Let $n(r)$ be the non-degenerate groundstate density of N electrons in the potential $v_1(r)$, corresponding to the groundstate Ψ_1, and the energy E_1. Then,

$$E_1 = (\Psi_1, H_1\Psi_1) = \int v_1(r)n(r)dr + (\Psi_1, (T+U)\Psi_1) \ , \tag{4.1}$$

where H_1 is the total Hamiltonian corresponding to v_1, and T and U are the kinetic and interaction energy operators. Now assume that there exists a second potential $v_2(r)$, *not* equal to $v_1(r)$ + constant, with groundstate Ψ_2, necessarily $\neq e^{i\Theta}\Psi_1$, which gives rise to the same $n(r)$. Then

$$E_2 = \int v_2(r)n(r)dr + \int(\Psi_2, (T+U)\Psi_2). \tag{4.2}$$

Since Ψ_1 is assumed to be non-degenerate, the Rayleigh-Ritz minimal principle for Ψ_1 gives the inequality,

$$E_1 < (\Psi_2, H_1\Psi_2) = \int v_1(r)n(r)dr + (\Psi_2, (T+U)\Psi_2)$$
$$= E_2 + \int (v_1(r) - v_2(r))n(r)dr. \qquad (4.3)$$

Similarly

$$E_2 \leq (\Psi_1, H\Psi_1) = E_1 + \int (v_2(r) - v_1(r_1))n(r)dr \quad , \qquad (4.4)$$

where we use ≤ since the non-degeneracy of Ψ_2 was not assumed. Adding (4.3) and (4.4) leads to the contradiction

$$E_1 + E_2 < E_1 + E_2. \qquad (4.5)$$

We conclude by *reductivo ad absurdum* that the assumption of the existence of a second potential $v_2(r)$, which is unequal to $v_1(r)$ + constant and gives the same $n(r)$, must be wrong. The lemma is thus proved for a non-degenerate groundstate.

Since $n(r)$ determines both N and $v(r)$ (ignoring an irrelevant additive constant) it gives us the full H and N for the electronic system. Hence $n(r)$ determines implicitly *all* properties derivable from H through the solution of the time-independent or time-dependent Schroedinger equation (even in the presence of additional perturbations like electromagnetic fields), such as: the many-body eigenstates $\Psi^{(0)}(r_1,...,r_N)$, $\Psi^{(1)}(r_1,...,r_N)$, ... the 2-particle Green's function $G(r_1t_1, r_2t_2)$, the frequency dependent electric polarizability $\alpha(\omega)$, and so on. We repeat that all this information is implicit in $n(r)$, the groundstate density.

Remarks:

1. The requirement of non-degeneracy can easily be lifted[12].
2. Of course the lemma remains valid for the special case of non-interacting electrons.
3. Lastly we come to the question whether *any* well-behaved positive function $n(r)$, which integrates to a positive integer N, is a possible groundstate density corresponding to *some* $v(r)$. Such a density is called v-representable (VR). On the positive side it is easy to verify that, in powers of λ, any nearly uniform, real density of the form $n(r) = n_0 + \lambda\Sigma\, n(q)\, e^{iq.r}$ is VR, and that for a single particle any normalized density $n(r) = |\psi(r)|^2$ is also VR. On the other hand Levy[13] and Lieb [14] have shown by an example which involves degenerate groundstates, that there do exist well-behaved densities which are *not* VR. The topology of the regions of v-representability in the abstract space of all $n(r)$ continues to be studied. But this issue has so far not appeared as a limitation in practical applications of DFT.

B. The Hohenberg-Kohn Variational Principle

The most important property of an electronic groundstate is its energy E. By wavefunction methods E could be calculated either by direct approximate solution of the Schroedinger Equation $H\Psi = E\Psi$ or from the Rayleigh Ritz minimal principle,

$$E = min_{\tilde{\Psi}}(\tilde{\Psi}, H\tilde{\Psi}) \ , \tag{4.6}$$

where $\tilde{\Psi}$ is a normalized trial function for the given number of electrons, N.

The formulation of the minimal principle in terms of trial densities, $\tilde{n}(r)$, rather than trial wavefunctions $\tilde{\Psi}$ was first presented in ref. 1. Here we shall follow the more succinct derivation due to Levy[13] and Lieb[14], called the *constrained search method.*

Every trial function $\tilde{\Psi}$ corresponds to a trial density $\tilde{n}(r)$ obtained by integrating $\tilde{\Psi}^* \tilde{\Psi}$ over all variables except the first and multiplying by N. One may carry out the minimization of (4.6) in two stages. First fix a trial $\tilde{n}(r)$ and denote by $\tilde{\Psi}^{\alpha}_{\tilde{n}}$ the class of trial functions with this $\tilde{n}$. We define the constrained energy minimum, with $\tilde{n}(r)$ fixed, as

$$\begin{aligned} E_v[\tilde{n}(r)] &\equiv \min{}_{\alpha}(\tilde{\Psi}^{\alpha}_{\tilde{n}}, H\tilde{\Psi}^{\alpha}_{\tilde{n}}) \\ &= \int v(r)\tilde{n}(r)dr + F[\tilde{n}(r)], \end{aligned} \tag{4.7}$$

where

$$F[\tilde{n}(r)] \equiv \min{}_{\alpha}(\tilde{\Psi}^{\alpha}_{\tilde{n}(r)}, (T+U)\Psi^{\alpha}_{\tilde{n}(r)}) \ . \tag{4.8}$$

$F[\tilde{n}(r)]$ requires no explicit knowledge of $v(r)$. It is a universal functional of the density $\tilde{n}(r)$, (whether the latter is VR or not). In the second step minimize (4.7) over all $\tilde{n}$,

$$E = min_{\tilde{n}(r)} E_v[\tilde{n}(r)] = min_{\tilde{n}(r)}\{\int v(r)\tilde{n}(r)dr + F[\tilde{n}(r)]\}. \tag{4.9}$$

For a non-degenerate groundstate, the minimum is attained when $\tilde{n}(r)$ is the groundstate density; and, for the case of a degenerate groundstate, when $\tilde{n}(r)$ is any one of the groundstate densities. The HK minimum principle (4.9) may be considered as the formal exactification of Thomas-Fermi theory.

The formidable problem of finding the minimum of $(\tilde{\Psi}, H\tilde{\Psi})$ with respect to the 3N-dimensional trial function $\tilde{\Psi}$ has been transformed into the *seemingly* trivial problem of finding the minimum of $E_v[\tilde{n}(r)]$ with respect to the 3-dimensional trial function $\tilde{n}(r)$.

Actually the definition (4.8) of $F[\tilde{n}(r)]$ leads us right back to minimization with respect to 3N-dimensional trial wavefunctions. Nevertheless, significant *formal* progress has been made: *the strict formulation* of the problem of groundstate densities and energies entirely in terms of the density distribution $\tilde{n}(r)$ and of a well-defined, though, not explicitly known, functional of the density, $F[\tilde{n}(r)]$, which represents the sum of kinetic energy and interaction energy $(T+U)$, associated with $\tilde{n}$ (see Eq. (4.8)).

One can now easily re-derive the Thomas-Fermi (TF) theory by making the approximations

$$T = \int n(r) \frac{3}{10} k_F^2(n(r)) dr \,, \tag{4.10}$$

$$U = \frac{1}{2} \int \frac{n(r) n(r')}{| r - r' |} dr dr' \,, \tag{4.11}$$

where $k_F(n)$ is the Fermi wave-vector of a uniform electron gas of density n and $\frac{3}{10} k_F^2(n)$ is the mean kinetic energy per electron of such a gas. The expression for U is the classical (or mean field) approximation. Various previously known corrections, of TF theory for exchange, correlation and density gradients can also be easily re-derived.

The main remaining error is due to the seriously inadequate representation of the kinetic energy, T, by Eq.(4.10) or its gradient-corrected forms. This deficiency is largely remedied by the self-consistent, so-called Kohn-Sham equations, discussed in the following Section 4C.

A second interesting class of systems $n(r) = n_0 + n_1(r)$, where $n_1(r) \ll n_0$, could also be treated using the n_0-dependent density-density response function, $K(|r - r'|)$.

C. The Self Consistent Kohn-Sham Equations

Soon after the publication of the TF theory, Hartree[8] proposed a set of self-consistent single particle equations for the approximate description of the electronic structure of atoms[8]. The concept was physically very simple. Every electron was regarded as moving in an effective single particle potential

$$v_H(r) = -\frac{Z}{r} + \int \frac{n(r')}{| r - r' |} dr' \,, \tag{4.12}$$

where the first term represents the potential due to a nucleus of atomic number Z and the second the potential due to the average density distribution $n(r)$. (The negative charge of the electron has been allowed for). Thus each electron obeys the single particle Schroedinger equation

$$\{ -\frac{1}{2} \nabla^2 + v_H(r) \} \varphi_j(r) = \epsilon_j \varphi_j(r) \,, \qquad (4.13$$

where j denotes both spatial as well as spin quantum numbers. The mean density is given by

$$n(r) = \sum_{j=1}^{N} | \varphi_j(r) |^2 \,, \tag{4.14}$$

where, in the groundstate, the sum runs over the N lowest eigenvalues, to respect the Pauli exclusion principle. Equations (4.12)–(4.14) are called the self-consistent Hartree equations. One may start from a first approximation for $n(r)$, (e.g. from TF theory), construct $v_H(r)$, solve (4.13) for the φj; and recalculate $n(r)$ from Eq. (4.14), which should be the same as the initial $n(r)$. If it is not one iterates appropriately until it is.

In the winter of 1964, I returned from France to San Diego, where I found my new post-doctoral fellow, Lu Sham. I knew that the Hartree equations described atomic groundstates much better than TF theory. The difference between them lay in the different treatments of the kinetic energy T (See Eqs. (4.10) and (4.13). I set ourselves the task of extracting the Hartree equations from the HK variational principle for the energy, Eqs. (4.9), (4.7), (4.8), which I knew to be formally *exact* and which therefore had to have the Hartree equations *and* improvements "in them". In fact it promised a Hartree-like formulation, which – like the HK minimal principle – would be formally exact.

The Hartree differential equation (4.13) had the *form* of the Schroedinger equation for *non-interacting* electrons moving in the external potential v_{eff}. Could we learn something useful from a DFT for non-interacting electrons moving in a given external potential $v(r)$?. For such a system, the HK variational principle takes the form

$$E_{v(r)}[\tilde{n}] \equiv \int v(r)\tilde{n}(r)dr + T_s[\tilde{n}(r)] \tag{4.15}$$

$$\geq E\,, \tag{4.16}$$

where (assuming that $\tilde{n}(r)$ is *VR* for non-interacting electrons),

$T_s[\tilde{n}(r)] \equiv$ kinetic energy of the groundstate of *non-interacting* electrons with density distribution $\tilde{n}(r)$. (4.17)

The Euler-Lagrange equations, embodying the fact that the expression (4.14) is stationary with respect to variations of $\tilde{n}(r)$ which leave the total number of electrons unchanged, is

$$\delta E_v[\tilde{n}(r)] \equiv \int \delta\tilde{n}(r)\,\{v(r) + \frac{\delta}{\delta\tilde{n}(r)}T_s[\tilde{n}(r)]\mid_{\tilde{n}=n} - \epsilon\}dr = 0\;, \tag{4.18}$$

where $\tilde{n}(r)$ is the exact groundstate density for $v(r)$. Here ε is a Lagrange multiplyer to assure particle conservation. Now in this soluble, non-interacting case we know that the groundstate energy and density can be obtained by calculating the eigenfunctions $\varphi_j(r)$ and eigenvalues ε_j of non-interacting, single-particle equations

$$(-\frac{1}{2}\nabla^2 + v(r) - \epsilon_j)\varphi_j(r) = 0\,, \tag{4.19}$$

yielding

$$E = \sum_{j=1}^{N}\epsilon_j\;;\; n(r) = \sum_{j=1}^{N}\mid\varphi_j(r)\mid^2 . \tag{4.20}$$

(Here j labels both orbital quantum numbers and spin indices, ± 1).

Returning now to the problem of *interacting* electrons, which had previously been addressed approximately by the single-particle-like Hartree equations, we deliberately wrote the functional $F[\tilde{n}(r)]$ of Eq. (4.8) in the form

$$F[\tilde{n}(r)] \equiv T_s[\tilde{n}(r)] + \frac{1}{2}\int \frac{\tilde{n}(r)\tilde{n}(r')}{|r-r'|} dr dr' + E_{xc}[\tilde{n}(r)] \ , \qquad (4.21)$$

where $T_s[\tilde{n}(r)]$ is the kinetic energy functional for *non-interacting* electrons, Eq. (4.15). The last term, $E_{xc}[\tilde{n}(r)]$, the so-called exchange-correlation energy functional is then *defined* by Eq. (4.21). The HK variational principle for *interacting* electrons now takes the form,

$$E_v[\tilde{n}(r)] \equiv \int v(r)\tilde{n}(r)dr + T_s[\tilde{n}(r)] + \frac{1}{2}\int \frac{\tilde{n}(r)\tilde{n}(r')}{|r-r'|} dr dr' + E_{xc}[\tilde{n}(r)] \qquad (4.22)$$

$$\geq E \, .$$

The corresponding Euler-Lagrange equations, for a given total number of electrons has the form

$$\delta E_v[\tilde{n}(r)] = \int \delta\tilde{n}(r)\{v_{eff}(r) + \frac{\delta}{\delta\tilde{n}(r)} T_s[\tilde{n}(r)]\ |_{\tilde{n}(r)=n(r)} - \epsilon\}\, dr = 0 , \qquad (4.23)$$

where

$$v_{eff}(r) \equiv v(r) + \int \frac{n(r')}{|r-r'|} dr' + v_{xc}(r) \qquad (4.24)$$

and

$$v_{xc}(r) \equiv \frac{\delta}{\delta\tilde{n}(r)} E_{xc}[\tilde{n}(r)]\ |_{\tilde{n}(r)=n(r)} . \qquad (4.25)$$

Now the *form* of Eq. (4.23) is identical to that of Eq. (4.18) for non-interacting particles moving in an effective external potential v_{eff} instead of $v(r)$, and so we conclude that the minimizing density $n(r)$ is given by solving the single-particle equation

$$(-\frac{1}{2}\nabla^2 + v_{eff}(r) - \epsilon_j)\ \varphi_j(r) = 0 \ , \qquad (4.26)$$

with

$$n(r) = \sum_{j=1}^{N} |\varphi_j(r)|^2 \ , \qquad (4.27)$$

$$v_{eff}(r) = v(r) + \int \frac{n(r')}{|r-r'|} dr' + v_{xc}(r) , \qquad (4.28)$$

where $v_{xc}(r)$ is the *local* exchange-correlation potential, depending functionally on the entire density distribution $\tilde{n}(r)$, as given by Eq. (4.25). These self-consistant equations are now called the Kohn-Sham (KS) equations.

The groundstate *energy* is given by

$$E = \sum_j \epsilon_j + E_{xc}[n(r)] - \int v_{xc}(r)n(r)dv - \frac{1}{2}\int \frac{n(r)n(r')}{|r-r'|} . \quad (4.29)$$

If one neglects E_{xc} and v_{xc} altogether, the KS equations (4.26)-(4.29), reduce to the self-consistent Hartree equations.

The KS theory may be regarded as the formal exactification of Hartree theory. With the *exact* E_{xc} and v_{xc} all many body effects are in principle included. Clearly this directs attention to the functional $E_{xc}[\tilde{n}(r)]$. The practical usefulness of groundstate DFT depends entirely on whether approximations for the functional $E_{xc}[\tilde{n}(r)]$ could be found, which are at the same time sufficiently simple and sufficiently accurate. The next section V briefly describes the development and current status of such approximations.

Remarks:

1. The exact effective single particle potential $v_{eff}(r)$ of KS theory, Eq. (4.28) can be regarded as that unique, fictitious external potential which leads, for non-interacting particles, to the same physical density $n(r)$ as that of the interacting electrons in the physical external potential $v(r)$. Thus if the physical density $n(r)$ is independently known (from experiment or–for small systems–from accurate, wavefunction-based calculations) $v_{eff}(r)$ and hence also $v_{xc}(r)$ can be directly obtained from the density $n(r)$ [15].
2. Because of their linkage to the exact physical density $n(r)$, the KS single particle wavefunctions $\varphi_j(r)$ may be considered as "density-optimal", while, of course, the Hartree-Fock HF wavefunctions $\varphi_j^{HF}(r)$ are "total energy-optimal" in the sense that their normalized determinant leads to the lowest groundstate energy attainable with a single determinant. Since the advent of DFT the term "exchange energy" is often used for the exchange energy computed with the exact KS $\varphi_j(r)$, and not with the HF φ_j^{HF}. (For the uniform electron gas the two definitions agree; typically the differences are very small).
3. Neither the exact KS wavefunctions φ_j nor energies ε_j have any known, directly observable, strict meaning, except for a) the connection (4.27) between the φ_j and the true, physical density $n(r)$; and b) the fact that the magnitude of the highest occupied ε_j, relative to the vacuum equals the ionization energy [16].

In concluding this Section we remark that most practical application of DFT use the KS equations, rather than the generally less accurate HK formulation.

V. APPROXIMATION FOR $E_{xc}[n(r)]$: FROM MATHEMATICS TO PHYSICAL SCIENCE

So far DFT has been presented as a formal *mathematical framework* for viewing electronic structure from the perspective of the electron-density $n(r)$. This mathematical framework has been motivated by physical considerations, but to make concrete use of it we require effective approximations for $F[n(r)]$ in the HK formulation, and for $E_{xc}[n(r)]$ in the KS formulation. *These approximations reflect the physics of electronic structure and come from outside of DFT.* In this account I limit myself to the much more extensively used functional E_{xc}.

The most important approximations for $E_{xc}[n(r)]$ have a quasi-local form. As will be discussed in Section 5B, $E_{xc}[n(r)]$ can be written in the form

$$E_{xc}[n(r)] = \int e_{xc}(r; [n(\tilde{r}])n(r)dr \,, \tag{5.1}$$

where $e_{xc}(r;[n(\tilde{r})])$ represents an exchange-correlation (xc) energy/particle at the point r, which is a functional of the density distribution $n(\tilde{r})$. It depends primarily on the density $n(\tilde{r})$ at points $\tilde{r}$ near r, where "near" is a microscopic distance such as the local Fermi wavelength $\lambda_F(r) \equiv [3\pi^2 n(r)]^{-1/3}$ or TF screening length, typically of similar magnitude. The general form of Eq. (5.1), representing the total E_{xc} as an integral over all space of a suitable integrand is similar to the treatment of kinetic energy in Thomas-Fermi theory, Eq. (4.10). All components of the KS energy, can be expressed in terms of the 1- and 2- particle density matrices of the interacting and non-interacting system $n_1(r_1;r_1')$, $n_2(r_1,r_2;r_1', r_2')$ and $n_1^0(r_1;r'_1)$, $n_2^0(r_1r_2;r'_1r'_2)$, all corresponding to and uniquely defined by the *same* physical $n(r)$; their calculation involves these Green's functions primarily for arguments, such as (r_1,r_1') and $(r_1,r_2;r_1'r_2')$, which are microscopically close to one another; furthermore, for given r_1, these Green's functions depend only on the form of $n(\tilde{r})$ for $\tilde{r}$ near r_1 – the property of "nearsightedness" previously mentioned[10]. This leads immediately to the form (5.1) for $E_{xc}[n(r)]$, where e_{xc} is a nearsighted functional of $n(\tilde{r})$.

We now briefly discuss several implementations of this quasi-local approach.

A. The Local Density Approximation (LDA)

The simplest, and at the same time remarkably serviceable, approximation for $E_{xc}[n(r)]$ is the so-called local density approximation (LDA),

$$E_{xc}^{LDA} \equiv \int e_{xc}(n(r))\, n(r)dr, \tag{5.2}$$

where $e_{xc}(n)$ is the exchange-correlation energy of a *uniform* electron gas of density n[2]. The exchange part is elementary and given, in atomic units, by

$$e_x(n) \equiv -\frac{0.458}{r_s}\,, \tag{5.3}$$

where r_s, is the radius of a sphere containing one electron and given by $(4\pi/3)r_s^3 = n^{-1}$. The correlation part was first estimated by E.P. Wigner[17]

$$\epsilon_c(n) = -\frac{0.44}{r_s + 7.8}\,, \tag{5.4}$$

and more recently with a high precision of about ± 1 % by D.M. Ceperly[18] using Monte Carlo methods.

Remarks:

1. The LDA, obviously exact for a uniform electron gas, was a priori expected to be useful only for densities varying slowly on the scales of the local Fermi wavelength λ_F and TF wavelength, λ_{TF}. In atomic systems these conditions are rarely well satisfied and very often seriously violated. Nevertheless the LDA has been found to give extremely useful results for most applications. This has been at least partly rationalized by the observation that the LDA satisfies a sum rule which expresses the normalization of the exchange correlation hole. In other words, given that an electron is at r, the conditional electron density $n(r,r')$ of the other electrons is depleted near r in comparison with the average density $n(r')$ by the hole distribution $n_h(r';r)$ which integrates to 1.
2. The solution of the KS equation in the LDA is minimally more difficult than the solution of the Hartree equation and very much easier than the solution of the HF equations. Its accuracy for the exchange energy is typically within O(10 %), while the normally much smaller correlation energy is generally overestimated by up to a factor of 2. The two errors typically cancel partially.
3. Experience has shown that the LDA gives ionization energies of atoms, dissociation energies of molecules and cohesive energies with a fair accuracy of typically 10–20 %. However the LDA gives bond-lengths and thus the geometries of molecules and solids typically with an astonishing accuracy of ~ 1 %.
4. The LDA (and the LSDA, its extension to system with unpaired spins) can fail in systems, like heavy fermion systems, so dominated by electron-electron interaction effects that they lack any resemblance to non-interacting electron gases.

B. Beyond the Local Density Approximation

The LDA is the "mother" of almost all approximations currently in use in DFT. To discuss more accurate approximations we now introduce the concept of the *average xc hole* distribution around a given point r. The *physical xc hole* is given by

$$n_{xc}(r,r') = g(r,r') - n(r') \quad , \tag{5.5}$$

where $g(r,r')$ is the conditional density at r' given that one electron is at r. It describes the "hole" dug into the average density $n(r')$ by the electron at r. This hole is normalized

$$\int n_{xc}(r,r')dr' = -1 \quad , \tag{5.6}$$

which reflects a total "screening" of the electron at r, and generally is localized due to the combined effect of the Pauli principle and the electron-electron interaction. Of course, like everything else, it is a functional of the density distribution $n(\tilde{r})$. To define the *average xc* hole one introduces a fictitious, λ-depentent Hamiltonian, H_λ for the many body system, $0 \leq \lambda \leq 1$, which differs from the physical Hamiltonian, $H_{\lambda=1}$, by the two replacements

$$\frac{e^2}{\mid r_i - r_j \mid} \longrightarrow \frac{\lambda e^2}{\mid r_i - r_j \mid} , \tag{5.7}$$

$$v(r) \longrightarrow v_\lambda(r) , \tag{5.8}$$

where the fictitious $v_\lambda(r)$ is so chosen that for all λ in the interval (0,1) the corresponding density equals the physical density, $n(r)$:

$$n_\lambda(r) \equiv n_{\lambda=1}(r) = n(r) . \tag{5.9}$$

The procedure (5.2), (5.3) represents an interpolation between the KS system ($\lambda = 0$) and the physical system ($\lambda = 1$). The average *xc* hole density $\bar{n}(r,r')$ is then defined as

$$\bar{n}_{xc}(r, r') = \int_0^1 d\lambda n_{xc}(r, r'; \lambda) . \tag{5.10}$$

Its importance stems from the exact result, proved independently in three important publications[19], that

$$E_{xc} = \frac{1}{2} \int dr dr' \frac{n(r)\bar{n}_{xc}(r, r')}{\mid r - r' \mid} . \tag{5.11}$$

An equivalent expression is[20]

$$E_{xc} = -\frac{1}{2} \int dr n(r) \bar{R}_{xc}^{-1}(r, [n(\tilde{r})] , \tag{5.12}$$

where

$$\bar{R}_{xc}^{-1}(r, n(\tilde{r})) \equiv \int dr' \frac{(-\bar{n}_{xc}(r, r'[n(\tilde{r})])}{\mid r - r' \mid} , \tag{5.13}$$

is the moment of degree (–1) of $\bar{n}_{xc}(r,r')$, i.e., minus the inverse radius of the λ-averaged *xc* hole. Comparison of Eqs. (5.12) and (5.1) gives the very physical, formally exact relation

$$e_{xc}(r; [n(\tilde{r})]) = -\frac{1}{2} R_{xc}^{-1}(r; [n(\tilde{r})]) . \tag{5.14}$$

Gradient Expansion and Generalized Gradient Approximation

Since $R_{xc}^{-1}(r)$ is a functional of $n(\tilde{r})$, expected to be (predominantly) short-sighted, we can formally expand $n(\tilde{r})$ around the point r which we take to be the origin:

$$n(\tilde{r}) = n + \sum n_i \tilde{r}_i + \frac{1}{2} \sum n_{ij} \tilde{r}_i \tilde{r}_j + ..., \tag{5.15}$$

where $n \equiv n(0)$, $n_i \equiv \nabla_i\, n(r)\,\big|_{r=0}$ etc., and then consider $R_{xc}(r)$ as a function of the coefficients $n, n_i, n_{ij}, ...$. Ordering in powers of the differential operators and respecting the scalar nature of R_{xc}^{-1} gives

$$\begin{aligned} R_{xc}^{-1}(r) &= F_0(n(r)) + F_{21}(n(r))\, \nabla^2 n(r) + F_{22}(n(r)) \\ &\quad \times \sum (\nabla_i n(r))(\nabla_i (n(r)) + ... \end{aligned} \tag{5.16}$$

When this is substituted into Eq. (5.12) for E_{xc} it leads (after an integration by parts) to the gradient expansion

$$E_{xc} = E_{xc}^{LDA} + \int G_2(n)(\nabla n)^2 dr + \int [G_4(n)(\nabla^2 n)^2 + ...]\, dr + ..., \tag{5.17}$$

where $G_2(n)$ is a universal functional of $n^{[2]}$. In application to real systems this expansion has generally been disappointing, indeed often worsened the results of the LDA.

The series (5.15) can however be formally resummed to result in the following sequence

$$E_{xc}^0 = \int \epsilon(n(r)) n(r) dr \quad (LDA), \tag{5.18}$$

$$E_{xc}^{(1)} = \int f^{(1)}(n(r), |\nabla n(r)|) n(r) dr \quad (GGA), \tag{5.19}$$

$$E_{xc}^{(2)} = \int f^{(2)}(n(r), |\nabla n(r)|\ \nabla^2 n(r))\, dr, \tag{5.20}$$

E_{xc}^0 is the (LDA), requiring the independently calculated function of one variable, $x \equiv n$. $E_{xc,}^{(1)}$ the so-called generalized gradient approximation (GGA) requires the independently calculated function of two variables, $x \equiv n$, $y \equiv |\nabla n|$ etc.

Thanks to much thoughtful work important progress has been made in deriving successful GGA's of the form (5.19). Their contruction has made use of sum rules, general scaling properties, asymptotic behavior of effective potentials and densities in the tail regions of atoms and their aggregates. In addition, A. Becke in his work on GGAs, introduced some numerical fitting parameters which he determined by optimizing the accuracy of atomization energies of standard sets of molecules. This subject was recently reviewed[21]. We mention here some of the leading contributors: A.D. Becke, D.C. Langreth, M. Levy, R.G. Parr, J.P. Perdew, C. Lee, W. Yang.

In another approach A. Becke introduced a successful *hybrid* method

$$E_{xc}^{hyb} = \alpha E_x^{KS} + (1-\alpha) E_{xc}^{GGA}, \tag{5.21}$$

where E_x^{KS} is the exchange energy calculated with the exact KS wavefunctions, E_{xc}^{GGA} is an appropriate GGA, and α is a fitting parameter[22]. The *form* of this linear interpolation can be rationalized by the λ-integration in Eq. (5.10), with the lower limit corresponding to pure exchange.

Use of GGAs and hybrid approximations instead of the LDA has reduced errors of atomization energies of standard sets of small molecules, consisting of light atoms, by factors of typically 3-5. The remaining errors are typically ± (2–3) kg moles per atom, about twice as high as for the best current wave-function methods. This improved accuracy, the ease of calculation, together with the previously emphasized capability of DFT to deal with systems of very many atoms, has, over a period of relatively few years beginning about 1990, made DFT a significant component of quantum chemistry.

For other kinds of improvements of the LDA, including the weighted density approximation (WDA) and self-interaction corrections (SIC) we refer the reader to the literature, e.g.[21].

Before closing this section I remark that the treatments of *xc*-effects in the LDA and all of its improvements, mentioned above, is completely inappropriate for all those systems or subsystems for which the starting point of an electron gas of slowly varying density $n(r)$ is fundamentally incorrect. Examples are a) the electronic Wigner crystal; b) Van der Waals (or polarization) energies between non-overlapping subsystems; c) the electronic tails evanescing into the vacuum near the surfaces of bounded electronic systems. However this does not preclude that DFT with appropriate, *different* approximations could successfully deal with such problems (See Sec. VII).

VI. GENERALIZATIONS AND QUANTITATIVE APPLICATIONS

While DFT for non-degenerate, non-magnetic systems has continued to progress over the last several decades, the DFT paradigm was also greatly extended and generalized in several directions. The purpose of this section is to give the briefest mention of these developments. For further details we refer to two monographs[23], [24] and a recent set of lecture notes[21]

A. Generalizations

a. Spin DFT for spin polarized systems: $v(r)$, $B_z(r)$; $n(r)$, $(n_\uparrow(r) - n_\downarrow(r))$.
b. Degenerate groundstates: $v(r)$; $n_\nu(r)$ $\nu = 1,...M$; E_0.
c. Multicomponent systems (electron hole droplets, nuclei): $v_\alpha(r)$; $n_\alpha(r)$; E_0.
d. Ensemble DFT for M degenerate groundstates: $v(r)$; $n(r)$ $(\equiv M^{-1}$ $(Tr$ $n_\nu(r))$; E_0.

e. Free energy at finite temperatures T: $v(r)$; $n(r)$, Ω (grand potential).
f. Superconductors with electronic pairing mechanisms: $v(r)$, $\Delta(r)$ (gap function); $n_{norm}(r)$, $n_{super}(r)$, E_0.
g. M excited states equi-ensembles $v(r)$, $\bar{n}(r) \equiv M^{-1}\sum_1^M n_m(r))$, $\bar{E} \equiv M^{-1}\sum_1^M, E_m$.
h. Relativistic electrons.
i. Current-density functional theory diamagnetism: $v(r)$, *curl* $A(r)$; $n(r)$, *curl* $j(r)$; E_0.
j. Time-dependent phenomena: $v(r,t)$; $n(r,t)$, and excited states $v(r)e^{-i\omega t}$; $n(r)e^{-i\omega t}$; $E_j - E_i = \omega$.
k. Bosons (instead of fermions) $v(r)$; $n(r)$; E_0.
l. Combination of DFT with molecular dynamics or Monte Carlo methods (especially for determinations of structures). (Car-Parrinello method).
m. Combination of the LDA with Hubbard on-site repulsion parameter U ("LDA + U").

This incomplete list is only intended to give a general sense of the great diversity of contexts in which the basic concept of DFT has been found useful.

B. Applications

To do any kind of justice to the many thousands of applications of DFT to physical and chemical systems is entirely impossible within the framework of this lecture. So I will, quite arbitrarily, choose one example, the spin susceptibility of the alkali metals (Table 1)[25].

Table I. Spin Susceptibility of the Alkali Metals

	χ/χ_0	
Metal	Variational Theory	Experiment
Li	2.66	2.57
Na	1.62	1.65
K	1.79	1.70
Rb	1.78	1.72
Cs	2.20	2.24

After S.H. Vosko *et al.*,[25]. χ_0 is the Pauli susceptibility of a free electron gas.

This is an early, completely parameter-free calculation. It uses only the independently calculated external pseudo-potential $v(r)$ and the exchange correlation energy of a spatially uniform, magnetized electron gas (the so-called local spin density approximation, LSDA). The only input specific to each metal is the atomic number *Z*. Note how accurately the theoretical results agree with the rather irregular sequence of experimental data. The deviations of the ratio (χ/χ_0) from 1, are due, in comparable degree, to the combination of the effects of the non-uniform, periodic potentials and the electron-electron interactions.

Of course these metals have, over most of space, fairly uniform densities, which makes them favorable test-cases for local spin density calculations. For other classes of systems and their properties the accuracies can be considerably poorer, with the exception of the already mentioned very accurate results for structures, with typically a 1% error (which is still somewhat astonishing to me).

Inclusion of gradient corrections and/or hybrid schemes have improved calculated energies for large classes of chemical applications by typically almost an order of magnitude; in physical applications the improvement is usually less dramatic. Accuracies of geometric parameters remain at the 1 % level.

VII. CONCLUDING REMARKS

DFT has now been widely accepted by both physicists and chemists. For periodic solids it is sometimes referred to as the standard model. In chemistry DFT complements traditional wave-function based methods, particularly for systems with very many atoms($\gtrsim O(10)$).

In cases where DFT currently works still rather poorly (e.g. long range polarization energies; regions of evanescent electron densities; partially filled electronic shells; reaction barriers) it often provides clues of how our present understanding of electronic structure in *real space coordinates* needs to be modified.

Looking into the future I expect that wavefunction-based and density-based theories will, in complementary ways, continue not only to give us quantitatively more accurate results, but also contribute to a better physical/chemical understanding of the electronic structure of matter.

REFERENCES

[1] P. Hohenberg and W. Kohn, *Phys. Rev.*, **136**, B864 (1964).

[2] W. Kohn and L. J. Sham, *Phys. Rev.* **140**, A1133 (1965).

[3] E. Schroedinger, *Am. Physik* **79**, 361 (1926).

[4] L.H. Thomas, *Proc. Camb. Phil. Soc.* **23**, 542 (1927).

[5] E. Fermi, Atti. *Accad. Nazl. Lincei* **6**, 602 (1927).

[6] W. Heitler and F. London, Zs. f. Physics **44**, 455 (1927).

[7] R.S. Mullikan, *Phys. Rev.* **32**, 186 (1928).

[8] D.R. Hartree, *Proc. Camb. Phil. Soc.* **24**, 89 (1928), F. Fock, Zs. f. Phys., **61**, 126 (1930).

[9] H.M. James and A.S. Coolidge, *J. of Chem. Phys.* **1**, 825(1933).

[10] W. Kohn, *Phys. Rev.* **76**, 17 (1996).

[11] J.H. Van Vleck, *Phys. Rev.* **49**, 232 (1936).

[12] W. Kohn, *Proc. Int'l School of Physics*, Enrico Fermi, Course LXXXIX, p. 4 (1985).

[13] M. Levy, *Phys. Rev. A* **26**, 1200 (1982).

[14] E. Lieb, in Physics as Natural Philosophy: Essays in Honor of Laszlo Tisza on his 75th Birthday, Ed. by A. Shimony and H. Feshbach (Cambridge, Mass., 1982), p. 111.

[15] Y. Wang and R. Parr, *Phys. Rev. A* **47**, 1591 (1993).

[16] C.O. Almbladh and U. von Barth, *Phys. Rev. B* **31**, 3231 (1985).

[17] E.P. Wigner, *Trans. Faraday Soc.* **34**, 678 (1938).

[18] D.M. Ceperley, *Phys. Rev. B* **18**, 3126 (1978); D.M. Ceperly and B.J. Alder, *Phys. Rev. Lett.* **45**, 566 (1980).

[19] J. Harris and R.O. Jones, *J. Phys. F* **4**, 1170 (1974); D.C. Langreth and J.P. Perdew, *Sol. State Comm.* **17**, 1425 (1975); O. Gunnarsson and B.I. Lundquist, *Phys. Rev. B* **13**, 4274 (1976).

[20] W. Kohn and A.E. Mattsson, *Phys. Rev. Lett.* **81**, 16 (1998).

[21] J.P. Perdew and S. Kurth in Density Functionals: Theory and Applications, p. 8, D. Joubert (Ed.), Springer, Berlin (1998).

[22] A.D. Becke, *J. Chem. Phys.* **104**, 1040 (1996).
[23] R.G. Parr and W. Yang, Density Functional Theory of Atoms and Molecules, Oxford University Press, Oxford, 1989.
[24] R.M. Dreizler and E.K.V. Gross, Density Functional Theory, Springer, Berlin, 1990.
[25] S.H. Vosko *et al., Phys. Rev. Lett* **35**, 1725 (1975).

John A. Pople

JOHN A. POPLE

My early life was spent in Burnham-on-Sea, Somerset, a small seaside resort town (population around 5000) on the west coast of England. I was born on October 31, 1925 and lived there with my parents until shortly after the end of the Second World War in 1946. No member of my family was involved in any scientific or technical activity. Indeed, I was the first to attend a university.

My father, Keith Pople, owned the principal men's clothing store in Burnham. In addition to selling clothes in the shop, he used to drive around the surrounding countryside with a car full of clothes for people in remote farms and villages. He was resourceful and made a fair income, considering the economic difficulties during the depression of the 1930s. My great-grandfather had come to Burnham around 1850 and set up a number of local businesses. He had a large family and these were split up among his children. As a result, I had relatives in many of the other businesses in the town. My grandfather inherited the clothing shop and this passed to my father when he returned from the army at end of the First World War.

My mother, Mary Jones, came from a farming background. Her father had moved from Shropshire as a young man and had farmed near Bath for most of his life. I suspect that he would have preferred to be a teacher, for he had a large collection of books and encyclopedias. He wanted my mother to be a schoolteacher, but this did not happen. Instead, she became a tutor to children in a rich family and, later, a librarian in the army during the first war. Most of her relatives were farmers in various parts of Somerset and Wiltshire so, as small children, my younger brother and I spent much time staying on farms.

Both of my parents were ambitious for their children; from an early age I was told that I was expected to do more than continue to run a small business in this small town. Education was important and seen as a way of moving forward. However, difficulties arose in the choice of school. There was a good preparatory school in Burnham but, as part of the complex English class system, it was not open to children of retail tradesmen, even if they could afford the fees. The available alternative was unsatisfactory and my parents must have agonized over what to do. Eventually, they decided to send us to Bristol Grammar School (BGS) in the nearest big city thirty miles away. BGS was the prime day school for boys, catering mainly to middle class families resident in the city, although it received a government grant for accepting about thirty boys a year from the state elementary schools. I went there in the spring of 1936 at the age of ten. Some arrangement had to be made for boarding and

I used to return home by train each weekend. This I found unappealing and eventually I persuaded my parents to allow me to commute daily – two miles by bicycle, twenty-five miles by train and one mile on foot. I continued to do this during the early part of the war, a challenging experience during the many air attacks on Bristol. Often, we had to wend our way past burning buildings and around unexploded bombs on the way to school in the morning. Many classes had to be held in damp concrete shelters under the playing fields. In spite of all these difficulties, the school staff coped well and I received a superb education.

At the age of twelve, I developed an intense interest in mathematics. On exposure to algebra, I was fascinated by simultaneous equations and rapidly read ahead of the class to the end of the book. I found a discarded textbook on calculus in a wastebasket and read it from cover to cover. Within a year, I was familiar with most of the normal school mathematical curriculum. I even started some research projects, formulating the theory of permutations in response to a challenge about the number of possible batting orders of the eleven players in a cricket team. For a very short time, I thought this to be original work but was mortified to find $n!$ described in a textbook. I then attempted to extend $n!$ to fractional numbers by various interpolation schemes. Despite a lot of effort, this project was ultimately unsuccessful; I was angry with myself when I learned of Euler's solution some years later. However, these early experiences were valuable in formulating an attitude of persistence in research.

All this mathematical activity was kept secret. My parents did not comprehend what I was doing and, in class, I often introduced deliberate errors in my exercises to avoid giving an impression of being too clever. My grades outside of mathematics and science were undistinguished so I usually ended up several places down in the monthly class order. This all changed suddenly three years later when the new senior mathematics teacher, R. C. Lyness, decided to challenge the class with an unusually difficult test. I succumbed to temptation and turned in a perfect paper, with multiple solutions to many of the problems. Shortly afterwards, my parents and I were summoned to a special conference with the headmaster at which it was decided that I should be prepared for a scholarship in mathematics at Cambridge University. During the remaining two years at BGS, I received intense personal coaching from Lyness and the senior physics master, T.A. Morris. Both were outstanding teachers. The school, like many others in Britain, attached great importance to the placement of students at Oxford or Cambridge. Most such awards were in the classics and I think that the mathematics and science staff were very anxious to compete. Ironically, during the last two years at BGS, I abandoned chemistry to concentrate on mathematics and physics. In 1942, I travelled to Cambridge to take the scholarship examination at Trinity College, received an award and entered the university in October 1943.

In the middle of the war, most young men of my age were inducted into the armed forces at the age of seventeen. However, a small group of students in mathematics, science and medicine was permitted to attend university before taking part in wartime research projects such as radar, nuclear explosives, code-breaking and the like. This was a highly successful project and many of my predecessors in earlier years made important contributions to the war effort. The plan was to complete all degree courses in only two years, followed by secondment to a government research establishment. In my case, I completed Part II of the mathematical tripos in May 1945, just as the European war was ending. In fact, it was hard to concentrate on the examinations because of the noisy celebrations going on in the streets outside. The government no longer had need for my services and the university was under great pressure to make room for the deluge of exservicemen as they were demobilized from the armed forces. So, I had to leave Cambridge and take up industrial employment for a period. This was with the Bristol Aeroplane Company, close to where I had attended school. There was little to do there and I had a period of enforced idleness as changing employment was illegal at the time (part of the obsession for a planned economy in postwar Britain).

In 1945, I had little idea of what my future career might be. My interest in pure mathematics began to wane; after toying with several ideas, I finally resolved to use my mathematical skills in some branch of science. The choice of a particular field was postponed, so I devoted much of my time to pestering government offices for permission to return to Cambridge and resume my studies. In the late summer of 1947, I finally received a letter informing me that an unexpectedly large number of students had failed their examinations and a few places were available. So, in October 1947, I returned to Cambridge to begin a career in mathematical science.

Cambridge in 1947 had greatly changed since 1943. The university was crowded with students in their late twenties who had spent many years away at the war. In addition, the lectures were given by the younger generation who had also been away on research projects. There was a general air of excitement as these people turned their attention to new scientific challenges. I remained as a mathematics student but spent the academic year 1947–8 taking courses in as many branches of theoretical science as I could manage. These included quantum mechanics (taught in part by Dirac), fluid dynamics, cosmology and statistical mechanics. Most of the class opted for research in fundamental areas of physics such as quantum electrodynamics which was an active field at the time. I felt that challenging the likes of Einstein and Dirac was overambitious and decided to seek a less crowded (and possibly easier) branch of science. I developed an interest in the theory of liquids, particularly as the statistical mechanics of this phase had received relatively little attention, compared with solids and gases. I approached Fred Hoyle, who was giving the statistical mechanics lectures (following the death of R.H. Fowler). However, his current interests were in the fields of astrophysics and cosmology, which I found rather remote from everyday experience. I next approached Sir John Lennard-Jones (LJ), who had published important papers on a theory of liquids in 1937. He held the chair of theoretical chemistry at Cambridge and was lecturing on molecular orbital theory at the time. When I approached him, he told me that his interests were currently in electronic structure but he would very possibly return to liquid theory at some time. On this basis, we agreed that I would become a research student with him for the following year. Thus, after the examinations in June 1948, I began my career in theoretical chemistry at the beginning of July. I had almost no chemical background, having last taken a chemistry course at BGS at the age of fifteen.

Other important events took place in my life at this time. In late 1947, I was attempting to learn to play the piano and rented an instrument for the attic in which I lived in the most remote part of Trinity College. The neighbouring room was occupied by the philosopher Ludwig Wittgenstein, who had retired to live in primitive and undisturbed conditions in the same attic area. There is some evidence that my musical efforts distracted him so much that he left Cambridge shortly thereafter. In the following year, I sought out a professional teacher. The young lady I contacted, Joy Bowers, subsequently became my wife. We were married in Great St. Mary's Church, Cambridge in 1952, after a long courtship. Like many other Laureates, I have benefitted immeasurably from the love and support of my wife and children. Life with a scientist who is often changing jobs and is frequently away at meetings and on lecture tours is not easy. Without a secure home base, I could not have made much progress.

The next ten years (1948–1958) were spent in Cambridge. I was a research student until 1951, then a research fellow at Trinity College and finally a lecturer on the Mathematics Faculty from 1954 to 1958. Cambridge was an extraordinarily active place during that decade. I was a close observer of the remarkable developments in molecular biology, leading up to the double-helix papers of Watson and Crick. At the same time, the X-ray group of Perutz and Kendrew (introduced to the Cavendish Laboratory by Lawrence Bragg) were achieving the first definitive structures of proteins. Elsewhere, Hoyle, Bondi and Gold were arguing their case for a cosmology of continuous creation, ultimately disproved but vigorously presented. Looking through the list of earlier Nobel laureates, I note a large number with whom I became acquainted and with whom I interacted during those years as they passed through Cambridge.

In the theoretical chemistry department, LJ was professor and Frank Boys started as lecturer in September 1948. I began research with some studies of the water molecule, examining the nature of the lone pairs of electrons. This was an initial step towards a theory of hydrogen bonding between water molecules and a preliminary, rather empirical study of the structure of liquid water. This fulfilled my initial objective of dealing with properties of liquids and gained me a Ph.D. and a research fellowship at Trinity College. This highly competitive stage accomplished, I was able to relax a bit and formulate a more general philosophy for future research in chemistry. The general plan of developing mathematical models for simulating a whole chemistry was formulated, at least in principle, some time late in 1952. It is the progress towards those early objectives that is the subject of my Nobel lecture.

At that early date, of course, computational resources were limited to hand calculators and very limited access to motorized electric machines. So my early notes show attempts to simplify theories enough to turn them into practical possibilities. The work paralleling studies of Pariser and Parr led to what became known as PPP theory. This was not a complete model but rather one applicable to systems with only one significant electron per atom. It did fit the general form of conjugated hydrocarbons and achieved some notoriety. In 1953, Bob Parr came to Cambridge to spend a year with Frank Boys. We shared an office and had many valuable discussions; he was to have a major influence on my future. I talked about PPP theory when I began to speak at international meetings in 1955.

In addition to the PPP work, I started theoretical work on other topics in physical chemistry. I began supervision of research students in 1952, beginning with David Buckingham, who completed a masterly thesis on properties of compressed gases. He was the first of a long list of remarkably able and dedicated students who have worked with me over the years. In 1954, LJ was succeeded as professor of theoretical chemistry by Christopher Longuet-Higgins, who was joined by Leslie Orgel shortly afterwards. I continued to spend a lot of time in the chemistry department, although by then I had undertaken new teaching responsibilities as a lecturer in mathematics. The department was crowded and active in those years. Among the many visitors were Linus Pauling, Robert Mulliken, Jack Kirkwood, Clemens Roothaan and Bill Schneider. Frank Boys was also managing a lively group of students.

At the end of 1955, I developed an interest in nuclear magnetic resonance, which was then emerging as a powerful technique for studying molecular structure. At the urging of Bill Schneider, I agreed to spend two summers (1956 and 1957) at the National Research Council in Ottawa, Canada, working on the theoretical background of NMR. This was extremely stimulating for, at that time, we were measuring the spectra and interpreting the nuclear spin behaviour of many standard chemicals for the first time. My time there with Bill and Harold Bernstein led to a book, *High Resolution Nuclear Magnetic Resonance,* which was well received. This area was the main emphasis of my research during the final years in Cambridge.

By 1958, I had become dissatisfied with my mathematics teaching position at Cambridge. I had clearly changed from being a mathematician to a practicing scientist. Indeed, I was increasingly embarassed that I could no longer follow some of the more modern branches of pure mathematics, in which my undergraduate students were being examined. I therefore resolved to seek a new job with greater scientific content. After some hesitation, I accepted a position as head of the new Basics Physics Division at the National Physical Laboratory near London. This involved direction of experimental work and a considerable amount of administration. When I took the job, I hoped that the adminstrative burden would not be large enough to interfere with my research programme. Although I was given plenty of help, this turned out not to be so and I had a rather fallow period while I was there.

In the spring of 1961, I organized an international conference in Oxford, along with Charles Coulson and Christopher Longuet-Higgins. Bob Parr was an invited speaker and, during a break, he urged me to come and spend a sabbatical year at Carnegie Institute of Technology in Pittsburgh. This was an attractive suggestion and I arranged to come for the academic year 1961–2 with my family. By this time, Joy and I had three children and were expecting a fourth. We arrived in September, accompanied by a charming young Swedish *au pair,* Elisabeth Fahlvik. One of the most delightful side-effects of winning the Nobel Prize is the opportunity to meet her again after a gap of over thirty-six years.

By the time we arrived in Pittsburgh, Bob Parr had decided to leave for Johns Hopkins University and he did, in fact, leave in January. Nevertheless, we had a delightful year, travelling as a family over much of the eastern part of the U.S.A. During this period, I made up my mind to abandon my administrative job and seek an opportunity to devote as much time as possible to chemical research. I was approaching the age of forty, with a substantial publication record, but had not yet held any position in a chemistry department. When we returned to England in June, 1962, it was not clear where we might go for there were opportunities both in the U.K. and the U.S.A. Eventually, after much debate, we decided to return to Pittsburgh in 1964. Leaving England was a painful decision and we still have some regrets about it. However, at that time, the research environment for theoretical chemistry was clearly better in the U.S.

On my return to Pittsburgh, I resolved to go back to the fundamental problems of electronic structure that I had contemplated abstractly many years earlier. Prospects of really implementing model chemistries had improved because of the emerging development of high-speed computers. I was late in recognizing the role that computers would play in the field – I should not have been, for Frank Boys was continually urging the use of early machines back in Cambridge days. However, by 1964, it was clear that the development of an efficient computer code was one of the major tasks facing a practical theoretician and I learned the trade with enthusiasm. Mellon Institute, where I had an adjunct appointment, acquired a Control Data machine in 1966 and my group was able to make rapid progress in the dingy deep basement of that classic building. In 1967, Carnegie Tech and Mellon Institute merged to become Carnegie-Mellon University (CMU) and I remained on the faculty there until 1993. Almost all of the work honoured by the Nobel Foundation was done at CMU. That institution deserves much of the credit for their continuing support and encouragement over many years.

The scientific details of the Pittsburgh work are related, in part, in the accompanying lecture. Over the years, we were able to keep abreast with the rapid developments in computer technology. Around 1971, the work was moved to a Univac 1108 machine and then, in 1978, we were fortunate enough to acquire the first VAX/780 minicomputer from the Digital Equipment Corporation for use entirely within the chemistry department. This became a valuable workhorse as we began to distribute programs to the general chemical community. In more recent years, of course, the techniques have become available on small workstations and personal computers. The astonishing progress made in computer technology has had profound consequences in so many branches of theoretical science.

Our children were mostly brought up and educated in the Churchill suburb east of Pittsburgh. Each summer, we took them back to England for an extended period. By 1979, all had gone away and Joy and I decided to move again to Illinois, where our daughter had settled. In 1981, we set up house in Rogers Park, Chicago and then moved to Wilmette in 1988. Our family is now scattered in Chicago, Houston, Pittsbugh and Cork, Ireland. We have been blessed with ten grandchildren (an eleventh expected), who greatly enrich our lives in many ways.

From 1981 to 1993, I continued to run my research group in Pittsburgh, commuting frequently and communicating with my students by telephone and modem. Northwestern University kindly offered me an adjunct appointment and I became a full member of their faculty in 1993. I am very grateful to them for the opportunity to continue my research programme and interact with other members of the chemistry department.

I have had many opportunities to visit universities all over the world in the past fifty years. Among the most rewarding have been frequent trips to Australia and New Zealand, where Joy and I have wintered no fewer than nine times since 1982. The campus of the Australian National University, where Leo Radom became Professor after spending time with me as a post-doctoral fellow from 1968 to 1972, has become a second academic home – a great place for relaxed contemplation.

Israel and Germany are other countries with which I have become closely associated, having visited and collaborated many times. In the 1980s, I held a von Humboldt Award, which allowed me to spend some time in Erlangen, where I collaborated with Paul Schleyer on a large number of applications of the theory. In Israel, I have visited and lectured at all universities, including a period as Visiting Professor at the Technion, Haifa. In 1992, I was fortunate enough to receive the Wolf Prize in Chemistry at a ceremony in the Knesset.

I must emphasize that my contribution to quantum chemistry has depended hugely on work by others. The international community in our field is a close one, meeting frequently and exchanging ideas freely. I am delighted to have had students, friends and colleagues in so many nations and to have learned so much of what I know from them. This Nobel Award honours them all.

QUANTUM CHEMICAL MODELS

Nobel Lecture, December 8, 1998

by

JOHN A. POPLE

Department of Chemistry, Northwestern University, 2145 Sheridan Road, Evanston, Illinois 60208, USA

INTRODUCTION

The fundamental underpinnings of theoretical chemistry were uncovered in a relatively short period at the beginning of the present century. Rutherford's discovery of the nucleus in 1910 completed the identification of the constituent subparticles of atoms and molecules and was followed shortly thereafter by the Bohr treatment of electronic orbits in atoms, the "old quantum theory". The relation between the positive nuclear charge, atomic number and position of an atom in the periodic table was uncovered by 1913. It proved difficult to extend Bohr's orbits to a polyatomic situation and the next advance had to await the development of the wave theory of matter and the associated quantum mechanics in the early 1920s. By 1926, Heisenberg had developed matrix mechanics and Schrödinger had proposed the basic nonrelativistic wave equation governing the motion of nuclei and electrons in molecules. The latter,

$$H\Psi = E\Psi \tag{1}$$

is a differential eigenvalue equation for the energy E and wavefunction Ψ of a particular state. H is the Hamiltonian operator and Ψ depends on cartesian and spin coordinates of the component particles. The only further restrictions are the permutational symmetry requirements for Ψ (antisymmetry for fermions such as electrons and symmetry for bosons). A relativistic generalization of this equation was proposed a short time later by Dirac.

The Schrödinger equation is easily solved for the hydrogen atom and found to give results identical to the earlier treatment of Bohr. With inclusion of relativistic corrections via the Dirac equation, almost perfect agreement was found with experimental spectroscopic data. However, exact solution for any other system was not found possible, leading to a famous remark by Dirac in 1929:

"The fundamental laws necessary for the mathematical treatment of a large part of physics and the whole of chemistry are thus completely known, and the difficulty lies only in the fact that application of these laws leads to equations that are too complex to be solved"

This was a cry both of triumph and of despair. It marked the end of the process of fundamental discovery in chemistry but left a collossal mathematical task of implementation. In retrospect, the implied finality of the claim seems excessively bold. In 1929, there had only been one preliminary approximate quantum mechanical calculation on the hydrogen molecule by Heitler and London, leading to a value of the bond energy of only about 70% of the experimental value. Nevertheless, the physicists were highly confident and most moved on to study the internal structure of the nucleus during the 1930s. In fact, their boldness was apparently justified, for no significant failure of the full Schrödinger-Dirac treatment has ever been demonstrated.

This was the challenge presented to the early quantum chemists by 1930. Given the hopelessness of exact solution, how would it be possible to develop approximate mathematical procedures that could (a) assist the qualitative interpretation of chemical phenomena; and (b) provide predictive capability. Attempts to approach this problem by a model approach is the topic addressed here.

FEATURES OF THEORETICAL MODELS

A theoretical model for any complex process is an *approximate but well-defined* mathematical procedure of simulation. When applied to chemistry, the task is to use input information about the number and character of component particles (nuclei and electrons) to derive information and understanding of resultant molecular behavior. Five stages may be distinguished in the development and use of such a model:

Target

A target accuracy must be selected. A model is not likely to be of much value unless it is able to provide clear distinction between possible different modes of molecular behavior. As the model becomes quantitative, the target should be that data is reproduced and predicted within experimental accuracy. For energies, such as heats of formation or ionization potentials, a global accuracy of 1 kcal/mole would be appropriate.

Formulation

The approximate mathematical procedure must be precisely formulated. This should be *general and continuous* as far as possible. Thus, particular procedures for particular molecules or particular symmetries should be avoided. If this can be done, the procedure becomes a *full theoretical model chemistry,* which can be explored in detail as far as available resources permit.

Implementation

The formulated method has to be implemented in a form, which permits its application in reasonable times and at reasonable cost. In recent times, this stage involves the development of efficient and easily used computer programs. It is closely comparable to the stage of building equipment in an experimental investigation.

Verification
The next step is to test the model against known chemical facts to determine whether the target has been achieved. If quantitative accuracy is being sought, this can be done by various statistical criteria such as the root-mean-square difference between the results of the theoretical model and experimental data. In selecting such a data-set, it is important to make it as broad as possible, while limiting it to experimental facts known to be of high quality. If the results of such a comparison do meet the target requirements, the model may be said to be *validated.*

Prediction
Finally, if the model has been properly validated according to some such criterion, it may be applied to chemical problems to which the answer is unknown or in dispute. If the experimental data-set is sufficiently broad, there is a reasonable expectation that the results will be accurate to something like the target accuracy. This stage, of course, is the one of most interest to the larger chemical community.

One further aspect of theoretical models is the introduction of empirical parameterization. Models which utilize only the fundamental constants of physics are generally termed *ab initio*; if some parameters are introduced which are determined by fitting to some experimental data, the methods are *semi-empirical*. Clearly, there is a wide range of possible empiricism, as will be noted in subsequent parts of this article.

HARTREE-FOCK MODELS

During the 1930s, most work was of a qualitative nature, treating the electrons as moving in independent *molecular orbitals.* However, the foundations of the the orbital theory of many-electron systems was laid by Hartree, Fock and Slater. If the $2n$ electrons in a closed-shell molecule are assigned to a set of n molecular orbitals ψ_i ($i = 1,\ldots n$), the corresponding many-electron wavefunction can be written

$$\Psi = (n!)^{-1/2} det[(\psi_1\alpha)(\psi_1\beta)(\psi_2\alpha)...] \tag{2}$$

Here the ψ_i are taken to be orthonormal and α and β are spin functions. This single-configuration wavefunction is usually described as a *Slater determinant.*

If the molecular orbitals ψ_i are varied to minimize the energy, calculated as the expectation value of the full Hamiltonian H,

$$E = < \Psi|H|\Psi > \tag{3}$$

then the energy E is fully defined and, according to the variational principle, is an upper bound for the exact Schrödinger energy from the full wave equation (1). This procedure leads to a set of coupled differential equations for the ψ_i, as first derived by Fock. The method is known as *Hartree-Fock* theory, early applications having been made (to atoms) by Hartree.

Following the break due to World-War II, work on quantum chemistry resumed in a number of countries. In Cambridge, Lennard-Jones and his group (of which I became a member in 1948) reexamined the Hartree-Fock equations with a view to transforming the orbitals ψ_i into localized or equivalent orbitals, representing bonding and lone electron pairs, concepts widely used in the qualitative description of molecular structure. However, the coupled 3-dimensional differential equations appeared intractable and little progress was made towards their solution.

A major advance occurred in 1951 with the publication from Chicago of the Roothaan equations [1]. (Actually, these had been circulated in a report some time earlier.) Roothaan considered molecular orbitals that were restricted to be linear combinations of a set of prescribed 3-dimensional 1-electron functions $\chi_\mu(\mu = 1,2,..N, N>n)$.Thus

$$\psi_i = \sum_{\mu=1}^{N} c_{\mu i}\chi_\mu \tag{4}$$

Variation of the total energy (3) was then carried out with respect to the coefficients $c_{\mu i}$. This leads to a set of *algebraic* equations which can be written in matrix form (using real functions and atomic units throughout),

$$\mathrm{FC} = \mathrm{SCE} \tag{5}$$

where

$$F_{\mu\nu} = H_{\mu\nu} + \sum_{\lambda\sigma} P_{\lambda\sigma}[(\mu\nu|\lambda\sigma) - (\mu\lambda|\nu\sigma)/2] \tag{6}$$

$$H_{\mu\nu} = \int \chi_\mu H \chi_\nu d\tau \tag{7}$$

$$S_{\mu\nu} = \int \chi_\mu \chi_\nu d\tau \tag{8}$$

$$E_{ij} = \epsilon_i \delta_{ij} \tag{9}$$

$$P_{\mu\nu} = 2\sum_{i}^{n} c_{\mu i}c_{\nu i} \tag{10}$$

$$(\mu\nu|\lambda\sigma) = \int\int \chi_\mu(1)\chi_\nu(1)(1/r_{12})\chi_\lambda(2)\chi_\sigma(2)d\tau_1 d\tau_2 \tag{11}$$

In these and subsequent equations, we follow a useful practice of using roman suffixes for molecular orbitals ψ and greek for the expansion functions χ. H is the core Hamiltonian, describing motion of a single electron moving in the bare field of the nuclei. The eigenvalues ε_i are the one-electron *Fock energies,* the lowest n corresponding to the occupied molecular orbitals 1,2,...n.

These nonlinear equations provide a complete mathematical model if the prescribed functions χ_μ are uniquely specified by the nuclear positions. They are often referred to as *self consistent field* (SCF) equations. In the earliest versions of molecular orbital theory, the χ_μ were chosen to be the atomic orbitals of the component atoms, in which case the theory was described as LCAOSCF for 'linear combination of atomic orbitals'. More generally, the set $\{\chi_\mu\}$ is referred to as the *basis set.* Normal practice is to choose basis functions which are centered at the nuclei and depend only on the atomic number (positive charge) of that nucleus.

The Roothaan-type of equations can be extended to electron configurations in which some orbitals are doubly occupied and some singly. Another extension is one in which electrons of α-spin and β-spin are assigned to *different* molecular orbitals ψ^α and ψ^β. This is usually referred to as a *spin-unrestricted* configuration. There will be two sets of coefficients $c^\alpha_{\mu i}$ and $c^\beta_{\mu i}$. The corresponding generalization of the Roothaan equations was published by the author and Nesbet in 1954 [2]. These are usually denoted as Unrestricted Hartree-Fock or UHF, and the option of double and single occupation as Restricted Open Hartree-Fock or ROHF.

The introduction of basis set expansions played a major role in the development of quantum chemistry. It changed the mathematical task from the numerical solution of coupled differential equations (following the atomic work of Hartree) to the double challenge of evaluation of the 3- or 6-dimensional integrals (7),(8) and (11), followed by solution of the algebraic SCF equations (5). If analytic integration were possible, the model could become *precise* in the sense that good arithmetic correctness would be possible, even though the underlying approximations (use of a single configuration determinant and a finite basis) might still be unsatisfactory.

During the 1950s, integral evaluation was regarded as the main barrier to progress. The best choice of basis functions for LCAOSCF theory appeared to be Slater-type atomic orbitals (STO), which have exponential radial parts by analogy to the hydrogen atom. The one- and two-electron integrals (7),(8) and (11) can then be evaluated analytically in the two-center case. However, great difficulties were encountered for the three- and four-center cases. It was common to describe this impasse as "the nightmare of the integrals".

There were two responses to the integral difficulties. One was to make approximations for the more difficult integrals and to introduce parameters for others, with values obtained by empirical fits to experimental data. This practice became known as *semi-empirical.* The alternative of proceeding without approximation or empirical parameterization was, at the time, necessarily limited to very small molecules and became known as the *ab initio* approach. The most widely used semi-empirical methods were based on the zero-differential-overlap approximation, in which products of different atomic oritals $\chi_\mu\chi_\nu$ are neglected in most integrals. This approximation, when applied to the π-electrons of conjugated organic molecules, became known as the Pariser-Parr-Pople (PPP) theory [3-5]. It was later generalized to the treatment of all valence electrons in the CNDO and INDO theories [6] (1964-6) and then pursued at a more empirical level by the group of M.J.S.Dewar. The CNDO/INDO methods were genuine chemical models in the sense that they could be used to study many molecules, vary structure to determine equilibrium geometries and generate potential surfaces. However, they were limited by uncertainty over the consequences of the massive integral approximations and the large number of empirical parameters.

Within the *ab initio* community, a truly major development was the introduction of gaussian-type basis functions. In 1950, S.F. Boys [7], working in Cambridge, had demonstrated that all integrals in SCF theory could be evaluated analytically if the radial parts had the form $P(x,y,x)\,exp(-r^2)$, where $P(x,y,z)$ is any polynomial in the cartesian coordinates x,y,z. Initially, this appeared to be of limited value, since single gaussian functions were poor approximations to atomic orbitals, but it was clear that prospects would improve if larger numbers of basis functions could be handled. For several years, there was competition between proponents of Slater-type and gaussian-type basis sets.

The 1950s also saw the introduction of computers into quantum chemistry. By the time of the 1959 meeting, there were already several groups developing *ab initio* programs, using both Slater and gaussian bases. Early codes for 2-center integrals with Slater basis functions were developed in Chicago and used by Ransil in the first full LCAOSCF treatment of diatomic hydrides. At the same meeting, Boys presented several prescient papers describing simple SCF calculations using gaussians. During the early 1960s, other general purpose programs were developed, notably the gaussian packages POLYATOM and IBMOL, leading to a number of individual computations of molecular orbitals at the LCAO or minimal basis level.

My own research group began *ab initio* work in 1968 with the development of the GAUSSIAN program. At that time, the relative cost of *ab initio* LCAOSCF and CNDO computations on small organic molecules was over 1000. The original intention was to use full *ab initio* results to test various integral approximations that were less severe than the use of zero differential overlap. However, in the course of developing the program, Warren Hehre and I were able to generate a new integral algorithm that improved efficiency for highly contracted gaussian basis sets by more than two orders of magnitude [8]. This was based on a method of axis rotation inside inner loops, thereby limiting the number of arithmetic operations in the innermost sections of the program. Using a procedure of least-squares fitting Slater-type basis functions by a fixed contraction of K gaussians, we were able to reproduce the results of earlier full Slater results on a series of small molecules. The choice K=3 proved adequate and led to the STO-3G basis and the general theoretical model HF/STO-3G. This was published in 1969 [9] and the code was made generally available as GAUSSIAN70 shortly thereafter.

Investigation of the minimal HF/STO-3G model quickly showed major failures. Comparison of some isomeric species (e.g. propene and cyclopropane) showed too much stability for single bonds, relative to multiple bonds. This can be traced to the failure of the minimal basis to describe *anisotropic* atoms. In acetylene, for example, the carbon $2p\sigma$ atomic orbitals should be much tighter than $2p\pi$; this effect cannot be properly simulated by the isotropic structure implied by a minimal basis with identical $2p$ functions in all three directions. This difficulty can be overcome by using two basis functions per valence atomic orbital instead of one. Such a basis is 6-31G, which has a single contracted 6-gaussian basis function for the inner shell, a set of inner 3-contracted and a set of outer uncontracted gaussians for the valence shell of each atom. This is an example of a *split-valence* basis. Another similar commonly used type of basis set is *double-zeta,* in which there are two basis functions per atomic orbital for *all* atomic shells.

There are several notable failures for split-valence bases. In the first place, such bases tend to favor structures of high symmetry. For example, the ammonia molecule NH_3 is predicted to have a trigonal structure which is too close to planarity. This deficiency can reasonably be attributed to the fact that, in a planar structure, the lone pair of electrons are assigned to a nitrogen orbital that is pure p-type, which cannot mix with higher angular momentum d-type functions, whereas, in a non- planar structure, the lone-pair orbital is a *sp* mixture, for which further stabilization by d-mixing is possible. A second deficiency in Hartree-Fock studies at the split-valence level is an exaggeration of polarity, as measured by electric dipole moments. This can also be attributed to restriction of lone-pair orbitals to pure p-type. The $3p\pi$ lone-pair orbitals in *HCl,* for example, will probably be polarized towards hydrogen if mixing with $d\pi$ basis is allowed, thereby reducing the predicted dipole moment.

Considerable improvement is found in Hartree-Fock models if a single set of uncontracted d-functions are added on each heavy (non-hydrogen) atom. Such a basis is 6-31G*, or 6-31G(d) [10,11]. If a single set of uncontracted p-functions is added on each hydrogen, the basis is denoted by 6-31G** or 6-31G(d,p). These additional basis functions are termed *polarization* functions. The full model with the 6-31G* basis is then described as HF/6-31G*. Other important basis set extensions are the introduction of higher polarization functions (as in 6-31G(2df,p) which contains two sets of d-functions and a set of f-functions on heavy atoms and a single set of p-functions on hydrogen) and the use of diffuse functions, which are particularly important for anions and electronic states. The latter are denoted by a '+' as in 6-31+G(d).

The Hartree-Fock model HF/6-31G* has proved quite effective in the description of molecular conformations. Its overall performance in this and other regards has been documented elsewhere [12]. It is notably successful in giving differences of different isomeric forms of organic molecules, where no major changes of bond lengths are involved. Rotational potentials about single bonds were successfully explored using this level of theory [13]. A particular example is the anomeric effect in carbohydrate chemistry, which was not properly understood until the interaction of rotational potentials about geminal C-O single bonds was investigated using HF/6-31G* theory [14].

CORRELATED MODELS

The major fault implicit in all Hartree-Fock models is neglect of electron correlation between the motions of electrons of antiparallel spin ($\alpha\beta$ correlation). In the very early days of quantum chemistry, it was recognized that neglect of correlation led to severe underestimation of bond dissociation energies. This may be understood qualitatively by considering the process of complete homolytic dissociation of a bond in which one electron ends up on one center and one on the other. If the motion of the two electrons is uncorrelated, there will be a finite possibility of both electrons ending up on the same center.

Neglect of $\alpha\beta$ electron correlation is implicit in the use of a single-determinant wavefunction; improved wavefunctions necessarily involve the use of many determinants. Most practical correlation procedures start with the Hartree-Fock determinant and form linear combinations with other determinants. It is particularly convenient to form additional determinants from the unoccupied or virtual molecular orbitals, which are the higher eigenfunctions of the Fock operator. If a finite basis is used, with $2n$ electrons and N cartesian basis functions, there will be N-n virtual orbitals, which may be occupied by α or β electrons.

At this point, it is convenient to change the notation somewhat and use *spinorbital basis functions* which are products of the cartesian basis functions and the α or β spin functions. N is now the size of this spinorbital basis (twice the number of cartesian basis functions) and n is the total number of electrons. This notation enables us to use a common notation for both spin-restricted and spin-unrestricted cases. If labels $i,j,k,...$ are used for occupied spinorbitals and labels $a,b,c,...$ for virtual, then single-determinant functions using Fock orbitals may be classified as unsubstituted (i.e. Hartree-Fock) Ψ_0, singly substituted Ψ_i^a, doubly substituted Ψ_{ij}^{ab} and so forth. A general multi-determinant wavefunction can then be written

$$\Psi = a_0\Psi_0 + \sum_{ia} a_i^a\Psi_i^a + \sum_{ijab} a_{ij}^{ab}\Psi_{ij}^{ab} + \ldots \qquad (12)$$

The a-coefficients can be determined by variation to minimize the calculated energy. This is the method of *configuration interaction* (CI). If only singles are mixed in, no energy lowering follows, since the occupied orbitals are already optimized. The simplest effective form of CI allows for doubles only in (12). This is usually denoted by CID. If singles are also included, the method is CISD. These configuration interaction techniques were first implemented as iterative schemes around 1970 and are still often used in practical computations. If all possible substitutions are included in the expansion (a large but finite set if a finite basis set is used), the method is described as full configuration or FCI. The FCI procedure, although desirable in principle, is usually too costly to apply except for very small systems.

Although CID and CISD are well-defined models, given a standard basis set, they suffer some serious disadvantages. These have to do with *size-con-*

sistency. If a method such as CID is applied to a pair of completely separated systems, the resulting energy is *not* the sum of the energies obtained by applying the same theory to the systems separately. If CID is applied to two separated helium atoms, for example, the wavefunction does not allow for *simultaneous* excitation of pairs in each atom, this being strictly a quadruple excitation. This failure of CID and CISD models is likely to lead to poor descriptions of large molecules and interacting systems.

A second general method of incorporating electron correlation is to treat its effects by perturbation theory. Suppose we define a perturbed Hamiltonian as

$$H(\lambda) = F_0 + \lambda\{H - F_0\} \tag{13}$$

where F_0 is the Fock Hamiltonian (for which the single determinants in (12) are exact eigenfunctions), then Ψ_0 is the appropriate wavefunction if $\lambda = 0$ and the exact(FCI) Ψ is obtained if $\lambda = 1$. The perturbation procedure used is to expand the computed energy in powers of λ,

$$E(\lambda) = E_0 + \lambda E_1 + \lambda^2 E_2 + \lambda^3 E_3 + \dots \tag{14}$$

cut the series off at some level and then put $\lambda = 1$. This perturbation method was first introduced by Moeller and Plesset [15] and is often denoted by MPn if terminated at order n. The MP1 energy $(E_0 + E_1)$ is identical to the Hartree-Fock value. MP2 is the simplest practical perturbative procedure for electron correlation and incorporates only effects of double substitutions. At third order, MP3 also involves only double substitutions. At the fourth order level, MP4 includes a description of the (indirect) effects of singles, the leading contributions of triples and some treatment of certain quadruple substitutions.

Moeller-Plesset theory is size-consistent if the computations are carried out completely at any given order. Difficulties are that the terms become algebraically complicated at higher orders and also are increasingly costly to apply. In fact, Hartree-Fock theory (with no integral approximations) scales as N^4, MP2 as N^5, MP3 as N^6 and MP4 as N^7. The triple contributions in the MP4 energy are the most expensive and generally limit the applicability of Moeller-Plesset theory to this level. The MP2,MP3 and MP4 models were implemented by several groups in the 1970s and incorporated into the GAUSSIAN program [16,17].

A third general approach to correlation theory is the use of *coupled cluster* methods, originally introduced into quantum chemistry by Cizek [18]. If the configuration interaction CID wavefunction is written in the form

$$\Psi = (1 + T_2)\Psi_0 \tag{15}$$

where T_2 is an operator specifiying all double substitutions, with undetermined coefficients, then the corresponding coupled- cluster function (CCD) is

$$\Psi = exp(T_2)\Psi_0 \tag{16}$$

The CCD coefficients are determined, not by the variational method, but by requiring zero projection of $(H-E)\Psi$ onto Ψ_0 and all Ψ_{ij}^{ab}. This method was first implemented in 1978 [19–21]. Single substitutions are incorporated by using the operator $exp(T_1 + T_2)$ instead of $exp(T_1)$. This then defines a CCSD model [22,23].

Unlike CISD, the CCSD method is size-consistent. The cost is of order N^6, as for CISD. Being non-variational, the resulting total energy is no longer an upper bound for the exact result, but it is generally thought that the achievement of size-consistency is a matter of greater importance. Another, slightly simpler, method is *quadratic configuration* denoted QCISD. This is also size-consistent and can be regarded as an approximation intermediate between CISD and CCSD.

The QCISD and CCSD methods take no account of the effects of triple substitutions, known to be important by studies at the MP4 level. A useful way to take account of triples is to carry out an iterative QCISD or CCSD computation and the do a single computation of the effects of triples, using the single and double amplitudes already found. These are the QCISD(T) and CCSD(T) methods [24,25]. A third related method is the Brueckner-doubles method, BD(T) [26], which alters the underlying occupied orbitals so that ther is no singles mixing. All three of these methods are superior to MP4 in that, when the energy is expanded in a Moeller-Plesset series, complete agreement with a FCI expansion is obtained up to fourth order and many other terms at higher order are also included [27]. In fact, the QCISD, CCSD and BD methods have the further advantage of being completely correct for composite two-electron systems such as a set of isolated helium atoms.

The cost of QCISD(T) or CCSD(T) scales as iterative N^6, followed by a single computation at N^7. They represent the most sophisticated correlation methods that are simple enough to be incorporated into general theoretical models at the present time.

GENERAL ENERGY MODELS

In recent years, progress has been made in developing models which reproduce chemical energies to an accuracy approaching that achieved in good experimental work. The description of model features in the two previous sections indicates that two main features are involved, basis set and level of correlation. The options available are usefully summarized in a two-dimensional *model chart* as shown in Figure 1. The various correlation methods are displayed horizontally in order of increasing sophistication from left to right. Basis sets are displayed vertically, becoming more flexible from top to bottom. At the far right, full configuration interaction (FCI) represents complete solution *within the finite space defined by the basis.* At the bottom of the table, we have (in principle but not in practice), the results of applying a complete basis set. At the bottom right, application of a complete basis set with full configuration interaction corresponds to full solution of the non-relativistic Schrödinger equation.

Basis	HF	MP2	MP3	MP4	QCI	FCI
STO-3G						
6-31G						
6-31G(d)						
6-31+G(d)						
6-311+G(d)						
6-311+G(2df)						
∞						S-eqn

Figure 1. General Model Table. (QCI refers to QCISD(T)).

Each empty box in this chart represents a well-defined size-consistent theoretical model as specified in Section 2. Clearly, we may test each level to find how far we have to proceed from top-left to bottom-right for acceptable agreement between theory and experiment. Eventually, adequate performance will be achieved, if the underlying assumptions of quantum mechanics are correct.

In practice, full models usually have to make some compromises to achieve a wide range of applicability. If the prediction of energies is most important, a common practice is to carry out a geometry optimization (to an equilibrium structure, for example) at some lower level of theory and then make a final, more expensive, computation at a higher level. A useful notation for this type of composite model is "model-1//model-2", meaning single-point calculations using model-1 at geometrical structures determined by model-2.

To illustrate these ideas, we give a partial description of the G3 model for molecular energies, recently published [28]. This is a refinement of previous energy models G1 and G2 which have been under development for more than a decade [29,30]. The main computational steps are summarized in Figure 2.

Basis	HF	MP2	MP4	QCI
6-31G(d)	freq	opt-1	2	3
6-31+G(d)		4	5	
6-31G(2df,p)		6	7	
G3large		8		?

Figure 2. G3 Model Table. (QCI refers to QCISD(T)).

In addition to the standard type of basis sets already described, a large basis (G3large), which permits a flexible description of the whole space with inner shells, is added. This basis is so large that only MP2 computations are reasonably possible. Geometrical structures in the G3 model are determined at the MP2/6-31G(d) level [31]. This is followed by a sequence of single-point calculations which aim to estimate the results of a potential QCI/G3 large energy, by assuming that effects of some of the improvement steps can be treated additively. The actual formula used is:

$$? = 2 + (3 - 2) + (5 - 2) + (7 - 2) + (8 - 1) - (4 - 1) - (6 - 1) \qquad (17)$$

Earlier studies [32] had indicated that this kind of addititivy was reasonably accurate. (It should be noted that all correlation computations except the full MP2/G3 large are carried out in the "frozen core" approximation, only interactions between valence electrons being treated).

An important contribution to total molecular energies is the zero-point vibrational energy. This is estimated in G3 theory by using harmonic frequencies calculated at the HF/6-31G(d) level and then empirically scaled by a factor 0.8929 (HF theory being known to systemically overestimate the magnitudes of frequencies [33]). In addition, a small correction is added for the spin-orbit splitting in isolated atoms, obtained from experimental data [34].

The computations as described up to this level give a reasonable account of significant energy differences, such as dissociation energies and ionization potentials. However, there is a significant systematic error, all binding energies being slightly too low. This can be reasonably interpreted as being due mostly to the limitation in the basis sets being used. An accurate description of the wavefunction cusp at the point where electrons of opposite spin come to the same point in space requires basis sets involving high angular momentum. Another reason is that, in molecules, the symmetry is lower than in atoms and again neglect of the effects of higher angular momentum basis functions will favor atomic energies relative to molecular.

These difficulties can be partly overcome by adding a *small* empirical correction, depending on the number of electrons and distinguishing between atoms and molecules. The theory therefore becomes semi-empirical or, perhaps "slightly empirical" since the parameters are small and their origin is partly understood. This higher-level correction (HLC) is $-An_\beta - B(n_\alpha - n_\beta)$ for molecules and $-Cn_\beta - D(n_\alpha - n_\beta)$ for atoms (including atomic ions). n_α and n_β are the numbers of α and β electrons, respectively, with $n_\alpha \geq n_\beta$. This completes the specification of a total G3 energy for any atom or molecule.

The parameters A,B,C,D are determined as part of the validation process. This is carried out using a large set of 299 experimental energy differences, involving molecules up to the size of benzene (42 electrons). These data include 148 heats of formation, derived from heats of atomization, 85 ionization potentials, 58 electron affinities and 8 proton affinities. All of these experimental results are believed known to an accuracy of 1 kcal/mole or better. Values of A,B,C,D are obtained by minimization of the mean absolute deviation between theory and experiment. These are (in millihartrees) 6.386, 2.977,6.219 and 1.185. The resulting mean deviation is 1.02 kcal/mole, close to the target accuracy. The corresponding root-mean-square deviation, which lays more emphasis on the poorer levels of agreement, is 1.45 kcal/mole. However, nearly 88 % of the G3 deviations fall in the range, –2.0 to +2.0 kcal/mole.These results are significantly better than the prior G1 and G2 models, which use a smaller database of experimental facts.

The poorest results are worthy of note. Largest absolute deviations are 4.9 kcal/mole (C_2F_4) for heats of formation, 7.0 kcal/mole (B_2F_4) for ionization potentials, 4.2 kcal/mole (*NH*) for electron affinities and 1.8 kcal/mole (PH_3 and SH_2) for proton affinities.

CONCLUSIONS

The current status of *ab initio* quantum chemical models is that some success has been achieved in approaching experimental accuracy in predictive power. The target of 1 kcal/mole is not far away for small molecules containing up to about fifty electrons. However, the G3 model has a number of remaining deficiencies that merit further attack.

1. The use of an empirical correction, which depends only on the number of electrons, is undesirable. One consequence is that the model becomes discontinuous in some manner. For example, if a bond is broken, the electron count of paired versus unpaired electrons has to change at some point, thereby providing a discontinuity in the potential surface. The same criticism can be applied to the use of different parameters for atoms and molecules. Some form of extrapolation is probably necessary, but it would be much better if this could be carried out in a continuous and differentiable manner.
2. The G3 model is based on MP2/6-31G(d) geometries, which are known to show considerable errors [31]. Some of the failures of the final energies can be attributed to this; clearly a method which would reproduce known bond lengths and angles more accurately would be preferable.
3. No account is taken of relativity in the G3 model. The total energy of a molecule is known to depend significantly on relativistic corrections, particularly for inner-shell electrons. However, considerable cancellation of errors occurs in processes such as bond dissociation. Nevertheless, some inclusion of relativistic contributions to chemical processes is clearly desirable.
4. The applicability of the G3 model to large systems is presently limited by the very expensive treatment of the triples terms, where computational cost scales as the seventh power of the size of the system. The magnitude of these terms is small, but not insignificant. A simpler treatment of three-electron effects would be desirable.

Finally, some brief comment should be made about theoretical models based on density functional theory (DFT). Such methods do not handle the two-electron interactions explicitly but rather allow for them using properties of the one- electron density. This leads to lower cost and therefore a wider range of applicability. Recent forms of DFT have also introduced a considerable amount of empirical parameterization, sometimes using the same set of experimental data. At the present time, the principal limitation of DFT models is that there is no clear route for convergence of methods to the correct answer, comparable to the *ab initio* chart shown in Figure 1. Interaction between these two groups of theoretical chemists is a hopeful direction for future progress.

REFERENCES

1. C. C .J. Roothaan, Rev. Mod. Phys. **23**, 61 (1951).
2. J. A. Pople and R. K. Nesbet, J. Chem.Phys. **22**, 571 (1954).
3. R. G. Parr, J. Chem. Phys. **20**, 1499 (1952).
4. R. Pariser and R. G. Parr, J. Chem. Phys. **21**, 466,767 (1953).
5. J. A. Pople, Trans. Faraday Soc. **49**, 1375 (1953).
6. J. A. Pople, D. P. Santry and G. A. Segal, J. Chem. Phys. **43**, 5129,5136 (1965).
7. S. F. Boys, Proc. Roy. Soc. **A200**, 542 (1950).
8. J. A. Pople and W. J. Hehre, J. Comput. Phys. **27**, 161 (1978).
9. W. J. Hehre, R. F. Stewart and J. A. Pople, J. Chem. Phys. **51**, 2657 (1969).
10. P.C.Hariharan and J.A.Pople, Theor. Chim. Acta, **28**, 213 (1971).
11. M. M. Francl, W. J. Pietro, W. J. Hehre, J. S. Binkley, M. S. Gordon, D. J. DeFrees and J. A. Pople, J. Chem. Phys **77**, 3654 (1982).
12. W. J. Hehre, L. Radom, P. v. R. Schleyer and J. A. Pople, "Ab Initio Molecular Orbital Theory", John Wiley, (New York 1986).
13. L. Radom, W. J. Hehre and J. A. Pople, J. Amer. Chem.Soc. **54**, 724 (1971).
14. G. A. Jeffrey, J. A. Pople and L. Radom, Carbohydrate Res. **25**, 117 (1972).
15. C. Moeller and M. S. Plesset, Phys. Rev. **46**, 618 (1934).
16. R. Krishnan and J. A. Pople, Int. J. Quantum Chem. **14**, 91 (1978).
17. R. Krishnan, M. J. Frisch and J. A. Pople, J. Chem. Phys. **72**, 4244 (1980).
18. J. Cizek, J. Chem. Phys. **45**, 4256 (1966).
19. J. A. Pople, R. Krishnan, H. B. Schlegel and J. S. Binkley, Int. J. Quantum Chem. **14**, 545 (1978).
20. R. J. Bartlett and G. D. Purvis, Int. J. Quantum Chem. **14**, 561 (1978).
21. P. R. Taylor, G. B. Bacskay, N. S. Hush and A. C. Hurley, Chem.Phys.Lett **41**, 444 (1976).
22. G. D. Purvis and R. J. Bartlett, J. Chem. Phys. **76**, 1910 (1982).
23. G. E. Scuseria, A. C. Schneiner, T. J. Lee, J. E. Rice and H. F. Schaefer, J. Chem.Phys. **86**, 2881 (1987).
24. J. A. Pople, M. Head-Gordon and K. Raghavachari, J. Chem. Phys. **87**,5968 (1987).
25. K. Raghavachari, G. W .Trucks, M.Head-Gordon and J. A. Pople, Chem. Phys. Lett. **157**, 479 (1989).
26. N. C. Handy, J. A. Pople, M. Head-Gordoni, K. Raghavachari and G. W. Trucks, Chem.Phys.Lett. **164**, 185 (1989).
27. K. Raghavachari, J. A. Pople, E. S. Replogle and M. Head-Gordon, J. Chem. Phys. **94**, 5579 (1990).
28. L. A. Curtiss, K. Raghavachari, P. C. Redfern, V. Rassolov and J. A. Pople, J. Chem. Phys. **109**, 7764 (1998).
29. L. A. Curtiss, C. Jones, G. W. Trucks, K. Raghavachari and J. A. Pople, J. Chem. Phys. **93**, 2537 (1990).
30. L. A. Curtiss, K. Raghavachari, G. W. Trucks and J. A. Pople, J. Chem. Phys. **94**, 7221 (1991).

31. D. J. DeFrees, B. A. Levi, S. K. Pollack, W. J. Hehre, J. S. Binkley and J. A. Pople, J. Amer. Chem. Soc. **101**, 4843 (1979).
32. L. A. Curtiss, J. E. Carpenter, K. Raghavachari and J. A. Pople, J. Chem. Phys. **96**, 9030 (1992).
33. J. A. Pople, H. B. Schlegel, K. Raghavachari, D. J. Defrees, J. S. Binkley, M. J. Frisch, R. A. Whiteside, R. F. Hout and W. J. Hehre, Int J. Quantum Chem. **15**, 269 (1981).
34. C. Moore, Natl. Bur. Stand. (U.S.) Circ 467 (1952).

Chemistry 1999

AHMED H. ZEWAIL

"for his studies of the transition states of chemical reactions using femtosecond spectroscopy"

THE NOBEL PRIZE IN CHEMISTRY

Speech by Professor Bengt Nordén of the Royal Swedish Academy of Sciences. Translation of the Swedish text.

Your Majesties, Your Royal Highness, Ladies and Gentlemen,

We chemists want to understand molecules and their intrinsic essence, and to be able to predict what happens when molecules meet – do they attach weakly to each other or do they react passionately to form new molecules? Not least, we want to understand the complicated chemistry called life. Through a revolution in knowledge, molecules today take center stage in all fields, from biology and medicine through environmental sciences, and technology.

The heart of chemistry is the chemical reaction, meaning the breaking and formation of chemical bonds between atoms. How then do chemical reactions occur? We all know that they can proceed at different rates – compare the time it takes a nail to rust with explosion of dynamite! Alfred Nobel knew that reaction rates are important; dynamite reacts too rapidly to be used in cannons – they would blow up. He also knew that chemical reactions proceed at greater speed at higher temperatures, but he did not see why. This was, however, realized by the docent of physical chemistry in Uppsala, Svante Arrhenius. Inspired by the Dutch scientist Jacobus van't Hoff (the first Nobel laureate in chemistry, 1901), Arrhenius presented the first theory on reaction rates and an equation for their temperature dependence that has been used for more than a hundred years now. Arrhenius was himself awarded the third Nobel Prize in Chemistry (1903), but for different achievements.

Science has always strived to see smaller and smaller things and faster and faster events. Since the time of Arrhenius a number of methods have been developed to measure increasingly faster reaction rates, many of them rewarded with Nobel prizes. However, no one had, until recently, been able to observe what actually happens to the reacting molecule as it passes through its so-called transition state, a metaphor for a kind of intermediate state of the reaction in which bonds are broken and formed. This remained a misty no-man's land.

The molecule passes the transition state as fast as the atoms in the molecule move. They move at a speed of the order of 1000 m/second – about as fast as a rifle bullet – and the time required for the atoms to move slightly within the molecule is typically tens of femtoseconds (1 fs = 10^{-15} seconds). Only few believed that such fast events would ever be possible to see.

This, however, is exactly what Ahmed Zewail has managed to do. Twelve years ago he published results that gave birth to the scientific field called femtochemistry. This can be described as using the fastest camera in the world to film the molecules during the reaction and to get a sharp picture of the transition state. His "camera" is a laser technique with light flashes of only a few tens of femtoseconds in duration. The reaction is initiated by a strong laser flash and is then studied by a series of subsequent flashes to follow the events. The key to his success was that the first femtosecond flash or starting shot, excited all molecules in the sample at once, causing their atoms to swing in rhythm. The first experiments demonstrated in slow motion how bonds were stretched and broken in rather simple reactions, but soon studies of more complex reactions followed. The results were often surprising, and the dance of the atoms during the reaction was found to differ from what was expected. Zewail's use of the fast laser technique can be likened to Galilei's use of his telescope, which he directed towards everything that lit up the vault of heaven. Zewail tried his femtosecond laser on literally everything that moved in the world of molecules. He turned his telescope toward the frontiers of science.

Ahmed Zewail is being awarded the Nobel Prize in Chemistry because he was the first to conduct experiments that clearly show the decisive moments in the life of a molecule – the breaking and formation of chemical bonds. He has been able to see the reality behind Arrhenius' theory.

It is of great importance to be able in detail to understand and predict the progress of a chemical reaction. Femtochemistry has found applications in all branches of chemistry, but also in adjoining fields such as material science (*future electronics?*) and biology. The retinal molecule is an example – a substance that you are all making use of at this very moment, namely to see with. It has been found that light causes this molecule to twist like a hinge around a well-greased bond, which sends a nerve signal to the brain. The reaction takes only 200 fs, which explains the eye's sensitivity to light.

Femtochemistry has radically changed the way we look at chemical reactions. A hundred years of mist surrounding the transition state has cleared.

Professor Zewail,

I have tried to explain how your pioneering work has fundamentally changed the way scientists view chemical reactions. From being restricted to describe them only in terms of a metaphor, the transition state, we can now study the actual movements of atoms in molecules. We can speak of them in time and space in the same way that we imagine them. They are no longer invisible.

May I convey to you my warmest congratulations on behalf of the Royal Swedish Academy of Sciences and ask you to come forward to receive the Nobel Prize in Chemistry of 1999 from the hands of His Majesty the King.

Professor Kohn,

You were awarded last year's Nobel Prize together with Professor John Pople for your contributions to Computational Quantum Chemistry. They have been of fundamental importance also in the context of this year's Nobel Prize, as calculations of energies and structures of the molecules in their various states are crucial for the interpretation of the experiments.

On behalf of the Royal Swedish Academy of Sciences, may I warmly congratulate you and ask you to come forward to receive the Nobel Prize in Chemistry of 1998 from the hands of His Majesty the King.

Ahmed Zewail

احمد زويل

AHMED H. ZEWAIL

On the banks of the Nile, the Rosetta branch, I lived an enjoyable childhood in the City of Disuq, which is the home of the famous mosque, Sidi Ibrahim. I was born (February 26, 1946) in nearby Damanhur, the "City of Horus", only 60 km from Alexandria. In retrospect, it is remarkable that my childhood origins were flanked by two great places – Rosetta, the city where the famous Stone was discovered, and Alexandria, the home of ancient learning. The dawn of my memory begins with my days at Disuq's preparatory school. I am the only son in a family of three sisters and two loving parents. My father was liked and respected by the city community – he was helpful, cheerful and very much enjoyed his life. He worked for the government and also had his own business. My mother, a good-natured, contented person, devoted all her life to her children and, in particular, to me. She was central to my "walks of life" with her kindness, total devotion and native intelligence. Although our immediate family is small, the Zewails are well known in Damanhur.

The family's dream was to see me receive a high degree abroad and to return to become a university professor – on the door to my study room, a sign was placed reading, "Dr. Ahmed," even though I was still far from becoming a doctor. My father did live to see that day, but a dear uncle did not. Uncle Rizk was special in my boyhood years and I learned much from him – an appreciation for critical analyses, an enjoyment of music, and of intermingling with the masses and intellectuals alike; he was respected for his wisdom, financially well-to-do, and self-educated. Culturally, my interests were focused – *reading, music, some sports and playing backgammon.* The great singer Um Kulthum (actually named Kawkab Elsharq – a superstar of the East) had a major influence on my appreciation of music. On the first Thursday of each month we listened to Um Kulthum's concert – "waslats" (three songs) – for more than three hours. During all of my study years in Egypt, the music of this unique figure gave me a special happiness, and her voice was often in the background while I was studying mathematics, chemistry ... etc. After three decades I still have the same feeling and passion for her music. In America, the only music I have been able to appreciate on this level is classical, and some jazz. Reading was and still is my real joy.

As a boy it was clear that my inclinations were toward the physical sciences. Mathematics, mechanics, and chemistry were among the fields that gave me a special satisfaction. Social sciences were not as attractive because in those days much emphasis was placed on memorization of subjects, names and the like, and for reasons unknown (to me), my mind kept asking "how" and "why". This characteristic has persisted from the beginning of my life. In my teens, I recall feeling a thrill when I solved a difficult problem in mechanics, for instance, considering all of the tricky operational forces of a car going uphill or downhill. Even though chemistry required some memorization, I was intrigued by the "mathematics of chemistry". It provides laboratory phenomena which, as a boy, I wanted to reproduce and understand. In my bedroom I constructed a small apparatus, out of my mother's oil burner (for making Arabic coffee) and a few glass tubes, in order to see how wood is *transformed* into a burning gas and a liquid substance. I still remember this vividly, not only for the science, but also for the danger of burning down our house! It is not clear why I developed this attraction to science at such an early stage.

After finishing high school, I applied to universities. In Egypt, you send your application to a central Bureau (Maktab El Tansiq), and according to your grades, you are assigned a university, hopefully on your list of choice. In the sixties, Engineering, Medicine, Pharmacy, and Science were tops. I was admitted to Alexandria University and to the faculty of science. Here, luck played a crucial role because I had little to do with Maktab El Tansiq's decision, which gave me the career I still love most: science. At the time, I did not know the depth of this feeling, and, if accepted to another faculty, I probably would not have insisted on the faculty of science. But this passion for science became evident on the first day I went to the campus in Maharem Bek with my uncle – I had tears in my eyes as I felt the greatness of the university and the sacredness of its atmosphere. My grades throughout the next four years reflected this special passion. In the first year, I took four courses, mathematics, physics, chemistry, and geology, and my grades were either excellent or very good. Similarly, in the second year I scored very highly (excellent) in Chemistry and was chosen for a group of seven students (called "special chemistry"), an elite science group. I graduated with the highest honors – "Distinction with First Class Honor" – with above 90% in all areas of chemistry. With these scores, I was awarded, as a student, a stipend every month of approximately £13, which was close to that of a university graduate who made £17 at the time!

After graduating with the degree of Bachelor of Science, I was appointed to a university position as a demonstrator ("Moeid"), to carry on research toward a Masters and then a Ph.D. degree, and to teach undergraduates at the University of Alexandria. This was a tenured position, guaranteeing a faculty appointment at the University. In teaching, I was successful to the point that, although not yet a professor, I gave "professorial lectures" to help students after the Professor had given his lecture. Through this experience I discovered an affinity and enjoyment of explaining science and natural phenomena in the clearest and simplest way. The students (500 or more) enriched this sense with the appreciation they expressed. At the age of 21, as a Moeid, I believed that behind every universal phenomenon there must be beauty and simplicity in its description. This belief remains true today.

On the research side, I finished the requirements for a Masters in Science in about eight months. The tool was spectroscopy, and I was excited about developing an understanding of how and why the spectra of certain molecules change with solvents. This is an old subject, but to me it involved a new level of understanding that was quite modern in our department. My research advisors were three: The head of the inorganic section, Professor Tahany Salem and Professors Rafaat Issa and Samir El Ezaby, with whom I worked most closely; they suggested the research problem to me, and this research resulted in several publications. I was ready to think about my Ph.D. research (called "research point") after one year of being a Moeid. Professors El Ezaby (a graduate of Utah) and Yehia El Tantawy (a graduate of Penn) encouraged me to go abroad to complete my Ph.D. work. All the odds were against my going to America. First, I did not have the connections abroad. Second, the 1967 war had just ended and American stocks in Egypt were at their lowest value, so study missions were only sent to the USSR or Eastern European countries. I had to obtain a scholarship directly from an American University. After corresponding with a dozen universities, the University of Pennsylvania and a few others offered me scholarships, providing the tuition and paying a monthly stipend (some $300). There were still further obstacles against travel to America ("Safer to America"). It took enormous energy to pass the regulatory and bureaucratic barriers.

Arriving in the States, I had the feeling of being thrown into an ocean. The ocean was full of knowledge, culture, and opportunities, and the choice was clear: I could either learn to swim or sink. The culture was foreign, the language was difficult, but my hopes were high. I did not speak or write English fluently, and I did not know much about western culture in general, or American culture in particular. I remember a "cultural incident" that opened my eyes to the new traditions I was experiencing right after settling in Philadelphia. In Egypt, as boys, we used to kid each other by saying "I'll kill you", and good friends often said such phrases jokingly. I became friends with a sympathetic American graduate student, and, at one point, jokingly said "I'll kill you". I immediately noticed his reserve and coolness, perhaps worrying that a fellow from the Middle East might actually do it!

My presence – as the Egyptian at Penn – was starting to be felt by the professors and students as my scores were high, and I also began a successful course of research. I owe much to my research advisor, Professor Robin Hochstrasser, who was, and still is, a committed scientist and educator. The diverse research problems I worked on, and the collaborations with many able scientists, were both enjoyable and profitable. My publication list was increasing, but just as importantly, I was learning new things literally every day – in chemistry, in physics and in other fields. The atmosphere at the Laboratory for Research on the Structure of Matter (LRSM) was most stimulating and I was enthusiastic about researching in areas that crossed the disciplines of physics and chemistry (sometimes too enthusiastic!). My courses were enjoyable too; I still recall the series 501, 502, 503 and the physics courses I took with the Nobel Laureate, Bob Schrieffer. I was working almost "day and night," and doing several projects at the same time: The Stark effect of simple molecules; the Zeeman effect of solids like NO_2^- and benzene; the optical detection of magnetic resonance (ODMR); double resonance techniques, etc.

Now, thinking about it, I cannot imagine doing all of this again, but of course then I was "young and innocent".

The research for my Ph.D. and the requirements for a degree were essentially completed by 1973, when another war erupted in the Middle East. I had strong feelings about returning to Egypt to be a University Professor, even though at the beginning of my years in America my memories of the frustrating bureaucracy encountered at the time of my departure were still vivid. With time, things change, and I recollected all the wonderful years of my childhood and the opportunities Egypt had provided to me. Returning was important to me, but I also knew that Egypt would not be able to provide the scientific atmosphere I had enjoyed in the U.S. A few more years in America would give me and my family two opportunities: First, I could think about another area of research in a different place (while learning to be professorial!). Second, my salary would be higher than that of a graduate student, and we could then buy a big American car that would be so impressive for the new Professor at Alexandria University! I applied for five positions, three in the US, one in Germany and one in Holland, and all of them with world-renowned professors. I received five offers and decided on Berkeley.

Early in 1974 we went to Berkeley, excited by the new opportunities. Culturally, moving from Philadelphia to Berkeley was almost as much of a shock as the transition from Alexandria to Philadelphia – Berkeley was a new world! I saw Telegraph Avenue for the first time, and this was sufficient to indicate the difference. I also met many graduate students whose language and behavior I had never seen before, neither in Alexandria, nor in Philadelphia. I interacted well with essentially everybody, and in some cases I guided some graduate students. But I also learned from members of the group. The obstacles did not seem as high as they had when I came to the University of Pennsylvania because culturally and scientifically I was better equipped. Berkeley was a great place for science – the BIG science. In the laboratory, my aim was to utilize the expertise I had gained from my Ph.D. work on the spectroscopy of pairs of molecules, called dimers, and to measure their coherence with the new tools available at Berkeley. Professor Charles Harris was traveling to Holland for an extensive stay, but when he returned to Berkeley we enjoyed discussing science at late hours! His ideas were broad and numerous, and in some cases went beyond the scientific language I was familiar with. Nevertheless, my general direction was established. I immediately saw the importance of the concept of coherence. I decided to tackle the problem, and, in a rather short time, acquired a rigorous theoretical foundation which was new to me. I believe that this transition proved vital in subsequent years of my research.

I wrote two papers with Charles, one theoretical and the other experimental. They were published in *Physical Review*. These papers were followed by other work, and I extended the concept of coherence to multidimensional systems, publishing my first independently authored paper while at Berkeley. In collaboration with other graduate students, I also published papers on energy transfer in solids. I enjoyed my interactions with the students and professors, and at Berkeley's popular and well-attended physical chemistry seminars. Charles decided to offer me the IBM Fellowship that was only given to a few in the department. He strongly felt that I should get a job at one of the top universities in America, or at least have the experience of going to the interviews; I am grateful for his belief in me. I only applied to a few places and thought I had no chance at these top universities. During the process, I contacted Egypt, and I also considered the American University in Beirut (AUB). Although I visited some places, nothing was finalized, and I was preparing myself for the return. Meanwhile, I was busy and excited about the new research I was doing. Charles decided to build a picosecond laser, and two of us in the group were involved in this hard and "non-profitable" direction of research (!); I learned a great deal about the principles of lasers and their physics.

During this period, many of the top universities announced new positions, and Charles asked me to apply. I decided to send applications to nearly a dozen places and, at the end, after interviews and enjoyable visits, I was offered an Assistant Professorship at many, including Harvard, Caltech, Chicago, Rice, and Northwestern. My interview at Caltech had gone well, despite the experience of an exhausting two days, visiting each half hour with a different faculty member in chemistry and chemical engineering. The visit was exciting, surprising and memorable. The talks went well and I even received some undeserved praise for style. At one point, I was speaking about what is known as the FVH picture of coherence, where F stands for Feynman, the famous Caltech physicist and Nobel Laureate. I went to the board to write the name and all of a sudden I was stuck on the spelling. Half way through, I turned to the audience and said, "you know how to spell Feynman". A big laugh erupted, and the audience thought I was joking – I wasn't! After receiving several offers, the time had come to make up my mind, but I had not yet heard from Caltech. I called the Head of the Search Committee, now a colleague of mine, and he was lukewarm, encouraging me to accept other offers. However, shortly after this, I was contacted by Caltech with a very attractive offer, asking me to visit with my family. We received the red carpet treatment, and that visit did cost Caltech! I never regretted the decision of accepting the Caltech offer .

My science family came from all over the world, and members were of varied backgrounds, cultures, and abilities. The diversity in this "small world" I worked in daily provided the most stimulating environment, with many challenges and much optimism. Over the years, my research group has had close to 150 graduate students, postdoctoral fellows, and visiting associates. Many of them are now in leading academic, industrial and governmental positions. Working with such minds in a *village of science* has been the most rewarding experience – Caltech was the right place for me.

My biological children were all "made in America". I have two daughters, Maha, a Ph.D. student at the University of Texas, Austin, and Amani, a junior at Berkeley, both of whom I am very proud. I met Dema, my wife, by a surprising chance, a fairy tale. In 1988 it was announced that I was a winner of the King Faisal International Prize. In March of 1989, I went to receive the award from Saudi Arabia, and there I met Dema; her father was receiving the same prize in literature. We met in March, got engaged in July and married in September, all of the same year, 1989. Dema has her M.D. from Damascus University, and completed a Master's degree in Public Health at UCLA. We have two young sons, Nabeel and Hani, and both bring joy and excitement to our life. Dema is a wonderful mother, and is my friend and confidante.

The journey from Egypt to America has been full of surprises. As a Moeid, I was unaware of the Nobel Prize in the way I now see its impact in the West. We used to gather around the TV or read in the newspaper about the recognition of famous Egyptian scientists and writers by the President, and these moments gave me and my friends a real thrill – maybe one day we would be in this position ourselves for achievements in science or literature. Some decades later, when President Mubarak bestowed on me the Order of Merit, first class, and the Grand Collar of the Nile ("Kiladate El Niel"), the highest State honor, it brought these emotional boyhood days back to my memory. I never expected that my portrait, next to the pyramids, would be on a postage stamp or that the school I went to as a boy and the road to Rosetta would be named after me. Certainly, as a youngster in love with science, I had no dreams about the honor of the Nobel Prize.

Since my arrival at Caltech in 1976, our contributions have been recognized by countries around the world. Among the awards and honors are:

Special Honors

King Faisal International Prize in Science (1989).
First Linus Pauling Chair, Caltech (1990).
Wolf Prize in Chemistry (1993).
Order of Merit, first class (Sciences & Arts), from President Mubarak (1995).
Robert A. Welch Award in Chemistry (1997).
Benjamin Franklin Medal, Franklin Institute, USA (1998).
Egypt Postage Stamps, with Portrait (1998); the Fourth Pyramid (1999).
Nobel Prize in Chemistry (1999).
Grand Collar of the Nile, Highest State Honor, conferred by President Mubarak (1999).

Prizes and Awards

Alfred P. Sloan Foundation Fellow (1978-1982).
Camille and Henry Dreyfus Teacher-Scholar Award (1979–1985).
Alexander von Humboldt Award for Senior United States Scientists (1983).
National Science Foundation Award for especially creative research (1984; 1988; 1993).
Buck-Whitney Medal, American Chemical Society (1985).
John Simon Guggenheim Memorial Foundation Fellow (1987).
Harrison Howe Award, American Chemical Society (1989).
Carl Zeiss International Award, Germany (1992).
Earle K. Plyler Prize, American Physical Society (1993).

Medal of the Royal Netherlands Academy of Arts and Sciences, Holland (1993).
Bonner Chemiepreis, Germany (1994).
Herbert P. Broida Prize, American Physical Society (1995).
Leonardo Da Vinci Award of Excellence, France (1995).
Collége de France Medal, France (1995).
Peter Debye Award, American Chemical Society (1996).
National Academy of Sciences Award, Chemical Sciences, USA (1996).
J. G. Kirkwood Medal, Yale University (1996)
Peking University Medal, PU President, Beijing, China (1996).
Pittsburgh Spectroscopy Award (1997).
First E. B. Wilson Award, American Chemical Society (1997).
Linus Pauling Medal Award (1997).
Richard C. Tolman Medal Award (1998).
William H. Nichols Medal Award (1998).
Paul Karrer Gold Medal, University of Zürich, Switzerland (1998).
E. O. Lawrence Award, U. S. Government (1998).
Merski Award, University of Nebraska (1999).
Röntgen Prize, (100th Anniversary of the Discovery of X-rays), Germany (1999).

Academies and Societies
American Physical Society, Fellow (elected 1982).
National Academy of Sciences, USA (elected 1989).
Third World Academy of Sciences, Italy (elected 1989).
Sigma Xi Society, USA (elected 1992).
American Academy of Arts and Sciences (elected 1993).
Académie Européenne des Sciences, des Arts et des Lettres, France (elected 1994).
American Philosophical Society (elected 1998).
Pontifical Academy of Sciences (elected 1999).
American Academy of Achievement (elected 1999).
Royal Danish Academy of Sciences and Letters (elected 2000).

Honorary Degrees
Oxford University, UK (1991): M.A.,h.c.
American University, Cairo, Egypt (1993): D.Sc.,h.c.
Katholieke Universiteit, Leuven, Belgium (1997): D.Sc.,h.c.
University of Pennsylvania, USA (1997): D.Sc.,h.c.
Université de Lausanne, Switzerland (1997): D.Sc.,h.c.
Swinburne University, Australia (1999): D.U.,h.c.
Arab Academy for Science & Technology, Egypt (1999): H.D.A.Sc.
Alexandria University, Egypt (1999): H.D.Sc.
University of New Brunswick, Canada (2000): Doctoris in Scientia, D.Sc.,h.c.
Universitá di Roma "La Sapienza", Italy (2000): D.Sc.,h.c.
Université de Liège, Belgium (2000): Doctor *honoris causa*, D.,h.c.

FEMTOCHEMISTRY

ATOMIC-SCALE DYNAMICS OF THE CHEMICAL BOND USING ULTRAFAST LASERS

Nobel Lecture, December 8, 1999

by

AHMED H. ZEWAIL

California Institute of Technology, Pasadena, USA.

Over many millennia, humankind has thought to explore phenomena on an ever shorter time scale. In this race against time, femtosecond resolution (1 fs = 10^{-15} second) is the ultimate achievement for studies of the fundamental *dynamics of the chemical bond.* Observation of the very act that brings about chemistry – the making and breaking of bonds on their actual time and length scales – is the wellspring of the field of *Femtochemistry*, which is the study of molecular motions in the hitherto unobserved ephemeral *transition states* of physical, chemical and biological changes. For molecular dynamics, achieving this atomic-scale resolution using *ultrafast lasers* as strobes is a triumph, just as x-ray and electron diffraction, and, more recently, STM and NMR, provided that resolution for static molecular structures. On the femtosecond time scale, matter wave packets (particle-type) can be created and their *coherent* evolution as a *single-molecule trajectory* can be observed. The field began with simple systems of a few atoms and has reached the realm of the very complex in isolated, mesoscopic and condensed phases, and in biological systems such as proteins and DNA. It also offers new possibilities for the control of reactivity and for structural femtochemistry and femtobiology. This anthology gives an overview of the development of the field from a personal perspective, encompassing our research at Caltech and focusing on the evolution of techniques, concepts, and new discoveries.

TABLE OF CONTENTS

I. PROLOGUE

In the history of human civilization, the measurement of time and recording the order and duration of events in the natural world are among the earliest endeavors that might be classified as science. The development of calendars, which permitted the tracking of yearly flooding of the Nile valley in ancient Egypt and of the seasons for planting and harvesting in Mesopotamia, can be traced to the dawn of written language. Ever since, time has been an important concept **[1]** and is now recognized as one of the two fundamental dimensions in science. The concept of time encapsulates an awareness of its duration and of the passage from *past* to *present* to *future* and surely must have existed from the very beginning with humans searching for the meaning of birth, life and death and, in some cultures, rebirth or recurrence.

My ancestors contributed to the beginning of the science of time, developing what Neugebauer **[1]** has described as *"the only intelligent calendar which ever existed in human history"*. The "Nile Calendar" was an essential part of life as it defined the state of yearly flooding with three seasons, the *Inundation* or *Flooding*, *Planting*, and *Harvesting*, each four months long. A civil year lasting 365 days was ascertained by about 3000 BC or before, based on the average time between arrivals of the flood at Heliopolis, just north of Cairo; nilometers were used in more recent times, and some are still in existence today. By the time of the First Dynasty of United Egypt under Menes in ca. 3100 BC, the scientists of the land introduced the concept of the "Astronomical Calendar" by observing the event of the helical rising of the brilliant star Sothis (or

Sirius). Inscribed on the Ivory Tablet, dating from the First Dynasty and now at the University Museum in Philadelphia, were the words, "Sothis Bringer of the Year and of the Inundation" **[1, Winlock]**. On the Palermo Stone, the *annals* of the kings and their *time-line* of each year's chief events were documented from pre-dynastic times to the middle of the Fifth Dynasty **[1, Breasted]**. Thus, as early as 3100 BC, they recognized a definite natural phenomenon for accurate timing of the coming flood and recounted the observed reappearance of the star as the New Year Day – real-time observation of daily and yearly events with the zero-of-time being well defined![1]

In about 1500 BC, another major contribution to this science was made, the development of *Sun-Clocks*, or *Sundials*, using moving shadows **[Fig. 1]**. Now in Berlin, the sun-clock bearing the name of Thutmose III (or Thothmes after the Egyptian God Thoth of wisdom and enlightenment), who ruled at Thebes from 1447–1501 BC, shows the graduation of hours for daytime measurements. This clock with uneven periods for hours was man-made and transportable. For night time, the water-clock was invented, and the device provided even periods for timing. With these developments, the resolution of time into periods of *year, month, day* and *hour* became known and has now been used for more than three millennia. The durations of minutes and seconds followed, using the Hellenistic sexagesimal system, and this division, according to Neugebauer, *"is the result of a Hellenistic modification of an Egyptian practice combined with Babylonian numerical procedures."* About 1300 AD, the mechanical clock **[1, Whitrow]** was advanced in Europe, ushering in a revolution in precision and miniaturization; Galileo began studies of pendulum motions and their clocking, using his heartbeats (seconds), in 1582 AD. Our present time standard is the cesium atomic clock[2], which provides precision of about $1{:}10^{13}$, i.e., the clock loses or gains one second every 1.6 million years. For this work, Norman Ramsey shared the 1989 Nobel Prize in Physics.

Until 1800 AD, the ability to record the timing of individual steps in any process was essentially limited to time scales amenable to direct sensory perception – for example, the eye's ability to see the movement of a clock or the

[1] Recognizing the incommensurability of the lunar month and the solar year, they abandoned the lunar month altogether and used a 30-day month. Thus the year is made of 12 months, 30 days each, with 5 feast days at the end of the year. The "Civil Calendar" was therefore 365 days per year and differs from the astronomical calendar of Sothis by approximately a 1/4 day every year. The two calendars must coincide at intervals of 365 × 4=1460 years, and historians, based on recorded dates of the reappearance of Sothis in dynastic periods, give the four dates for the coincidence of both calendars: 139 AD, 1317 BC, 2773 BC, and 4233 BC (+139-3 × 1460 ≡ 4241 BC). Even though the Egyptians discovered the astronomical calendar, 365-1/4 days, they decided, presumably for "bookkeeping", to use the civil calendar of 365 without leap years. They also divided the day into two periods of 12 hours each for day and night time. The remarkable calendar of years, months, days, and hours was adapted throughout history and formed the basis for the 365.25-day Julian (46 BC, Alexandria school) and 365.2422-day Gregorian (from 1582 AD, Pope Gregory XIII) calendars. In the words of the notable Egyptologist J. H. Breasted **[1]**, *"It has thus been in use uninterruptedly over six thousand years."* Many historians regard the date around 4241 BC as the beginning of *history* itself as it defines the period of written records; anything earlier is *prehistory*.

[2] Since 1967 one second has been *defined* as the time during which the cesium atom makes exactly 9, 192,631,770 oscillations.

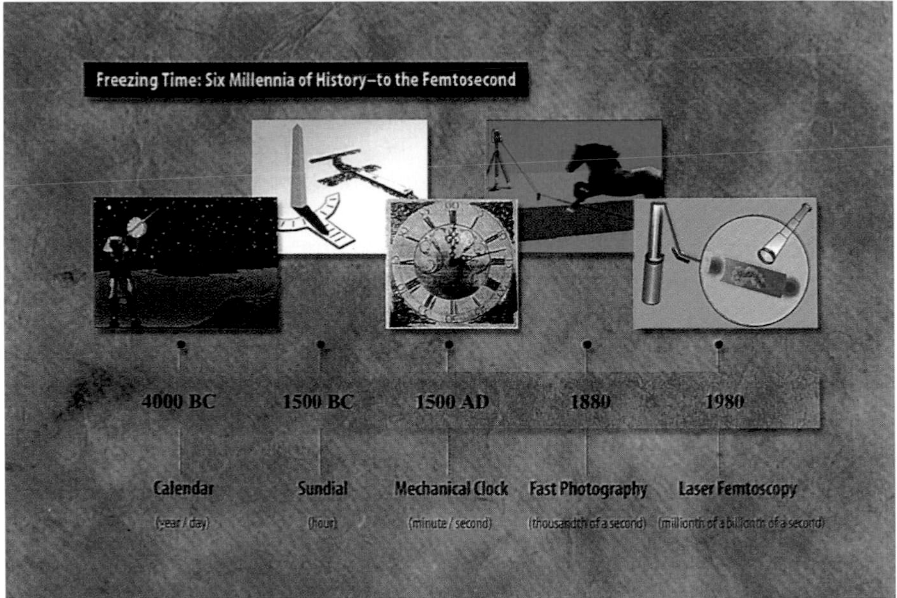

Figure 1. Timeline of some events in the history of measurements of time, from yearly calendars to the femtosecond regime (see text).

ear's ability to recognize a tone. Anything more fleeting than the blink of an eye (~0.1 second) or the response of the ear (~0.1 millisecond) was simply beyond the realm of inquiry. In the nineteenth century, the technology was to change drastically, resolving time intervals into the sub-second domain. The famous motion pictures by Eadweard Muybridge (1878) of a galloping horse, by Etienne-Jules Marey (1894) of a righting cat, and by Harold Edgerton (mid-1900's) of a bullet passing through an apple and other objects are examples of these developments, with millisecond to microsecond time resolution, using snapshot photography, chronophotography and stroboscopy, respectively **[2]**. By the 1980's, this resolution became ten orders of magnitude better [see **Section II2**], reaching the femtosecond scale[3], the scale for atoms and molecules in motion.

The actual atomic motions involved in chemical reactions had never been observed in real time despite the rich history of chemistry over two millennia **[3]**, as khem became khemia, then alchemy, and eventually chemistry **[3,4]** – see the Stockholm Papyrus in **Fig. 2**. Chemical bonds break, form, or geometrically change with awesome rapidity. Whether in isolation or in any other phase, this ultrafast transformation is a dynamic process involving the mechanical motion of electrons and atomic nuclei. The speed of atomic motion

[3] The prefix milli comes from Latin (and French), micro and nano from Greek, and pico from Spanish. Femto is Scandinavian, the root of the word for "fifteen"; nuclear physicists call femtometer, the unit for the dimensions of atomic nuclei, fermi. Atto is also Scandinavian.

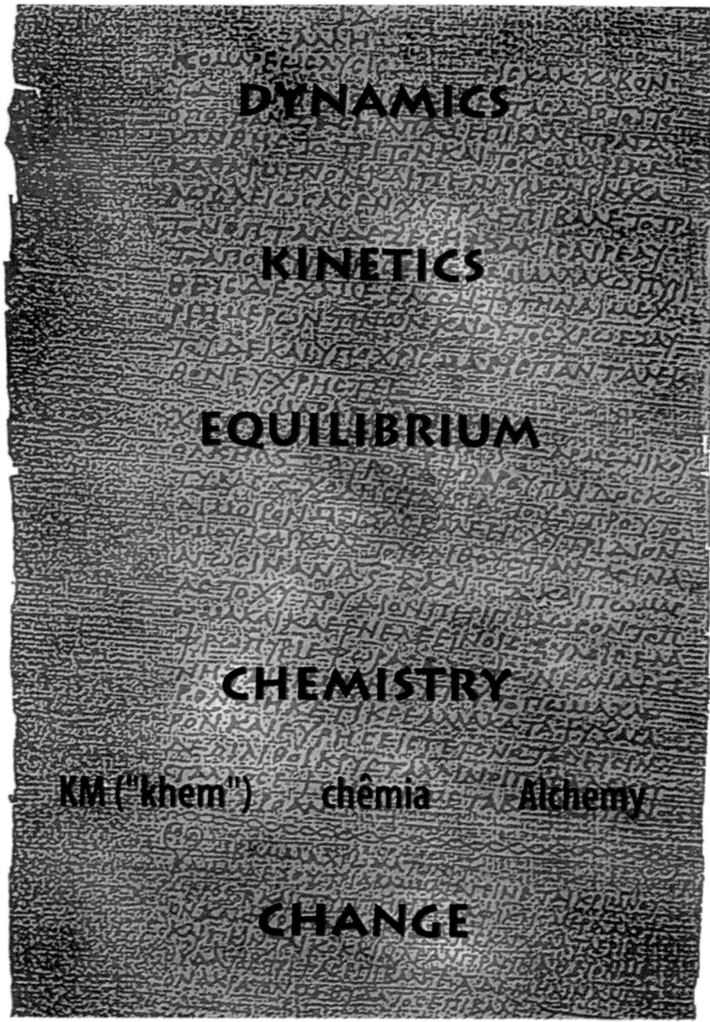

Figure 2. A page of the Stockholm papyrus (300 AD) describing the recipe for "making" (actually imitating) emerald. Note that *change* has been at the heart of chemistry from its millennia-old definition KM, to Chêmia and Alchemy, and to Chemistry. [Ref. B26, and references therein] Equilibrium, kinetics and dynamics are the foundation for the describtion of chemical changes.

is ~ 1 km/second and, hence, to record atomic-scale dynamics over a distance of an ångström, the average time required is ~ 100 fs. The very act of such atomic motions as reactions unfold and pass through their transition states is the focus of the field of femtochemistry. With fs time resolution we can "freeze" structures far from equilibrium and prior to their vibrational and rotational motions, or reactivity. The pertinent questions about the dynamics of the chemical bond are the following: How does the energy put into a reactant molecule redistribute among the different degrees of freedom, and how fast does this happen?; What are the speeds of the chemical changes connecting individual quantum states in the reactants and products?; What are the detai-

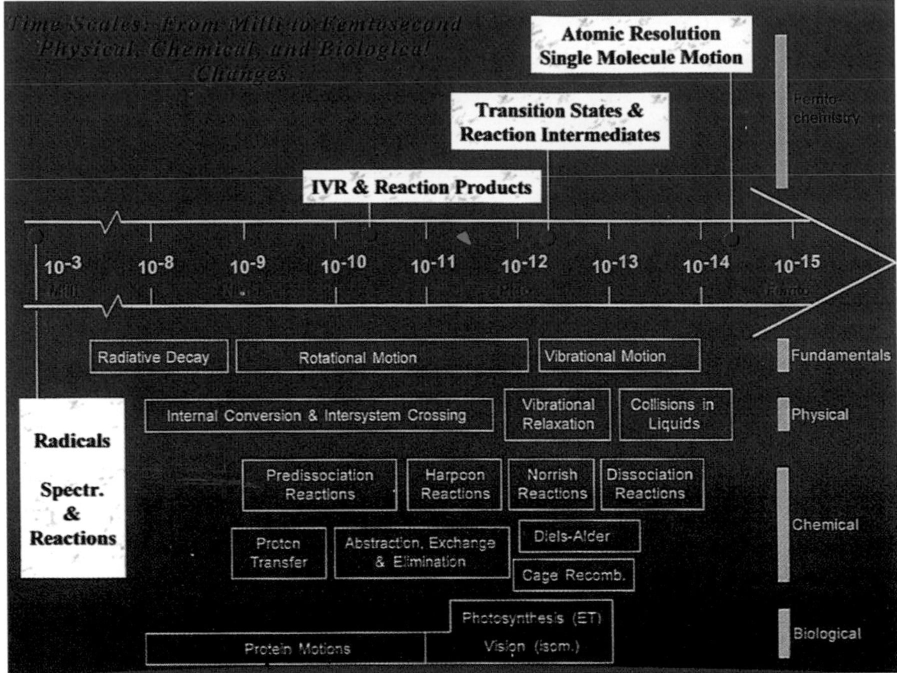

Figure 3. Time scales. The relevance to physical, chemical, and biological changes. The fundamental limit of the vibrational motion defines the regime for femtochemistry. Examples are given for each change and scale. [Ref. B4, B10]

led nuclear motions that chart the reaction through its transition states, and how rapid are these motions? As pointed out by Jim Baggott, *"the entire history of chemical reaction dynamics and kinetics has been about providing some approximate answers to these three questions* **[4]**."

In femtochemistry, studies of physical, chemical or biological changes are at the fundamental time scale of molecular vibrations: the actual nuclear motions **[Fig. 3]**. In this sense, femtoscience represents the end of the race against time, or, as the report of Ref. **[5]** puts it, "... *reaching the end of the road*". For this same reason, Martens stated, "*all chemistry is femtochemistry* **[5]**". The ephemeral transition states, denoted in the past by a bracket $[TS]^{\ddagger}$ for their elusiveness, can now be clocked as a molecular species $TS^{\ddagger}$. Moreover, on this time scale, the time-dependent description of a *coherent*, single-molecule trajectory represents the classical nuclear "motion picture" of the reaction as its wave packet proceeds from the initial state, through transition states, and on to final products – the language of the actual dynamics! The fs time scale is unique for the creation of such coherent matter waves on the atomic scale of length, a basic problem rooted in the development of quantum mechanics and the duality of matter. Figure (4) highlights some steps made in the description of the duality of light-matter and time scales, and the Appendix discusses the importance of coherence for localization.

Matter Waves
Particle-type Control & Dynamics

de Broglie (1924)

Einstein's light wave/particle

$E = h\nu \qquad E = c\,p$

$\therefore \lambda = h/p$

Similarly, matter particle/wave

Schrödinger (1926)

The Wave Equation – Stationary waves

$H\Psi = E\Psi$

Schrödinger (1926)

Micro- to Macro-mechanics

Quantum to Newton Mechanics

Ψ to wave group

Femtochemistry & Quantum Limit (h):
Particle-type

$\lambda_{\text{de Broglie}}$ (initial localization) = h/p

uncertainty in time measurement

$\Delta x\, \Delta p \geq h/(2\pi) \qquad \Delta t\, \Delta E \geq h/(2\pi)$

for force free

$\Delta x = p/m\, \Delta t \equiv v\, \Delta t$

$\Delta t \sim 10$ fs $\quad \Delta x \sim 0.1$ Å

$\Delta t \sim 10$ ps $\quad \Delta x \sim 100$ Å

Figure 4. Matter waves and particle-type limit of dynamics. The atomic-scale de Broglie wavelength in the coherent preparation of a quantum system and in the uncertainty of probe measurements are both reached on the fs time scale. See also the Appendix and its Figure.

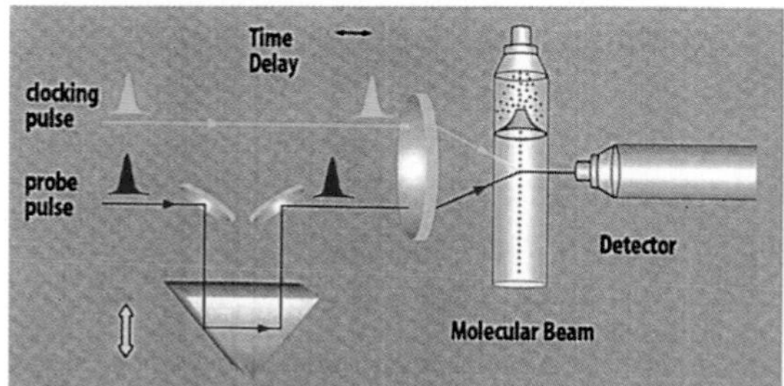

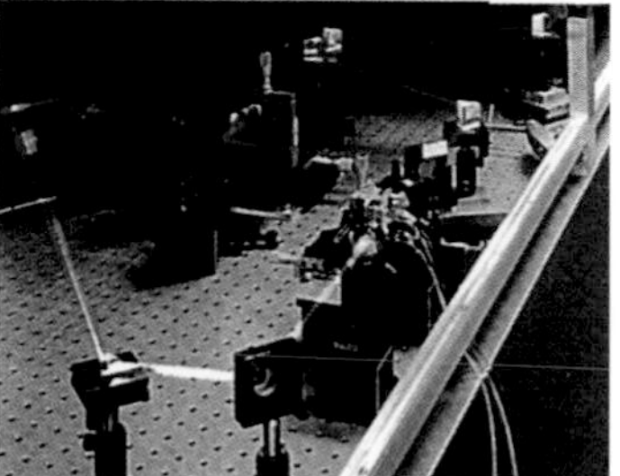

Figure 5. A femtochemistry apparatus typical of early Femtolands. (A, above) The laser system. (top) the first CPM oscillator used in Femtoland I; (bottom) the continuum generation to the right and the experimental layout for clocking to the left. (B, next page) The molecular beam apparatus of Femtoland III, together with a view of the beam/laser arrangement. [Ref. B1, B16, B28, 49]

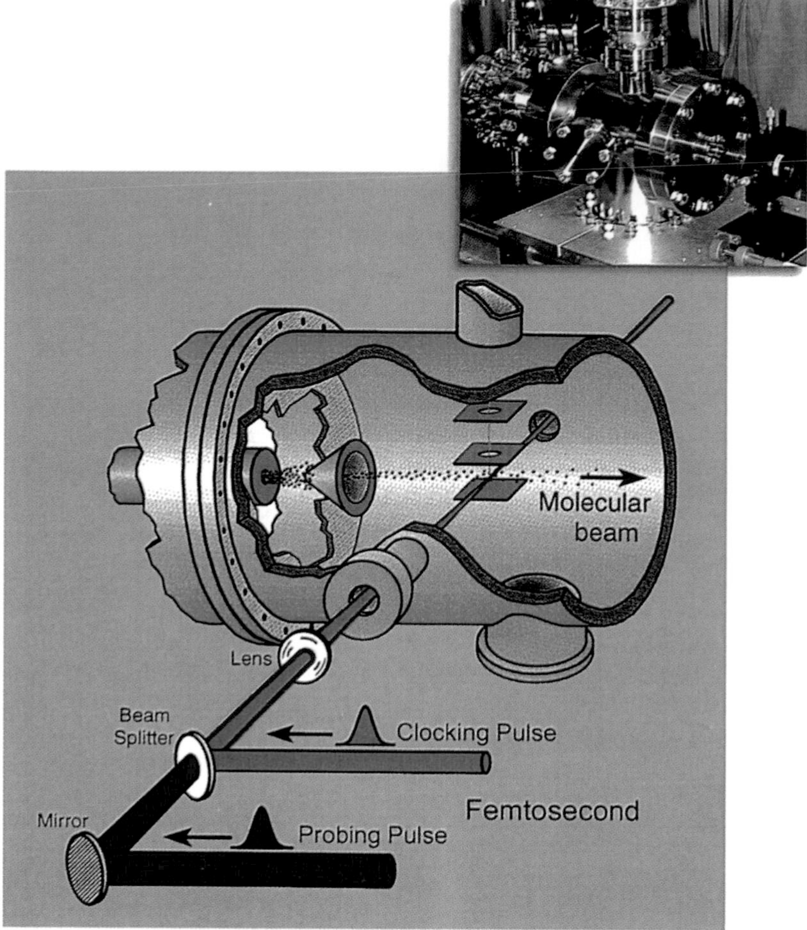

This powerful concept of *coherence* lies at the core of femtochemistry and was a key advance in observing the dynamics at atomic resolution. The realization of its importance and its detection by selectivity in both preparation and probing were essential in all studies, initially of states and orientations, and culminating in atomic motions in reactions. With these concepts in mind, the marriage of ultrafast lasers with molecular beams **[Fig. 5A, B]** proved to be essential for the initial development. Laser-induced fluorescence was the first probe used, but later we invoked mass spectrometry and non-linear optical techniques. Now numerous methods of probing are known **[Fig. 6]** and used in laboratories around the world; Coulomb explosion is the most recent powerful probe developed by Will Castleman **[5]** for arresting reactive intermediates, as mentioned in Section III.

Applications of femtochemistry have spanned the different types of chemical bonds – covalent, ionic, dative and metallic, and the weaker ones, hydrogen and van der Waals bonds. The studies have continued to address the varying complexity of molecular systems, from diatomics to proteins and DNA.

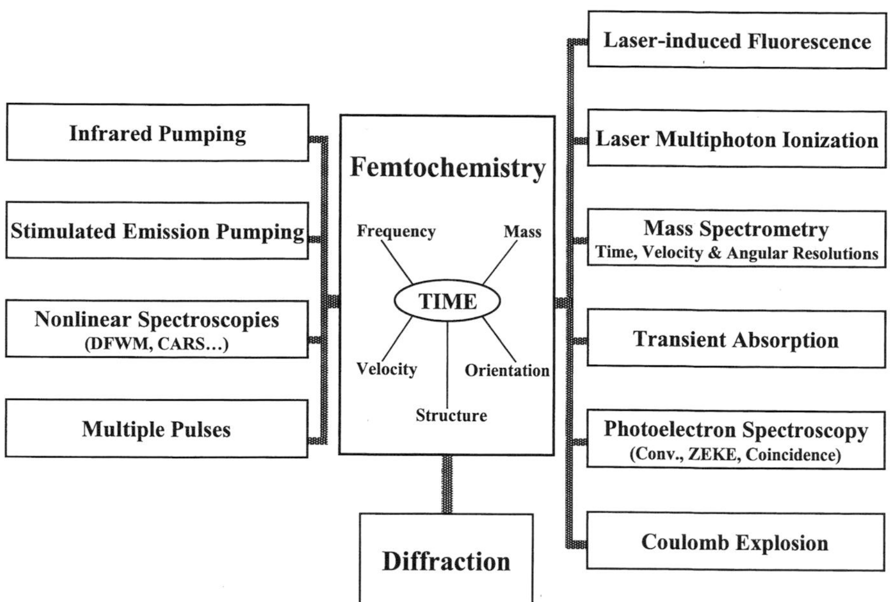

Figure 6. Techniques for probing in femtochemistry. Both excited and ground states have been probed by these methods. The correlations of time with frequency, mass, velocity and orientation were essential in the studies of complex systems. Diffraction represents the new effort for probing structures (see text and Fig. 29).

Studies have also been made in all phases of matter: gases and molecular beams; mesoscopic phases of clusters, nanostructures, particles and droplets; condensed phases of dense fluids, liquids, solids, surfaces and interfaces; and in sibling fields of femtoscience such as femtobiology **[Fig. 7]**.

Twenty-four centuries ago, the Greek philosopher Democritus and his teacher Leucippus gave birth to a new way of thinking about matter's invisible and elementary entity, the atom. Richard Feynman once asked, if you had only one sentence to describe the most important scientific knowledge we possess, what would that sentence be? He said, *"everything is made of atoms."* Democritus' atomism, which was rejected by Aristotle, was born on a purely philosophical basis, surely without anticipating some of the twentieth century's most triumphant scientific discoveries. Atoms can now be seen, observed in motion, and manipulated **[6]**. These discoveries have brought the microscopic world and its language into a new age, and they cover domains of length, time and number. The *length* (spatial) resolution, down to the scale of atomic distance (ångström), and the *time* resolution, down to the scale of atomic motion (femtosecond), have been achieved. The trapping and spectroscopy of a single ion (electron) and the trapping and cooling of neutral atoms have also been achieved. All of these achievements have been recognized by the awarding of the Nobel Prize to STM (1986), to single-electron and -ion trapping and spectroscopy (1989), to laser trapping and cooling (1997), and to laser femtochemistry (1999).

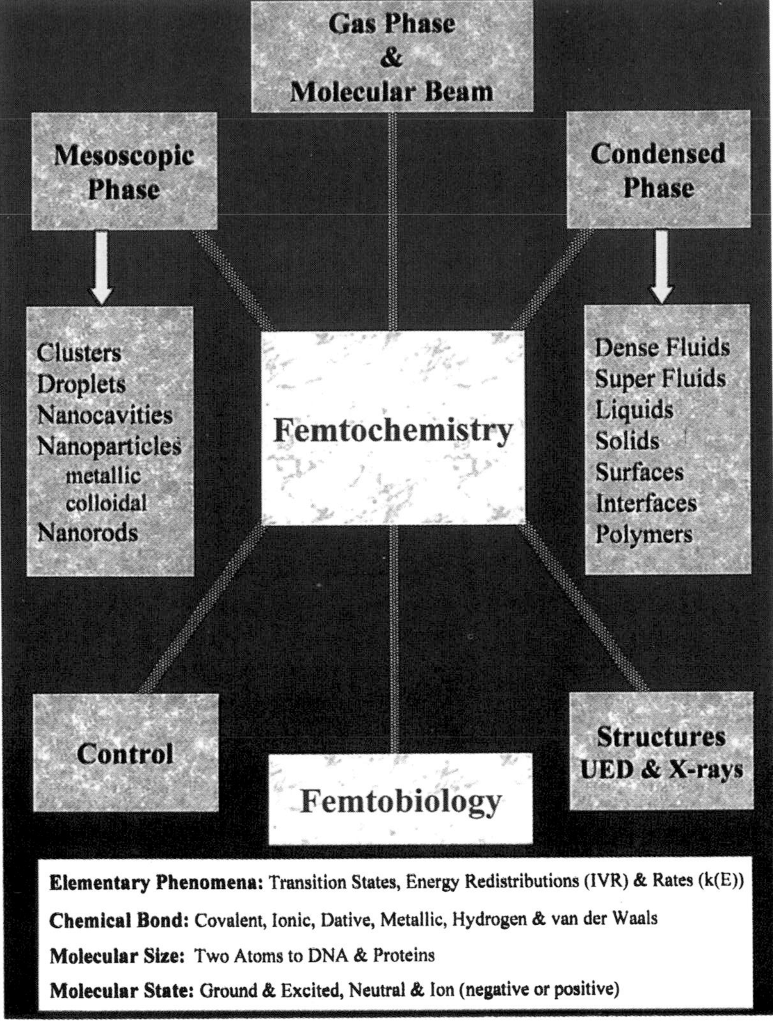

Figure 7. Femtochemistry and the scope of applications.

II. DYNAMICS AND ARROW OF TIME

(1) Origins – From Kinetics to Dynamics

The Arrhenius Seminal Contribution

At the turn of the 20th century, the study of reactivity was dominated by the question[4]: *How do reactions proceed and what are their kinetic rates?* Svante Arrhenius **[7]** gave the seminal description of the change in rates of chemical

[4] The main focus of prior studies was on the thermodynamics and equilibrium characteristics. Interestingly, chemical equilibrium was already established a century before (1798) by Claude Berthollet, during a visit to the Natron Lakes in Napoleonic times, through his studies of the reaction $Na_2CO_3 + CaCl_2 \leftrightarrow 2NaCl + CaCO_3$ [S. W. Weller, *Bull. Hist. Chem.* **24**, 61 (1999)].

reactions with temperature and formulated in 1889 the familiar expression for the rate constant,

$$k = A \exp(-E_a/RT) \tag{1}$$

which, as Arrhenius acknowledged, had its roots in van't Hoff's (1884) equations **[7]**. For any two reactants, the rate constant (k) depends on temperature (T) according to an energy of activation (E_a), which is different from the thermodynamic net energy change between reactants and products, and the dependence is exponential in form. (R is the universal gas constant). If we think of the reaction as a finite probability of reactant A colliding with B, then the rate is simply the collision frequency times the fraction of successful collisions with an energy equal to or more than E_a. Besides the value of equation (1) through the well-known plots of "ln k *vs.* 1/T" to obtain the energy of activation E_a, Arrhenius introduced a "hypothetical body", now known as the "activated complex", a central concept in the theory of reaction rates: the reaction, because of collisions or other means, proceeds only if the energy is sufficient to exceed a barrier whose energy is defined by the nature of the complex. Since then, various experimental data for different temperatures T were treated with equation (1), yielding E_a and the pre-exponential factor A.

A few years after Arrhenius' contribution, Bodenstein (1894) **[7]** published a landmark paper on the hydrogen/iodine system, which has played an important role in the development of gas-phase chemical kinetics, with the aim of understanding *elementary* reaction mechanisms. In the twenties, Lindemann (1922), Hinshelwood (1926), Tolman (1920), and others **[7&8]** developed, for unimolecular gas-phase reactions, elementary mechanisms with different steps describing activation, energy redistribution, and chemical rates. By 1928, the Rice-Ramsperger-Kassel (RRK) theory was formulated, and Marcus, starting in 1952, blended RRK and transition state theory in a direction which brought into focus the nature of the initial and transition-state vibrations in what is now known as the RRKM theory **[8]**.

The rate constant, k(T), does not provide a detailed molecular picture of the reaction. This is because k(T), which was obtained from an analogy with van't Hoff's description of the change with T of the equilibrium constant K (thermodynamics), is an average of the microscopic, reagent-state to product-state rate coefficients over all possible encounters. These might include different relative velocities, mutual orientations, vibrational and rotational phases, and impact parameters. A new way was needed to describe, by some quantitative measure, the process of the chemical reaction itself: How reagent molecules approach, collide, exchange energy, sometimes break bonds and make new ones, and finally separate into products. Such a description is the goal of molecular reaction dynamics **[9]**.

The London, Eyring and Polanyi Contributions

For some time, theory was ahead of experiment in studies of *microscopic* molecular reaction dynamics. The effort started shortly after the publication of the Heitler-London quantum-mechanical treatment (1927) of the hydrogen molecule **[10]**, a breakthrough in thinking not only about the stable structure of the chemical bond, but also about how two atoms can interact at different separations. One year later (1928), for Sommerfeld's Festschrift (60th birthday), London **[10]** presented an approximate expression for the potential energy of triatomic systems, e.g., H_3, in terms of the coulombic and exchange energies of the "diatomic" pairs. In 1931 Henry Eyring and Michael Polanyi **[10]**, using the London equation, provided a semiempirical calculation of a potential energy surface (PES) of the $H + H_2$ reaction describing the journey of nuclei from the reactant state of the system to the product state, passing through the crucial transition state of activated complexes. The birth of "reaction dynamics" resulted from this pioneering effort and, for the first time, one could think of the PES and the trajectories of dynamics on it – *in those days, often, expressed in atomic units of time!* But no one could have dreamed in the 1930's of observing the transient molecular structures of a chemical reaction, since the time scale for those *far from equilibrium* activated complexes in the transition state was estimated to be less than a picosecond (ps).

The time scale was rooted in the theory developed for the description of reaction rates. Building on Arrhenius' work and the work of Polanyi and Wigner (1928) **[11]**, in 1935, Eyring, and independently Evans and Polanyi, formulated *transition-state theory*, which gave an explicit expression for Arrhenius' pre-exponential factor **[11]**:

$$k = \frac{\mathbf{k}T}{h} K^{\ddagger} = \frac{\mathbf{k}T}{h} \frac{Q^{\ddagger}}{Q_A Q_B} \exp(-E_0 / \mathbf{k}T) \qquad (2)$$

where **k** is Boltzmann's constant, h is Planck's constant and Q is the partition function. The key idea here is to assume the *equilibration* of the population between reactants and the transition state: $A + B \leftrightarrow [TS]^{\ddagger} \rightarrow$ products. Thus, the rate constant can be related to the equilibrium constant for formation of the transition state, $K^{\ddagger}$, and hence $\Delta G^{\ddagger}$, $\Delta H^{\ddagger}$ and $\Delta S^{\ddagger}$; physically, $\Delta G^{\ddagger}$ ($\Delta H^{\ddagger}-T\Delta S^{\ddagger}$) in the exponent gives the barrier energy (through $\Delta H^{\ddagger}$) along the reaction coordinate, and the pre-exponential entropic term which reflects the change in vibrational modes perpendicular to the reaction coordinate. By comparing (2) with Arrhenius' equation (1), A can be identified; E_a, the activation energy, and E_o, the barrier energy, with zero-point energy corrections, are related [8, Laidler]. Kramers' (1940) classic work **[11]** modified the pre-exponential factor to include friction from the surrounding medium, with transition state theory giving an upper limit, and Casey Hynes, in the 1980's, provided a dynamical theory of friction with emphasis on time scales in the transition-state region and for solvent interaction **[11]**. In the 1950's, Marcus **[11]** obtained a transition-state type expression for reactions of electron transfer in solutions with the Gibbs free energy of activation expressing the dependence on solvent reorganization energies. This work was awarded the 1992 Nobel Prize.

According to transition state theory, the fastest reaction is given by **k**T/h, which is basically the "frequency" for the passage through the transition state

[Equation 2]. At room temperature this value is 6×10^{12} second^{-1}, corresponding to ~ 170 fs; the time scale of molecular vibrations is typically 10–100 fs. This estimate is consistent with knowledge of the speeds of nuclei and the distance change involved in the reaction. In 1936 the first classical trajectory from Hirschfelder-Eyring-Topley molecular dynamics simulations of the $H+H_2$ reaction showed the fs steps needed to follow the reaction profile, albeit on the wrong PES. Later, Karplus, Bunker and others showed a range for the time scales, ps to fs, depending on the reaction and using more realistic PES's **[see 9 and 26]**.

Transition State and Definition

In general, for an elementary reaction of the type,

$$A + BC \rightarrow [ABC]^{\ddagger} \rightarrow AB + C \tag{3}$$

the whole journey from reagents to products involves changes in internuclear separation totaling ~10Å. If the atoms moved at 10^4–10^5 cm/sec then the entire 10Å trip would take 10^{-12}–10^{-11} sec. If the 'transition state', $[ABC]^{\ddagger}$, is defined to encompass *all configurations of ABC significantly perturbed from the potential-energy of the reagents A + BC or the products AB + C,* then this period of 1–10 ps is the time available for its observation. To achieve a resolution of ~ 0.1 Å, the probe time window must be 10-100 fs.

The above definition of the transition state follows the general description given by John Polanyi and the author **[12],** namely the full family of configurations through which the reacting particles evolve *en route* from reagents to products. This description may seem broad to those accustomed to seeing the TS symbol, ‡, displayed at the crest of the energy barrier to a reaction. As stated in **Ref. 12,** even if one restricts one's interest to the over-all rates of chemical reactions, one requires a knowledge of the family of intermediates sampled by reagent collisions of different collision energy, angle and impact parameters. The variational theory of reaction rates further extends the range of TS of interest, quantum considerations extend the range yet further, and the concern with rates to yield products in specified quantum states and angles extends the requirements most of all. A definition of the TS that embraces the entire process of bond-breaking and bond-making is therefore likely to prove the most enduring. This is specially important as we address the energy landscape of the reaction, as discussed in Section IV.

The cardinal choice of the transition state at the saddle point, of course, has its origin in chemical kinetics – calculation of rates – but it should be remembered that this is a mathematical 'single-point' with the division made to define the speed of a reaction **[8]**. Even in thermal reactions, there is enough of an energy distribution to ensure many types of trajectories. Furthermore, transition state theory is not a quantum theory, but a classical one, because of the assertion of a *deterministic* point. Quantum uncertainty demands some delocalization **[8, Hase]** and, as mentioned above, the theory invokes the equilibration of reactants and transition state populations, using statistical thermodynamics. In fact, the term transition state is used ambiguously to refer both to the quasi-equilibrium *state* of the reaction and to the molecular *structure* of the saddle point. As discussed in **Ref. 13 [Williams, Doering, Baldwin]**, the molecular species at the saddle point perhaps could be referred to as the *transition structure;* the *activated complex* is even more descriptive.

The location at the saddle point provides the highest energy that must be reached, defining the exponential probability factor; the dynamics (forces and time scales) are governed by the nature of the TS region. Provided that the energy landscape of the reaction is controlled by a narrow region, the structure of the transition state becomes important in structure-reactivity correlations; rates *vs.* Gibbs energy between the TS and the ground state. (It is also useful in designing TS analogs as enzyme inhibitors and TS complements as catalysts **[13]**). The position of the TS and its energy relative to that of reactants and products along the reaction path becomes relevant. In the mid 1930's, the Bell-Evans-Polanyi principle gave a predictive correlation between changes in barrier heights and the enthalpies of reactions, especially for a series of related reactions; the TS for an exoergic or an endoergic reaction is very different. In 1955, this led to Hammond's postulate which characterizes a reactant-like TS (so called "early" TS) for exoergic reactions, product-like TS (so called "late" TS) for endoergic reactions, and "central" TS for energy neutral reactions **[8, Shaik]**. John Polanyi, the son of Michael, formulated some concepts regarding energy disposal in relation to the position of the TS on the PES **[14]**, using molecular dynamics simulations and experimental studies of chemiluminescence.

It should be recognized that selectivities, efficiencies, and stereochemistries are quantified only when the dynamics on the global energy landscape are understood. For example, the picture for simple reactions of a few (strong) bonds being made and broken is changed when the energy surface is nearly flat or there is a significant entropic contribution. Many transition states will exist in the region as in the case of protein folding **[13, Fersht]**. Finally, in *the transition state,* chemical bonds are in the process of being made and broken. In contrast, for *intermediates,* whose bonds are fully formed, they are in potential wells, typically "troughs" in the TS region. However, the time scale is crucial. In many cases, the residence time in intermediates approaches the fs regime characteristic of TS structures and the distinction becomes fuzzy. Moreover, for a real multidimensional PES, the non-reactive nuclear motions can entropically lock the system even though there is no well in the energy landscape. And the presence of shallow wells in the energy landscape does not guarantee that trajectories will visit such wells.

Transition State and Spectroscopy

Various techniques have been advanced to probe transition states more directly, especially for elementary reactions. Polanyi's analogy **[14]** of transition-state spectroscopy, from "spectral wing emission", to (Lorentz) collisional line broadening studies, made earlier by A. Gallagher and others **[see 12]**, set the stage for the use of CW spectroscopic methods as a probe. (In this way, only about one part in a million of the population can be detected.) Emission, absorption, scattering and electron photodetachment are some of the novel methods presented for such time-integrated spectroscopies, and the groups of Jim Kinsey, Philip Brooks and Bob Curl, Benoit Soep and Curt Wittig, Dan Neumark, and others, have made important contributions to this area of research. The key idea was to obtain, as Kinsey **[15]** puts it, *short-time dynamics from long-time experiments.* (The renaissance of wave packet dynamics in spectroscopy, pioneered by Rick Heller, will be highlighted in **Section III4 and III6**). With these spectroscopies in a CW mode, a distribution of spectral frequencies provides the clue to the desired information regarding the distribution of the TS over successive configurations and potential energies. Recently, this subject has been reviewed by Polanyi and the author and details of these contributions are given therein **[12]**, and also in **[26]**.

(2) Arrow of Time

In over a century of development, time resolution in chemistry and biology has witnessed major strides, which are highlighted in **Fig. 8 [16]**. As mentioned before, the Arrhenius equation (1889) for the speed of a chemical reaction gave information about the time scale of rates, and Eyring and Michael Polanyi's (1931) microscopic theoretical description made chemists think of the atomic motions through the transition state and on the vibrational time scale. But the focus naturally had to be on what could be measured in those days, namely the slow rates of reactions. Systematic studies of reaction velocities were hardly undertaken before the middle of the 19th century; in 1850 Ludwig Wilhelmy reported the first quantitative rate measurement, the hydrolysis of a solution of sucrose to glucose and fructose **[8, Laidler].** In 1901, the first Nobel Prize for chemistry was awarded to van't Hoff for, among other contributions, the theoretical expressions (chemical dynamics) which were precursors to the important work of Arrhenius on rates. Arrhenius too received the Prize in 1903 for his work on electrolytic theory of dissociation.

A major advance in experiments involving sub-second time resolution was made with flow tubes in 1923 by H. Hartridge and F. J. W. Roughton for solution reactions. Two reactants were mixed in a flow tube, and the reaction products were observed at different distances. Knowing the speed of the flow, one could translate this into time, on a scale of tens of milliseconds. Such measurements of non-radiative processes were a real advance in view of the fact that they were probing the "invisible", in contrast with radiative glows which were seen by the naked eye and measured using phosphoroscopes. Then came the stopped-flow method (B. Chance, 1940) that reached the millisecond scale. The stopped-flow method is still used today in biological kinetics.

Around 1950, a stride forward for time resolution in chemistry came about when Manfred Eigen in Germany and R. G. W. Norrish and George Porter in England developed techniques reaching the microsecond time scale **[17].** For this contribution, Eigen and Norrish & Porter shared the 1967 Nobel Prize. The method of flash photolysis was developed by Norrish and Porter a few years after World War II, using electronics developed at the time. They pro-

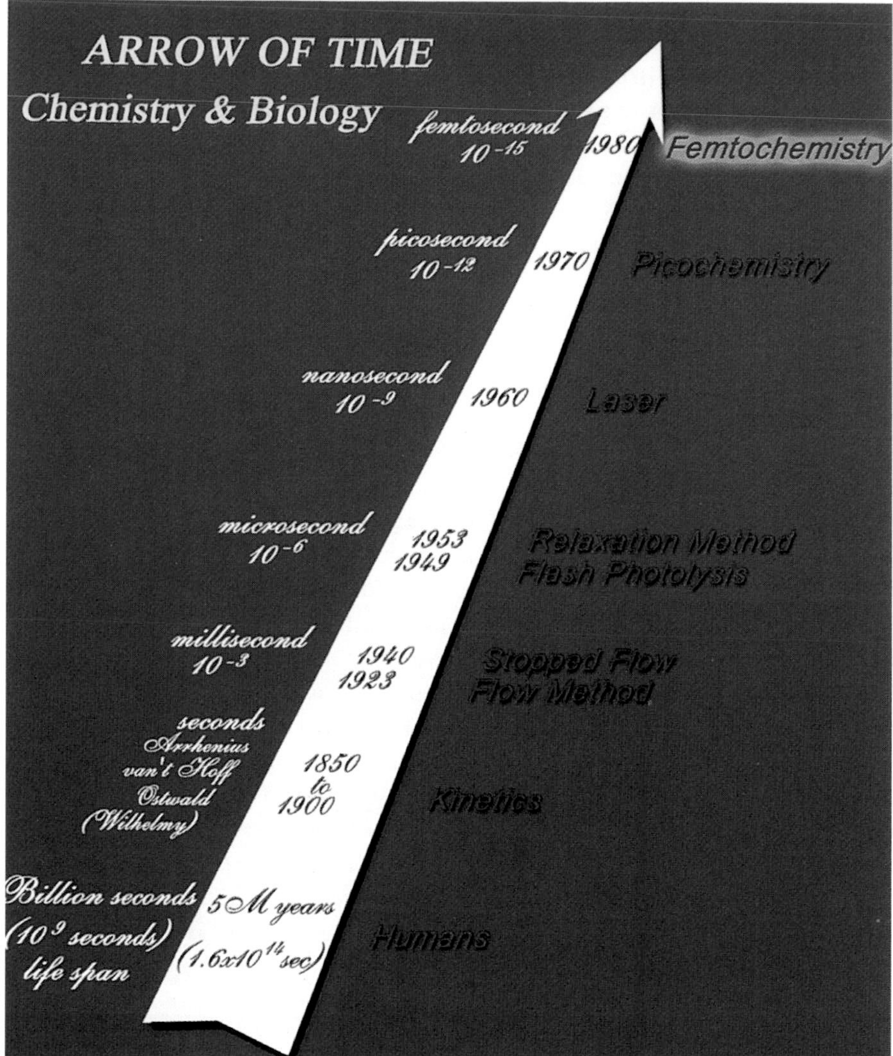

Figure 8. Arrow of Time in chemistry and biology, some steps over a century of development (see text). [Ref. B1]

duced an intense burst of light and created radicals in the sample, and, using other light, they recorded the spectra of these radicals. They achieved kinetics on this time scale and observed some relatively stable intermediates.

Before the turn of the 20th century, it was known that electrical sparks and Kerr cell shutters could have response times as short as ten nanoseconds. In an ingenious experiment, Abraham and Lemoine (1899) **[18]** in France demonstrated that the Kerr response of carbon disulfide was faster than ten nanoseconds; it has now been measured to be about two picoseconds (with femtosecond response) **[18]**. They used an electrical pulse which produced a spark and simultaneously activated a Kerr shutter. Light from the spark was collimated through a variable-delay path and through the Kerr cell (polarizer,

CS_2 cell and analyzer). The rotation of the analyzer indicated the presence of birefringence in the cell for short optical delays; this birefringence disappeared for pathlengths greater than several meters, reflecting the total optical/electrical response time of 2.5 ns. In this landmark "pump-probe" experiment, they demonstrated in 1899 the importance of synchronization. The setting of time delays was achieved by varying the light path. Bloembergen **[18]** has recently given a historical perspective of short-pulse generation and Shapiro has reviewed the early developments, including mechanical, streak, spark, stroboscope, and other high-speed photography methods **[see Refs. 2 & 18]**. As pointed out in these references **[18]**, flash photolysis utilized the above approach but one of the flashes was made very strong to generate high concentrations of free radicals and hence their utility in chemical and spectroscopic applications.

Eigen developed "the relaxation method", which reached the microsecond and close to the nanosecond scale. By disturbing the equilibrium of a solution by either a heat jump, a pressure jump, or an electric field, the system shifts from equilibrium. This is the point of time zero. Then the system equilibrates, and its kinetics can be followed. (At about the same time, shock-tube methods were used to provide kinetics on similar time scales.) Eigen called these reactions "immeasurably fast" in his Nobel lecture. There was a feeling that this time resolution was the fastest that could be measured or that needed to be measured for relevance to chemistry **(Section IV)**. The invention of the laser has changed the picture.

Shortly after the realization of the first (ruby) laser by Maiman (1960), the generation of giant and short pulses became possible: nanoseconds by Q-switching (Hellwarth, 1961) and picoseconds (De Maria, *et al* 1966) by mode-locking (1964). Sub-picosecond pulses from dye lasers (Schäfer & Sorokin, 1966) were obtained in 1974 by Chuck Shank and Eric Ippen at Bell Labs, and in 1987 a six fs pulse was achieved **[19]**. In 1991, with the discovery of femtosecond pulse generation from solidstate Ti-sapphire lasers by Sibbett and colleagues **[19]**, dye lasers were rapidly replaced and femtosecond pulse generation became a standard laboratory tool; the state-of-the-art **[19]**, once 8 fs, is currently about 4 fs and made it into the Guinness Book of World Records (Douwe Wiersma's group **[19]**). The tunability is mastered using continuum generation **[19, Alfano & Shapiro]** and optical parametric amplification.

In the late sixties and in the seventies, ps resolution made it possible to study *non-radiative* processes, a major detour from the studies of conventional *radiative* processes to infer the non-radiative ones. As a beginning student, I recall the exciting reports of the photophysical rates of internal conversion and biological studies by Peter Rentzepis **[20]**; the first ps study of chemical reactions (and orientational relaxations) in solutions by Ken Eisensthal **[21]**; the direct measurement of the rates of intersystem crossing by Robin Hochstrasser **[22]**; and the novel approach for measurement of ps vibrational relaxations (in the ground state of molecules) in liquids by Wolfgang Kaiser and colleagues **[23]**. The groups of Shank and Ippen **[19]** have made important contributions to the development of dye lasers and their applications in the ps and into the fs regime. Other studies of chemical and biological non-radiative processes followed on the ps time scale, the scale coined by G. N. Lewis as the "jiffy" – the time needed for a photon to travel 1 cm, or 33 picoseconds **[24]**.

At about the same time in the sixties, molecular-beam studies of reactions were being developed, and, although I was not initially a member of this community, beams later became part of our effort in femtochemistry. Molecular collisions occur on a shorter time scale than a ps and real time studies were not possible at the time. Crossed molecular beams and chemiluminescence techniques provided new approaches for examining the dynamics of single collisions using the post-attributes of the event, the reaction products. The contributions by Dudley Herschbach, Yuan Lee and John Polanyi **[14, 25]** were acknowledged by the 1986 Nobel Prize. From state and angular distributions of products, information about the dynamics of the collision was deduced and compared with theoretical calculations of the PES and with molecular dynamics simulations; the goal was to find self-consistency and to deduce an estimate of the lifetime of the collision complex. Crossed molecular beam-laser studies have probed dynamics via careful analyses of product internal energy (vibrational and rotational) distributions and steady-state alignment and orientation of products. The contributions to this important area are highlighted in the article by Dick Zare and Dick Bernstein **[25]** and in the book by Raphy Levine and Bernstein **[9]**. An overview of femtochemistry (as of 1988) in connection with these other areas is given in a feature article **[26]** by Zewail and Bernstein.

III. FEMTOCHEMISTRY: DEVELOPMENT OF THE FIELD

The development of the field is highlighted in this section, from the early years of studying coherence to the birth of femtochemistry and the explosion of research, or, as the report of **Ref. 5** puts it, "*... the revolution in chemistry and adjacent sciences.*" On the way, there were conceptual and experimental problems to overcome and many members of our Caltech group have made the successful evolution possible. The review article published in the Journal of Physical Chemistry **[B14]** names their contributions in the early stages of development. Here, references are given to the work and explicitly in the Figure Captions.

(1) The Early Years of Coherence

When I arrived in the US as a graduate student in 1969, nine years after the invention of the first laser, I had no idea of what lasers were about. When appointed to the Caltech faculty as an assistant professor in 1976, I was not thinking or dreaming of femtosecond time resolution. But I had the idea of exploring *coherence* as a new concept in dynamics, intra- and inter-molecular. This proved to be vital and fruitful.

New Techniques for Molecules

The Caltech offer included start-up funds of $50,000 for capital equipment ($15K for shop services), an empty laboratory of two rooms and an office next to it. A few months before moving to Caltech in May of 1976, I made the decision not to begin with the type of picosecond research I was doing at Berkeley as a postdoctoral fellow. Instead, the initial effort was focused on two directions: (i) studies of coherence in disordered solids, and (ii) the development of a new laser program for the studies of the phenomena of (optical) coherence. Prior to the final move, I came down from Berkeley for several visits in order to purchase the equipment and to outline the laboratories' needs for electricity, water, gases and so on; my feeling was that of a man left out in a desert with the challenge to make it fertile. We spent a significant fraction of the $50K on setting up the apparatus for optical detection of magnetic resonance (ODMR) to study disordered solids. My experience at Penn with Professor Robin Hochstrasser and at Berkeley with Professor Charles Harris was to culminate in these experiments. The key questions I had in mind were: What is the nature of energy migration when a crystal is systematically disordered? Is it coherent, incoherent or partially coherent? Is there a relationship between optical and spin coherence?

While the ODMR apparatus was being built, I was thinking intensely about new laser techniques to probe the coherence of optical transitions (so-called optical coherence). I had the intuitive feeling that this area was rich and at the time had in mind several issues which were outlined in my research proposal to Caltech. Laser experiments were designed with objectives focused on the same issues outlined in the proposal. First, in the work I published with Charles Harris, and alone at Berkeley, the coherence probed had been that of spin (triplet excitons) and I felt that we should directly probe the optical coherence, i.e., the coherence between the excited electronic state and the ground state, not that between two spin states of the same excited state **[Fig. 9]**. Second, I did not believe that *the time scale* of spin coherence was the same as that of optical coherence. Later, I wrote a paper on the subject which was published in the Journal of Chemical Physics (1979) with the title: *Are the homogeneous linewidths of spin resonance and optical transitions related?*

I was convinced that essentially all molecular optical transitions in solids are inhomogeneous, that is, they do not reflect the true dynamics of a homogeneous ensemble, but rather the overlapping effects of sub-ensembles. This was the key point outlined in my research proposal to Caltech. If the new set of laser experiments at Caltech proved successful, we should be able to find the answer to these important (to me) questions. However, the funds remaining were insufficient to realize our dream for the new experiments. Fortunately, Spectra Physics, started in 1961 and now a huge laser company, was proud of a new product and was interested in helping us demonstrate the usefulness of one of the first single-mode dye lasers they produced. At Caltech, we had the pump argon-ion laser, from Wilse Robinson's Laboratory, and added the dye laser we obtained from Spectra Physics with the idea that we would purchase it if the experiments were successful. David Evans of

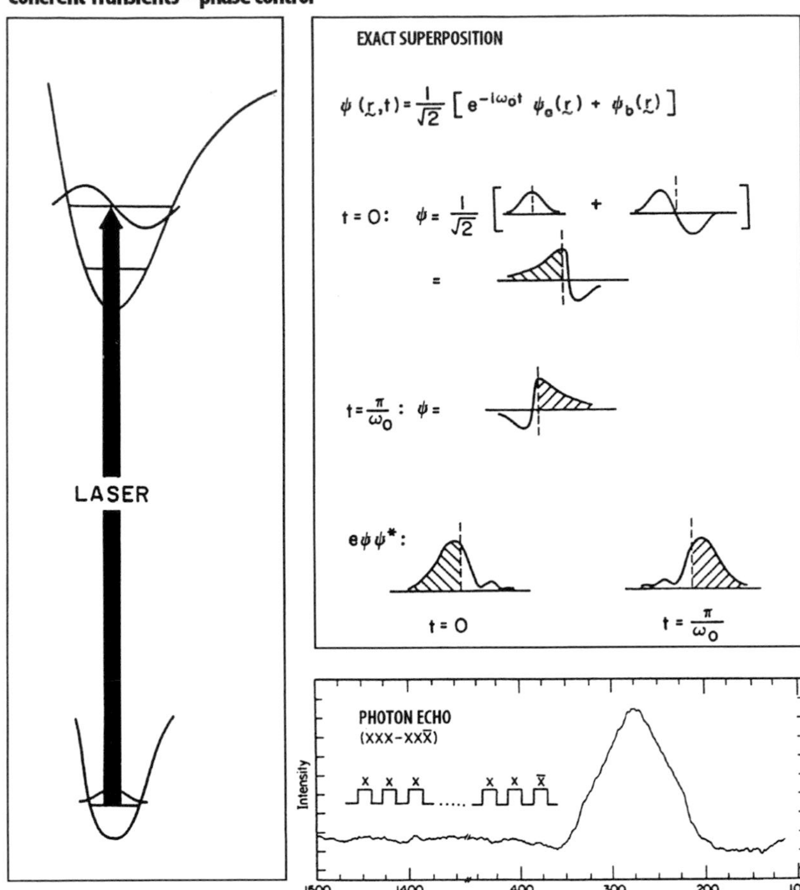

Figure 9. Molecular coherence and dephasing. Coherent transients from the superposition of states, and control of pulse phase (x or x̄) in multiple pulse experiments, the optical analog of NMR spectroscopy. The photon echo of iodine gas was observed on the spontaneous emission using the described pulse sequence. [Ref. B9, B13, B22, B23, 50]

Spectra Physics was instrumental in helping us achieve this goal. What was left were the low-temperature cryostat and electronics. We could not afford a real metal cryostat, so we custom-made a glass Dewar that could go down to 1.8 Kelvin. Most of the electronics were obtained on loan for several months. To generate laser pulses we used switching methods developed at JILA (Joint Institute for Laboratory Astrophysics) and IBM. The work at IBM by Dick Brewer's group triggered our interest in using electro-optic switching methods.

We succeeded in making the laser perform according to specifications and began the first experiments with phenazine crystals. This system, phenazine, has unique properties which I learned about in Robin Hochstrasser's laboratory. At Berkeley I published a paper outlining the nature of coherence in multidimensional systems and used phenazine as a prototype experimental system. Our first laser transient at Caltech was beautiful. Unfortunately, the transient was from the electronics, not from the crystal, but we soon realized this! We decided to abandon this particular system for a while and to try an impurity crystal of pentacene in a host of terphenyl. We also decided to study gases, and some success came our way.

With the theoretical knowledge acquired in handling coherence effects, which requires expertise with density matrix formalism and its manipulation in geometrical frames, I had a novel idea: we should be able to detect *coherence* on the *incoherent* emission at optical frequencies. From many discussions with Alex Pines and Charles Harris at Berkeley, I knew the power of "adding" pulses in NMR and ESR. This idea was successful and indeed we were able to observe the photon echo on the spontaneous emission **[Fig. 9]** using three optical pulses. Only months after my arrival in May of 1976, we published our first scientific paper from Caltech. This success gave us confidence in the approach and in our understanding of the principles of coherence and its probing in molecular systems. We applied it to larger molecules with success but also encountered some disappointments. This work was followed by a variety of extensions to studies in gases and solids and also in a home-made (from glass) effusive molecular beam. The small group in our laboratory of optical coherence became productive and we had an exciting time. Our group and that of Douwe Wiersma in Holland were then the two most active in these areas of chemical research.

In the meantime, the work on disorder in solids began to yield interesting results and, surprisingly, we observed an unusual change in the degree of energy transfer with concentration, which we published as evidence of Anderson localization, a hot topic in the 1970's. This was followed by detailed studies of several systems and I wrote my first research proposal to the National Science Foundation (NSF). The research was funded! The then program director at NSF, Fred Stafford, was supportive of the effort and we have maintained our support from the Division of Materials Research to this day; the NSF's Chemistry Division continues to support our research. Knowledge of energy transfer was also helpful in another area. Terry Cole, a visiting scholar at Caltech, and I initiated work on the studies of Luminescent Solar Concentrators (LSC) using energy transfer between dyes as a key principle. This idea, too, was successful and funding for this research came in from SERI and from ARCO. The work on LSC resulted, ultimately, in a patent (with Sam Batchelder; issued in 1980) and in several publications. My research group was rapidly expanding.

After a year-and-a-half at Caltech, I was pleased to learn that my colleagues were considering my case for tenure. Tenure was granted a few months later, two years after my arrival at the Institute. I was both appreciative and pleased. We continued research in four areas: optical coherence phenomena and dephasing; disorder in solids; picosecond spectroscopy; and LSC. In my own department, some colleagues were not too excited about 'this stuff of coherence and dephasing' – thinking that it was *not relevant to chemistry*! Many chemists on the outside were also unsure what this was all about. In fact, a notable chemist once said publicly at a conference I attended that coherence and dephasing had nothing to do with chemistry! On the physics side, I was invited to numerous conferences, including one in which the Nobel Laureate Willis Lamb asked me to have dinner to discuss our research. This was a special experience!

I was not convinced by these doubts and my faith helped us to continue along with the development. The concept of coherence turned out to be fundamental in femtochemistry, and it is now well accepted that coherence is a key process in the probing and controlling of molecular dynamics. With the success we had with observations and studies of coherence in different systems, I wrote an Account of Chemical Research article, published in 1980, with the title: *Optical Molecular Dephasing – Principles of and Probings by Coherent Laser Spectroscopy.* I felt that the nanosecond time scale we had mastered should be extended to the picosecond time scale, but did not wish to repeat the Berkeley experience with glass lasers. Fortunately, the design for the first sub-picosecond dye laser was reported in 1974 and we decided to build one to study the phenomena of coherence – but now on a shorter time scale.

Optical Analogue of NMR: Controlling the Phase

From the studies of optical transients, we learned that coherence can be probed directly in real time in gases (and solids) and that incoherent decay (e.g. fluorescence) can be used to monitor such coherences provided that the laser pulse(s) is capable of forming a *coherent* superposition of states. For two states of a transition (say ψ_a and ψ_b), the coherent state can be written as

$$\Psi_{coherent}(t) = a(t)\,\psi_a + b(t)\,\psi_b \qquad (4)$$

where the coefficients, a(t) and b(t) contain in them the familiar quantum-mechanical phase factors, $\exp(-iE_at/\hbar)$ and $\exp(-iE_bt/\hbar)$, respectively **[see Fig. 9]**.

With pulse sequences, we could directly monitor the behavior of the ensemble-averaged coefficients of $\Psi \cdot \Psi^*$, $\langle a(t)\, b^*(t)\rangle$, which contain information on the coherence decay time (optical T_2); they are the off-diagonal elements of a density matrix, ρ_{ab}. The term $\langle a(t)\, a^*(t)\rangle$ is the population of state ψ_a and represents the diagonal density-matrix element, ρ_{aa}; $\langle a(t)\, a^*(t)\rangle$ decays with optical T_1. We were thus able to demonstrate the power of the optical analogue of NMR pulse techniques in learning about coherence and the origin of optical dephasing in molecular systems of interest to chemical dynamics. This advance changed the thinking of many with the recognition that it was impossible to deduce T_1 and T_2 from measurements of the line width of inhomogeneous transitions.

One feature of this work which later helped us in the study of molecular reaction dynamics was the realization of the importance of the *pulse phase (shape)* in studies of coherence. With the acousto-optic modulation techniques we developed earlier, it became possible to make optical pulse sequences with well-defined phases. This development took us into the domain of selective and prescribed pulse sequences which could then be used to enhance coherences or suppress them – *the optical analogue of NMR multiple pulse spectroscopy*. We published several papers on phase control and extended the applications to include photon locking. We were eager to extend these techniques to the picosecond time domain in order to study solids, but, for several reasons, our attention was diverted to gas-phase molecular dynamics.

By this time, our group's efforts were narrowing on two major areas. (Dick Bernstein, who was on the Visiting Committee for our Division, hinted that I was doing too much in too many areas!) The work on disorder and LSC was gradually brought to completion. Picosecond spectroscopy of rotational diffusion and energy transfer in liquids were similarly handled. I felt that the latter area of research was too crowded with too many scientists, a characteristic I do not enjoy when venturing into a new area. I must add that I was not too thrilled by the exponential (or near exponential) decays we were measuring and by the lack of molecular information. Our effort began to emphasize two directions: (i) the studies of coherence and dynamics of isolated molecules in supersonic beams and (ii) the development of the optical analogue of NMR spectroscopy. The low-temperature facility was put to use to study the dephasing and polarization of highly vibrationally-excited molecules in the ground state. Coherence in chemical dynamics was occupying my thinking, and I made a detour in the applications that turned out to be significant.

(2) The Marriage with Molecular Beams

The Bell Labs design for the dye laser (passively mode-locked, CW, and cavity dumped) was too restrictive for our use and, even though we published several papers on studies in the condensed phase with 0.6 ps resolution, we decided to change to a new system. The synchronously pumped mode-locked (CW) dye laser allowed for tunability and also for photon-counting detection techniques. The power of single-photon-counting became apparent, and a new laser system, a synchronously pumped, cavity-dumped dye laser was constructed for studies of gas-phase molecular dynamics, but now with the benefit of all the expertise we had gained from building the first system used for studies in the condensed phase and for probing the torsional rigidity of DNA.

Stimulated by the work on coherence, and now with the availability of picosecond pulses, I thought of an interesting problem relating to the question of intramolecular *vs.* intermolecular dephasing. In large, isolated molecules (as opposed to diatomics), there are the so-called heat bath modes which can be a sink for the energy. The question arose: Could these bath modes in isolated large molecules dephase the optically-excited initial state in the same way that phonons of a crystal (or collisions in gases) do? This problem has some roots in the question of state preparation, and I was familiar with its relationship to the description of radiationless transitions through the work of Joshua Jortner, Wilse Robinson and others. There was much theoretical activity about dephasing, but I felt that they were standard extensions and they did not allow for surprises. We decided on a new direction for the studies of coherence in a supersonic molecular beam.

Rick Smalley came to Caltech in May of 1980 and gave a talk entitled "Vibrational Relaxation in Jet-Cooled Polyatomics". He spoke about his exciting work on the naphthalene spectra. From the line width in the excitation spectra, he inferred the "relaxation time". At the time, the work by Don Levy, Lennard Wharton, and Smalley on CW (or nanosecond) laser excitation of molecules in supersonic jets was providing new ways to examine the spectroscopy of molecules and van der Waals complexes. Listening to Rick and being biased by the idea of coherence, I became convinced that the way to monitor the homogeneous dynamics was not through the apparent width but by using coherent laser techniques. This was further kindled by the need for direct measurement of energy redistribution rates; we were encouraged by Charlie Parmenter, after he had reported on a chemical timing method using collisions as a "clock" to infer the rate of energy redistribution. The first "real" supersonic molecular beam was huge. We did not know much about this kind of technology. However, in a relatively short time, it was designed and built from scratch, thanks to the effort of one graduate student who must have consumed kilos of coffee! The molecular beam and picosecond system were interfaced with the nontrivial addition of a spectrometer to resolve fluorescence in *frequency* and *time*. This was crucial to much of the work to come.

The Anthracene Discovery: A Paradigm Shift

Our goal in the beginning was to directly measure the rate of IVR (Intramolecular Vibrational-energy Redistribution), expecting to see a decrease with time (exponential decay) in the population of the initially-excited vibrational state and to possibly see a rise in population in the state after the redistribution, thus obtaining T_1 directly. What we saw in these large systems was contrary to the popular wisdom and unexpected. The population during IVR was oscillating coherently back and forth **[Fig. 10]** with well-defined period(s) and phases! We were very excited because the results revealed the significance of coherence at its best in a complex molecular system, now *isolated* in a molecular beam, with many degrees of freedom. I knew this would receive attention and skepticism. We had to be thorough in our experimental tests of the observation and three of us went to the laboratory to see how robust the observation was. We published a Communication in the *Journal of Chemical Physics* (1981). Earlier there had been attempts by another group to observe such a "quantum coherence effect" in large molecules, but the observation turned out to be due to an artifact. Some scientists in the field were skeptical of our new observation and the theorists argued that the molecule is too big to see such quantum coherence effects among the *vibrational* states. Furthermore, it was argued that rotational effects should wash out such an observation.

We followed the initial publication with several others and the effect became even more pronounced with shorter time resolution. Physicists appreciated the new results. We published a *Physics Review Letter* on the nature of non-chaotic motion in isolated systems, and Nico Bloembergen and I wrote a review (1984) on the relevance to laser-selective chemistry **[B21]**. We and other groups subsequently showed the prevalence of this phenomenon in large molecules. As is often the case in science, after the facts and in retrospect, the phenomenon was clear and was soon accepted; to some it even became obvious! Looking back, this novel and unexpected observation was a *paradigm shift* of critical importance, for a number of reasons:

First, the observation was the first to clearly show the presence of "quantum coherence effect" in isolated complex chemical systems and only among selected vibrational states. In other words, out of the expected *chaotic motion* in the vibrational and rotational phase space we could see *ordered and coherent motion* despite the presence of numerous vibrational degrees of freedom (from S_0 and S_1 states). This point was theoretically appreciated by only a few scientists. In fact, at one point Stuart Rice and I drafted a paper on the subject, thinking of clarifying the point. At the time, researchers in high-resolution spectroscopy were observing complex spectra and attributing this complexity to chaotic vibrational motion in the molecule. Stuart and I argued that spectral complexity does not mean chaos, and the anthracene experiment was a clear demonstration in real time.

Second, the observation demonstrated that coherence had not previously been detected in complex systems, not because of its absence but due to the inability to devise a proper probe. Detection of total absorption or total emission (at all wavelengths) from molecules gives a non-selective window on the dynamics and in this way coherence cannot be detected. This was a key point for the success of the anthracene experiment for which both *time* and *wavelength* were resolved and correlated. For all subsequent work on wave packet dynamics, nuclear motion in chemical reactions, and femtochemistry, this concept of "window probing" was essential. The concept was further elucidated by resolving the phase character. By probing at two different wavelengths, we found that the quantum oscillations exhibit identical periods, but were phase shifted by exactly 180° (i.e., they are out-of-phase) **[Fig. 10]**. The two wavelengths resolved were those corresponding to emission of the initial vibrational state and to that of the vibrational state to which the population goes by IVR. Thus, if "total detection" was invoked, the in-phase and out-of-phase oscillations will add up to cancel each other and coherence would have remained undiscovered.

Third, observation of phase coherent dynamics gave us a new dimension. The phase shift indicates a true transfer of population, in contrast with conventional quantum beats, and by analyzing the phases we could understand the nature of IVR: *"concerted"*, i.e. going at the same time to all states, or *"non-concerted"* i.e., going in a sequential redistribution of vibrational energy. We could also obtain the time scale and the effect of molecular rotations on coherence of the vibrational motion.

Fourth, the observation illustrated the importance of the "preparation of non-stationary states" in molecules. This issue was of fundamental importance in radiationless transition theories involving multiple electronic states, and experiments by Jan Kommandeur and by Doug McDonald have shown this interstate coupling. The question of interest to us was: what nuclear states do we

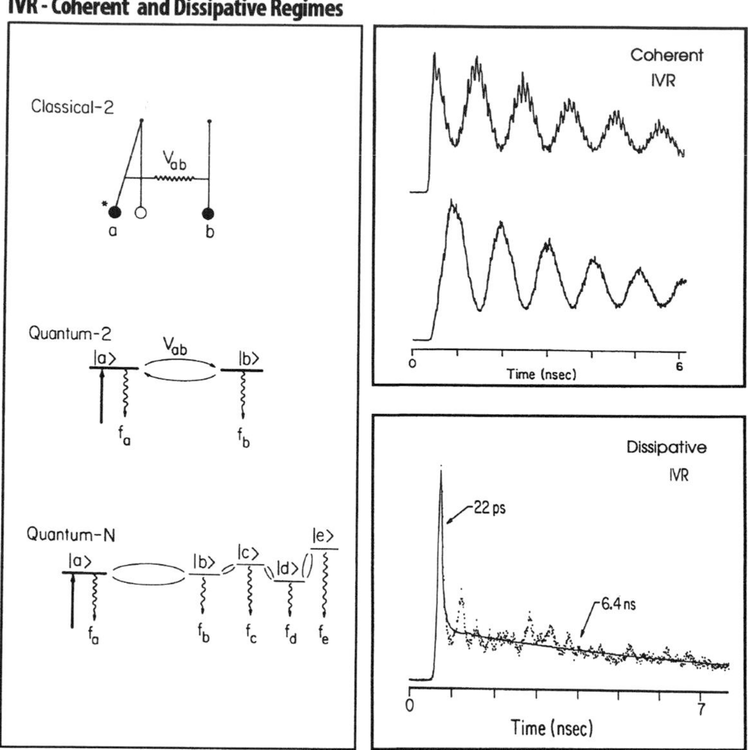

Figure 10. Dynamics of IVR, intramolecular vibrational-energy redistribution. The coherent, restricted and the dissipative regimes. Note the exact in-phase and out-of-phase oscillatory behavior between the vibrational states of the system (anthracene in a molecular beam). The theory for classical and quantum pictures (to the left) has been discussed in detail in the literature. [Ref. 51]; See also ref B20, B21.

prepare in the isolated molecule on a single surface? The anthracene experiment taught us that a coherent source spanning the stationary states can prepare a non-stationary state which evolves with time. Moreover, we can prepare molecules in-phase at time zero to observe the subsequent coherent dynamics. This concept indicates that the description, in terms of Schrödinger's molecular stationary states, is not cardinal and that the time-dependent picture is real and directly relevant to dynamics. Most textbooks describe dynamics in terms of stationary states and it took some time for this concept of a time-dependent description to be appreciated. I recently found a theoretical article by Roy Gordon published in the 1960's touching on similar issues. In femtochemistry, the concept of time-domain dynamics is what describes elementary motions.

Fifth, by directly probing coherence and its extent in isolated, complex molecular systems we advanced some concepts regarding the nature of IVR and its regions. We divided the regions of IVR into three basic ones: *no IVR, restricted IVR,* and *dissipative IVR.* We also established that the IVR picture of one vibrational state coupled to a continuum of vibrational levels is not adequate. Instead it is a multi-tier coupling among vibrational states.

This work and its implications were published in two series of papers and reviewed in two book chapters.

The Successful 036 Laboratory

The laboratory known as 036 was in the sub-basement of Noyes. In this laboratory, the initial work on IVR was followed by fruitful applications spanning (i) studies of IVR in other systems, (ii) radiationless transitions, and (iii) reaction rates of a variety of systems. One of our first studies of reactions on the ps time scale, isomerization of stilbene, was stimulated by discussion with Robin Hochstrasser about his work on stilbene vapor at room temperature. He felt that if we could resolve the low-frequency modes in the molecular beam, we would derive a great deal of information on the torsional potential. We resolved these torsional modes. Furthermore, we decided to study the rates as a function of energy and in the process found the barrier for twisting around the double bond and observed coherent IVR in reactions, the first such observation. Even now, stilbene remains a member of our molecular family and continued studies have been pursued by us and others, also on the fs time scale.

The following list highlights some of the work **[Fig. 11]** done in this initial period from 1981 to 1983: (1) IVR in anthracene and stilbene; (2) trans-cis isomerization of stilbene; (3) quantum beats and radiationless transitions in pyrazine; (4) intramolecular hydrogen bonding in methyl salicylate; (5) intramolecular electron transfer in donor-bridge-acceptor systems; (6) IVR and dissociation of intermolecular hydrogen-bonded complexes; and (7) isomerization of diphenylbutadiene and styrene.

Over the years, in the same laboratory (036 Noyes), members of our group have made new extensions covering the following topics: isomerization in isolated molecules vs. in bulk solutions; nonchaotic multilevel vibrational energy flow; mode-specific IVR in large molecules; IVR dynamics in alkyl-anthracenes; isotope effects on isomerization of stilbene; charge transfer and exciton dynamics in isolated bianthracene; isotope effects on the intramolecular dephasing and molecular states of pyrazine; IVR dynamics in alkylanilines (the "ring + tail" system); mode-specific (non-RRKM) dynamics of stilbene-rare gas vdW complexes; solvation effects on intramolecular charge transfer; IVR dynamics in p-difluorobenzene and p-fluorotoluene (real time vs. chemical timing); IVR dynamics in deuterated anthracenes; dynamics of interstate coupling in chromyl chloride; dynamics of IVR and vibrational predissociation in anthracene-Ar_n ($n = 1,2,3$); structural effects on the dynamics of IVR and isomerization in stilbenes; and dynamics of IVR and vibrational predissociation in n-hexane solvated *trans*-stilbene. The research resulted in a series of publications.

Changing A Dogma: Development of RCS

The success with the anthracene experiment made us ask a similar question, but now regarding the coherent *rotational motion* of isolated, complex molecules. There were theories around which discarded its possibility because of

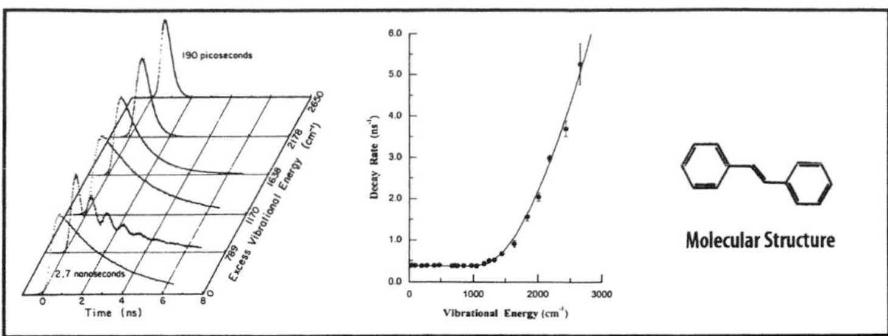

Electron Transfer Reaction

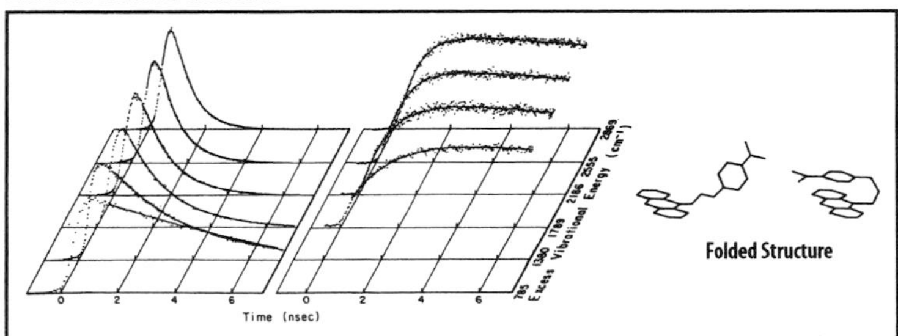

Microscopic Solvation

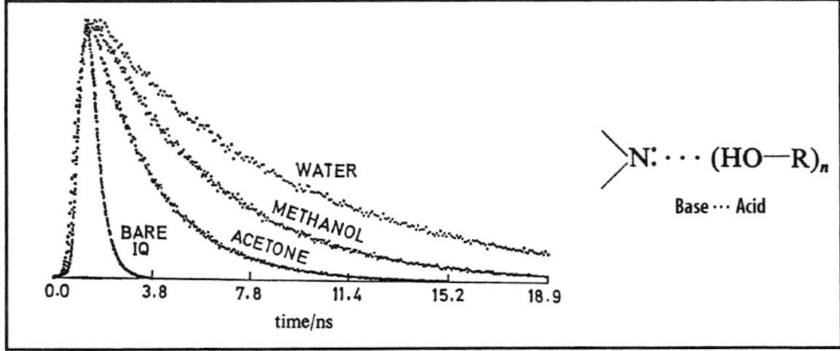

Figure 11. Some examples of studies in the 036 Laboratory. The isomerization of stilbene, intramolecular electron transfer, and solvation in clusters are examples of the studies made in the early 1980's (see text); IQ = isoquinoline. [Ref. B3, 52].

the general belief that Coriolis interactions, anharmonicity and other interactions would destroy the coherence. We worked out the theoretical implications and the results suggested possibly another surprise: if we could align the molecules with a *polarized* picosecond pulse and probe (polarization-selective) the rotating molecules, we should be able to observe rotational recurrences which would give the full period of rotations of the isolated (large) molecule. Classically, it is as if the molecule rotated back to its initial configuration. This rotation period gives the moment of inertia and, since the masses of the atoms are known, we can deduce distances, and hence obtain information on molecular structures of very large molecules.

Indeed the recurrences in stilbene were observed with high precision, and its molecular structure deduced. Coherence in rotational motion was clearly evident and could be probed in a manner similar to what we had done with vibrational coherence. The approach was again met with some skepticism regarding its generality as a molecular structure technique. However, it is now accepted by many as a powerful Doppler-free technique; more than 120 structures have been studied this way. The method **[Fig. 12]** is termed *"Rotational Coherence Spectroscopy (RCS)"* and is successfully used in many laboratories. Some book chapters and review articles have been published on the subject [see **Bibliography**].

Out of this first marriage between ultrafast lasers and molecular beams came the developments and concepts discussed above. We were now poised to study molecules and reactions with even shorter time resolution. We could study their vibrational and rotational dynamics and align ("orient") them by controlling time.

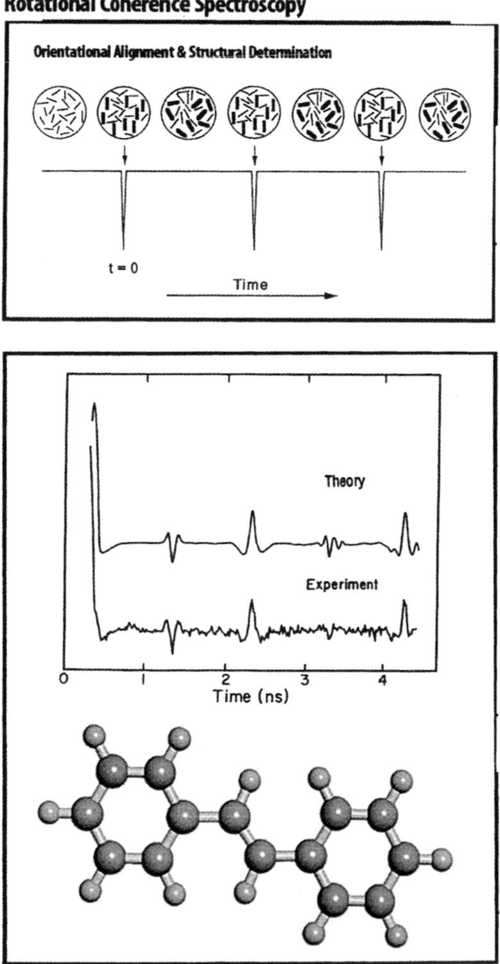

Figure 12. Rotational Coherence Spectroscopy (RCS), the concept and first experimental observation made in a beam of trans-stilbene. [Ref. B7, B13, 53]

(3) The Transition to the Sub-Picosecond Regime

By the early 1980's, our laser time resolution for studying molecules in supersonic jets was 15 picoseconds and detection was made using a microchannel plate (~ 40 ps). With this resolution, we had already studied reactions such as the isomerization (twisting) of stilbene, charge-transfer, and intra- and intermolecular proton and hydrogen atom transfer. How could we improve the time resolution and study, in a general way, the elementary steps of reactions? The only approach I knew of was to use two pulses, one to "pump" and the other to "probe". Unlike liquid-state studies, where the approach was proven successful, in this case, the density of a molecular beam is very low. Furthermore, it was not clear how to establish the zero of time *in situ* in the molecular beam and how to avoid temporal broadening due to propagation effects.

A New Beam Machine: Pump-Probe Mass Spectrometry

I thought we should build a second generation beam apparatus to house a time-of-flight mass spectrometer. From the physics literature, it was clear that single-atom detection could be observed using ionization techniques with lasers, and such detection had already been successful with nanosecond lasers. Unlike many nanosecond studies, we should propagate the two picosecond (and later femtosecond) pulses in the *same* direction, otherwise we would lose the ultrashort time resolution! The same beam machine was equipped with optics for laser-induced fluorescence detection. We began a new direction of research in a separate laboratory of our group. The new beam was built and integrated with two independently tunable dye lasers. This proved to be a precursor to the femtochemistry work as this taught us to master pump-probe picosecond and sub-picosecond experiments on chemical reactions. In this same laboratory, we studied with a resolution of a few ps: (1) dissociation reactions; (2) ground-state, overtone-initiated reactions; (3) van der Waals reactions and others **[Fig. 13]**. We wrote a series of papers on *state-to-state microcanonical rates*, k(E), and addressed theoretical consequences and deviations from the statistical regime.

It was in two of these systems (reactions of NCNO and ketene) that we found that the statistical phase-space theory, although successful in describing product-state distributions, failed in describing the microcanonical rates k(E) as a function of energy. Moreover, we made careful studies of the effect of rotational population on k(E), and the effect was dramatic near the threshold. Rudy Marcus, stimulated by these studies of k(E), applied variational RRKM theory and we published some papers in a collaborative effort. The key point here is that the TS "moves" to different (shorter) distances along the reaction coordinate at different energies; the cardinal definition is relaxed [see above, **Section II**]. In another system (H_2O_2), we studied the *ground-state* ("thermal") reaction for the first time in real time by initiating the reaction with direct excitation of the overtones of the OH stretch vibration **[Fig. 13]**. The coupling between theory and experiment stimulated my interest in

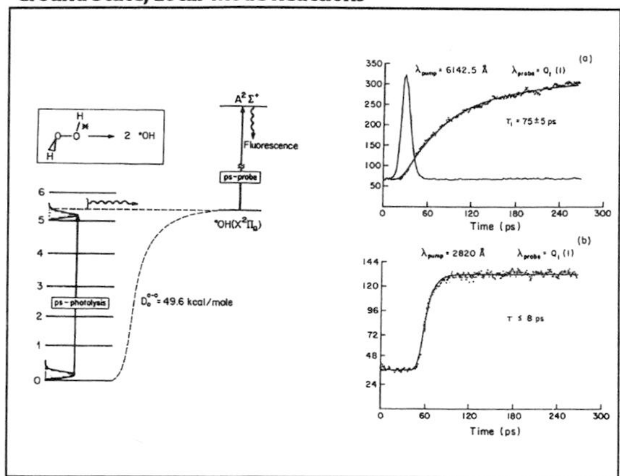

Reaction Rates of NCNO as a Function of Energy

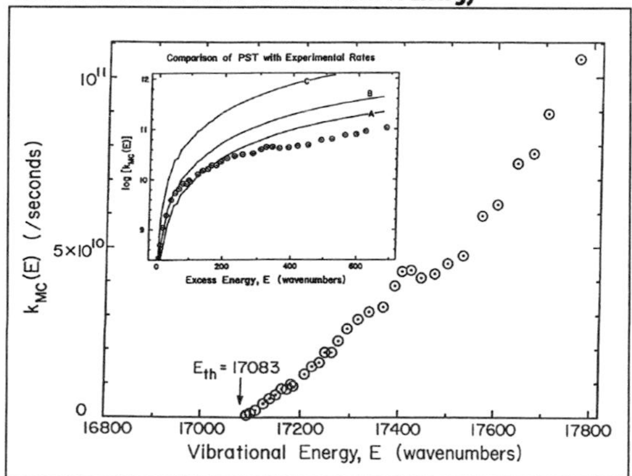

Figure 13. Microcanonical Rates of Reactions: (top) Ground-state reaction of H_2O_2 initiated by local-mode excitation; (bottom) the dissociation of NCNO as a function of energy, showing the breakdown of conventional phase-space theory at energies above threshold. [Ref. B1, 54]

the nature of transition states which generally live for less than a picosecond. The thirst for even shorter time resolution became real!

The First ICN Experiment: Sub-picosecond Resolution

In the early 1980's, the technology of pulse compression became available and we ordered, from Spectra Physics, a pulse compressor – a fiber optic arrangement to reduce the laser pulse width to sub-picosecond. The company indicated that it would take them several months to build one, and that the only one available was at Purdue University in the laboratory of Professor Duane Smith, one of the first two graduate students I had at Caltech. I mentioned to Duane my excitement about the experiment, which was intended to *directly* monitor the elementary bond breakage in a molecule, and asked if it was possible to borrow his compressor. The triatomic molecule ICN was chosen because the CN radical could be conveniently monitored by laser-induced fluorescence; we had been encouraged by the positive experience we had had with CN from NCNO and with earlier ps results on ICN. Also, ICN had been central to studies of dissociation reactions and to photofragment spectroscopy. Previous work, without time resolution, had provided a measurement of the so-called anisotropy parameter β and, hence, inference of the time scale.

All that we needed was a factor of 10-100 improvement in time resolution from what we currently had in the group. Duane shipped the compressor and joined us for two weeks. We observed the first ICN *sub-picosecond* transient, thus establishing the new methodology. In the same year, we wrote a paper which was accepted and published (December 1985) in the *Journal of Physical Chemistry*. We did not resolve the transition states of this reaction, but only detected the rise of the product. The last paragraph in this paper summarized what it would be possible to do if the time resolution could be improved by a further order of magnitude: "*Since the recoil velocity is ~ 2 x 10^5 cm/s, the fragment separation is ~ 10Å on the time scale of the experiment (~500 fs). With this time resolution, we must, therefore, consider the proximity of fragments at the time of probing, i.e., the evolution of the transition state to final products.*" I wrote this sentence having in mind that the fs resolution is the ideal one and that our next step in research should be in this direction.

Several factors influenced the fast entry into femtochemistry. The development in 1981 by Richard Fork, Ben Greene and Chuck Shank of the colliding pulse mode-locked (CPM) ring dye laser took the pulse duration into the 90-fs regime. By 1985, when we were involved in the ICN experiment, 27-fs pulses were generated with the help of intracavity pulse compression. Soon after, in 1987, 6-fs pulses were obtained by amplification and extracavity compression **[19]**. With such short pulses and with the help of the earlier development of continuum generation by Bob Alfano and Stanley Shapiro **[19],** continuously tunable fs pulses became available and only (!) required expertise in ultrafast lasers and nonlinear optics.

The interaction with several colleagues was a stimulating force in the initial effort. With Rudy Marcus and Vince McKoy, my colleagues at Caltech, I discussed many experiments and theories, especially on our way to lunch at the Athenaeum, Caltech's faculty club. John Polanyi came to Caltech in 1982 as a Fairchild Scholar. John saw the importance of (CW) transition-state spectroscopy, and his paper in a book I edited (proceedings of the conference in Alexandria) was on this subject, for the reactions H + H_2 and NaI dissociation (wing emission). For some reason, we did not discuss fs transition-state spectroscopy at this time, but instead we (with John providing all the notes in writing) were interested in intense-laser-field stimulated emission in the NaI system and in the field "dressing" of the potentials. I do not know why I did not think of this system as the first one for femtosecond transition-state spectroscopy. This may have been due to the earlier experience we had with the ICN picosecond experiments. Later, the potentials John sent from Toronto were helpful in my thinking of the NaI experiment.

At nearby UCLA, the arrival of Dick Bernstein to the area was a real blessing. He was extremely excited about the developments and the possibilities for real-time studies of molecular reaction dynamics. It was at his house in Santa Monica that the word *femtochemistry* was coined, helped by a discussion in the company of his wife Norma and brother Ken. Dick also came to Caltech as a Fairchild Scholar in 1986, and in 1988 we wrote a feature article together (published in *Chemical & Engineering News* **[26]**). We had great fun writing this article and we learned an enormous amount about molecular reaction dynamics. We also had a genuine collaboration on bimolecular reactions and published a paper in 1987. Dick came to Caltech again in 1990, but sadly died before ending his sabbatical; a number of experiments, particularly the new direction of surface femtochemistry, were designed as part of our plan. Finally, the collaboration I had with Rich Bersohn, while he was at Caltech as a Fairchild Scholar, was enlightening. We discussed the classical picture of fs spectroscopy of dissociating molecules, and Rich and I wrote a theoretical paper on this subject.

(4) The Femtosecond Dream

A Piece of Good Fortune

To achieve the fs time resolution, we needed a new laser system. A piece of good fortune came our way at a time when funding was limited and when the establishment of fs lasers and molecular beam technologies required a "quantum jump" in support. Shaul Mukamel invited me to a workshop in Rochester (October 1985) on intramolecular vibrational redistribution and quantum chaos. I spoke about "IVR and chemical reactivity", and there in the audience were two program directors from the Air Force Office of Scientific Research: Larry Davis and Larry Burggraf. They requested a preliminary proposal immediately, and I sent one in October, followed by a complete proposal in January of 1986.

The proposal was funded and approved in August of 1986, to start in November of the same year. Larry Davis saw to it that we could order the equipment needed as soon as possible and made the necessary arrangements to do so; AFOSR continues to support our program. We did not have laboratory space to house the new equipment, but Caltech responded. Fred Anson, our Division Chairman at the time, arranged for the space (which once housed the X-ray machines of Linus Pauling) and Murph Goldberger, our President, provided funds for the renovation without delay – Murph appreciated the physics and Fred saw the importance of the new research to chemistry. By Thanksgiving (1986), we entered the new laboratory, and the CPM laser was operational at the "femtosecond party" on December 11, 1986. We focused again on the ICN reaction, but this time on the fs time scale in FEMTOLAND I.

The Classic Femtosecond ICN Discovery

The goal of the ICN experiment was to resolve in time the transition-state configurations *en route* to dissociation:

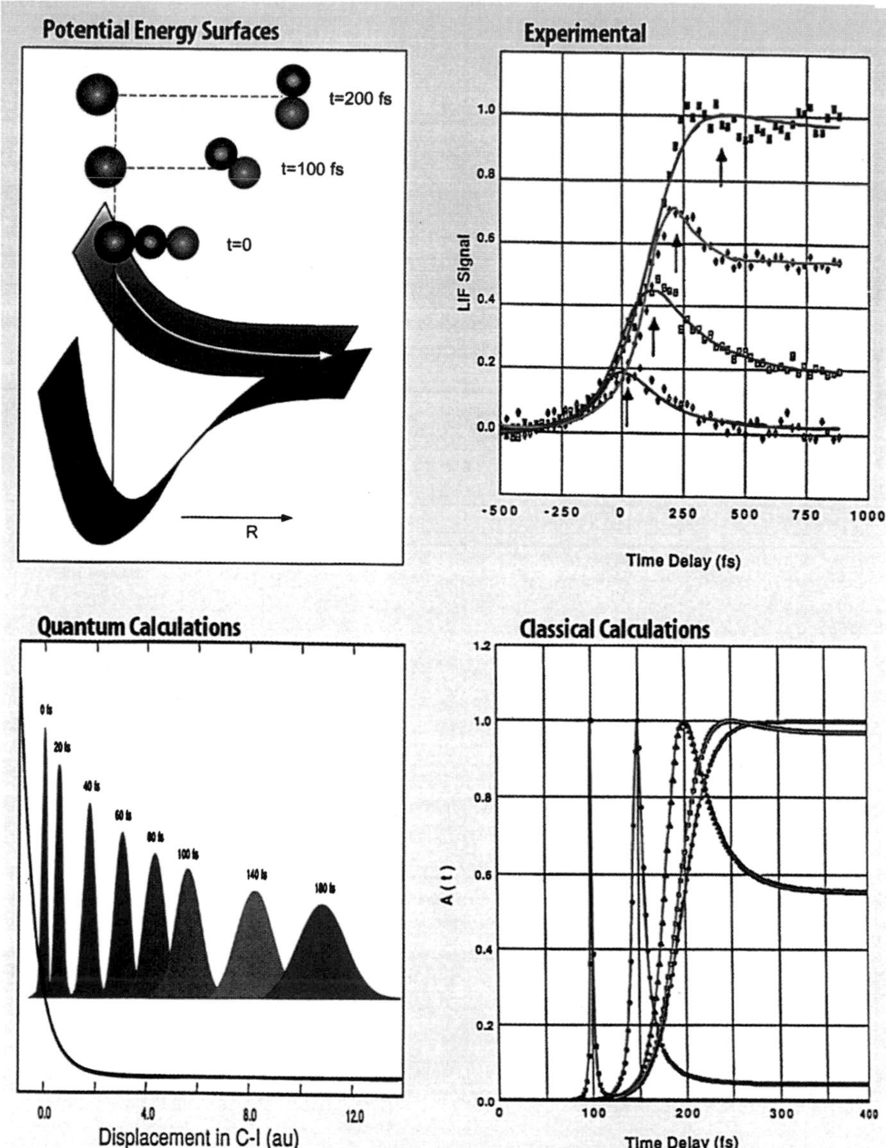

Figure 14. Femtochemistry of the ICN reaction, the first to be studied. The experimental results show the probing of the reaction in the transition state region (rise and decay) and the final CN fragment (rise and leveling) with precise clocking of the process; the total time is 200 fs. The I fragment was also detected to elucidate the translational energy change with time. Classical and quantum calculations are shown (see text). [Ref. B1, B4, B6, B14, B16, B18, B19, B28, 55]

$$ICN^* \rightarrow I \cdots CN^{*\ddagger} \rightarrow I + CN \quad (5)$$

Not only did we wish to monitor the final CN, free of the force field of iodine (which we did in 1985), but also the transitory species $I \cdots CN^{*\ddagger}$ **[Fig. 14]**. The first $I \cdots CN^{*\ddagger}$ transient surprised us, but after long and late hours of discussions and control experiments it became clear that, indeed, the transition configurations or the final products could be separately monitored in real time. We submitted our first communication to the *Journal of Chemical Physics* (received June 3, 1987), and it was accepted on June 15, 1987. The referee of this paper was not only prompt, but also, in retrospect, visionary. His report was ultrashort: *"It (the manuscript) has the smell that the authors are onto some very exciting new stuff... This manuscript meets all requirements for a communication. It may turn out to be a classic. Publish with all dispatch."*

Our thinking about the process of bond breakage was intuitive and relied on classical concepts, as discussed in the prologue. The basic observations made **[Fig. 14]** in the ICN experiment could be related to the fs nuclear dynamics; the delayed appearan-ce of the CN (on-resonance) and the build-up and decay of transition configurations (off-resonance) was understood using simple classical mechanics and even a helpful kinetic picture of A → B → C, describing the "elementary" steps of the reaction. Two papers (I and II of a series), published in the *Journal of Chemical Physics,* outlined the methodology of "femtosecond transition-state spectroscopy (FTS)" with applications to the ICN dissociation reaction. From these first experiments, we expressed the change in internuclear separation with time, i.e., the reactin trajectory, and the time of bond breakage:

$$\tau = \int_{R_0}^{R} \frac{dR'}{v(R')} \qquad \tau^{\ddagger} = \frac{\Delta V(R^{\ddagger})}{v(R^{\ddagger})|F(R^{\ddagger})|} \tag{6}$$

For a given potential of interaction V, the velocity is v and the force is F = $-\partial V/\partial R$.

We obtained the time of the motion to final products (τ) and during the transition state ($\tau^{\ddagger}$), as well as the distance of separation R(t). Thus, for a given total energy E, we expressed the distance of separation between I and CN and the time of bond breakage, and compared these with experimental results, hitherto unmeasured directly. Significantly, we were able, for the first time, to observe the passage through the transition state, potential energy window $\Delta V(R^{\ddagger})$, and measure its "lifetime" or transit time $\tau^{\ddagger}$ with fs resolution. This experiment on the dynamics of bond breaking and another one on the dynamics of bond making ($H + CO_2 \rightarrow OH + CO$) events were generally well received by colleagues in the scientific community worldwide. They even had impact on the public press with significant write-ups in popular newspapers and magazines such as New York Times, Los Angeles Times, Discover, ... etc. Isaac Asimov, H. C. von Baeyer, Gary Taubes, Philip Ball, and other notable science writers gave an exposition of this published work. Many books and textbooks referred and continue to refer to this 1987 discovery.

The quantum picture was intuitive too. On the basis of the experience outlined above, we could understand that coherent preparation can lead to non-stationary states which evolve with time (motion!) **[Section II]**. Extension to chemical reaction dynamics gives the following non-stationary wave packet:

$$\Psi(R,t) = \sum_i c_i \cdot \psi_i(R) \cdot \exp(-iE_i t/\hbar) \tag{7}$$

which evolves in time, similar to the two-level problem **[Equation 4]**, but now with spatial R localization because of the sum over many energy-states. This principle of superposition holds because of the linearity of the time-dependent Schrödinger equation. The phases in **[Equation 7]** give rise to the interferences (quantum coherence) and their fluctuations, due to intrinsic anharmonicities or interactions with the environment, lead to the delocalization. Since the packet can be synthesized easily when the sum criterion is satisfied, the fs pulse becomes the ideal initiator of the motion of nuclei in a reaction.

The concept of describing quantum systems using wave packets is fundamental and goes back to the 1920's **[Fig. 4]** when the connection between quantum mechanics and classical phenomena was the subject of discussion and correspondence among many notable scientists such as Schrödinger, Lorentz and others[5]; **Sections III4 and III6** highlight the theoretical developments since then. The ICN results demonstrated the experimental observation of wave packets in molecular systems, and since then they have been synthesized in atoms, complex molecules, and biological systems, as well as in the different phases of gases, liquids, clusters and solids. The behavior observed for ICN was found in other studies, the most recent is an elegant series of experiments of "bubbles in solids" by Majed Chergui's group in Lausanne.

The question then was: would quantum calculations reproduce the experimental results obtained for ICN? Dan Imre, being skeptical in the beginning, did the first of such calculations, and the results were important in showing the influence of the wave packet motion and spreading on the observed FTS transients. Horia Metiu addressed the role of rotations. We compared the quantum results with those obtained from the classical model of Bersohn and Zewail **[Fig 14]**. The model described the experimental trends quite well, just as did the quantum picture. This was followed by reports of trajectory calculations from Kent Wilson's group and a density-matrix description from Shaul Mukamel's group. The latter emphasized the different limits of dephasing and the time scale for nuclear motion. All theoretical results exhibited the general trends observed experimentally. In our early papers on ICN, we suggested that the δ-function limit of wave packet dynamics could be obtained if proper deconvolution was made, knowing the temporal response of the pulses. Very recently, Volker Engel and Niels Henriksen reported a quantum theoretical agreement with this simple picture and discussed its generality.

The NaI Discovery: A Paradigm for the Field

There were two issues that needed to be established on firmer bases: the issue of the uncertainty principle and the influence of more complex potentials on

[5] Schrödinger wrote a theoretical paper [*Naturwissenschaften* **14**, 664 (1926)], pointing out the transition from micro- to macro-mechanics using the superposition of eigenstates. There was a correspondence between Schrödinger and Lorentz on this problem and the difficulty of making wave groups or wave packets.

the ability of the technique (FTS) to probe reactions. The *alkali halide reactions* were thought of as perfect prototypes. Because they involve two potentials (covalent and ionic) along the reaction coordinate, I thought we would have fun with these systems. Moreover, their unique historical position in crossed molecular beam experiments ("The Alkali Age") made them good candidates for the "femto age". The reaction of NaI, unlike ICN, involves two electronic coordinates and one nuclear coordinate, the separation between Na and I. The resonance motion between covalent and ionic configurations is the key to the dynamics of bond breakage. *How could we probe such motion in real time?* We did the FTS experiments on NaI and NaBr, and the results, published in 1988, were thrilling **[Fig. 15]** and made us feel very confident about the ability of FTS to probe transition states and final fragments. The results also illustrated the importance of coherent wave packets in quasi-bound systems. The NaI experiment was a watershed event leading to an entirely new *paradigm* in the field of femtochemistry and establishing some new concepts for the dynamics of the chemical bond.

First, we could show experimentally that the wave packet was highly localized in space, ~ 0.1Å, thus establishing the concept of dynamics at *atomic-scale resolution. Second,* the spreading of the wave packet was minimal up to a few picoseconds, thus establishing the concept of *single-molecule trajectory,* i.e., the ensemble coherence is *induced* effectively, as if the molecules are glued together, even though we start with a random and noncoherent ensemble – the world of dynamics, *not* kinetics. *Third,* vibrational (rotational) coherence was observed during the entire course of the reaction (detecting products or transition states), thus establishing the concept of *coherent trajectories in reactions,* from reactants to products. *Fourth,* on the fs time scale, the description of the dynamics follows an *intuitive classical picture* (marbles rolling on potential surfaces) since the spreading of the packet is minimal. Thus, a time-evolving profile of the reaction becomes parallel to our thinking of the evolution from reactants, to transition states, and then to products. The emerging picture is physically and chemically appealing, and compellingly demonstrated that conversion from the energy space to the time domain is not needed.

Finally, the NaI case was the first to demonstrate the *resonance behavior,* in real time, of a bond converting from being covalent to being ionic along the reaction coordinate. From the results, we obtained the key parameters of the dynamics such as the time of bond breakage, the covalent/ionic coupling magnitude, the branching of trajectories, etc. In the 1930's, Linus Pauling's description of this bond was static at equilibrium; only now can the dynamics be described in real time by preparing structures far from equilibrium. I still reflect on the beauty of these NaI experiments and the rich number of concepts they brought to dynamics. Some of the concepts were not as clear when we first made the observations as they are now. The paradigm shift in our thinking is linked and similar in value to the work on IVR **[Section III2]**, but the difference is major – for IVR we studied coherence of states, but for reactions we observed coherence of the nuclear motion with atomic resolution.

After the initial set of experiments, we continued on this system for some

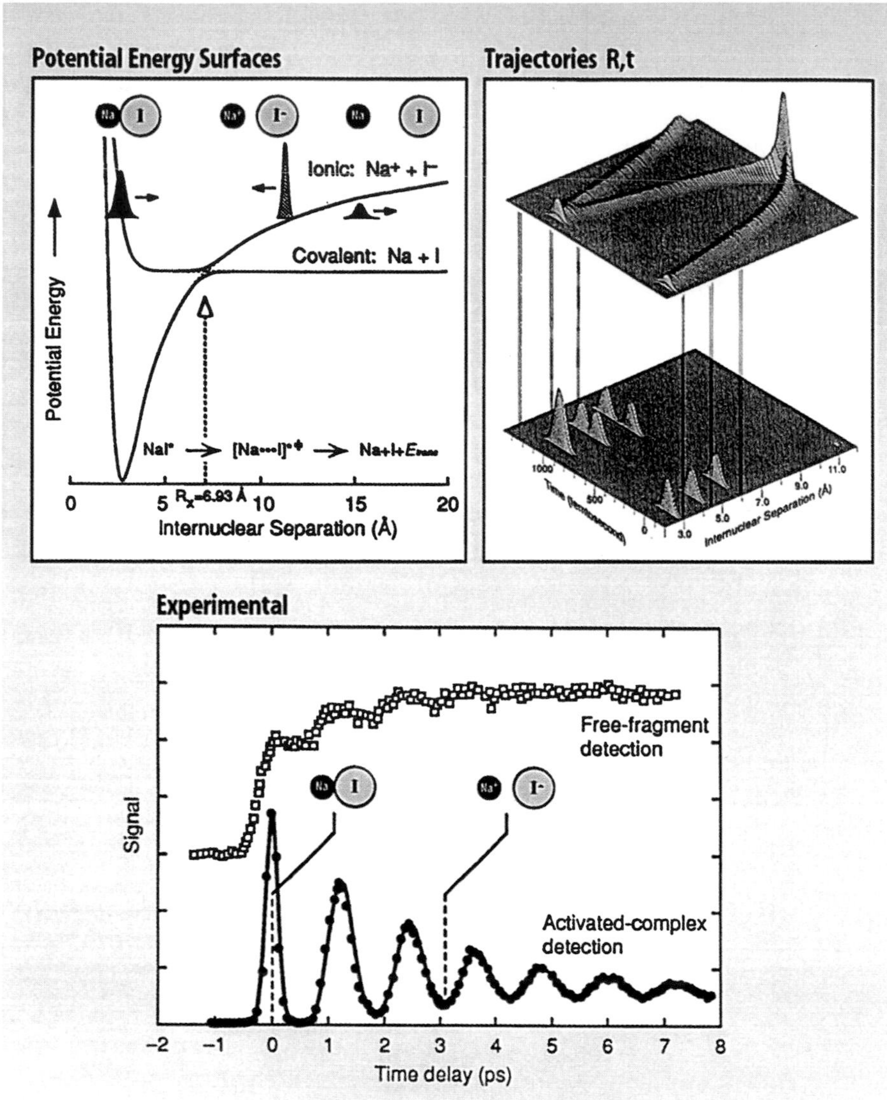

Figure 15. Femtochemistry of the NaI reaction, the paradigm case. The experimental results show the resonance motion between the covalent and ionic structures of the bond, and the time scales for the reaction and for the spreading of the wave packet. Two transients are shown for the activated complexes in transition states and for final fragments. Note the "quantized" behavior of the signal, not simply an exponential rise or decay of the ensemble. The classical motion is simulated as trajectories in space and time (top). [Ref. B1, B4, B12, B14, B17, B19, B28, 56]

time, exploring other phenomena of interest. The studies included: (1) direct observation of the reaction trajectory in R and t, and the resolution of the motion *into* and *from* the transition state (this work was first published in Nature, London); (2) direct observation of recurrences (echo-type), reflecting rephasing at long times (t = 20-40 ps), and their relationship to quantum interference effects due to the resonance behavior of the motion on the covalent and ionic potentials (this work was published in Chemical Physics Letters); (3) studies of the effect of the velocity of the nuclei on the crossing-to-products probability, providing the interaction matrix element for the coupling between the covalent and ionic potentials, and the classical and quantum treatment of the dynamics (these studies were published in the Journal of Physical Chemistry and Journal of Chemical Physics).

Numerous theoretical and experimental papers have been published by colleagues and the system enjoys a central role in femtodynamics. From the beginning we understood the major features of the dynamics from the point of view of classical mechanics. The "exact" quantum calculations were first made by Volker Engel and Horia Metiu and these were important in identifying the sensitivity of the observations to details of the motion and the potential. The agreement with the experimental results was remarkable. The same agreement was found for later theoretical studies involving classical, quantum, and semiclassical approaches.

The Saddle-Point Transition State

Our next goal was to examine reactions governed by multidimensional (nuclear) potentials, starting with "barrier reactions" which define a saddle-point transition state, the classic case of chemistry textbooks. If the reaction dynamics involve more than one nuclear coordinate, an interesting question arises: *can one observe in real time the reactive evolution from the TS at the saddle point to final products on the global PES?* The question was addressed by performing femtochemistry on ABA Systems. The IHgI system was the so-called "gift experiment" I suggested to a new postdoctoral fellow in our group, who joined us in 1988 from Ken Eisenthal's group. Stunning observations were made – the product HgI was *coherently* formed from the transition state **[Fig. 16]**. Also, the transition state, which absorbs a probe fs pulse in the red, as opposed to the HgI product which absorbs in the UV, was found to live for only ~200 fs and this state produces different coherent product states (different periods of vibrational oscillation). It was also in this system that we studied coherence of rotational motion (real-time alignment) and learned about the geometry of the (initially prepared) transition-state, activated complex $IHgI^{*\ddagger}$.

With simple theoretical PES's and molecular dynamics simulations we examined details of the motion, but the major features were evident in the experimental observations. Originally, we studied the TS and the evolution to HgI products by using laser-induced fluorescence. Later, we used mass spectrometry to also detect the I atoms and the translational energy; this effort triggered a great deal of theoretical work **[Fig. 16]** in our group addressing, in depth, the actual meaning of classical TS structure (see **Section IV**). Features of this reaction were similarly found in other classes of reactions, including those in condensed phases and biological systems.

The studies of this ABA system were published with an emphasis on the following points: *First,* coherent nuclear motion can be observed on *multi-dimensional surfaces* involving multiple-bond breakage (or formation). *Second,*

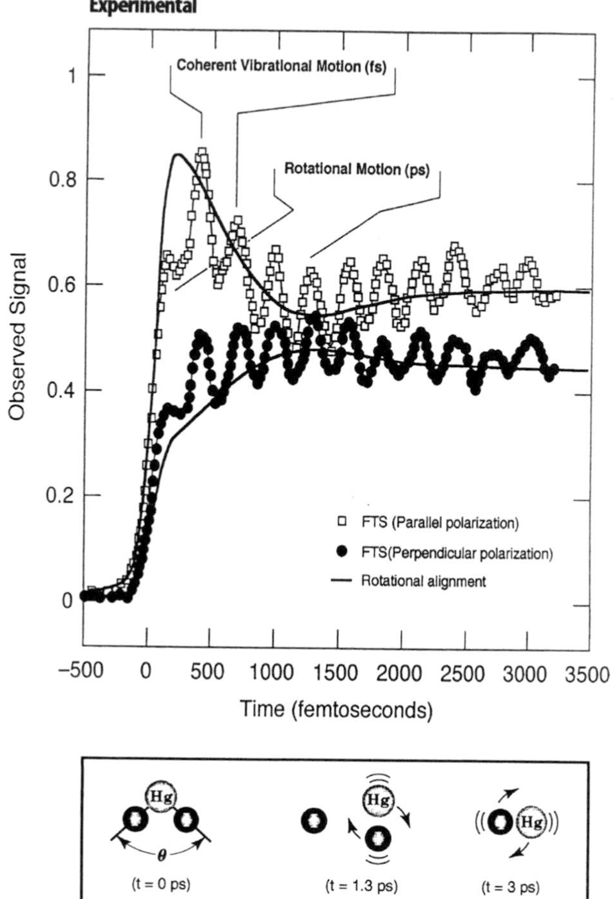

Figure 16. Femtochemistry of the IHgI reaction, the saddle-point transition state (barrier reactions). The experimental results show both the coherent vibrational and rotational motions of the reaction (A, above). The transition state $IHgI^{*\ddagger}$ and final fragment HgI were probed. We also probed the I fragment and the change of translational energy with time. The classical trajectory calculations are shown in (B, next page), together with experimental results for I detection; both theory and experiment illustrate the family of reaction trajectories on the global PES, in time and in kinetic energy distribution. Quantum calculations were also done (not shown). This ABA system is a prototype for saddle-point transition states. [Ref. B1, B4, B6, B10, 57]

coherence survives the entire reaction journey, even in multi-dimensional systems, and yields selective *coherence-in-products. Third,* for the first time, a *saddle-point TS* can be seen evolving in real time. *Fourth,* the *TS can be aligned (oriented)* at zero time and seen evolving into rotations of the diatom (AB) and the translation of the A and AB fragments – the vibrational (scalar) and rotational (vectorial) motions were easily separated using polarized fs pulses.

The Uncertainty Principle Paradox

At the time when I was giving lectures on the above examples of elementary reactions, some were raising a question about the "energy resolution" of the fs experiments: *How can a broad-energy pulse probe a sharp resonance*? In the con-

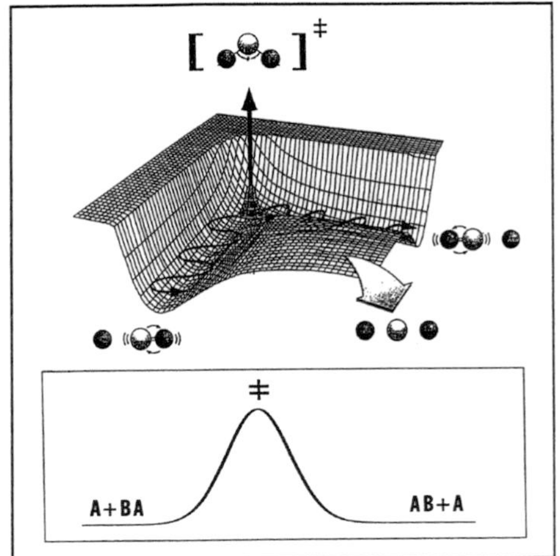

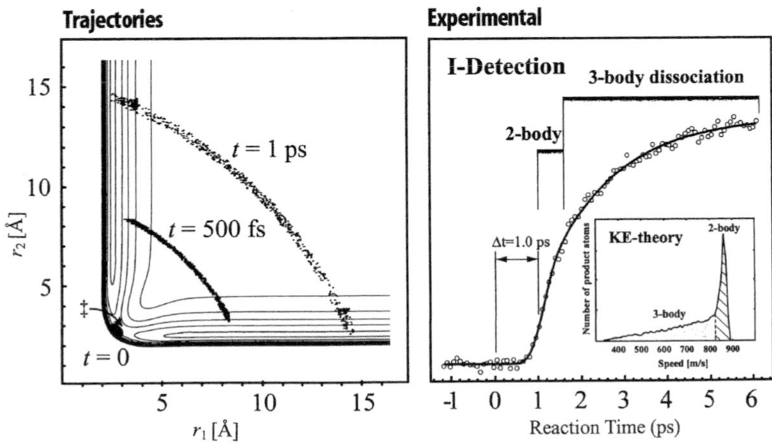

ventional teaching on spectroscopy of "eigenstates", one thinks of stationary states and their populations (diagonal elements of the density matrix [**Equation 4**]), but now we must think of coherent states. All information pertinent to eigenstates is in the wave packet structure. This point was elucidated by our experiments on a *bound nonreactive* system – the iodine system [**Fig. 17**]. Serendipity was at work. We were initiating studies of the FTS of CdI_2 (from the family of HgI_2) and instead observed the wave packet motion of I_2, made from the samples of CdI_2 without our knowledge! It turned out that when we heated CdI_2, we made I_2, which gave us striking oscillatory transients. The oscillations directly gave the periods of the nuclear (vibration) motion, and the data could be related to the change with time of the I-I separation and the rotation of I_2; the time scales were separated (fs *vs.* ps) and the vibrational (scalar) and rotational (vectorial) motion were clearly seen.

We used classical mechanical inversion methods and the RKR and quantum inversion methods to characterize the potential. This was followed by a

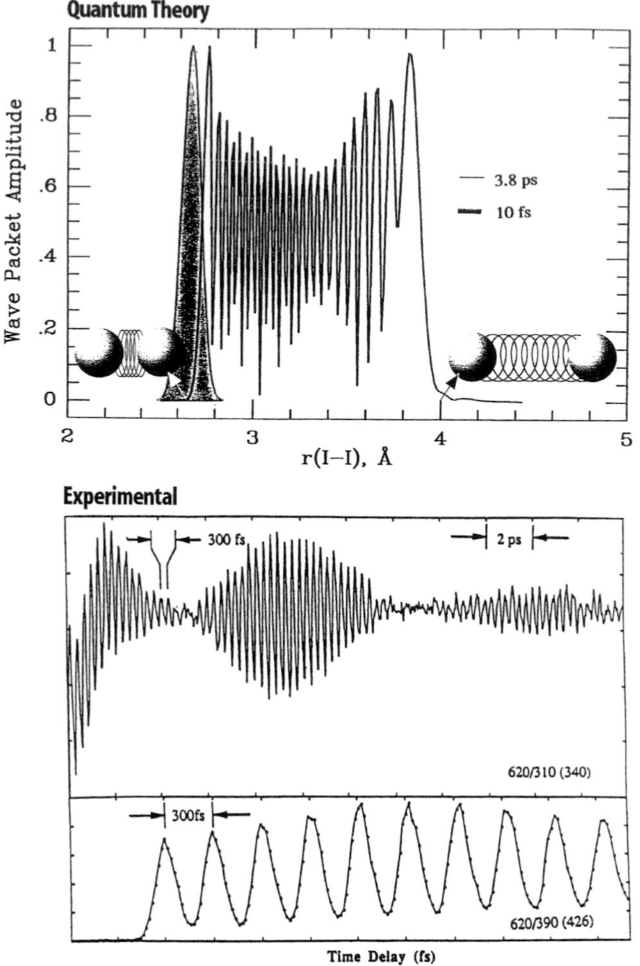

Figure 17. Femtosecond, real time observation of the vibrational (and rotational) motion of iodine. The experiments show the anharmonic nature of the bound motion. Quantum theory indicates the limit for creating a localized wave packet on the fs time scale. The localized wave packet describes the classical spring motion. [Ref. B1, B6, B17, B18, B25, 58]

study of the ICl system. It became evident that: (1) *the uncertainty in energy for short pulses works in our favor*, and the shorter the pulses, the better the localization of the wave packet; and (2) we could now observe *the vibrations and rotations of molecules in real time*, not from energy spectra. This work was first published in Nature (London) and detailed later in the Journal of Chemical Physics. Because the initial experiments were made on a diatomic molecule, the significance of this piece of work was originally missed, until later when many complex systems showed surprisingly similar behavior. The real message of this I_2 experiment was in its conceptual elucidation of the role of the uncertainty principle and the robustness of time, space, and energy resolutions because of *coherence*. Because $\Delta x\ \Delta p \sim \hbar$ and $\Delta t\ \Delta E \sim \hbar$, we can achieve localization with a very small de Broglie wavelength.

Bimolecular, Bond Making & Breaking: Bernstein's Passion

Bimolecular reactions were ready for a femtosecond treatment, to observe the "simultaneous" processes of bonds being broken and formed. With Dick Bernstein (1987), we had studied the IH/CO_2 system. The problem was that, for bimolecular reactions, the transient time for reactants to undergo a collision is generally nanoseconds to microseconds. By using van der Waals complexes, an idea introduced for the studies of product-state distributions by Benoit Soep and Curt Wittig, we could expand HI and CO_2 in a single molecular beam. But now, with the two reagents within angstroms of each other, we could examine the dynamics of the single collision. We used the first pulse to liberate the H atom with a given translational energy and a second pulse to probe the nascent OH product – the *zero of time* became well-defined and the collision was that of a limited impact parameter. The results were exciting and, in our joint paper, Dick termed this the "birth of OH from H + CO_2". Wittig's group improved the time resolution and studied the energy dependence of the rates. **Fig. 18** gives a summary for this system.

The H + CO_2 *ground-state* reaction proved to be important for a number of reasons. *First,* it showed how *Reactive Scattering Resonances* can be probed in real time during the collision and for a system of a complex number of degrees of freedom. *Second,* the experiments established that the *intermediate* $HOCO^{\ddagger}$ lives for $\tau \sim 1$ ps and that for this reaction the OH bond-making and the CO bond-breaking are made in a *nonconcerted* pathway. The nuclear motions of HOCO thus determine the reaction mechanism. If τ was found to be 10 -100 fs, the picture would have been entirely different; bond-making and breaking would occur as a result of the electron redistribution with the nuclei essentially "frozen" in configuration. Obtaining τ directly is critical for the nature of the *transition state/intermediate.* This is particularly true when τ is much longer than the vibrational and rotational periods, and all other methods will fail in deducing τ. *Third,* it provided a direct *test of theory* at the *ab initio* level. High quality *ab initio* calculations of the PES and dynamics have been made available by David Clary, George Schatz, John Zhang and many others. Theory compares favorably with experiments showing that resonances must be considered – the vibrations of HOCO bottleneck the trajectories. The reaction OH + CO → CO_2 + H is one of the key reactions in both combustion and atmospheric chemistry, and represents the most studied 4-atom reaction, both theoretically and experimentally **[see Ref. 27]**.

We constructed FEMTOLAND II, and I thought it would be interesting to examine "halogen bimolecular reactions". Precursors, of which the H-Br/I-I system is a prototype, were chosen to study bimolecular halogen atom + halogen molecule reactions. Upon breaking the HBr bond, the hydrogen goes many ångströms away from the field of the reaction (in femtoseconds) and we are left with the Br + I_2 collision. This halogen reaction had a history in crossed molecular beam experiments and comparison with real-time experimental results would be interesting. The Br + I_2 reaction was examined and found to occur through a sticky (~ 50 picoseconds) collision complex. It is a stable intermediate of BrII, and there is no other way we know of to deter-

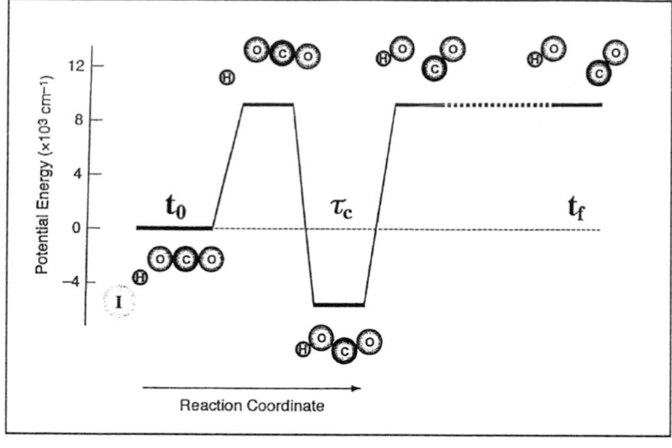

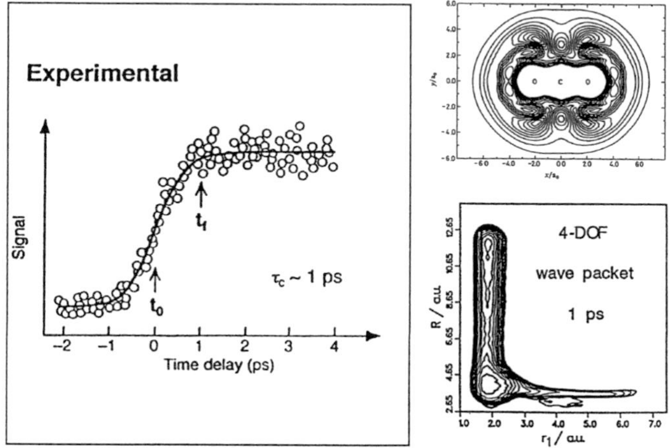

Figure 18. Femtochemistry of the bimolecular H+CO_2 reaction. The precursor in this molecular beam experiment is HI/CO_2 in a van der Waals complex. The initial experiments utilized ps pulses, but later sub-ps pulses were used (see text). Theoretical *ab initio* calculations of the PES and the dynamics (classical, semiclassical and quantum wave packet) have all been reported; the PES and wave packet calculations are from the work of D. Clary, G. Schatz and J. Zhang. [27] The transit species $HOCO^{\ddagger}$ lives for ~1ps. Similar studies were made of reactive Br+I_2 and of the inelastic collision between I and CH_3I. [Ref. B1, B5, B10, B16, B19, 59]

mine its lifetime and dynamics. We examined classical trajectories of motion and compared them with the experimental results. In more recent work, the McDonald group obtained similar times and discussed the possibility of the involvement of ground and excited halogen surfaces. In this study of an atom + diatom collision, we learned the effect of impact parameters, the influence of translational energy, and the interplay between bonding and dynamics. The analogy between full-collision (Br + I_2) and half-collision (hν + I_2) dynamics was based on the change in bonding, and we used frontier orbitals to describe such an analogy. Studies of bimolecular collisions in complex systems have continued in our and other laboratories; the most recent came from NIST (John Stephenson and colleagues) on the studies of $CH_4 + O \rightarrow CH_3 +$ OH, using $CH_4 \cdot O_3$ as a precursor.

Ultrafast Electron Diffraction (UED)

FEMTOLAND III was the home of our next effort, Ultrafast Electron Diffraction (UED). Our goal was to complement the detection schemes of spectroscopy and mass spectrometry and to use diffraction to follow structural changes, especially for large systems. I proposed the idea in 1991 in a Faraday Discussion paper **[B17]**, and we had our first success in 1992. We were able to record structures with an electron pulse duration of a few picoseconds, but with no time scan. This was followed by other studies, both on the theoretical and experimental fronts. In our group, rumor had it that UED was a "NO to the power 10 experiment!" We now have the third generation of UED machines, with a group of graduate students and post-doctoral fellows participating. In a Nature paper in 1997, we reported our state-of-the-art experimental development of the methodology. We also developed a "difference-method" which allows us to record the structure of radicals, carbenes and intermediates, and with higher sensitivity, as discussed below in **Section III6**.

Clusters, Dense Fluids & Liquids, and New Femtolands

With the above-mentioned spectroscopy, mass spectrometry and diffraction techniques, it was becoming possible to study many new systems. In addition to FEMTOLANDS I to III, PICOLANDS I and II were still operational in order to cover the different time scales of reaction dynamics. We are now up to FEMTOLAND VI. Two additional beam machines, equipped with mass spectrometry and spectroscopic detectors, were added. These new FEMTOLANDS were built to accommodate the expanding scope of research, from gas phase to clusters, to liquids and dense fluids, to the world of complex organic and inorganic chemistry, and to the very complex biological systems. I find complex reactions fascinating and we have devoted significant effort to this area, studying both uni- and bimolecular reactions. We also maintain strong theoretical efforts on *molecular structures* and *molecular dynamics* to compare theory with experiment on the relevant time scale.

Theoretical Femtochemistry

Our involvement had roots in the study of coherence and dephasing. This line of research was initiated by using pulsed lasers to form a coherent state, **[Equation 4]**. The evolution was followed in time to obtain the pure dephasing time (T_2'), which reflects the extent of phase interruptions, and the population or energy decay time (T_1). We invoked theoretical techniques such as density matrix formalism, Kubo relaxation theory and the Bloch equations to describe the optical analogue of NMR experiments. The 1956 Feynman, Vernon and Hellwarth paper **[29]** was an important contribution, as it pointed out the linkage between "spin" and "optical" coherence experiments: even in the absence of a magnetic field, used in the former, one can use a rotating frame picture to describe such optical experiments. All of this was known at the time. Our interest in theory was to relate T_1 and T_2 to molecular processes and to learn about their formal limits of applicability. We wrote a book chapter **[B9]** on the subject and published a number of papers relating these relaxation times to the anharmonicity of molecular vibrations, the phonon structure of solids, and collision dynamics in gases.

For collisionless, large molecules, the issue of intramolecular electronic-states coupling was, by the 1970's, well developed theoretically and heavily imbedded in theories of radiationless transitions formulated to explain the origin of non-radiative decays. Wilse Robinson, Joshua Jortner, Stuart Rice and many others were involved in the early stages of this development. The Bixon-Jortner model gave the description for such inter-electronic-states coupling and the important role of preparing a doorway state which "dephases" and "relaxes" depending on electronic coupling matrix element, Franck-Condon factors and the density of states **[30]**.

For IVR **[Section III 2]**, we developed a theoretical description for the coherent preparation of a set of vibrational eigenstates on a *single* potential surface, defining the preparation of a non-stationary *vibrational* packet, with the role of rotations and vibrational couplings explicitly expressed. The probability of being in the initial state ψ_0 is:

$$P(t) = |\langle \psi_0 | \Psi(t) \rangle|^2 = \sum_{i,j} \alpha(i,j) \quad \exp\left[-\left(i\omega_{ij} + \Gamma\right)t\right] \tag{8}$$

where the sum is over a product of coefficients α, and displays the interference of states i and j, together with their damping rate Γ. This treatment was useful because: (1) it gives a direct view of IVR, from the initial non-stationary state and as a function of time; (2) it indicates the critical role of selective probing – if all states were monitored, coherence would be obscured. On the other hand, selectivity provides rich information on the nature and extent of IVR, the level structure and the phase changes; and (3) it shows that P(t) is a measure of vibrational chaos, defining what we called restricted or nonchaotic IVR – coherence among vibrational states of a *single* electronic potential. Similarly, we considered the theory for rotational coherence using polarization-analyzed probes. This was treated in a series of papers with focus on the phenomenon of pure rotational coherence and its utility for molecular structural determination, and the rotational dephasing time.

At the time of the first femtochemistry experiment, I was thinking of the coherent state – wave packet – as an extension of the above picture. However, the span of states is now sufficient to create a localized, in R-space, atomic-scale wave packet **[Equation 7]**; see also **Section III6**. We needed a classical picture to connect with Newtonian mechanics, a simplified picture of the motion. The first of such models was published after we considered the theoretical treatment of absorption of fragments *during* reactions. We obtained the following expression:

$$A(t;R) = C\left\{\delta^2 + W^2\left(t, t^{\ddagger}\right)\right\}^{-1} \tag{9a}$$

where C is a constant and $W = V(t) - V(t^{\ddagger})$ is the potential (or more generally, the difference of the two potentials probed); δ is a half-width of the pulse (and damping). For exponential repulsion, V = E sech^2 (vt/2L) with L defining the length scale and v being the speed at the total energy E. Accordingly, the time for bond breakage can be related to FTS observables **[Section III4]**:

$$\tau_{BB} = (L/v)\, ln(4E/\delta) \tag{9b}$$

This expression defines bond breakage time when the potential drops to a value of δ. The model is basic and describes the reaction trajectory R(t) or τ (R). It provides a simple connection between observations and the dissociation time, transition-state lifetime, and the forces of the potential [see **Equation 6 and Fig. 14**]. While Dick Bernstein was at Caltech, we extended the model to obtain the potential using an inversion approach. We published two papers on the subject. Peter Sorokin and colleagues at IBM have addressed different limits of the classical regime in connection with their original studies of fs transient absorption of dissociation **[B50]**.

Next, we considered the treatment of the effect of alignment and orientation on femtochemical reaction dynamics. I considered the time evolution of alignment and coherence for a single rotational angular momentum and then averaged the different trajectories to define the coherence time; τ_c (in ps) becomes simple and equal to $2.2[B < E_R >]^{-1/2}$, where B is the rotational constant and $< E_R >$ is the average thermal rotational energy (cm^{-1}) produced in the reaction fragment(s). We applied this to reactions and I wrote a paper on the subject, published in 1989 **[B18]**. This was followed, in collaboration with Spencer Baskin, by a paper describing the details of the approach and its applications.

In quantum treatments, we have benefited greatly from the advances made in theoretical formalism and computation. A major step forward was made when Rick Heller **[31]** reformulated the time-dependent picture for applications in spectroscopy, and Jim Kinsey and Dan Imre **[31]** described their novel dynamical Raman experiments in terms of wave packet theory. Progress was significantly helped by advances made in the theoretical execution and speed of computation by Ronnie Kosloff **[31]** and, subsequently, by many others. In **Section III6**, we discuss the contributions made in the 1980's in connection with quantum control.

As mentioned above, the groups of Imre and Metiu did the first "exact" quantum calculations of femtochemical dynamics (ICN and NaI). The literature is now rich with numerous theoretical studies. There is a parallelism between the experimental diversity of applications in different areas and the impressive theoretical applications to many experiments and systems. This is summarized in the (1996 Nobel Symposium) book edited by Villy Sundström on Femtochemistry and Femtobiology **[28]**. Jörn Manz, who has played a significant role in this field, gave an overview of developments since Schrödinger's 1926 paper, with 1500 references. Jörn classifies the field into periods of *origins, sleeping beauty, renaissance, and revolution.* In this Nobel Symposium book (and another one **[30]**) Jortner provides a unifying overview of molecular dynamics in femtochemistry and femtobiology, and Mukamel gives an exposition of a general approach using the density matrix formalism. In the same volume, Clary presents the state-of-the-art in quantum theory of chemical reaction dynamics while Marcus and Casey Hynes review transition state theories for rates and dynamics.

Theory and experiment are now hand-in-hand, and many laboratories are doing both. For elementary chemical reactions, the above classical/quantum picture captured the essence of the observation and in many cases the comparison between theory and experiment was tested critically. For complex systems, our theoretical effort has taken on a different approach. With the help of molecular dynamics (MD) simulations, we compare theory with experiments. Then we use the MD simulations as a tool and vary parameters until we reduce the problem to identify the important key forces of dynamics. At this point, we can provide a microscopic dynamical picture with focus on the relative vibrational coordinates, time scales or system parameters. Two examples illustrate the point. The first was our study of the dynamics of a guest molecule in dense fluids with focus on the density dependence of microscopic friction, T_1 and T_2, and of bond breaking/remaking dynamics. The second is the study of numerous organic reaction mechanisms. For the latter, we also use advanced computational methods, such as Density Functional Theory (DFT), *ab initio* and CASSCF computations. For ground-state reactions, the theory can be compared in a critical way with experiment, while for excited states the situation is more challenging **[32]**.

Experimental Femtochemistry

The generation, amplification, and characterization of ultrashort pulses are a major part of femtochemistry experiments. Another is the reaction chamber: molecular-beam machine, gas cell, ultrahigh-vacuum (UHV) surface apparatus or the high-pressure/liquid cell. Here, we mention only the different systems designed for the studies presented in this anthology; further details can be found in the book chapter I wrote for the volumes edited by Manz and Wöste **[B6]** and in the two volumes of our collected works **[16]**.

At Caltech, over the years, we have constructed different types of lasers depending on the particular development and the resolution needed, picosecond to femtosecond [see **Fig. 19**]; one apparatus is shown in **Fig. 5**. Since 1976, and in evolutionary order, these are:

(1) Passive mode-locked, and cavity-dumped, dye laser (pumped by a CW argon ion laser); Ippen and Shank cavity design. Pulse characteristics: 615–625 nm, 2.4 ps (and 0.7 ps), 2 nJ, 100 kHz repetition rate.

(2) Synchronously pumped, mode-locked dye laser system (pumped by an actively mode-locked argon ion laser). Pulse characteristics: 550–600 nm, 3.1 ps, 1.8 nJ, 82 MHz repetition rate.

(3) Mode-locked argon ion laser. Pulse characteristics: 514.5 nm, 150 ps, 12 nJ, 82 MHz repetition rate.

(4) Synchronously pumped, cavity-dumped dye laser (pumped by a mode-

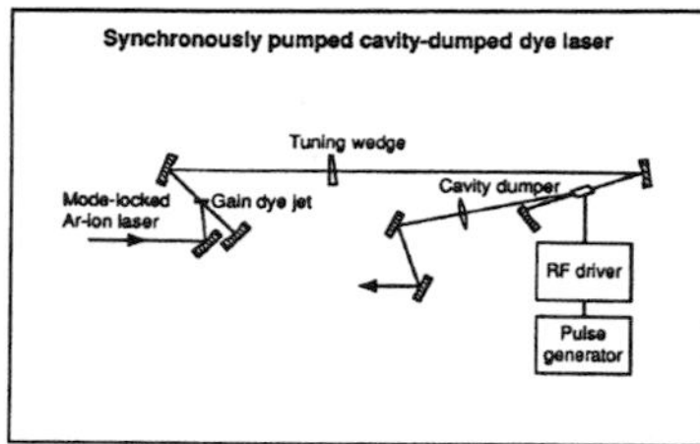

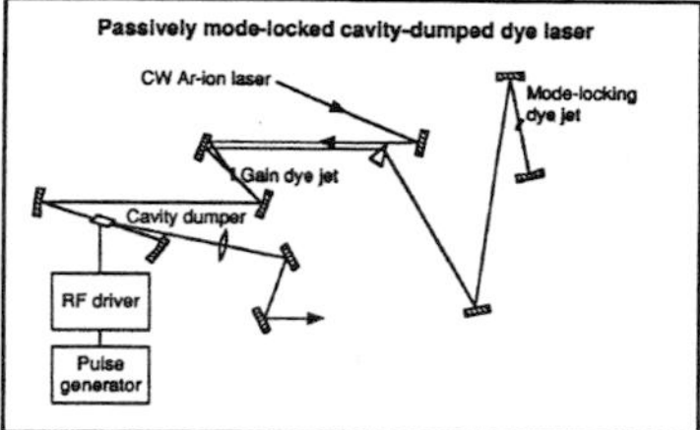

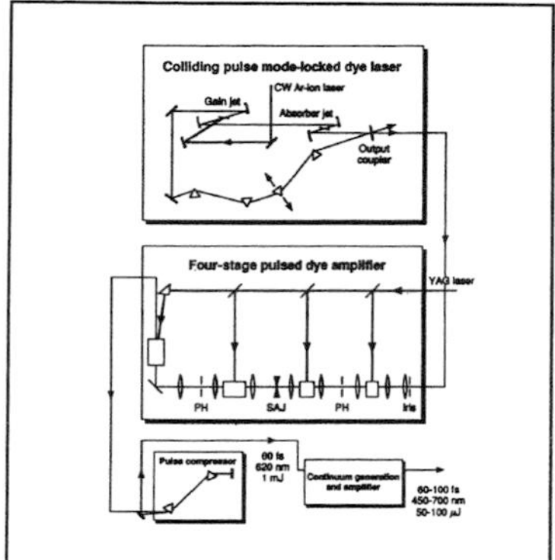

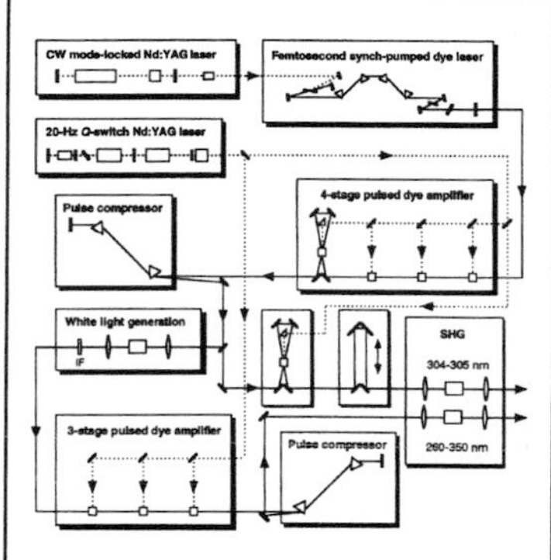

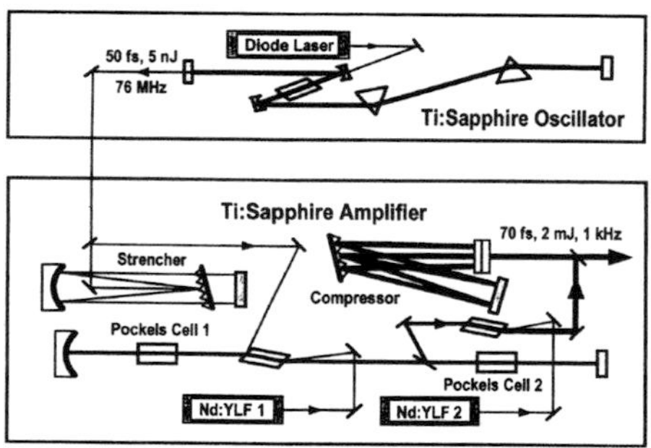

Figure 19. Examples of (A, above) the laser systems utilized in different laboratories (for more details see text and Ref. B6) (B, next page) the two-dimensional (time & wavelength) correlated pattern of a fs pulse (M. Chachisvilis, unpublished work from this laboratory).

locked argon ion laser). Pulse characteristics: 550–750 nm, 15 ps, 20 nJ, 4 MHz repetition rate.

(5) Mode-locked (CW) Nd:YAG laser which synchronously pumps two dye lasers, with two amplifiers pumped by a 20 Hz Q-switched Nd:YAG laser. Pulse characteristics: 550–750 nm, 3–5 ps, ~ 1 mJ, 20 Hz repetition rate.

(6) Dye lasers as described in (5) with an extra-cavity pulse compressor (using a fiber-grating optics arrangement) to obtain ~ 0.4 ps pulses.

(7) Colliding-pulse mode-locked (CPM) ring dye laser (pumped by a CW argon ion laser), and amplified in a four-stage dye amplifier pumped by a YAG laser. A compression at the output of the amplifier was also used. Pulse characteristics: 615-625 nm, 70 fs, ~ 0.5 mJ, 20 Hz repetition rate.

(8) Synchronously pumped, cavity-dumped dye lasers (two), pumped by a

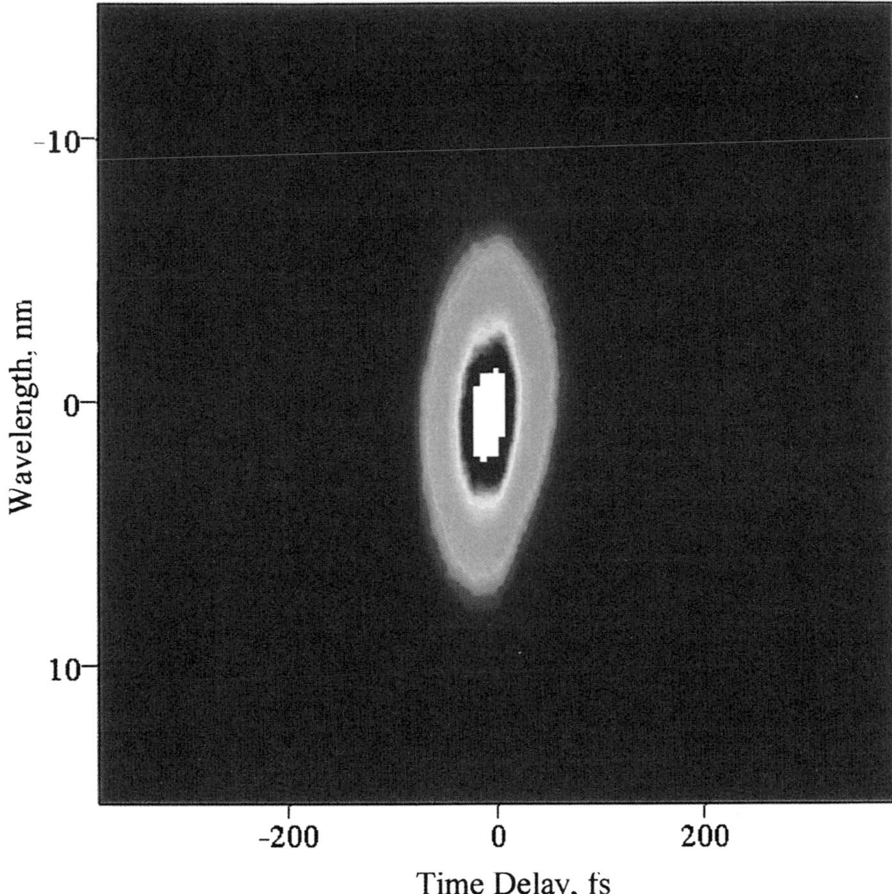

Q-switched, mode-locked Nd:YAG laser. Pulse characteristics: 550–750 nm, 50 ps, 10 μJ, ≤ 1000 Hz repetition rate.

(9) Passively mode-locked, dispersion-compensated tunable dye laser (synchronously pumped by a frequency-doubled, CW, mode-locked Nd:YAG laser), amplified in a four-stage dye amplifier. The compression is after the amplifier. Pulse characteristics: 500–700 nm, 150 fs, 1 mJ, 20 Hz repetition rate.

(10) CPM laser-amplifier system similar to (7), used for ultrashort electron pulse generation. Pulse characteristics: 615–625 nm, 60 fs, 1 mJ, 30 Hz repetition rate.

(11) Ti:sapphire laser system, Sibbett-type, argon-ion pumped; Ti:sapphire amplifier pumped by Nd:YAG laser. Pulse characteristics: 750–850 nm, 50 fs, 0.7 mJ, 1kHz repetition rate; one OPA system.

(12) Ti:sapphire laser system; oscillator, diode pumped and amplifier (Ti:sapphire) Nd:YLF (2) pumped – all solid state (no gas lasers). Pulse characteristics: 750 nm to 850 nm, 50 to 100 fs, 2 mJ, 1 kHz repetition rate. With two Optical Parametric Amplifiers (OPA), the tunability 1.1 μ to 2.6 μ, with nonlinear conversions, 200 nm to 2.6 μ.

(13) Ti:sapphire laser system, same as (12), but pulse width 120 fs and energy 3 mJ, 1 kHz repetition rate.

For recording and clocking in any study, the resolution must be determined accurately, as must the zero-of-time (t=0). The pulses were characterized using auto-correlation and cross-correlation techniques, typically by scanning the time delay between the two pulses (of the same or different colors) in an interferometer arrangement and observing the sum- or difference-frequency generation in a non-linear crystal. In this way, we can obtain the duration of the pulse. The central frequency of the pulse can be determined by passing the pulse through a calibrated spectrometer, while the shape of the pulse can be obtained from frequency-resolved-optical-gating (FROG) measurements, where the time and frequency components of the pulse are correlated and displayed as a 2-D image **[Fig. 19]**. In clocking experiments, the zero-of-time was precisely determined by an *in-situ* measurement, typically using ionization techniques in beam experiments, lensing techniques in diffraction experiments, or the solvent response in condensed phase experiments.

The detection probes are numerous. Initially, we used laser-induced fluorescence for selectivity and sensitivity, and there we had a *frequency-time correlation*. Later, we invoked mass spectrometry (multiphoton ionization) for *mass-time 2D correlations*. This was followed by *speed-time* and *angle-time correlations*. All these correlations proved important in the studies of complex systems; a prime example was the application of the latter two correlations to the study of electron transfer in isolated bimolecular reactions and in clusters. For absorption-type measurements, we introduced non-linear techniques such as degenerate-four-wave mixing. In a recent collaboration with the groups of Wolfgang Kiefer and Arnulf Materny, we also used CARS, with frequency-time correlations to study the dynamics of ground-state systems, in this case polymers. Other detection methods are: energy-resolved and ZEKE photoelectron spectroscopy, Coulomb explosion, ion-electron coincidence ionization techniques, absorption and photodetachment spectroscopy. The range of wavelength is from the IR to the far UV. Absorption, emission, reflection, ionization, and diffraction have all been involved **[Fig. 6]**.

(5) Femtocopia – Examples from Caltech

The range of applications to different systems and phases in many laboratories around the world is extensive and beyond the purpose of this report. In this section we limit ourselves to the examples studied by the Caltech group. The details are given in the original publications and are summarized in the reviews and books mentioned here. **Fig. 20** gives a summary of the different areas studied at Caltech with chronological flow from the south to the north!

Elementary Reactions and Transition States

The focus here was on the studies of elementary reactions. Some of these have already been discussed above. The dynamics are generally of three classes:

(i) Dynamics of Bond Breakage

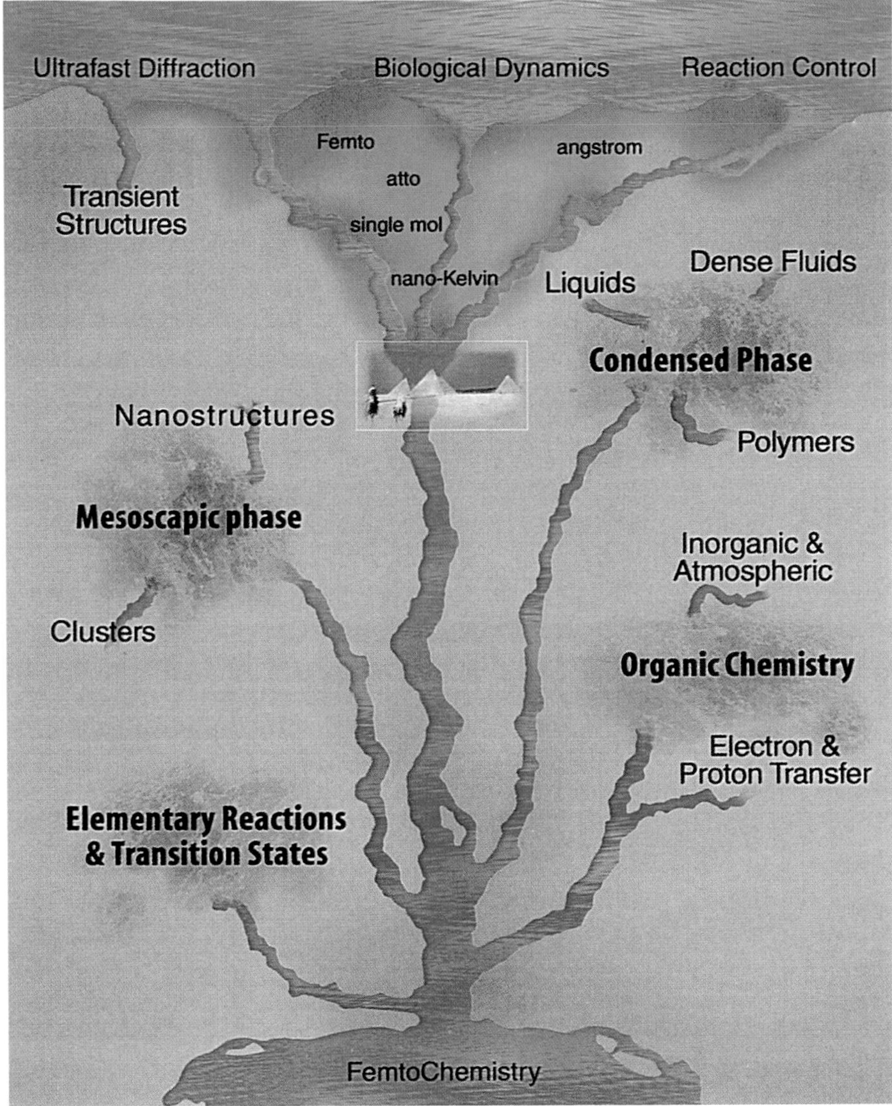

Figure 20. Femtochemistry branches, from the early days of studying elementary reactions and transition states ("southern part") to current activities ("northern part"). The flow has produced the different branches studied at Caltech.

(ii) Dynamics of the (saddle) Transition State
(iii) Dynamics of (bimolecular) Bond Breakage – Bond Formation

Organic Chemistry

With the integration of mass spectrometry into femtochemistry experiments, the field of organic reaction mechanisms became open to investigations of *multiple* transition states and reaction intermediates **[Fig. 21]**. The technique of femtosecond – resolved kinetic-energy-time-of-flight (KETOF) provided a new dimension to the experiment – correlations of *time, speed,* and *orientation*

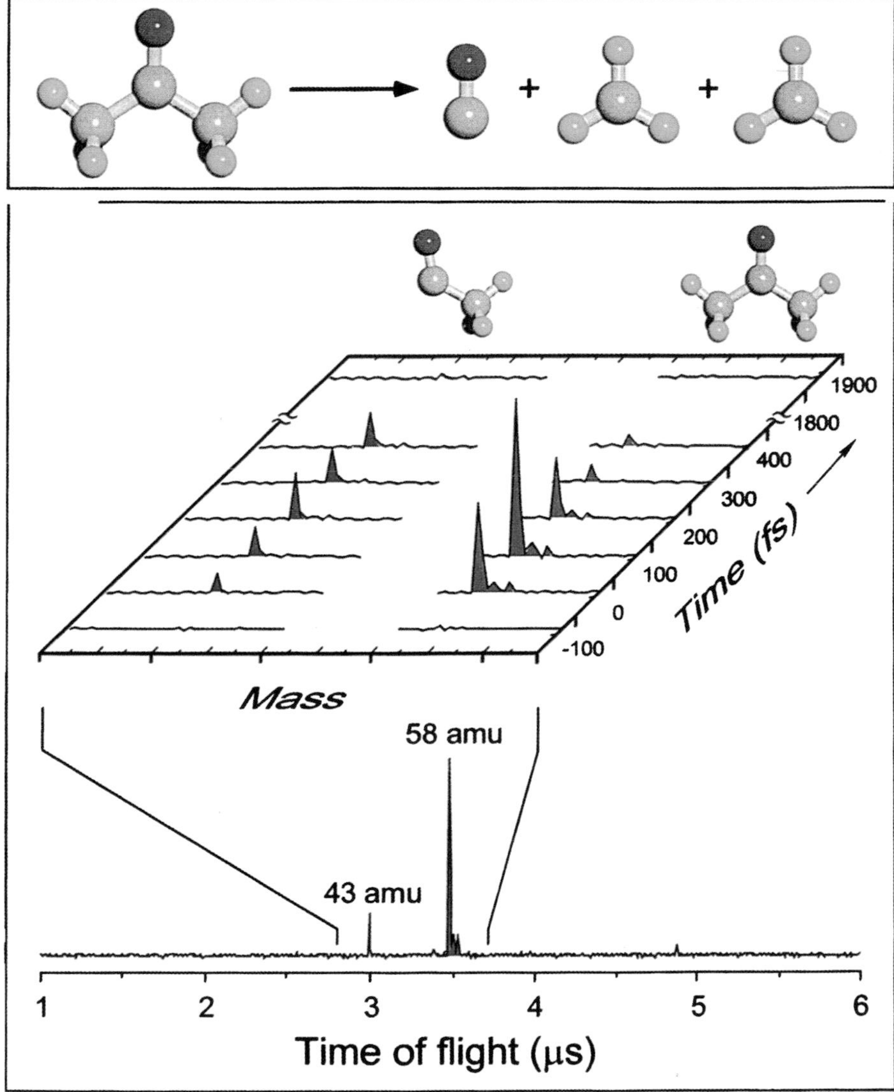

Figure 21. Femtosecond mass spectrometry, a 2D correlation important in the studies of reactive intermediates. The example given here is for the reaction of acetone (Norrish-type I) and its non-concerted behavior. [Ref. B10, 60]

which give *scalar* and *vectorial* dynamics. The examples of reactions include **[Figs. 21–26]**:

(i) Isomerization Reactions
(ii) Pericyclic Addition and Cleavage Reactions
(iii) Diels-Alder/Sigmatropic Reactions
(iv) Norrish-type I and II Reactions
(v) Nucleophilic Substitution (S_N) Reactions
(vi) Extrusion Reactions

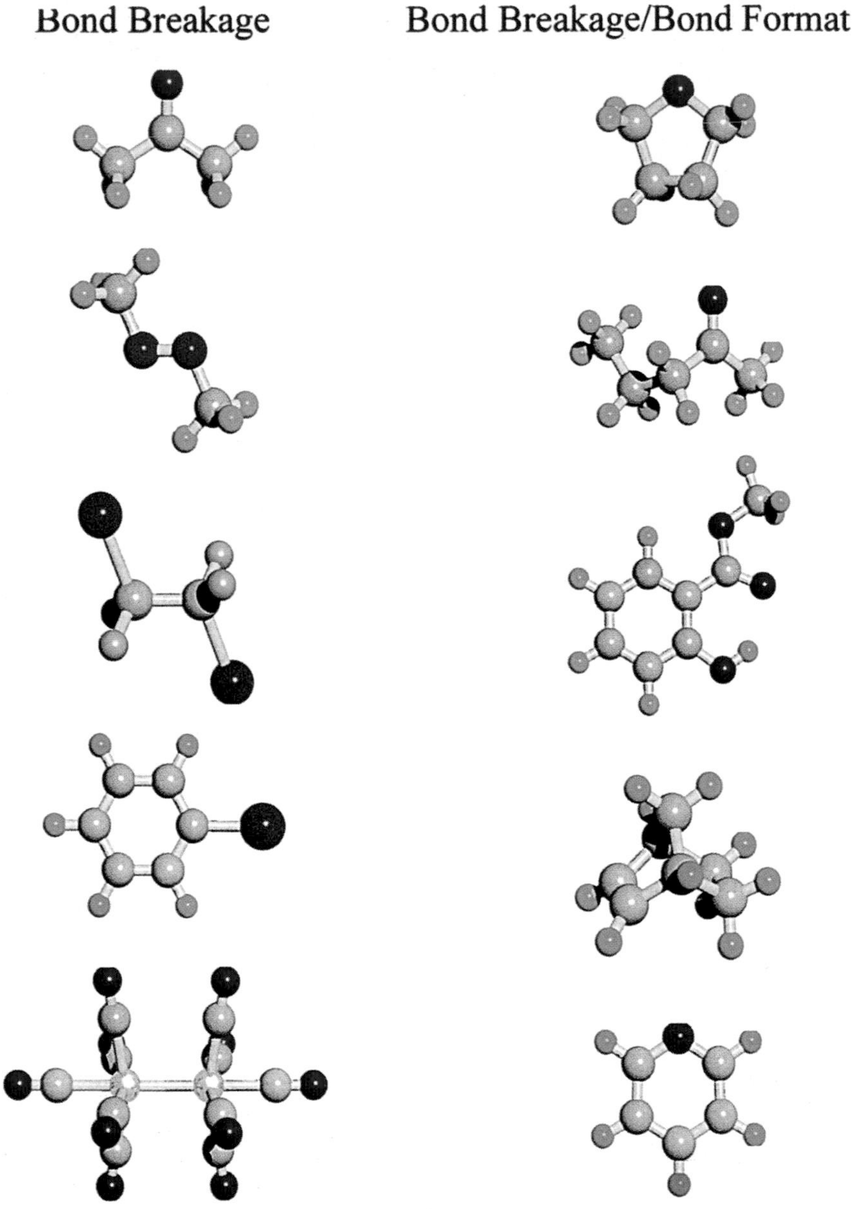

Bond Breakage

Bond Breakage/Bond Format

Figure 22. Molecular structures of different reactions studied, typical of the systems discussed in text for organic and organometallic femtochemistry: Acetone, Ref. 60; Azomethane, Ref. 61; Diiodoethane, Ref. 62; Iodobenzene, Ref. 63; $Mn_2(CO)_{10}$, Ref. 64; Cyclic ethers, Ref. 65; Aliphatic ketones for Norrish-II reactions, Ref. 66; Methyl salicylate, Ref. 67; One of the structures studied for addition and elimination reactions, Ref. 68; Pyridine for valence isomerization, Ref. 69.

(vii) β-Cleavage Reactions
(viii) Elimination Reactions
(ix) Valence Structure Isomerization
(x) Reactive Intermediates

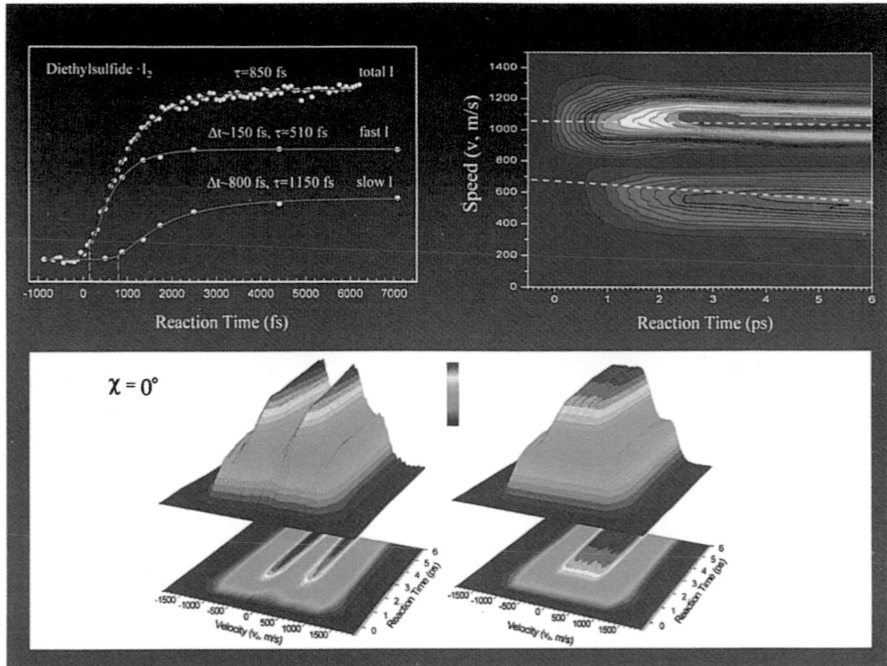

Figure 23. Femtochemistry of bimolecular electron-transfer reactions, the classic case of donors (e.g. benzene or diethylsulfide) and acceptors (e.g. iodine or iodomonochloride). (A, above) The experimental results clearly show the distinct velocity and time correlations, and thus the two-speed distributions and time scales of the reaction on the global PES. (B, below) Snapshots of the atomic motions according to experimental findings. [Ref. 70]

Electron & Proton Transfer

Here, we examined both bimolecular and intramolecular electron transfer reactions, and these studies were the first to be made under solvent-free conditions. We also studied the transfer in clusters and in solutions **[Figs. 11, 22, 23]**. For proton transfer, three classes of reactions were of interest, those of bi-

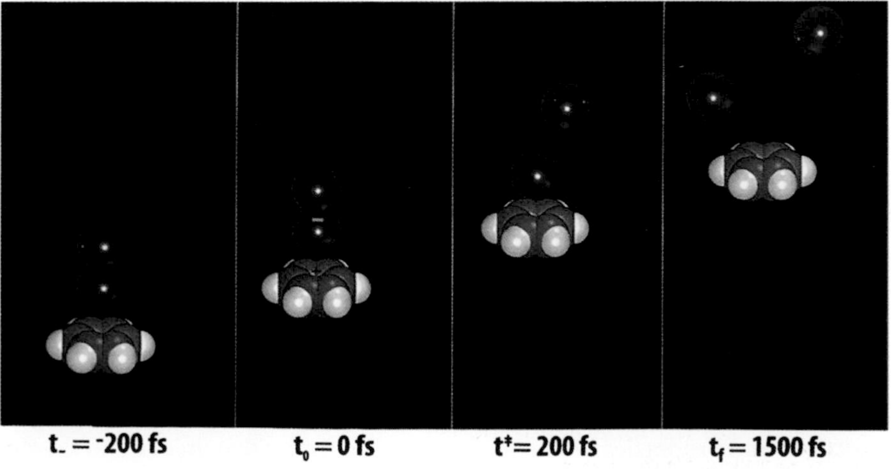

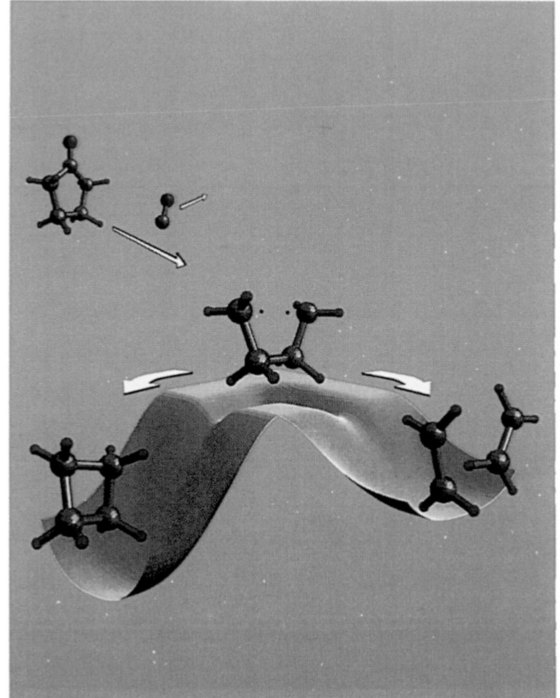

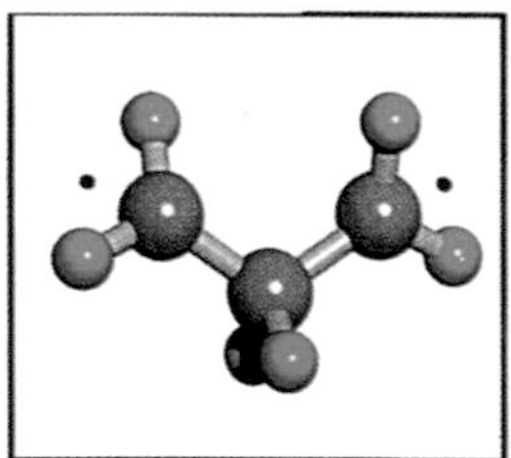
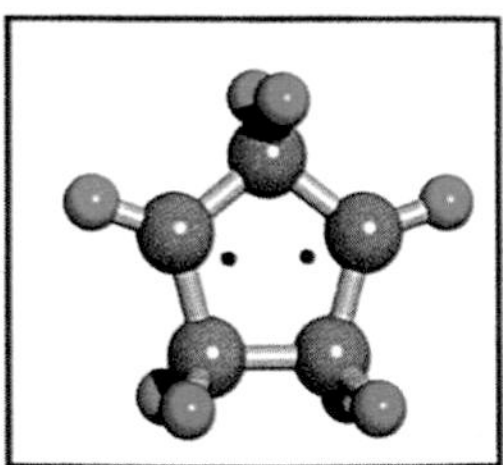
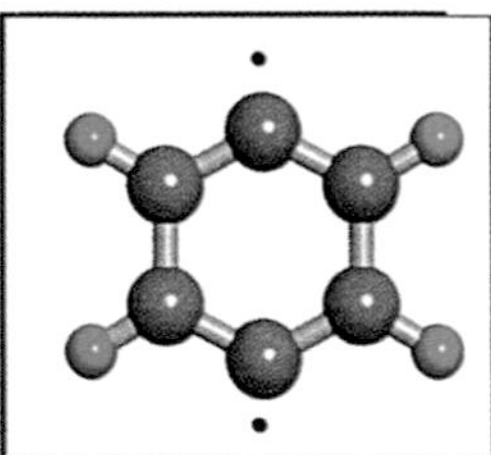

Figure 24. Reactive intermediates on the femtosecond time scale. Here, tetramethylene, trimethylene, bridged tetramethylene and benzyne are examples of species isolated on this time scale (see Fig. 22 for others). [Ref. 71]

molecular and intramolecular reactions, and those involving double proton transfer (base pair models):

(i) Bimolecular Electron Transfer Reactions
(ii) Intramolecular Electron Transfer and Folding Reactions
(iii) Acid-Base Bimolecular Reactions
(iv) Intramolecular Hydrogen-Atom Transfer
(v) Tautomerization Reactions: DNA Mimics

Inorganic & Atmospheric Chemistry

We extended the applications of femtochemistry to complex inorganic reactions of organometallics **[Fig. 22]**. Organometallic compounds have unique functions and properties which are determined by the dynamics of metal-me-

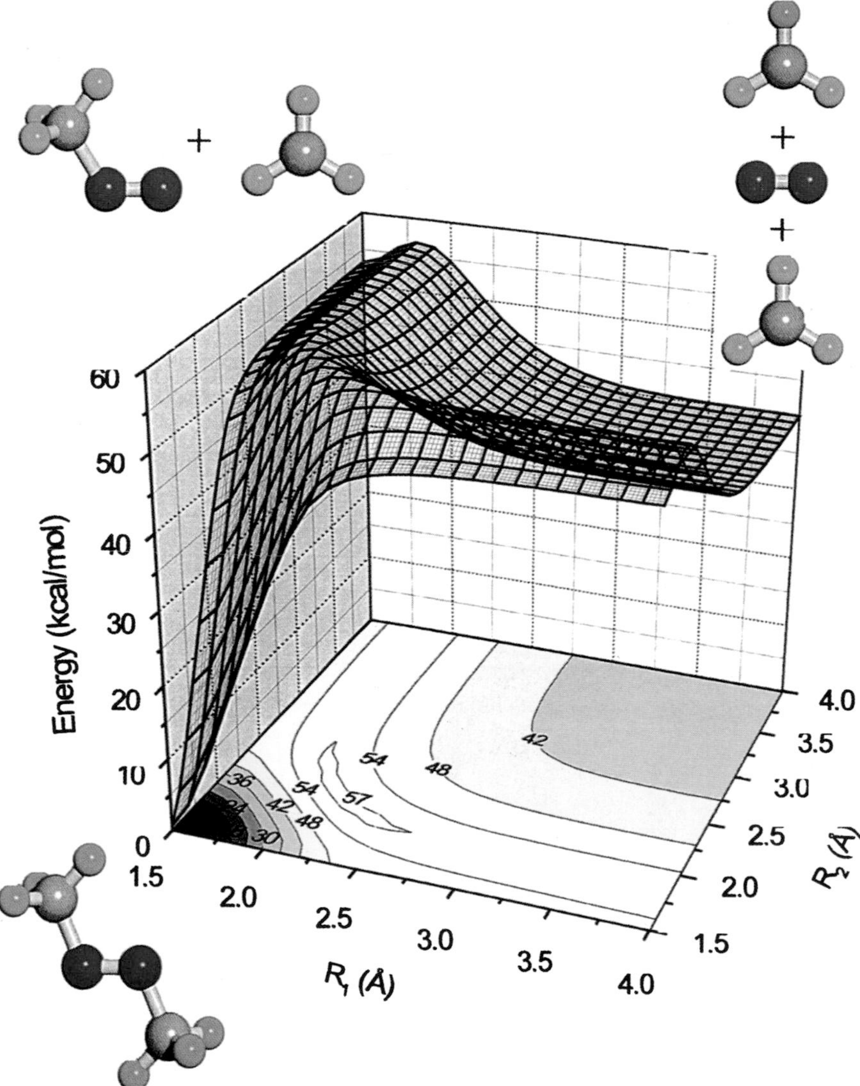

Figure 25. Reaction dynamics of azomethane, based on the experimental, femtosecond studies (Fig. 22, References). The *ab initio* PES was obtained from state-of-the-art calculations (Ref. 72) which show the two reaction coordinates (C-N) relevant to the dynamics. A third coordinate, which involves a twisting motion, was also studied. Note the concerted and non-concerted pathways.

tal (M-M) and metal-ligand (M-L) bonding. The time scales for cleavage of such bonds determine the product yield and the selectivity in product channels. They also establish the nature of the reactive surface: ground-state versus excited-state chemistry. Similarly, we studied the dynamics of chlorine atom production from OClO, a reaction of relevance to ozone depletion.

The Mesoscopic Phase: Clusters & Nanostructures

We have studied different types of reactions but under microscopic solvation conditions in clusters. These include **[Fig. 27]**:

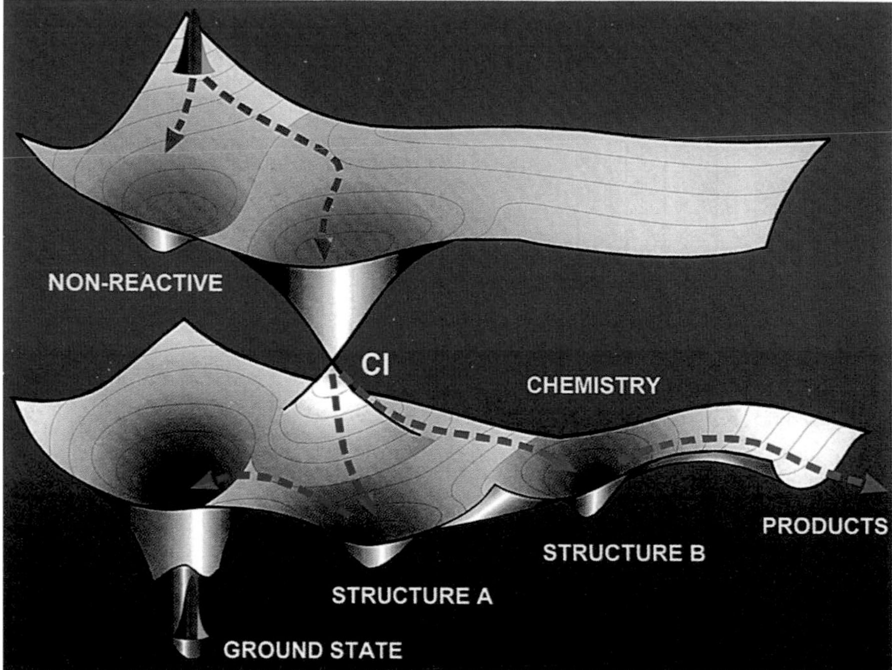

Figure 26. Generalized schematic showing reactive and non-reactive pathways (bifurcation) of wave packets. Both the photophysical and photochemical processes are shown with the conical intersection (CI) playing a crucial role. Examples are given in Ref. 73.

(i) Reactions of van der Waals complexes
(ii) Unimolecular Reactions
(iii) Bimolecular Reactions
(iv) Recombination, Caging Reactions
(v) Electron and Proton Transfer Reactions
(vi) Isomerization Reactions

The Condensed Phase: Dense Fluids, Liquids & Polymers

In this area of research, we have focused our efforts on the study of reactions in dense fluids and comparison with dynamics in liquids. By varying the solvent density, we could study the femtosecond dynamics from gas-phase conditions to the condensed phase of liquid-state density. Accordingly, we could observe the influence of solute-solvent collisions on reaction dynamics in real time. We also did studies in liquid solutions for some of the systems examined in the gas phase: bond breakage and caging; valence structure isomerization; and double proton transfer. Similarly, we studied systems of nanocavities and polymers. Some highlights include **[Fig. 28]**:

(i) Dynamics of the Gas-to-Liquid Transition Region (T_1 and T_2)
(ii) Dynamics of Bimolecular (one-atom) Caging
(iii) Dynamics of Microscopic Friction
(iv) Dynamics in the Liquid State

Snapshots of $I_2 \cdot Ar_{17}$

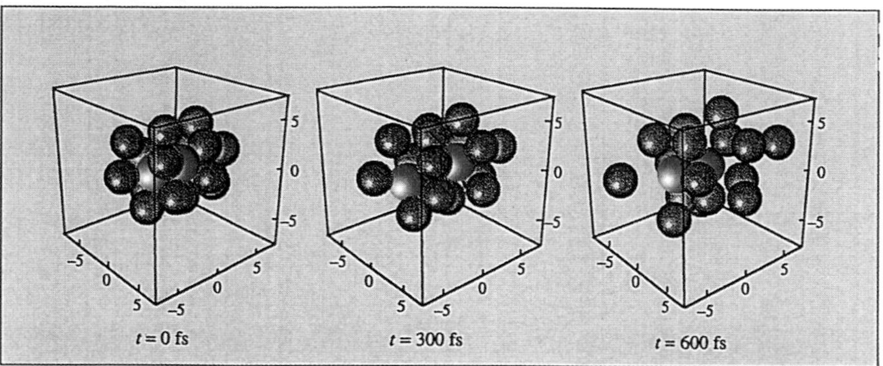

Snapshots of $I_2 \cdot benzene_5$

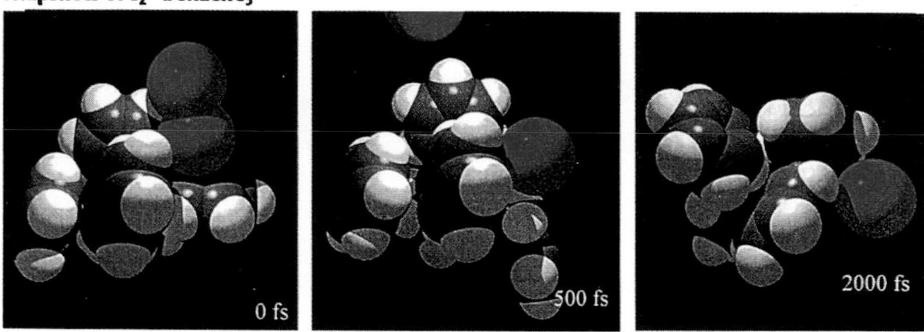

Figure 27. Femtosecond dynamics in the mesoscopic phase, reactions in solvent clusters. Several examples are given: (A, above) The coherent nuclear dynamics of bond breakage and recombination of iodine in argon (the cage effect), and the dynamics of the same solute but in polyatomic solvents (benzene). It was for the former that the first *coherent* bond breakage in the cage was observed and separated from the effect of vibrational relaxation. For the latter, the two atoms experience different force fields and the time scales are determined by the degree of solvation. (We also studied van der Waals complexes.) (B, next page) Shown is the study of acid-base reactions of naphthol with ammonia, changing the number of solvent molecules from 0 to 10. The isomerization of stilbene was similarly studied; the number of solvent hexane was varied from 0 to 6. [Ref. 74]

(v) Dynamics of Energy Flow in Polymers
(vi) Dynamics of Small & Large Molecules in Cyclodextrins

(6) Opportunities for the future

Three areas of study are discussed.

Transient Structures from Ultrafast Electron Diffraction

Electron diffraction of molecules in their ground state has been a powerful tool over the past 50 years, and both electron and x-ray methods are now being advanced in several laboratories for the studies of structural changes. We have reported in Nature (London) the latest advance in UED **[Fig. 29]**, by which major challenges were surmounted: the very low number densities of gas samples; the absence of the long-range order that is present in crystals, which enhances coherent interference; and the daunting task of determining

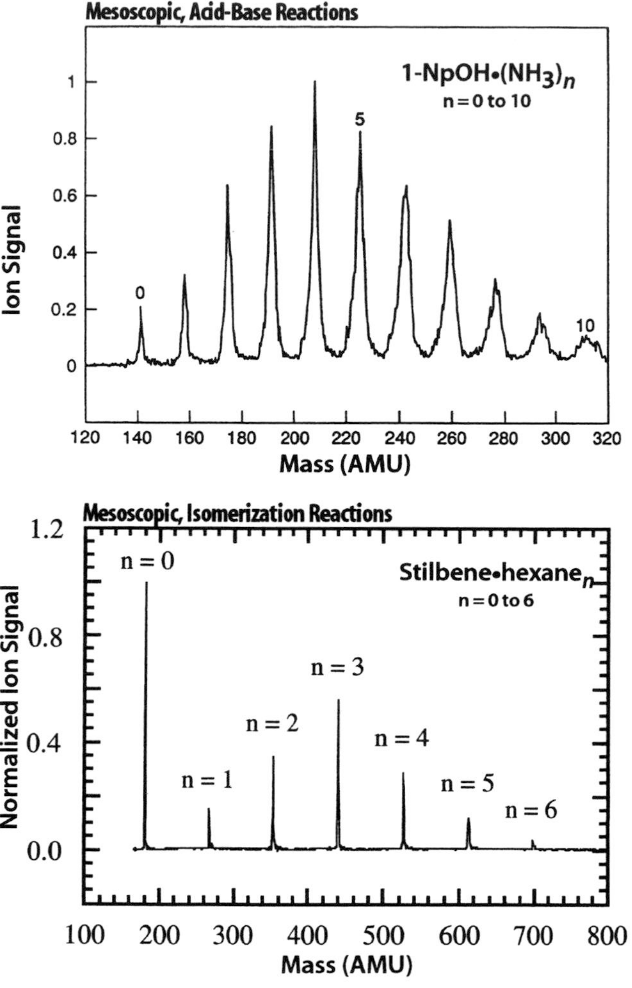

in situ the zero-of-time when diffraction changes are on the ps and sub-ps time scale.

With UED, we have been able to study molecular structures and branching ratios of final products on the ps time scale. The change in diffraction from before to after a chemical reaction was observed. However, the direct observation of transient structural changes in the course of a reaction was published only recently [in PNAS]. Specifically, we observed the transient intermediate in the elimination reaction of 1,2-diiodotetrafluoroethane ($C_2F_4I_2$) to produce the corresponding ethylene derivative by the breakage of two carbon-iodine bonds [see **Fig 29B** and **C**]. The evolution of the ground-state intermediate (C_2F_4I radical) was directly revealed in the population change of a single chemical bond, namely the second C-I bond. The elimination of two iodine atoms is nonconcerted, with the reaction time of the second C-I bond breakage being ~17 ps. The UED results for the short-lived C_2F_4I radical favor the classical structure over the bridged structure. *Ab initio* calculations were made to compare theory with experiments.

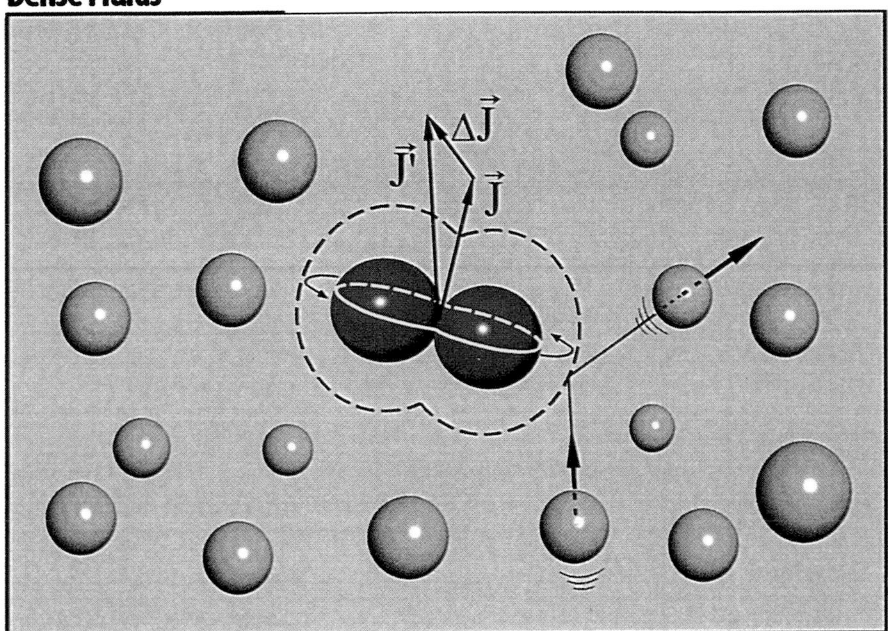

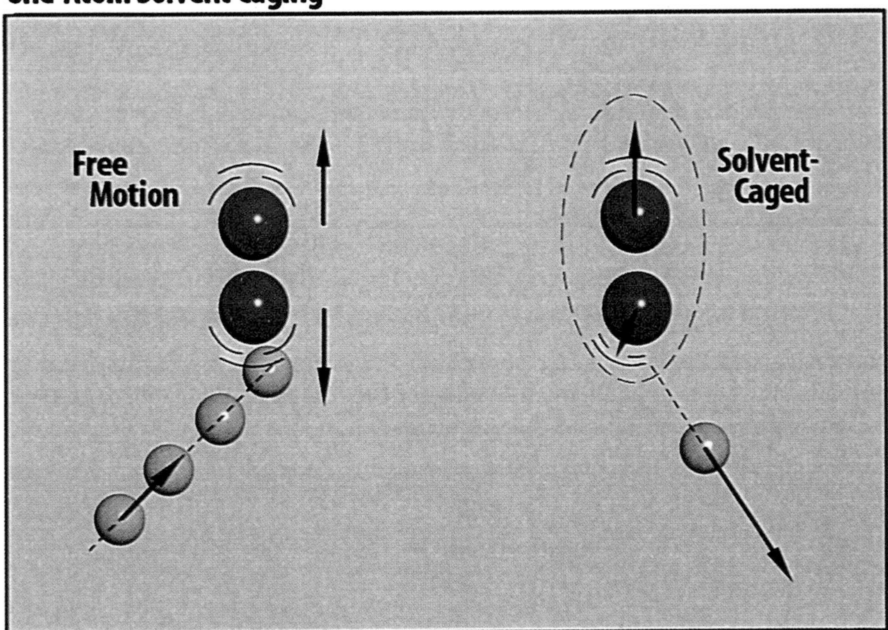

Figure 28. Femtosecond dynamics in the condensed phase: (A, above) dense fluids; (B, next page) the coherent vibrational and rotational motions observed in dense fluids as a function of density and down to the one-atom collision with iodine; (C, page 173) nanocavities of cyclodextrins and polymers of polydiacetylenes; liquids (not shown, but references are given [75]). Studies in these media include the one-atom coherent caging, J-coherence friction model, coherent IVR in polymer chains, anomalous T_2 behavior in dense fluids. [Ref. 75]

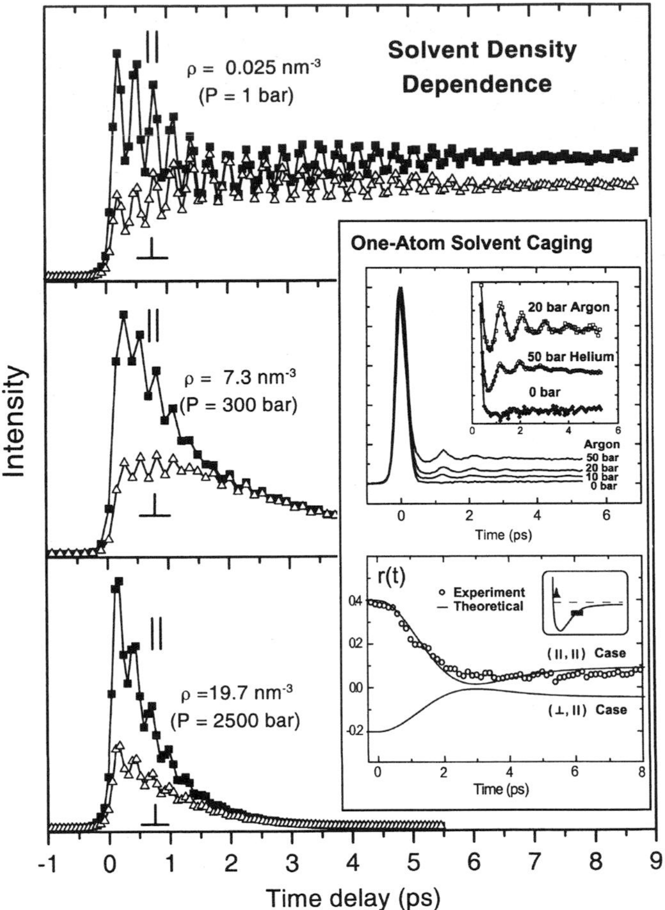

This leap in our ability to record structural changes on the ps and shorter time scales bodes well for many future applications to complex molecular systems, including biological systems. We have completed a new apparatus equipped with diffraction detection and also with mass spectrometry. This universal system is capable of studying complex systems in the gas and other phases. It holds great opportunities for the future.

Reaction Control

Our interest in this area goes back to the late 1970's when a number of research groups were reporting on the possibility of (vibrational) mode-selective chemistry with lasers. At the time, the thinking was directed along two avenues. One of these suggested that, by tuning a CW laser to a given state, it might be possible to induce selective chemistry. This approach was popularized enthusiastically, but it turned out that its generalization could not be made without knowing and controlling the time scales of IVR in molecules. Moreover, state-selective chemistry is quite different from bond-selective chemistry. The second avenue was that of IR multiphoton chemistry. In this case, it was shown that selectivity was lost in the quasi-continuum vibrational mani-

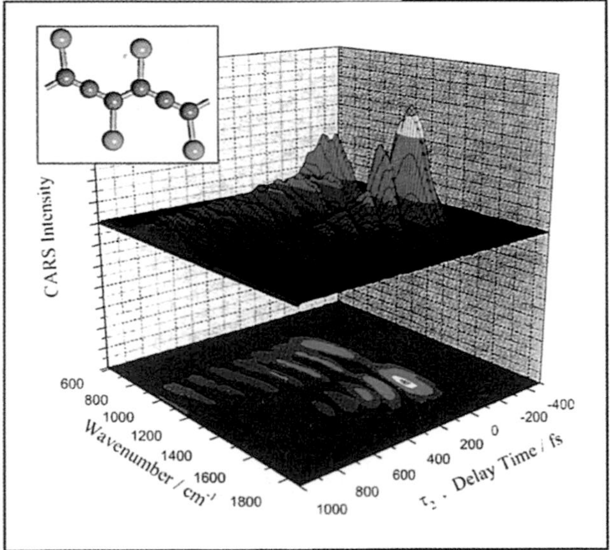

fold of molecules, but that the initial IR coherent pumping could be used for selective isotope separation. Such an approach has proven successful, even on the practical scale, and Letokhov has called the process "incoherent control" **[33]**.

The discovery **[Section III2]** of coherent and selective vibrational oscillations (in-phase and out-of-phase) in a large molecule such as anthracene made me think of the possibility of temporally controlling the state of the system. The key idea was coherence among the vibrational degrees of freedom and its observation (published in 1981) which triggered significant interest in the issue of chaotic *vs.* coherent "motion" of packets in isolated mo-

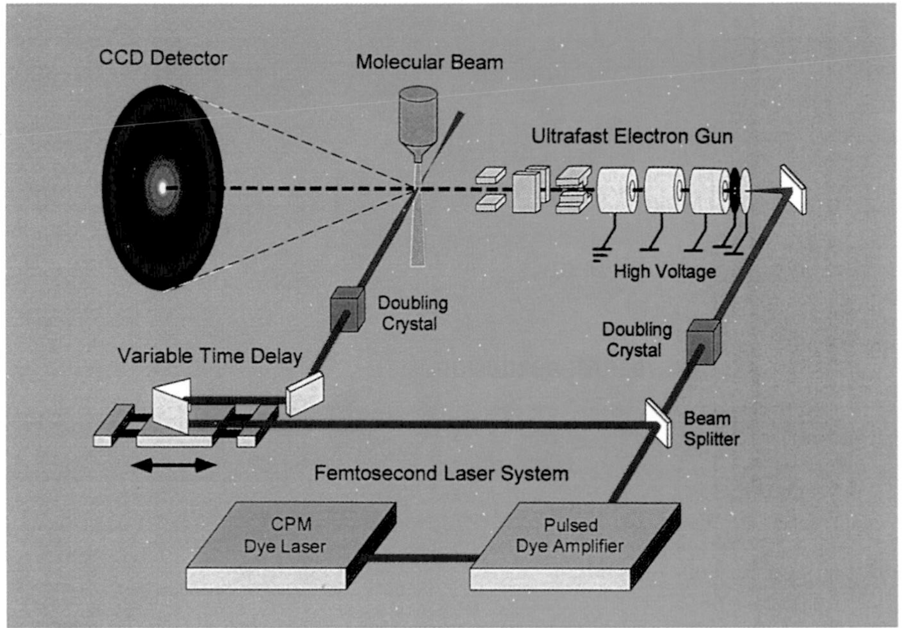

Figure 29. Ultrafast Electron Diffraction (UED). (A, above) The experimental arrangement; (B, below) A 2-D image (CCD) and the obtained molecular scattering sM(s) and radial distribution f(r) functions; (C, next page) The temporal change observed on a bond population, elucidates the structure of the reaction intermediate shown as two possibilities (see text). [Ref. B17, 76]

lecules. Some of us believed that, despite the complexity of the vibrational mode structure, coherence would be robust, provided it could be disentangled through proper preparation and probing. In fact, in a discussion with Richard Feynman (at Caltech) about the anthracene results, he informed me of a related problem noticed by Fermi, when both were at Los Alamos: A li-

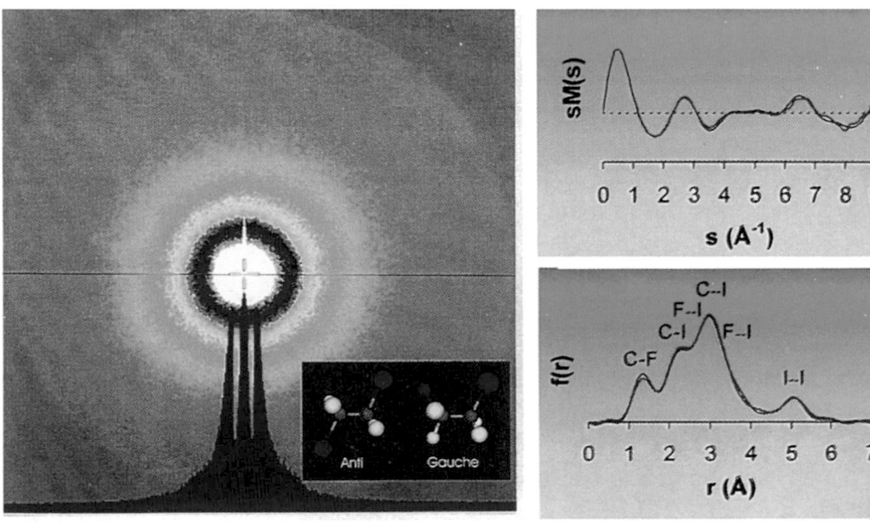

Reaction Intermediates

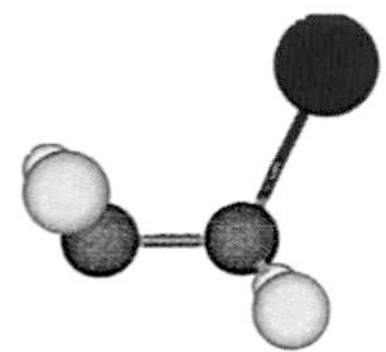

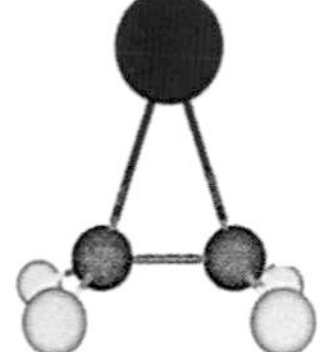

Ultrafast Electron Diffraction

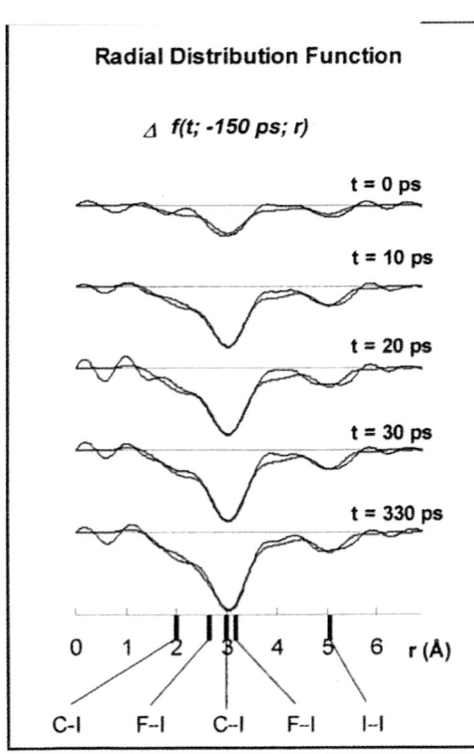

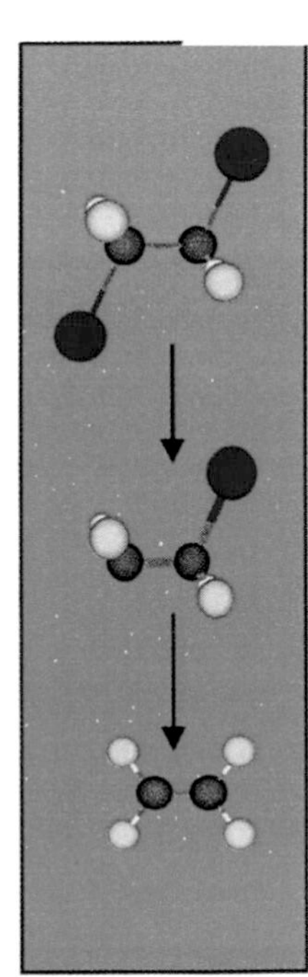

near chain of springs (vibrations) showed recurrences when the energy was initially localized (on the computer) in one spring. They expected dissipation and not recurrences. Nico Bloembergen and I wrote a feature article (1984) emphasizing this point of coherent motion and its significance to mode-selective chemistry **[B21]**. Stuart Rice believed strongly in the concept of coherence and, as I mentioned before, we even drafted a paper that was not finalized for publication.

Earlier in 1980, I wrote a *Physics Today* article in a special issue on laser chemistry suggesting the use of ultrashort pulses (not CW or long-time lasers) to control the outcome of a chemical reaction **[B24]**. The title of the paper was:

Laser Selective Chemistry – Is it Possible? The subtitle stated the message, *"With sufficiently brief and intense radiation, properly tuned to specific resonances, we may be able to fulfill a chemist's dream, to break particular selected bonds in large molecules."* In this article, I was concerned with the problem of IVR and so-called chaotic behavior. Stimulated by our work on IVR and coherence **[Section III]**, I thought that ultrashort pulses should be used to control the system in the desired configuration by proper choice of the time duration and delay and by the preparation of the packet which is controlled by the coherence width. Experimentally, we had already developed methods for the control of the phase of the field of optical pulses with the idea of using the *phase (pulse shaping)* to control molecular processes – collisions, inhomogeneous broadenings and even photon locking which could inhibit relaxation **[Section III]**; the time scale was ns and for the control of IVR, fs pulses were needed. Prior to this work, the optical pulse field,

$$E(t) = E_o A(t) \cos[\omega\tau + \phi(t)], \quad (10)$$

was simply defined by the envelope A(t) and the frequency ω; the phase $\phi(t)$ was unknown. By controlling $\phi(t)$ we were able to make sequences of phase-coherent multiple pulses and to tailor a composite "single" pulse with a prescribed $\phi(t)$. We published a series of papers demonstrating the power of the approach, as mentioned in **Section III**; see **Fig. 30**. In fact with composite shaped-pulses, a sequence of phase segments and tilt angles (in the rotating frame) of, e.g., 60_x–$300_{\bar{x}}$–60_x, we showed experimentally that the emission of a molecule can be made twice that as when a normal single pulse was used **[Fig. 30]**. Similarly, by choosing pulse sequences such as x-y-x($\bar{x}$) we experimentally locked the system and thus lengthened its relaxation time considerably. In theoretical papers, we examined the use of the approach for selectivity and control of molecular relaxations; in recent reviews **[34]**, Warren has discussed pulse shaping and its relevance to quantum control.

On the fs time scale, the theoretical work of Heller **[Section III]** stimulated the use of the time-dependent wave packet picture for absorption and emission. In 1985, David Tannor and Stuart Rice, using the wave packet picture, provided a two-photon scheme for the control of selectivity with pulse-sequence coherence being an important part of the evolution. This scheme was extended and, in their review article of 1988 **[35]**, they described phase sensitive experiments such as the ones we reported earlier. An important realization was the desire to optomize the yield of a given channel. With fs resolution, we began testing the idea of timing of pulses on small molecular systems **[Fig. 30]**. We first began with a single experiment on the control of the population in bound states (iodine). Then we reported results on the control of the yield in the reaction $Xe + I_2 \rightarrow XeI + I$ as a function of the delay time between pump and control. Although the mechanism is not fully resolved, the important point is that the yield of product *XeI* followed the temporal motion of the iodine wave packet. In a third experiment, we used pump-control-probe fs pulses to control the branching of the NaI reaction; these, together with the experiment by Gustav Gerber's group on $Na_2(Na_2^+ + e$ *vs.* $Na + Na^+ + e)$,

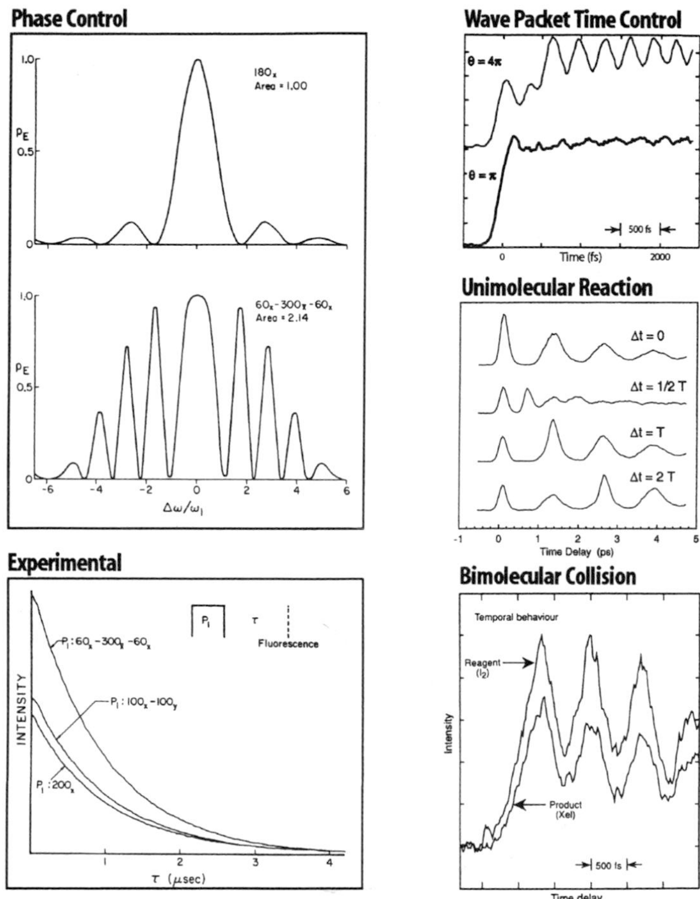

Figure 30. Control by the phase and/or the delay, or the duration of optical pulses. (A, above) (Left) The effect of a designed composite pulse on the fluorescence of a molecule (iodine), showing the large experimental enhancement for the labeled phase-controlled sequence. (Right) Control of the population (I_2), of unimolecular reactions (NaI), and of a bimolecular collision ($Xe+I_2$), see text. (B, next page) Localized wave packet control of the preparation at high energy on the femtosecond time scale, which is shorter than that of IVR. The series has the same reaction coordinate (C-C bond) but the molecular size has increased in complexity. The behavior is far from being statistical. [Ref. B20, B24, 77]

are prototypes for the Tannor-Rice-Kosloff scheme. Phase-locked pulses were extended to the fs resolution by Norbert Scherer and Graham Fleming in elegant studies of iodine.

Recently, we turned our attention to complex molecular systems, but this time using fs pulses to implement the 1980 idea. In a series of molecules of increasing complexity, but retaining the same reaction coordinate, we illustrated selectivity by beating IVR (and entering near the transition state); the rate of reaction was two to three orders of magnitude larger than the expected *statistical* limit. This work was published in Science **[Fig. 30]** and promises to be significant for achieving non-statistical chemistry at high energies. The concept suggests that control at high energies (chemical energies) is more rea-

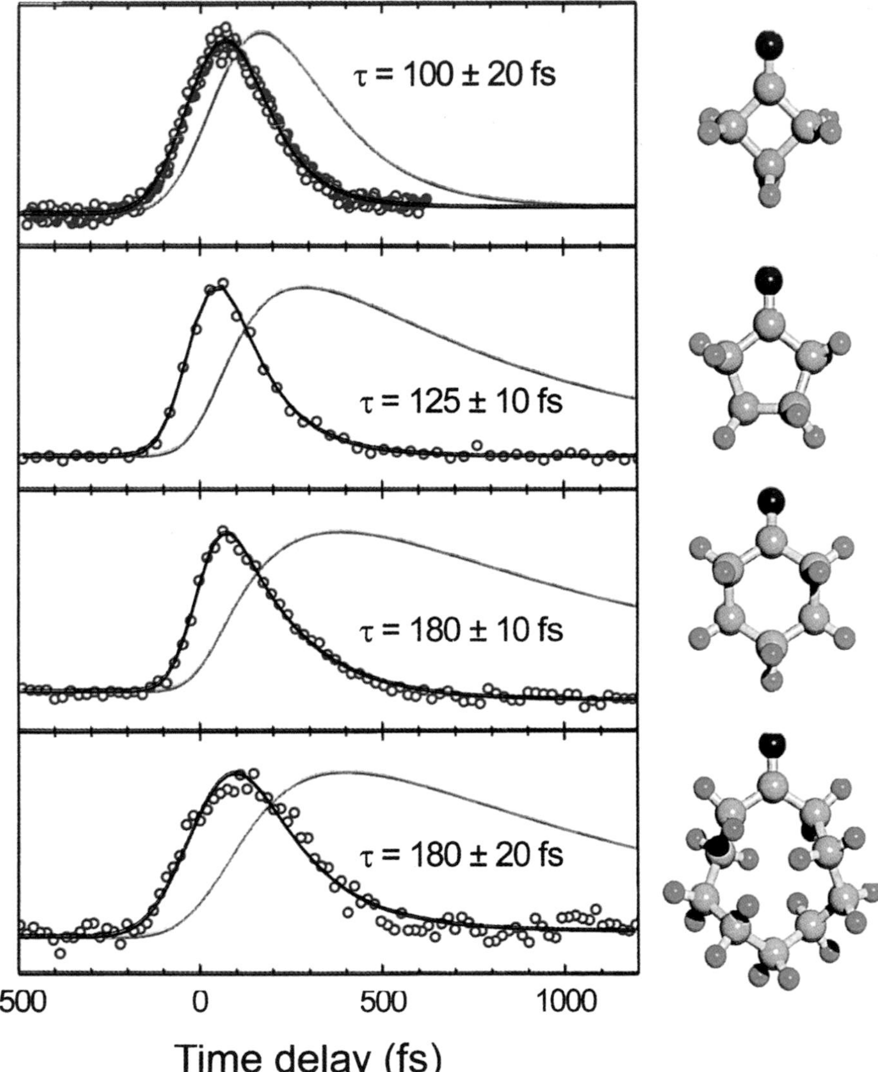

listic, in contrast with the conventional wisdom which asserts the need for low energies – time is of the essence! Another example of non-statistical femtochemistry comes from the work on surfaces **[36]**. Recently, the group in Berlin (Ertl and Wolf **[36]**) demonstrated, in an elegant experiment, the critical role of fs resolution in inducing oxidation (*vs.* desorption) of CO on Ru surfaces – the fs non-equilibrated electron distribution of the surface gives a selective chemistry different from that of equilibrated phonon distribution, or thermal heating.

In the future, there will be extensions and new directions in fs light-matter control based on the temporal coherence of light and its interference with matter waves. One area that holds promise is the use of fs pulses to induce selectivity by utilizing the three parameters of the pulse, the central frequency, the width and the chirp, in an iterative algorithm; the chirp is, in a way, similar to a composite pulse of the type described above. The technique of liquid-crystal-display developed by Andy Weiner for fs pulse shaping, combined with the evolutionary feedback idea of Herschel Rabitz, makes possible the generation of the desired complex E(t) field to achieve (combinatorial) control. This optimal control has been demonstrated nicely for a targeted second harmonic generation or a yield of chemical reaction as reported by Gerber's group in Würzburg **[37]**. Kent Wilson **[37]** showed the importance of chirped pulses in focusing and reflecting wave packets and, in a more recent contribution, he, with Warren Warren, used the evolutionary feedback approach to optimize the fluorescence of a molecule in solution, reminiscent of the composite pulse experiment we described above. In pulse shaping, the field is optimized and forms a complex pattern which is used through many generations to reach the fitted, desired population.

It should be noted that all of the above schemes change the coherent composition of the initial packet and hence the evolution in different channels – but we have not changed the evolution dictated by the natural forces of the atoms! Intense fields may do so. Paul Corkum, Thomas Baumert, and other colleagues have provided novel observations with intense fields **[38]**. Clearly these areas of control by ultrafast pulse timing (t), phase (ϕ) (shape), spatial localization (R) and intensity to alter the potential (V) offer new opportunities for the future. Many theoretical efforts have already been advanced ahead of current experiments and Manz' group is providing new possibilities, including deracemization by controlled pulses – timed and shaped [see his review of 1500 references in **Ref. 28**, to all work]. We did not discuss here the CW control scheme advanced by Paul Brumer and Moshe Shapiro, nor can we give references to all work done in this area.

Biological Dynamics

There have been important contributions to femtobiology and these include **[Refs. 28 & 39, and references therein]**: studies of the elementary steps of vision; photosynthesis; protein dynamics; and electron and proton transport in DNA. In proteins such as those of photosynthetic reaction centers and antennas, hemoglobins, cytochromes and rhodopsin, a femtosecond event, bond breaking, twisting or electron transfer occurs. There exist global and coherent nuclear motions, observed in these complex systems, and it is possible that the complexity is not as complicated as we think; for the chemistry and the efficiency to be unique, the system utilizes the organized structure around the "active center" with the necessary restraint on transition states and energy flow. Thus, in my view, the early fs events are critical to understanding the function of complex systems, as they reflect the important locality in biological dynamics.

Our efforts in this direction have so far focused on DNA twisting dynamics, electron transfer in DNA assemblies, DNA base-pair models, and on protein-ligand dynamics. The work on the torsional rigidity of DNA was published in 1980-1982, while that relating to proton transfer in model base pairs was reported in the last few years. With donors (D) and acceptors (A) covalently

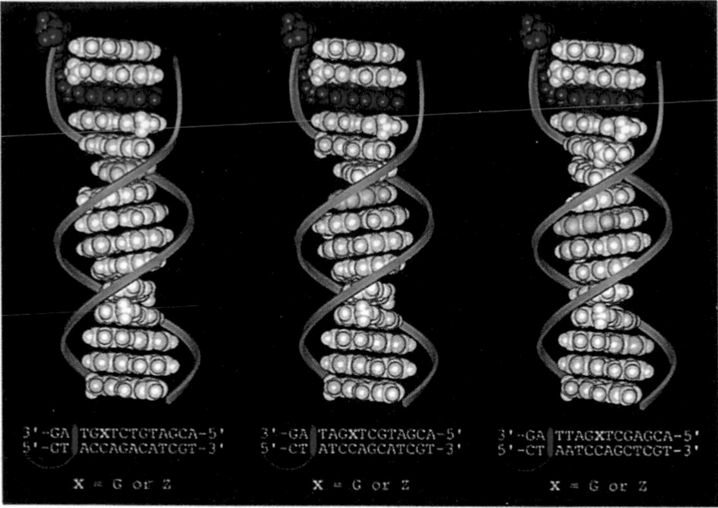

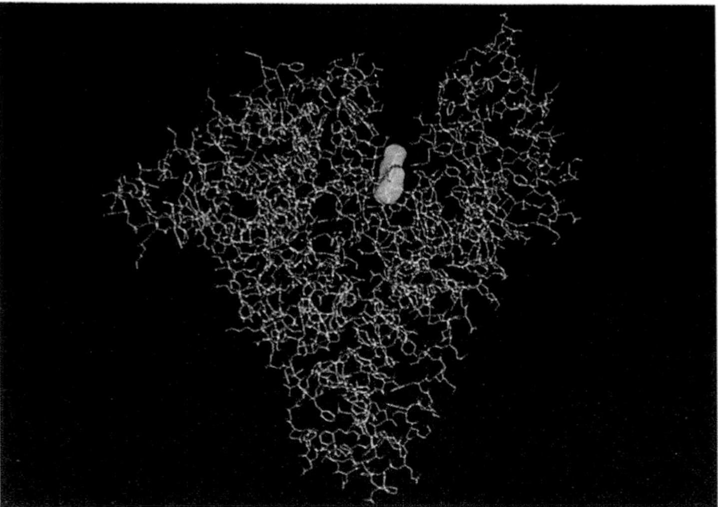

Figure 31. Femtosecond dynamics of biological systems, DNA assemblies and protein complexes. (A, above) The systems studied are the DNA with donors and acceptors at fixed distances (top) and protein HSA with the molecule HPMO shown in the interior. The focus of research is on electron transfer and molecular dynamics in the former and on probing solvation and ligand-recognition effects in the latter. (B, next page) Some illustrative femtosecond transients of DNA assemblies. Aminopurine was used for the initial excitation, and the effect of different bases G, C, A, T was studied at a fixed distance. The time scale of torsion dynamics was known from studies we and others made before. [Ref. 78, 79]

bonded to DNA, studies of ET on more well defined assemblies were made possible, and the effect of distance could be addressed. With fs resolution, we obtained the actual time scale of ET and related the rates to the distance between D and A. In collaboration with Jackie Barton's group, we published this work in PNAS. The time scale of orientational coherence and solvation was also examined, allowing us to elucidate the role of molecular motions, including the effect of DNA rigidity. The results reveal the nature of ultrafast ET

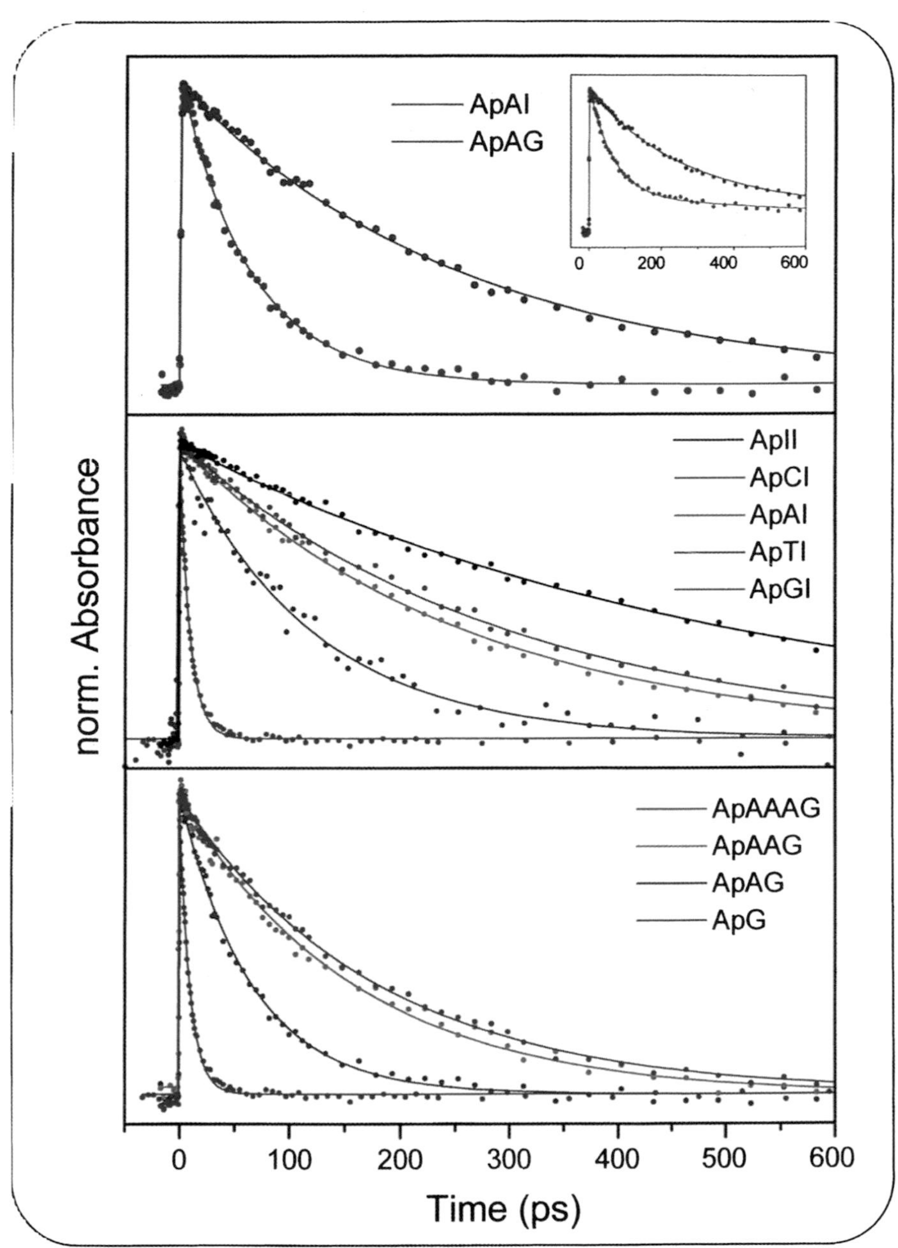

and its mechanism: in DNA, ET cannot be described as in proteins simply by a phenomenological parameter, β. Instead, the local involvement of the base pairs controls the time scale and the degree of coherent transport. Molecular dynamics are critical to the description of the transport. The measured rates **[Fig. 31]** and the distance range of the transfer suggest that DNA is not an efficient molecular wire.

For proteins, our current interest is in the studies of the hydrophobic forces and ET **[Fig. 31]**, and oxygen reduction in models of metallo-enzymes

[Fig. 32]. For the former, we have studied, with fs resolution, the protein Human Serum Albumin (HSA), probed with the small (ligand) molecule hydroxyphenyl methyloxazole (HPMO); this work is in collaboration with Abderrazzak Douhal. We also studied ET in hyperthermophilic proteins. For

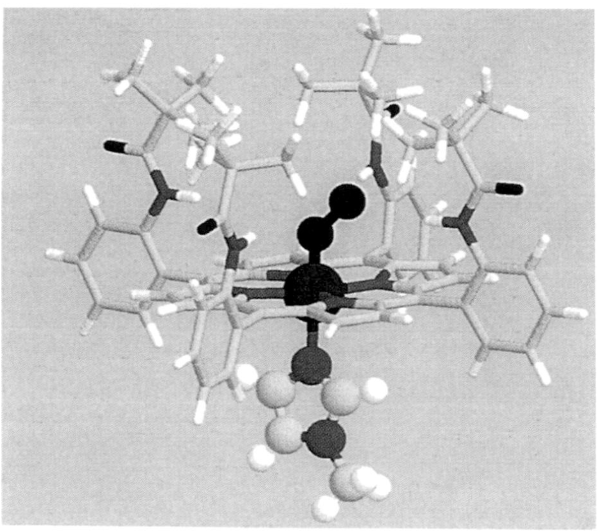

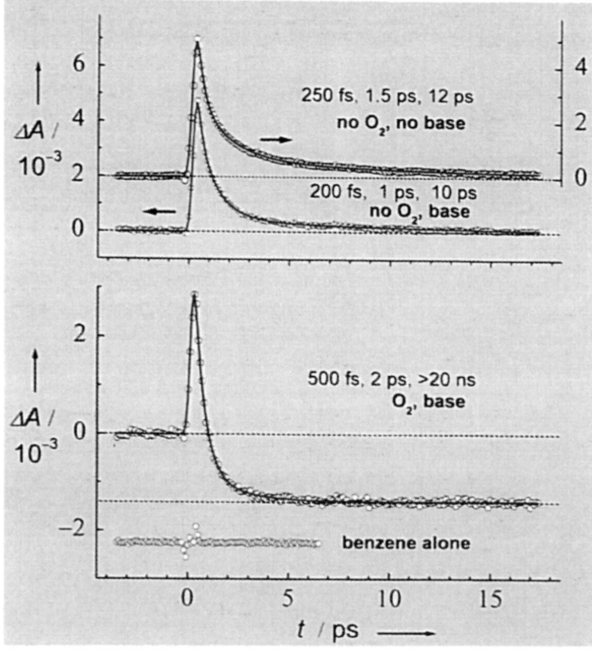

Figure 32. Femtosecond dynamics of model biological systems, hemoglobin and myoglobin, DNA base pairs, and photosynthetic assemblies. Shown here are the structure of Dioxygen-picket fence Cobalt Porphyrins and the fs transients which show the time scales involved and the release of O_2 in 1.9 ps at room temperature. These studies on this and the other model systems (not shown) are part of the continued effort in this area. [Ref. 80]

model enzymes, we examined novel picket-fence structures which bind oxygen to the central metal with ~ 85% efficiency at room temperature. In this system **[Fig. 32]**, we observed the release of O_2 in 1.9 ps and the recombination was found to occur on a much slower time scale. These are fruitful areas for future research, especially in that they provide prototype systems for O_2 reduction with complex metallo-porphyrins in the transition state, similar to the smaller systems of benzenes/halogens (discussed in **Fig. 23**) but at room temperature. We published our first report recently in Angewandte Chemie in collaboration with Fred Anson's group at Caltech.

In the future, new extensions are anticipated. The nature and control of enzymatic reactions, the catalytic function of the transition state, and the design of artificial biological functions seem to be areas of great promise for dynamical studies. Also, it is envisaged that the recording of a large biological structure changing with time and with atomic resolution may be realized. Already some success in studies of small chemical systems utilizing ultrafast electron diffraction have been reported. X-ray diffraction is another direction. The ultimate goal is the recording of all coordinates in space and time. The impact on problems such as protein folding and molecular recognition is clear.

IV. IMPACT AND CONCEPTS – A RETROSPECTIVE

In retrospect, the key to the explosion of research cited in **Ref. B60** can perhaps be traced to three pillars of femtochemistry:

Time Resolution – Reaching the Transition-State Limit

Three points are relevant: (i) The improvement of nearly ten orders of magnitude in time resolution, from the (milli)microsecond time scale (Eigen, Norrish & Porter) to present femtosecond resolution, opened the door to studies of new phenomena and to new discoveries; (ii) the transition state, the cornerstone of reactivity, could be clocked as a molecular species $TS^{\ddagger}$, providing a real foundation to the hypothesis of Arrhenius, Eyring and Polanyi for ephemeral species $[TS]^{\ddagger}$, and leading the way to numerous new studies. Extensions will be made to study transition state dynamics in complex systems, but the previous virtual status of the transition state has now given way to experimental reality **[B60]**; (iii) inferences deduced from "rotational periods" as clocks in uni- & bimolecular reactions can now be replaced by the actual clocking of the nuclear (vibrational) motion. This is particularly important when a chemical phenomenon such as concertedness is involved or the time scale of complexes or intermediates is many vibrational periods.

In the 1960's, there was some thought **[40]** that the relevant time scale for chemistry was the microsecond regime. Moreover, the uncertainty principle was thought to represent a severe limit of the utility of shorter time resolution; coherence was not part of the thinking in deciphering fs nuclear motion, as discussed in **Section III** (Development of Femtochemistry) and in what follows. The new vision is summarized in the following statement **[40]**: *"The study of chemical events that occur in the femtosecond time scale is the ultimate achievement in half a century of development and, although many future events will be run over the same course, chemists are near the end of the race against time."* Manfred Eigen, who gave the 1967 Nobel Lecture with the title "Immeasurably Fast Reactions" – "Die 'Unmessbar' Schnellen Reaktionen" – told me, when I teased him about the title, that nobody in the 1950s anticipated the laser and the short pulses they can provide.

Atomic-Scale Resolution

Two points are relevant: (i) The transition from *kinetics* to *dynamics*. On the femtosecond time scale, one can see the coherent nuclear motion of atoms – *oscillatory* or *quantized steps* instead of *exponential decays* or rises. This was proved to be the case for bound, quasi-bound or unbound systems and in simple (diatomics) and in complex systems (proteins). Because of coherence, we can speak of the motion classically and visualize it as the change actually occurs; (ii) the issue of *the uncertainty principle*. Many thought that the pulse was too short in time, thus broad in energy by the uncertainty principle $\Delta t\ \Delta E \sim \hbar$, but as discussed before, localization is consistent with the two uncertainty relationships **[Fig. 4]** and coherence is the key. The energy uncertainty ΔE should be compared with bond energies: ΔE is 0.7 kcal/mol for a 60 fs pulse [see details in **Ref. B17**]. In the condensed phase, localization may become shorter lived, but the basic picture is still valid **[41]**.

It took time for this concept of coherence to be appreciated, not only among some chemists, but also among notable physicists. In 1972, at a Welch Conference, picosecond time resolution was of concern because of the perceived fundamental limitation imposed on time and energy by Heisenberg's uncertainty principle. After his lecture on lasers in chemistry, the physicist Edward Teller had a lively exchange with another physicist and friend, Eugene Wigner. Even for picosecond resolution the question was asked, *is there a natural, real limit to the time...?* In the Welch Prize Address (1997), I highlighted these exchanges **[42]**. Jacob Bigeleisen **[24]**, although concerned about the uncertainty principle, asked why not venture into the "millijiffy" (femtosecond) range?

Generality of the Approach

Three points are relevant: (i) In retrospect, the femtosecond time scale was just right for observing the "earliest" dynamics at the actual time scale of the chemical bond, defining the earliest time possible; (ii) the methodology is versatile and general, as evidenced by the scope of applications in different phases and of different systems. Moreover, it has stimulated new directions of research in both experiment and theory in areas such as quantum control and ultrafast diffraction; (iii) the time resolution offers unique opportunities when compared with other methods.

First, processes often appear complex because we look at them on an extended time scale, during which many steps in the process are integrated. On the fs time scale, these steps are resolved, and the process breaks down into a series of simpler events. Second, only this time resolution can give the dynamics of transition states/intermediates in real time since for reactions neither the spectra of reactants nor those of products are directly relevant. This point was amply demonstrated in complex reactions, such as those of transient, reactive intermediates of organics. Even for simple, reactive systems this is still true. For example, the spectral bandwidth of the reactant, dissociative ICN has no information about the actual dynamics of the nuclear separation between I and CN fragments or about the transient configurations, as it only reflects the steepness of the potential at the initial nuclear configuration. For unreactive molecules, there is a different complexity; the spectra are usually inhomogenously broadened, especially in complex systems. Finally, with time resolution, we can observe the motion without resorting to a mathematical construct from eigenstates or other indirect methods.

It is worth noting that both *excited* and *ground state* reactions can be studied. It has been known for some time that the use of multiple pulses can populate the ground state of the system and, therefore, the population and coherence of the system can be monitored **[23]**. The use of CARS, DFWM, SRS, π-pulses or the use of direct IR excitation are some of the approaches possible. Two recent examples demonstrate this point: one invokes the use of IR fs pulses to study reactions involving hydrogen (bond) motions in liquid water, work done in France and Germany **[43]**; and the other utilizes CARS for the study of polymers in their ground state which we published in collaboration with the groups of Kiefer and Materny **[Fig. 28]**. Ground-state dynamics have also been studied by novel fs photodetachment of negative ions, and the subfield of fs dynamics of ions is now active in a number of laboratories **[44]**.

Some Concepts

New concepts and phenomena have emerged and include: Localization of wave packets; reaction path coherence, single-molecule trajectory; reaction landscapes *vs.* path; bifurcation; chemical *vs.* spectroscopic dynamics (time scales); concertedness; dynamical active space; non-statistical (non-ergodic) behavior; dynamical caging (by energy loss as opposed to barrier confinement); microscopic friction (energy *vs.* mechanical); and inhomogeneous dynamics of "soft matter" (e.g. biological) systems, with a whole range of time scales. These concepts have been discussed in the original publications, and below, only a few will be highlighted:

(1) Resonance (Non-equilibrium Dynamics): The concept of resonance in the structure of the chemical bond goes back to the era of Linus Pauling and the idea of interconversion between different electronic structures. The interconversion was a hypothesis, not an observable fact. Quantum mechanically, chemists usually speak of eigenstates of the system, which are stationary with

no time evolution. Resonance in dynamics is a concept which is not a stationary-state picture. With coherent preparation of molecules it is possible to prepare a non-stationary *(non-equilibrium)* state of a given nuclear structure and for the system to evolve in time. In our studies this was shown for vibrational redistribution, for rotational orientation, and for wave packet nuclear motions. Such non-stationary evolution does not violate the uncertainty principle and is fundamental to chemical dynamics.

(2) Coherence (Single-molecule-type Dynamics): Perhaps one of the most powerful concepts in femtochemistry is coherence of the *molecule*, of the *ensemble*, and of the *trajectory*. First, the coherence created by a femtosecond pulse is reflected in the motion of the wave packet; for a force-free motion the group velocity is that of a free particle (p/m), a classical motion **[45]**. Second is the ability to "transform" the ensemble's incoherent behavior to a coherent molecular trajectory. This is achieved because on the fs time scale the system can be promoted and localized in space with a localization length (ΔR) only limited by the uncertainty of the initial system, typically ~ 0.05 Å; all molecules which do not interact span this range. The chemical length scale (R) of interest is several ångströms and this is why the system behaves as a single-molecule trajectory **[Fig. 33]**. Because the initial state is promoted nearly intact on the fs time scale, the only dispersion is that which causes the different trajectories to spread under the influence of the new forces of the energy landscape or by external perturbations, such as solvation. If longer time pulses are used for the preparation, then ΔR is on the scale of R, and kinetics of the states are recovered. Put in time-domain language, the inhomogeneous dephasing time of the ensemble is relatively long for fs preparation *and* the homogeneous dynamics and the actual coherence of the packet become dominant, as amply demonstrated here and elsewhere. Third, the concept of coherence is crucial for achieving a coherent trajectory of reactions. Such control projects out the non-statistical behavior through the preparation of a localized configuration **[Fig. 33]**, as opposed to an incoherently-prepared configuration (by, for example, chemical activation) or a spatially-diffuse configuration (by long-time experiments). The concept is powerful and basic to many phenomena: atomic-scale motion; reaction path coherence (from TS to products); reaction landscape trajectories as opposed to a single reaction path; energetic *vs.* entropic structures near transition states and conical intersections; coherent caging by a solvent; bifurcation and others.

(3) Transition Structures (Landscape Dynamics): This concept became clear to us after studies of the elementary dynamics in simple reactions of three atoms and in complex reactions of organic systems. Traditionally, one uses a reaction path and makes a distinction between a TS and a reactive intermediate by the absence or presence of a potential well – if there is no well, bonds are not formed and thus we do not speak of a "real" structure. On the fs time scale, we can isolate a continuous trajectory of transition structures; none are in a potential well. Such structures are defined by the change in bond order

Concept of Coherence

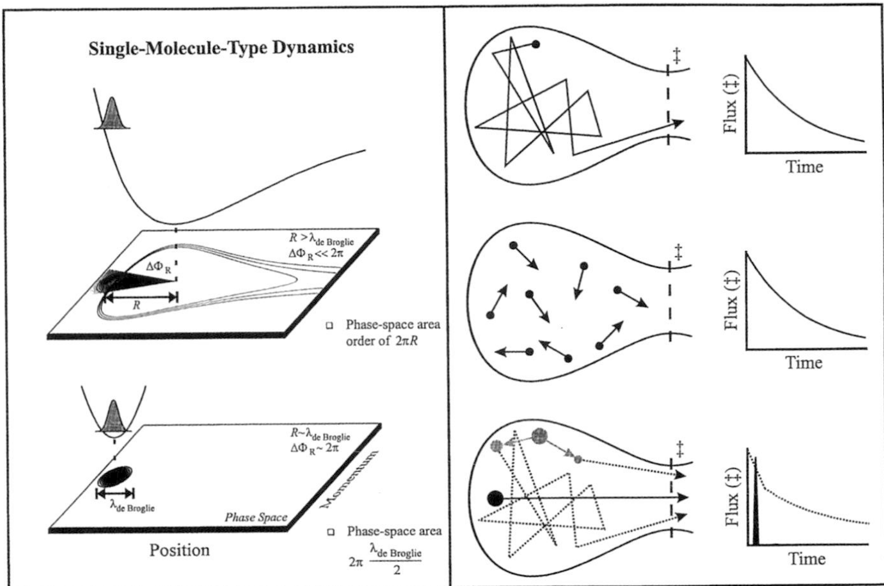

Figure 33. Concept of coherence, both in the dynamics at the atomic scale and in the control of non-statistical behavior. Shown is the phase space picture, describing the robustness of coherence (left); note the phase-space area of the initial state relative to that of the reaction. (Right) We present, for simplicity, a schematic of a configuration space made of the reactive coordinate and all nonreactive coordinates perpendicular to it (an equivalent phase-space picture can be made). Shown are three cases of interest: (top) the ergodic dynamics, (middle) the incoherent preparation and (bottom) the coherent wave packet preparation, showing the initial *localization*, spatially and temporally, and the *bifurcation* into direct and indirect reaction trajectories. Recent theoretical work (K. Møller, this laboratory) of the corresponding temporal behavior has elucidated the different regimes for the influence of the initial preparation, from a wave packet to a microcanonical limit.

and lead to a family of trajectories of reaction products. Thus, the two classical pathways of either a "concerted" or "sequential" process represent a very crude approximation for the actual landscape dynamics, even in a three-atom system. There is a whole distribution of reaction times and kinetic energy releases. The problem becomes even more severe if the landscape is complex and has near-flat energy regions or entropic changes, as discussed in **Section II** for the case of complex organic reactions and protein folding. The concept of transition structures and landscape dynamics is significant to issues addressing stereochemistry, product branching and selectivity, and to the real distinction between TS and intermediates in many reactions. Finally, the presence of such a family of trajectories on the energy landscape makes the restricted definition of TS – as only the saddle point – less clear. Concerted reactions, in the strict synchronous sense, essentially do not exist **[see Sections II & III]**.

(4) Reduced Space (Directed Dynamics): Another important concept in dynamics is the reduction of nuclear space to the sub-space critical to reactivity and

nonradiative behavior in complex systems. Here, the fs time scale allows one to project the *primary events* out of all processes possible. In complex systems with many degrees of freedom, the reduced space becomes the focus and the remaining space becomes a "continuum", thus moving the description from a multidimensional (3N - 6) PES to a few coordinates plus a weakly coupled continuum. This idea was central to our description of the dynamical isotope effect in elementary reactions, bifurcation to chemical and photophysical channels by conical intersections, concertedness and stereochemistry in organic reactions, non-statistical behavior, and reaction control in large systems at high energies. The consequences to photochemistry are significant: Reactions from high-energy states ($\pi\ \pi^*$, Rydberg, etc.) usually result in ground-state chemistry, and bifurcation into conical intersections is the key; for transitions involving σ^* orbitals, the time scale of rupture becomes comparable to that of the funneling through conical intersections, resulting in competitive chemical channels. It is possible that this same concept of reduced space is essential to biological dynamics. By reducing the space for dynamics, events occur efficiently and without "wasting" energy to all degrees of freedom possible. In addition, such designed local activity makes the system robust and immune to transferring "damage" over long distances. The ultrafast time scale is important because on it the system separates the important from the unimportant events – DNA bases quench their energy (nonradiatively) very rapidly, ET in DNA is locally ultrafast, and the first event of vision is very efficient and occurs in 200 fs.

V. EPILOGUE

As the ability to explore shorter and shorter time scales has progressed from the millisecond to the present stage of widely exploited femtosecond capabilities, each step along the way has provided surprising discoveries, new understanding, and new mysteries. In their editorial on the 10th anniversary of Femtochemistry, Will Castleman and Villy Sundström put this advance in a historical perspective **[46]**. The report in **Ref. 5** addresses with details the field and its position in over a century of developments. Developments will continue and new directions of research will be pursued. Surely, studies of transition states and their structures in chemistry and biology will remain active for exploration in new directions, from simple systems to complex enzymes and proteins **[47]**, and from probing to controlling of matter.

Since the current femtosecond lasers (4.5 fs) are now providing the limit of time resolution for phenomena involving nuclear motion, one may ask: Is there another domain in which the race against time can continue to be pushed? Sub-fs or attosecond (10^{-18} s) resolution may one day allow for the direct observation of the electron's motion. I made this point in a 1991 Faraday Discussion review **[B17]** and, since then, not much has been reported except for some progress in the generation of sub-fs pulses [**46,** Corkum, Harris]. In the coming decades, this may change and we may view electron rearrangement, say, in the benzene molecule, in real time. Additionally, there will be studies involving the combination of the "three scales" mentioned in the prologue, namely time, length and number. We should see extensions to studies of the femtosecond dynamics of *single molecules* and of *molecules on surfaces* (e.g. using STM). Combined time/length resolution will provide unique opportunities for making the important transition from molecular structures, to dynamics and to functions **(Section III6)**. We may also see that all of femtochemistry can be done at micro-to-nano Kelvin temperatures, utilizing lasers and other cooling techniques.

It seems that on the femtosecond to attosecond time scale we are reaching the "inverse" of the big bang time **[Fig. 34]**, with the human heartbeat "enjoying" the geometric average of the two limits. The language of molecular dynamics is even similar to that of cosmos dynamics. Cosmologists are speaking of energy landscapes and transition states for the big bang and universe inflation **[48]**. Perhaps we are approaching a universal limit of time!

Personally, I did not originally expect the rich blossoming in all of the directions outlined in this anthology; many more could, unfortunately, not be mentioned because of the limited space. What is clear to me is that my group and I have enjoyed the odyssey of discovery, seeing what was not previously possible, acquiring new knowledge and developing new concepts. Perhaps the best words to describe this feeling are those of the English archaeologist, Howard Carter, on November 25th 1922 when he got his first glimpse of the priceless contents of Tutankhamen's Tomb – *At first, I could see nothing, ... then shapes gradually began to emerge....* Lord Carnarvon, who financed the excava-

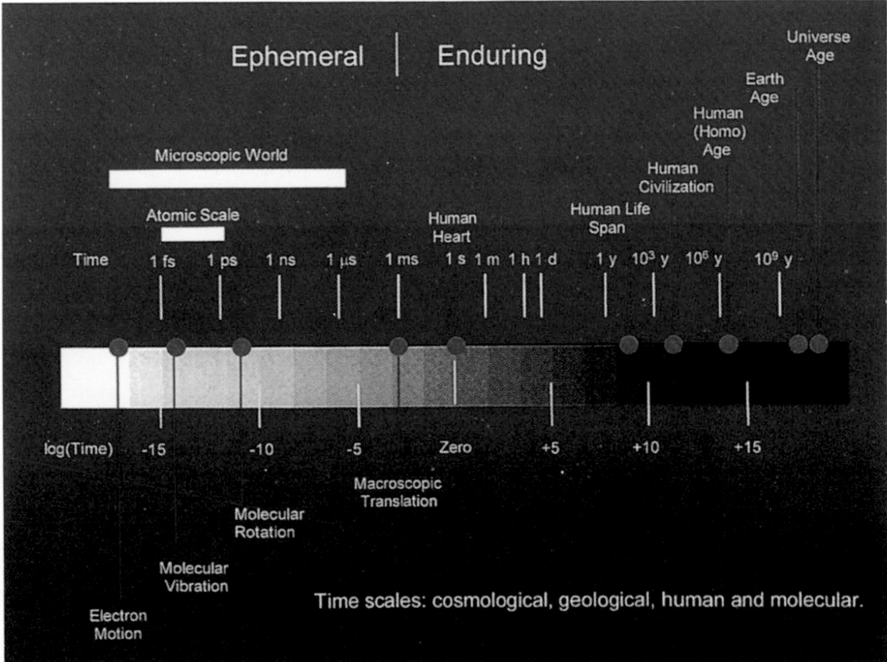

Figure 34. Time Scales of cosmological, geological, human and molecular events. Here, the time scale spans more than thirty orders of magnitude, from the big bang to the femto age. [Ref. B25]

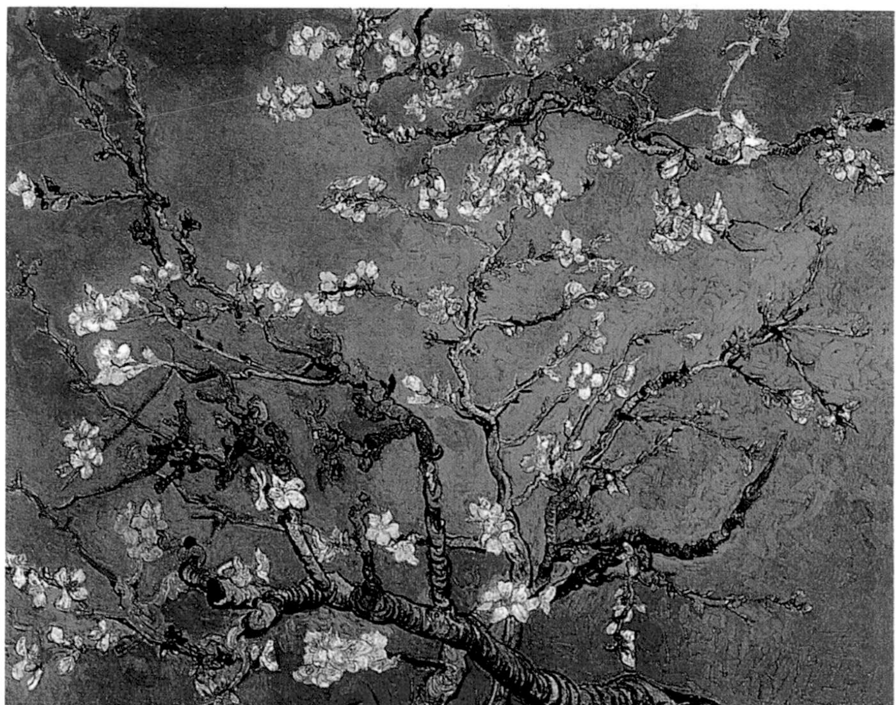

Figure 35. The Almond Blossom (st. Rémy, 1890) of Vincent van Gogh (1853–1890). With a good beginning, even if branching is unpredictable, the blossoms are rich and the big picture is beautiful. [L. A. County Museum of Art]

tion of this discovery, asked, when looking with Carter, *What do you see?* Carter replied, *"Beautiful Things", "Beautiful Things"*. This is the thrill of discovery in science, too. It seeks to unveil the hidden simplicity and beauty of Nature's truth. On a recent visit to the Los Angeles Art Museum with my wife, I stood in front of van Gogh's masterpiece – Almond Blossom **[Fig. 35]** – and wondered about the beauty of the big picture and the unpredictability of its details. That, also, is in the nature of scientific discoveries.

The future of femtoscience will surely witness many imaginative and unpredictable contributions. I hope that I will be able to enjoy the future as much as I have the past. Benjamin Franklin once wrote: *"The progress of human knowledge will be rapid and discoveries made of which we at present have no conception. I begin to be almost sorry I was born so soon since I cannot have the happiness of knowing what will be known in years hence."*

APPENDIX

A Primer for Femtoscopy, Coherence and Atoms in Motion

Pump-Probe Femtoscopy

In high-speed photography, a *continuous* motion is broken up into frames ("freezing") using a brief exposure time. For example, in Muybridge's experiment **(Section I)**, the shutter speed (exposure time) was ~2 milliseconds and the speed of the motion was ~10 m/s, resulting in a well-defined resolution (speed × exposure time) of 2 cm; the number of frames per second was about twenty since the cameras were 0.5 m apart. The huge contrast with molecular experiments is due to vast differences in speed (~1 km/s), resolution (~10^{-8} cm) and the number (millions) of molecules involved. Given the molecular speed and resolution, the ultrashort *strobes* must provide exposure time on the order of 100 femtoseconds, and in one second 10^{13} frames could be recorded! Ultrafast pulsed laser techniques have made direct exploration of this temporal realm a reality **(Sections II & III)**. Spectroscopy, mass spectrometry and diffraction play the role of ultra-high-speed photography in the investigation of molecular processes.

A femtosecond laser pulse provides the shutter speed for freezing nuclear motion with the necessary spatial resolution. The pulse probes the motion by stroboscopy, i. e. by pulsed illumination of the molecule in motion and recording or photographing the particular snapshot. A full sequence of the motion is achieved by using an accurately-timed series of these probe pulses, defining the number of frames per second. This method of probing, although different from Muybridge's, is in principle equivalent to his use of the cameras (with shutters) as probes. For molecules there exist three additional requirements in order to study the motion. First, we need to clock the motion by defining its zero of time, also accurate to tens of femtoseconds. Second, the motion must be synchronized since millions of molecules are typically used in the recording of molecular motion. Third, molecular coherence (see below) must be induced to localize the nuclei. These requirements are satisfied by using a femtosecond pump (*initiating*) laser pulse, in what is referred to as a pump-probe configuration.

With this methodology, the process to be studied is clocked from the instant that the substance under investigation absorbs radiation from the pump pulse. Passage of a probe pulse through the sample at some later point in time provides a snapshot of the status of the system at that time. For femtosecond studies, where femtosecond control of relative timing is needed, the laser pump and probe pulses are produced in synchrony, then the probe pulse is diverted through an adjustable optical path length **(Fig. 5)**. The finite speed of light translates the difference in path length into a difference in arrival time of the two pulses at the sample; 1μ corresponds to 3.3 fs. The individual snapshots combine to produce a complete record of the continuous time evolution – a motion picture, or a movie – in what may be termed femtoscopy.

Coherence and Atomic Motion

In a classical description, the motions of nuclei would be particle-like, i.e., they would behave as "marbles on a potential". At the scale of atomic masses and energies, however, the quantum mechanical wave/particle duality of matter comes into play, and the notions of position and velocity common to classical systems must be applied cautiously and in accord with the uncertainty principle, which places limits on the precision of simultaneous measurements. In fact, the state of any material system is defined in quantum mechanics by a spatially varying "wave function" with many similarities to light waves. Since the wave nature of light is a much more familiar concept than that of matter, we will use light to introduce the idea of wave superposition and interference, which plays an important role in atomic motion.

When light from two or more sources overlaps in space, the instantaneous field amplitudes (not intensities) from each source must be added together to produce the resultant light field. A well-known example is Young's two-slit experiment, in which light from a single source passes through two parallel slits in a screen to produce, in the space beyond, two phase-coherent fields of equal wavelength and amplitude. At points for which the distances to the two slits differ by $n + \frac{1}{2}$ wavelengths (for integer n) the two waves add to zero at all time, and no light is detected. Elsewhere, the amplitudes do not cancel. Thus, a stationary pattern of light and dark interference **(Fig. A,** next page, **inset)** fringes is produced ("light + light → darkness + more light!"). Knowledge of the wavelength of light and the spacing of fringes projected on a screen provide a measurement of the separation of the slits. In x-ray diffraction such interferences make it possible to obtain molecular structures with atomic resolution – the positions of the atoms replace the slits.

In studies of motion, we have exploited the concept of coherence among molecular wave functions to achieve atomic-scale resolution of dynamics – the change of molecular structures with time. Molecular wave functions are spatially diffuse and exhibit no motion. Superposition of a number of separate wave functions of appropriately chosen phases can produce a spatially localized and moving coherent superposition state, referred to as a wave packet **(Fig. A,** next page**)**; constructive and destructive interference (as in the interference of light waves) is the origin of such spatial localization. The packet has a well-defined (group) velocity and position which now makes it analogous to a moving classical marble, but at atomic resolution. The femtosecond light induces the coherence and makes it possible to reach atomic-scale spatial and temporal resolution, without violation of the uncertainty principle.

In the figure, a snapshot is shown of the position probability (in red) of a wave packet, formed from wave functions $n = 16$ to 20 with weighting according to the distribution curve at the left of the figure. This was calculated for the harmonic wave functions representing the vibrational states of a diatomic molecule, in this case iodine (n is the quantum number). As time advances, the wave packet moves back and forth across the potential well. As long as the wave packet (width ~0.04 Å) is sufficiently localized on the scale of all accessible space (~0.6 Å between the walls of the potential), as in the figure, a description in terms of the classical concepts of particle position and momentum is entirely appropriate. In this way, localization in time and in space are simultaneously achievable for reactive and nonreactive systems **(Section III4)**. Note that the width of the packet prepared by a 20 femtosecond pulse is very close to the uncertainty in bond distance for the initial $n = 0$ ground state. To prepare the packet at the inner turning point with the given phase composition, it must be launched vertically at the internuclear separation of ~2.4 Å; if

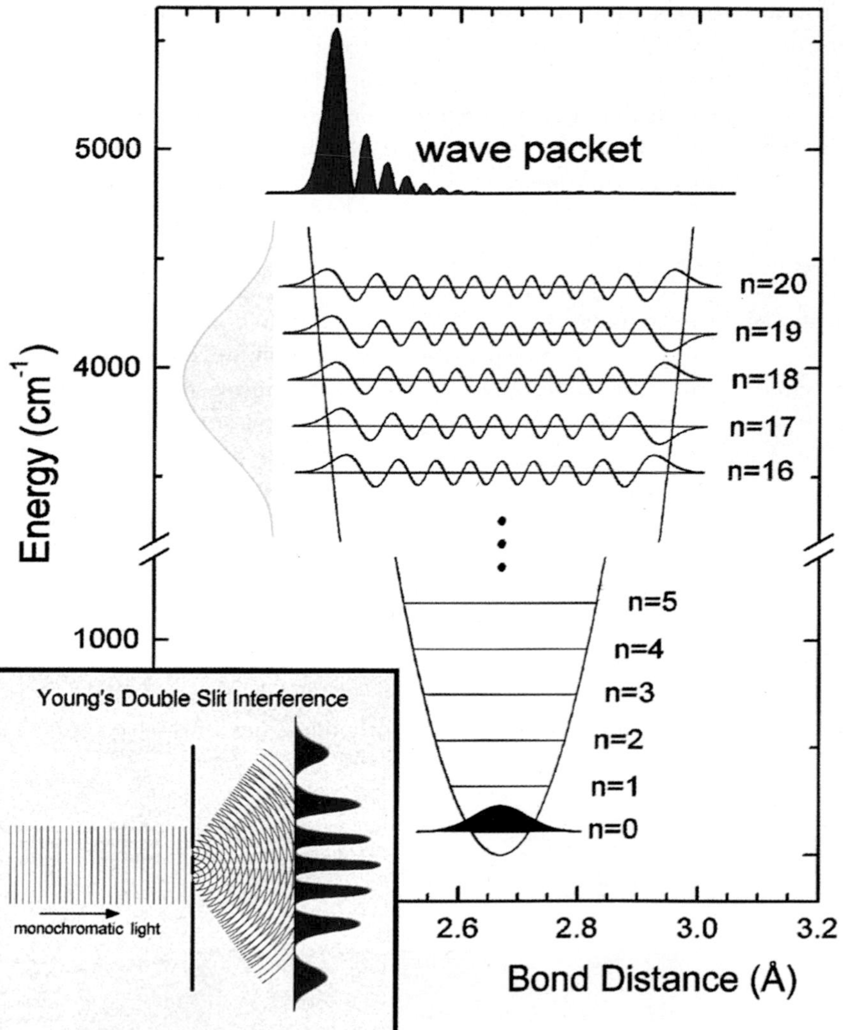

Figure A. The wave packet and wave function limits for molecular systems, in this case a diatomic molecule. The analogy with light interference is depicted – Thomas Young's experiment of 1801 [B25].

launched from the shown $n = 0$ distance, the packet will have a different phase composition and will be localized initially in the center of the well.

The observation of motion in a real system requires not only the formation of localized wave packets in each molecule, but also a small spread in position among wave packets formed in the typically millions of molecules on which the measurement is performed. The key to achieving this condition is generally provided by (a) the well-defined initial, equilibrium configuration of the studied molecules before excitation and (b) by the "instantaneous" femtosecond launching of the packet. The spatial confinement (in this case ~0.04

Å) of the initial ground state of the system ensures that all molecules, each with its own coherence among the states which form its wave packet, begin their motion on the excited potential in a bond-distance-range much smaller than that executed by the motion. The femtosecond launching ensures that this narrow range of bond distance is maintained during the entire process of preparation **(see Fig. 33)**. Unless the molecular and ensemble coherences are destroyed by intra- and/or inter-molecular perturbations, the motion is that of a classical *single-molecule trajectory*.

ACKNOWLEDGEMENTS

The story told here involves many dedicated students, post-doctoral fellows and research associates. Their contributions are recognized in the publications cited. I hope that by mentioning their work, they recognize the crucial role they have played in the journey of femtochemistry at Caltech. To me the exciting time with my research group represents the highlight of the story. For 20 years I have always looked forward to coming to work with them everyday and to enjoying the science in the truly international family **[Fig. 36]**. All members of the current research group have helped with the figures presented here. I particularly wish to thank Dongping Zhong for the special effort and care throughout the preparation and Ramesh Srinivasan for his devoted help with some of the figures. The artistic quality of many of the figures reflects the dedicated efforts of Wayne Waller and his staff.

There have been a number of friends and colleagues who have supported

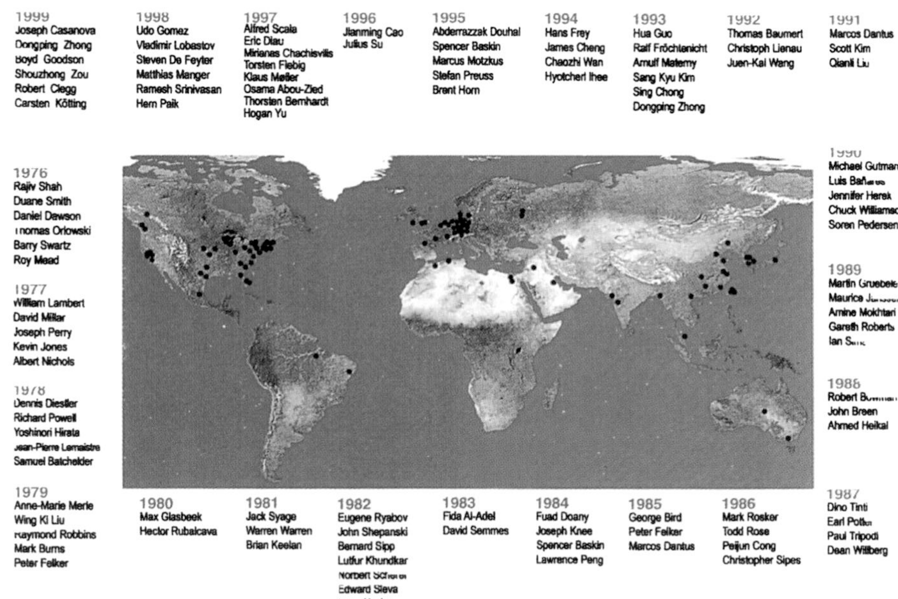

Figure 36. The International Roster of the Caltech group, from 1976–1999. The red circles are usually composite of many dots, reflecting the more than one hundred graduate students, post-doctoral fellows and visiting associates, who made up the research team over the years.

the field and made the experience enjoyable and humanly worthwhile – to them I wish the very best. All are with us today, except one: Dick Bernstein. Caltech proved to be the ideal scientific institution for me, not only because of the strong science it radiates, but also because of its culture, the "science village". I will never forget the impact of the "round table" at the Athenaeum. Vince McKoy, friend, colleague and neighbor at Caltech, has been the source of enjoyment and stimulation for over 20 years; Vince has taken an almost daily interest in our progress! Over the years I have enjoyed discussions with Spencer Baskin whose critical and wise judgment served as a sounding board for penetrating physics concepts.

Major support of this research came from the National Science Foundation, the US Air Force Office of Scientific Research, the Office of Naval Research, and it is a pleasure to acknowledge all of these organizations for making it possible. I take the opportunity to thank Carl and Shirley Larson for their support and friendship. Last, but not least, I wish to thank the staff of the Division and the Institute for their care and sincere efforts which have made our ship sail along with minimal resistance.

REFERENCES

1. J. H. Breasted, *A History of Egypt*, Scribner's Sons, New York (1909); Renewal (1937); A. Gardiner, *Egyptian Grammar*, Third Edition, Griffith Institute Ashmolean Museum, Oxford (1957), first published 1927 and last 1994; H. E. Winlock, *The Origin of the Ancient Egyptian Calendar, Proc. Am. Phys. Soc.* **83,** 447 (1940); O. Neugebauer, *The Exact Sciences in Antiquity*, Brown Univ. Press, Providence, RI (1957); R. K. Marshall, *Sundials*, McMillan, New York (1963); P. Tompkins, *The Magic of Obelisks*, Harper & Row, New York (1981); G. J. Whitrow, *Time in History*, Oxford Univ. Press, Oxford (1989); H. Selin (Ed.), *Encyclopedia of the History of Science, Technology, and Medicine in Non-Western Cultures*, Kluwer Academic Pub., The Netherlands (1997); *A Walk Through Time*, NIST publication (1998) http://physics.nist.gov/GenInt/Time/time.html; Mostafa M. Soliman, *History of Science and Technology in Ancient and Middle Ages* (in Arabic), Alhiaa Almasria Lelkitab, Cairo (1995); Duncan Steel, *Marking time – The Epic Quest to Invent the Perfect Calendar*, Wiley, New York (2000).
2. *Scientific American* , Volume XXXIX, Oct. 19 Issue, p. 241 (1878); T. K. Derry and T. I. Williams, *A Short History of Technology*, Dover, New York (1993); E. Jussim and G. Kayafas, *Stopping Time*, H. N. Abrams, Pub., New York (1987).
3. C. A. Ronan, *Science – Its History and Development Among the World's Cultures*, Facts on File, New York (1982); J. R. Partington, *A Short History of Chemistry*, Dover, New York (1989); J. W. Servos, *Physical Chemistry from Ostwald to Pauling*, Princeton (1990); C. van Doren, *A History of Knowledge*, Ballantine, New York (1991); W. H. Brock, *Chemistry*, Norton, New York (1992); C. H. Langford and R. A. Beebe, *The Development of Chemical Principles*, Dover (1995); R. Breslow, *Chemistry Today and Tomorrow*, American Chemical Society, Washington (1997); J. Read, *From Alchemy to Chemistry*, Dove, New York (1995); *Scientists – Random House Webster's Dictionary*, New York (1997);
4. A. H. Zewail, *Cambridge Rev.* **118**, No. 2330, 65 (1997); J. Baggott, *Chemistry in a New Light*, Cambridge University Press (2000), to be published.
5. B. Nordén, www.nobel.se/announcement-99/chemistry99.html; www.nobel.se/announcement-99/chemistry99.html#further reading; A. Donoso and C. C. Martens, *J. Phys. Chem.* A **102**, 4291 (1998) – see introduction of this reference; D. E. Folmer, W. S. Wisniewski, S. M. Hurley and A. W. Castleman, Jr., *Proc. Natl. Acad. Sci., USA* **96**, 12980 (1999), and references therein.
6. H. C. von Baeyer, *Taming the Atom*, Random House, New York (1992); P. Ball, *Designing the Molecular World*, Princeton University Press, Princeton (1994); P. W. Atkins, *Atoms, Electrons, and Change*, W. H. Freeman and Co., New York (1991).

7. S. Arrhenius, *Z. Phys. Chem.* (Leipzig) **4**, 226 (1889); J. H. van't Hoff, in *Etudes de Dynamiques Chimiques*, F. Muller and Co., Amsterdam, p. 114 (1884) (translation by T. Ewan, London, 1896); M. Bodenstein, *Z. Phys. Chem. (Munich)* **13**, 56 (1894); **22**, 1 (1897); **29**, 295 (1899); F. A. Lindemann, *Trans. Faraday Soc.* **17**, 598 (1922); C. N. Hinshelwood, *The Kinetics of Chemical Change in Gaseous Systems*, Calendron, Oxford (1926) (second printing, 1929; third printing, 1933); *Proc. R. Soc. London* **A113**, 230 (1926); R. C. Tolman, *J. Am. Chem. Soc.* **47**, 2652 (1925); **ibid. 42**, 2506 (1920); *Biographical memoirs*, National Academy of Sciences (USA), by J. G. Kirkwood, O. R. Wulf, and P. S. Epstein, Volume XXVII, p 139.
8. See, e.g., K. J. Laidler, *Chemical Kinetics*, 3rd ed., Harper Collins, New York (1987); J. I. Steinfeld, J. S. Francisco, and W. L. Hase, *Chemical Kinetics and Dynamics*, Prentice-Hall, New Jersey (1989); T. Baer and W. L. Hase, *Unimolecular Reaction Dynamics*, Oxford University Press, New York (1996); S. Shaik, H. Schlegel and S. Wolfe, *Theoretical Aspects of Phys. Org. Chem.*, Wiley, New York (1992); A. Pross, *Theoretical & Physical Principles of Organic Reactivity*, Wiley, New York (1995).
9. R. D. Levine and R. B. Bernstein, *Molecular Reaction Dynamics and Chemical Reactivity*, Oxford University Press, Oxford (1987) and references therein.
10. W. Heitler and F. London, *Z. Phys.* **44**, 455 (1927); F. London, *Probleme der Modernen Physik*, Sommerfeld Festschrift, p. 104 (1928); H. Eyring and M. Polanyi, *Z. Phys. Chem. B* **12**, 279 (1931); M. Polanyi, *Atomic Reactions*, Williams and Norgate, London (1932).
11. M. Polanyi and E. Wigner, *Z. Phys. Chem., Abt. A* **139**, 439 (1928); H. Eyring, *J. Chem. Phys.* **3**, 107 (1935); see also *J. Chem. Phys.* **3**, 492 (1935); M. G. Evans and M. Polanyi, *Trans. Faraday Soc.* **31**, 875 (1935); **33**, 448 (1937); H. A. Kramers, *Physica (Utrecht)* **7**, 284 (1940); R. F. Grote and J. T. Hynes, *J. Chem. Phys.* **73**, 2715 (1980) and the review in Ref. 28; R. A. Marcus, *J. Chem. Phys.* **24**, 966 (1956); **ibid.** p 974.
12. J. C. Polanyi and A. H. Zewail, *Acc. Chem. Res.* **28**, 119 (1995) and references therein.
13. I. H. Williams, *Chem. Soc. Rev.* **22**, 277 (1993); A. Williams, *Chem Soc. Rev.*, **23**, 93 (1994); W. v. E. Doering and W. Wang, *J. Am. Chem. Soc.* **121**, 10112 (1999); J. E. Baldwin, *J. Comp. Chem.* **19**, 222 (1998); P. G. Schultz and R. A. Lerner, *Science* **269**, 1835 (1995); A. R. Fersht, *Current Opinion in Structural Biology* **5**, 79 (1994).
14. J. C. Polanyi, *Science* **236**, 680 (1987) and references therein; J. C. Polanyi, *Faraday Discuss. Chem. Soc.* **67**, 129 (1979).
15. B. R. Johnson and J. L. Kinsey in: *Femtosecond Chemistry*, J. Manz and L. Wöste (Eds.), VCH Verlagsgesellschaft, Weinheim (1994).
16. A. H. Zewail, *Femtochemistry – Ultrafast Dynamics of the Chemical Bond*, World Scientific, Singapore (1994), Vols. I and II. Most of the Caltech publications in this field, to be referenced here (1976–1994), are collected in these two volumes. References will be given for the specific topics and will be updated in text and bibliography.
17. M. Eigen, *Discuss. Faraday Soc.* **17**, 194 (1954), in: *Techniques of Organic Chemistry*, Interscience, London (1963), Vol. VIII, Part II; M. Eigen, Immeasurable Fast Reactions, in: *Nobel Lectures (Chemistry)*, Elsevier, Amsterdam, p. 170 (1972), and references therein; R. G. W. Norrish and G. Porter, *Nature* **164**, 658 (1949); G. Porter in: *The Chemical Bond: Structure and Dynamics*, A. H. Zewail (Ed.), Academic Press, Boston, p. 113 (1992); and references therein.
18. H. Abraham and T. Lemoine, *Compt. Rend. (Paris)* **129**, 206 (1899). For reviews see: N. Bloembergen, *Rev. Mod. Phys.* **71**, S 283 (1999) and references therein; E. N. Glezer, in: *Spectroscopy and Dynamics of Collective Excitations in Solids*, Plenum Press, New York, p. 375 (1997); S. L. Shapiro (Ed.), *Ultrashort Light Pulses*, Springer, Berlin (1977); W. Kaiser (Ed.), *Ultrashort Laser Pulses & Applications*, Springer-Verlag, Berlin (1988); P. F. Barbara, W. H. Knox, G. A. Mourou, and A. H. Zewail (Eds.), *Ultrafast Phenomena IX*, Springer, New York (1994); T. Elsaesser, J. G. Fujimoto, D. A. Wiersma and W. Zinth (Eds.), *Ultrafast Phenomena XI*, Springer, New York (1998); and references therein; see also Ref. 2.
19. C. V. Shank and E. P. Ippen, *App. Phys. Lett.* **24**, 373 (1974); R. L. Fork, B. I. Greene, and C. V. Shank, *App. Phys. Lett.* **38**, 671 (1981); J. A. Valdmanis, R. L. Fork, and J. P. Gordon, *Opt. Lett.* **10**, 131 (1985); R. L. Fork, C. H. Brito Cruz, P. C. Becker, and C. V. Shank, *Opt. Lett.* **12**, 483 (1987); see also the reviews in Ref. 18; D. E. Spence, P. N. Kean, and W. Sibbett, *Opt. Lett.* **16**, 42 (1991); J. P. Zhou, G. Taft, C. P. Huang, M. M. Murnane, and H. C. Kaptyn, *Opt. Lett.* **19,** 1149 (1994); A. Baltuska, Z. Wei, M. S. Pshenichnikov, and D. Wiersma, *Opt. Letts.* **22**, 102 (1997); R. R. Alfano and S. L. Shapiro, *Phys. Rev. Lett.* **24**, 584 (1970).

20. P. M. Rentzepis, *Chem. Phys. Lett.* **2**, 117 (1968); T. L. Netzel, P. M. Rentzepis and J. S. Leigh, *Science* **182**, 238 (1973); K. J. Kaufmann and P. M. Rentzepis, *Acc. Chem. Res.* **8**, 407 (1975).
21. T. J. Chuang, G. W. Hoffmann, and K. B. Eisenthal, *Chem. Phys. Lett.* **25**, 201 (1974); K. B. Eisenthal and K. H. Drexhage, *J. Chem. Phys.* **51**, 5720 (1969); K. B. Eisenthal, *Acc. Chem. Res.* **8**, 118 (1975).
22. R. M. Hochstrasser, H. Lutz, and G. W. Scott, *Chem. Phys. Lett.* **24**, 162 (1974); R. Anderson, R. M. Hochstrasser, H. Lutz and G. W. Scott, *Chem. Phys. Lett.* **28**, 153 (1974); R. M. Hochstrasser in: *Advances in Laser Chemistry*, Springer Series in Chemical Physics, No. 3, p. 98 (1978).
23. D. von der Linde, A. Laubereau, and W. Kaiser, *Phys. Rev. Lett.* **26**, 954 (1971); A. Laubereau, D. von der Linde, and W. Kaiser, *Phys. Rev. Lett.* **28**, 1162 (1972); A. Laubereau and W. Kaiser, *Rev. Mod. Phys.* **50**, 607 (1978).
24. J. Bigeleisen, *Chemistry in a Jiffy*, *Chem. & Eng. News* **55**, 26 (1977). In a written correspondence to me, Jacob emphasized that he, in contrast with the 1960s view (Ref. 40), understood what could be learned from time scales below a nanosecond, but that he failed to fully appreciate how the two uncertainties would serve to localize the length scale.
25. D. R. Herschbach, *Angew. Chem., Int. Ed. Engl.* **26**, 1221 (1987) and references therein; Y. T. Lee, *Science* **236**, 793 (1987) and references therein; R. N. Zare and R. B. Bernstein, *Phys. Today* **33**, 11 (1980).
26. A. H. Zewail and R. B. Bernstein, *Chem. & Eng. News* (feature article), Nov. 7, pp. 24–43 (1988); also in *The Chemical Bond: Structure and Dynamics*, A. H. Zewail (Ed.), Academic Press, Boston, p. 223 (1992), (with an update section).
27. D. C. Clary, *Science* **279,** 1879 (1998); and D. C. Clary in Ref. 28, and references therein.
28. V. Sundström (Ed.) *Nobel Symposium Book: Femtochemistry and Femtobiology: Ultrafast Reaction Dynamics at Atomic-Scale Resolution*, World Scientific, Imperial College Press, London (1997).
29. R. P. Feynman, F. L. Vernon and R. W. Hellwarth, *J. Appl. Phys.* **28**, 49 (1957).
30. For recent reviews, see: J. Jortner and M. Bixon in Ref. 28; J. Jortner, *Phil. Trans. R. Soc. Lond. A* **356**, 477 (1998); and references therein for earlier work.
31. E. J. Heller, *Acc. Chem. Res.* **14**, 368 (1981); D. Imre, J. L. Kinsey, A. Sinha and J. Krenos, *J. Phys. Chem.* **88**, 3956 (1984); B. R. Johnson, C. Kittrell, P. B. Kelly and J. L. Kinsey, *J. Phys. Chem.* **100**, 7743 (1996); R. Kosloff, *J. Phys. Chem.* **92**, 2087 (1988) and references therein.
32. B. Roos, *Acc. Chem. Res.* **32**, 137 (1999), and references therein.
33. V. Letokhov in Ref. 28; his lecture in *Femtochemistry IV*, Leuven, Belgium (1999); and his articles in [B30 and B31].
34. W. S. Warren, *Science* **242**, 878 (1988); W. S. Warren, *Encyclopedia Magnetic Resonance*, Wiley, New York (1996).
35. D. J. Tannor and S. A. Rice, *J. Chem. Phys.* **83**, 5013 (1985); D. J. Tannor, R. Kosloff and S. A. Rice, *J. Chem.* Phys. **85**, 5805 (1986); D. J. Tannor and S. A. Rice, *Adv. in Chem.* Phys. **LXX**, 441 (1988).
36. H.-L. Dai and W. Ho, *Laser Spectroscopy and Photochemistry on Metal Surfaces, Part I & II*, World Scientific, Singapore (1995); see the articles by W. Ho and others; M. Bonn, S. Funk, Ch. Hess, D. N. Denzler, C. Stampfl, M. Scheffler, M. Wolf, and G. Ertl, *Science* **285**, 1042 (1999); R. J. Finlay, T.-H. Her, C. Wu, and E. Mazur, *Chem. Phys. Lett.* **274**, 499 (1997).
37. B. Kohler, J. L. Krause, F. Raksi, K. R. Wilson, V. V. Yakovlev, R. M. Whitnell, and Y. J. Yan, *Acc. Chem. Res.* **28**, 133 (1995); A. Assion, T. Baumert, M. Bergt, T. Brixner, B. Kiefer, V. Seyfried, M. Strehle, and G. Gerber, *Science* **282**, 919 (1998).
38. P. B. Corkum, M. Y. Ivanov and J. S. Wright, *Ann. Rev. Phys. Chem.* **48**, 387 (1997); T. Frohnmeyer, M. Hofmann, M. Strehle, and T. Baumert, *Chem. Phys. Lett.* **312**, 447 (1999); A. D. Bandrauk (Ed.), *Molecules in Laser Fields,* Marcel Dekker, New York (1994), and references therein.
39. R. M. Hochstrasser, *J. Chem. Ed.* **75**, No. 5, 559 (1998); V. Sundström, T. Pullerits, and R. van Grondelle, *J. Phys. Chem. B* **103**, 2327 (1999); S. Hahn and G. Stock, *J. Phys. Chem. B* **104**, 1146 (2000); J.-L. Martin and M. H. Vos, *Annu. Rev. Biophys. Biomol. Struct.* **21**, 199 (1992); L. Zhu, J. T. Sage and P. M. Champion, *Science* **266**, 629 (1994); G. R. Fleming and R. van Grondelle, *Phys. Today*, **February Issue**, 48 (1994); R. A. Mathies,

S. W. Lin, J. B. Ames, and W. T. Pollard, *Annu. Rev. Biophys. Biophys. Chem.* **20**, 491 (1991); R. J. D. Miller, *Annu. Rev. Phys. Chem.* **42**, 581 (1991); M. H. Vos and J.-L. Martin, *Biochim. et Biophys. Acta* **1411**, 1 (1999); U. Liebl, G. Lipowski, M. Négrerie, J.-C. Lambry, L.-L. Martin, and M. H. Vos, *Nature* **401**, 181 (1999); J. Jortner and M. Bixon (Eds.), *Electron Transfer – From Isolated Molecules to Biomolecules*, Wiley, New York (1999); see also references 28, 79 and 80.

40. S. Claesson (Ed.), *Nobel Symposium 5, Fast Reactions and Primary Processes in Chemical Kinetics,* Almqvist & Wiksell, Stockholm, p. 474 (1967); G. Porter in: *Femtosecond Chemistry,* J. Manz and L. Wöste (Eds.), VCH, Weinheim (Germany) **Vol. 1,** p. 3 (1995).
41. Z. Li, J-Y Fang and C. C. Martens, *J. Chem. Phys.* **104**, 6919 (1996).
42. A. H. Zewail, The 1997 Welch Prize Address, Femtochemistry, in: *Proceedings of the Robert A. Welch Foundation,* 41st Conference on Chemical Research, Houston, Texas (1997), p. **323**.
43. G. M. Gale, G. Gallot, F. Hache, N. Lascoux, S. Bratos, and J. C. Leicknam, *Phys. Rev. Lett.* **82**, 1068 (1999); C. Chudoba, E. T. J. Nibbering, and T. Elsaesser, *J. Phys. Chem.* A **103**, 5625 (1999); For references of the development of methods, see: J. N. Moore, P. A. Hansen, and R. M. Hochstrasser, *Chem. Phys. Lett.* **138**, 110 (1987); P. O. Stoutland, R. B. Dyer, and W. H. Woodruff, *Science* **257**, 1913 (1992); W. Kaiser and A. Laubereau in Ref. 28; and references therein for earlier work.
44. A. W. Castleman, Jr. and K. H. Bowen, Jr., *J. Phys. Chem. (Centennial* Issue) **100**, 12911 (1996); S. Wolf, G. Sommerer, S. Rutz, E. Schreiber, T. Leisner, L. Wöste, and R. S. Berry, *Phys. Rev. Lett.* **74**, 4177 (1995); P. Farmanara, W. Radloff, V. Stert, H. H. Ritze, and I. V. Hertel, *J. Chem. Phys.* **111**, 633 (1999); W. C. Lineberger, article in the book, Ref. 28; M. T. Zanni, B. J. Greenblatt, A. V. Davis, and D. M. Neumark, *J. Chem.* Phys. **111**, 2991 (1999); and references therein.
45. C. Cohen-Tannoudji, B. Diu, and F. Laloë, *Quantum Mechanics, Vol. I &*II, Wiley, New York (1977); J. L. Eisberg, *Fundamentals of Modern Physics*, Wiley, New York (1964); D. D. Fitts, *Principles of Quantum Mechanics*, Cambridge University Press (1999); L. Mühlbacher, A. Lucke and R. Egger, *J. Chem. Phys.* **110**, 5851 (1999); J. Manz, in Ref. 28 p 80; Michael A. Morrison, *Quantum Physics*, Prentice Hall, New Jersey (1990).
46. A. W. Castleman, Jr. and V. Sundström in: *J. Phys. Chem. Special Issue, "Ten Years of Femtochemistry",* A **102**, June 4, 4021 (1998); P. Corkum, *Nature* (London) **403**, 845 (2000); S. E. Harris and A. V. Sokolov, *Phys. Rev. Lett.* **81,** 2894 (1998).
47. A. R. Fersht, *Structure and Mechanism in Protein Science,* W. H. Freeman, New York (1999); W. A. Eaton, Proc. Natl. Acad. Sci., USA **96**, 5897 (1999); C. L. Brooks III, M. Gruebele, J. N. Onuchic and P. G. Wolynes, ibid, **95,** 11037 (1998); D. T. Leeson, F. Gai, H. M. Rodriguez, L. M. Gregoret, and R. B. Dyer, ibid, **97,** 2527 (2000).
48. J. Glanz, *Science* **284**, 1448 (1999) – which way to the Big Bang?; Stuart Clark, *Towards the Edge of the Universe,* 2nd Edition, Springer, New York (1999).
49. S. Pedersen, J. L. Herek, and A. H. Zewail, *Science* **266**, 1359 (1994).
50. A. H. Zewail, T. E. Orlowski, K. E. Jones, and D. E. Godar, *Chem. Phys. Lett.* **48**, 256 (1977); W. S. Warren and A. H. Zewail, *J. Chem. Phys.* **75**, 5956 (1981); W. S. Warren and A. H. Zewail, *J. Chem. Phys.* **78**, 2279 (1983); E. T. Sleva and A. H. Zewail, *Chem. Phys. Lett.* **110**, 582 (1984); E. T. Sleva, I. M. Xavier, Jr., and A. H. Zewail, *J. Opt. Soc. Amer.* **3**, 483 (1986).
51. W. R. Lambert, P. M. Felker, and A. H. Zewail, *J. Chem. Phys.* **75**, 5958 (1981); W. R. Lambert, P. M. Felker, and A. H. Zewail, *J. Chem. Phys.* **81**, 2209 & 2217 (1984); P. M. Felker and A. H. Zewail, *Phys. Rev. Lett.* **53**, 501 (1984); P. M. Felker and A. H. Zewail, *J. Chem. Phys.* **82**, 2961, **ibid.** 2975, **ibid.** 2994, (1985); P. M. Felker and A. H. Zewail, *Advances in Chemical Physics* **70**, 265 (1988); P. M. Felker and A. H. Zewail, in: *Jet Spectroscopy and Molecular Dynamics,* M. Hollas and D. Phillips (Eds.), Chapman and Hall, Blackie Academic, p. 222 (1995).
52. A. H. Zewail, *Faraday Discuss. Chem. Soc.* **75**, 315 (1983); J. A. Syage, W. R. Lambert, P. M. Felker, A. H. Zewail, and R. M. Hochstrasser, *Chem. Phys. Lett.* **88**, 266 (1982); J. A. Syage, P. M. Felker, and A. H. Zewail, *J. Chem. Phys.* **81**, 4706 (1984); P. M. Felker, J. A. Syage, W. R. Lambert, and A. H. Zewail, *Chem. Phys. Lett.* **92**, 1 (1982); J. A. Syage, P. M. Felker, and A. H. Zewail, *J. Chem. Phys.* **81**, 2233 (1984); P. M. Felker and A. H. Zewail, *Chem. Phys. Lett.* **94**, 454 (1983); P. M. Felker and A. H. Zewail, *J. Chem. Phys.* **78**, 5266 (1983).
53. J. S. Baskin, P. M. Felker, and A. H. Zewail, *J. Chem. Phys.* **84**, 4708 (1986); P. M. Felker,

J. S. Baskin, and A. H. Zewail, *J. Phys. Chem.* **90**, 724 (1986); P. M. Felker and A. H. Zewail, *J. Chem. Phys.* **86**, 2460 (1987); J. S. Baskin, P. M. Felker, and A. H. Zewail, *J. Chem. Phys.* **86**, 2483 (1987); N. F. Scherer, L. R. Khundkar, T. S. Rose, and A. H. Zewail, *J. Phys. Chem.* **91**, 6478 (1987); J. S. Baskin and A. H. Zewail, *J. Phys. Chem.* **93**, 5701 (1989).

54. L. R. Khundkar and A. H. Zewail, *Ann. Rev. Phys. Chem.* **41**, 15 (1990); N. F. Scherer, F. E. Doany, A. H. Zewail, and J. W. Perry, *J. Chem. Phys.* **84**, 1932 (1986); N. F. Scherer and A. H. Zewail, *J. Chem. Phys.* **87**, 97 (1987); L. R. Khundkar, J. L. Knee, and A. H. Zewail, *J. Chem. Phys.* **87**, 77 (1987); S. J. Klippenstein, L. R. Khundkar, A. H. Zewail, and R. A. Marcus, *J. Chem. Phys.* **89**, 4761 (1988).
55. N. F. Scherer, J. L. Knee, D. D. Smith, and A. H. Zewail, *J. Phys. Chem.* **89**, 5141 (1985); M. Dantus, M. J. Rosker, and A. H. Zewail, *J. Chem. Phys.* **87**, 2395 (1987); M. J. Rosker, M. Dantus, and A. H. Zewail, *Science* **241**, 1200 (1988); M. J. Rosker, M. Dantus, and A. H. Zewail, *J. Chem. Phys.* **89**, 6113 (1988); M. Dantus, M. J. Rosker, and A. H. Zewail, *J. Chem. Phys.* **89**, 6128 (1988); D. Zhong and A. H. Zewail, *J. Phys. Chem. A* **102**, 4031 (1998); R. Bersohn and A. H. Zewail, *Ber. Bunsenges. Phys. Chem.* **92**, 373 (1988); G. Roberts and A. H. Zewail, *J. Phys. Chem.* **95**, 7973 (1991); G. Roberts and A. H. Zewail, *J. Phys. Chem.* **99**, 2520 (1995); the quantum calculations in the last reference follows the original results by D. Imre (see text).
56. T. S. Rose, M. J. Rosker, and A. H. Zewail, *J. Chem. Phys.* **88**, 6672 (1988); T. S. Rose, M. J. Rosker, and A. H. Zewail, *J. Chem. Phys.* **91**, 7415 (1989); P. Cong, A. Mokhtari, and A. H. Zewail, *Chem. Phys. Lett.* **172**, 109 (1990); A. Mokhtari, P. Cong, J. L. Herek, and A. H. Zewail, *Nature* **348**, 225 (1990); P. Cong, G. Roberts, J. L. Herek, A. Mokhtari, and A. H. Zewail, *J. Phys. Chem.* **100**, 7832 (1996).
57. R. M. Bowman, M. Dantus, and A. H. Zewail, *Chem. Phys. Lett.* **156**, 131 (1989); M. Dantus, R. M. Bowman, M. Gruebele, and A. H. Zewail, *J. Chem. Phys.* **91**, 7437 (1989); M. Gruebele, G. Roberts, and A. H. Zewail, *Philos. Trans. Roy. Soc. London A* **332**, 35 (1990); D. Zhong and A. H. Zewail, *J. Phys. Chem. A* **102**, 4031 (1998); K. B. Møller and A. H. Zewail, *Chem. Phys. Lett.* **295**, 1 (1998).
58. M. Dantus, R. M. Bowman, and A. H. Zewail, *Nature* **343**, 737 (1990); M. Gruebele, G. Roberts, M. Dantus, R. M. Bowman, and A. H. Zewail, *Chem. Phys. Lett.* **166**, 459 (1990); R. B. Bernstein and A. H. Zewail, *Chem. Phys. Lett.* **170**, 321 (1990); M. H. M. Janssen, R. M. Bowman, and A. H. Zewail, *Chem. Phys. Lett.* **172**, 99 (1990); M. Gruebele and A. H. Zewail, *J. Chem. Phys.* **98**, 883 (1993).
59. N. F. Scherer, L. R. Khundkar, R. B. Bernstein, and A. H. Zewail, *J. Chem. Phys.* **87**, 1451 (1987); N. F. Scherer, C. Sipes, R. B. Bernstein, and A. H. Zewail, *J. Chem. Phys.* **92**, 5239 (1990); M. Gruebele, I. R. Sims, E. D. Potter, and A. H. Zewail, *J. Chem. Phys.* **95**, 7763 (1991); I. R. Sims, M. Gruebele, E. D. Potter, and A. H. Zewail, *J. Chem. Phys.* **97**, 4127 (1992); D. Zhong, P. Y. Cheng, and A. H. Zewail, *J. Chem. Phys.* **105**, 7864 (1996); An up-to-date experimental and theoretical work is reviewed in Ref. B5 (some are adapted here) and the review by D. Clary (see Ref. 27).
60. J. L. Knee, L. R. Khundkar, and A. H. Zewail, *J. Chem. Phys.* **82**, 4715 (1985); M. Dantus, M. H. M. Janssen, and A. H. Zewail, *Chem. Phys. Lett.* **181**, 281 (1991); T. Baumert, S. Pedersen, and A. H. Zewail, *J. Phys. Chem.* **97**, 12447 (1993); S. Pedersen, J. L. Herek, and A. H. Zewail, *Science* **266**, 1359 (1994); S. K. Kim, S. Pedersen, and A. H. Zewail, *J. Chem. Phys.* **103**, 477 (1995); S. K. Kim and A. H. Zewail, *Chem. Phys. Lett.* **250**, 279 (1996); S. K. Kim, J. Guo, J. S. Baskin, and A. H. Zewail, *J. Phys. Chem.* **100**, 9202 (1996).
61. E. W.-G. Diau, O. Abou-Zied, A. A. Scala, and A. H. Zewail, *J. Am. Chem. Soc.* **120**, 3245 (1998).
62. L. R. Khundkar and A. H. Zewail, *J. Chem. Phys.* **92**, 231 (1990); D. Zhong, S. Ahmad, and A. H. Zewail, *J. Am. Chem. Soc.* **119**, 5978 (1997); D. Zhong and A. H. Zewail, *J. Phys. Chem. A* **102**, 4031 (1998).
63. P. Y. Cheng, D. Zhong, and A. H. Zewail, *Chem. Phys. Lett.* **237**, 399 (1995).
64. S. K. Kim, S. Pedersen, and A. H. Zewail, *Chem. Phys. Lett.* **233**, 500 (1995).
65. A. A. Scala, E. W.-G. Diau, Z. H. Kim, and A. H. Zewail, *J. Chem. Phys.* **108**, 7933 (1998).
66. S. De Feyter, E. W.-G. Diau, and A. H. Zewail, *Angew. Chem. Int. Ed. Engl.* **112**, 266 (2000).
67. P. M. Felker, W. R. Lambert, and A. H. Zewail, *J. Chem. Phys.* **77**, 1603 (1982); J. L. Herek, S. Pedersen, L. Bañares, and A. H. Zewail, *J. Chem. Phys.* **97**, 9046 (1992).
68. B. A. Horn, J. L. Herek, and A. H. Zewail, *J. Am. Chem. Soc.* **118**, 8755 (1996); E. W.-G.

Diau, S. De Feyter, and A. H. Zewail, *Chem. Phys. Lett.* **304**, 134 (1999); S. De Feyter, E. W.-G. Diau, and A. H. Zewail, *Phys. Chem. Chem. Phys.* **2**, 877 (2000).

69. D. Zhong, E. W.-G. Diau, T. M. Bernhardt, S. De Feyter, J. D. Roberts, and A. H. Zewail, *Chem. Phys. Lett.* **298**, 129 (1998); M. Chachisvilis and A. H. Zewail, *J. Phys. Chem. A* **103**, 7408 (1999).
70. P. Y. Cheng, D. Zhong, and A. H. Zewail, *J. Chem. Phys.* **105**, 6216 (1996); D. Zhong and A. H. Zewail, Proc. Natl. Acad. Sci. **96**, 2602 (1999); D. Zhong, T. M. Bernhardt, and A. H. Zewail, *J. Phys. Chem. A* **103**, 10093 (2000).
71. S. Pedersen, J. L. Herek, and A. H. Zewail, *Science* **266**, 1359 (1994); S. De Feyter, E. W.-G. Diau, A. A. Scala, and A. H. Zewail, *Chem. Phys. Lett.* **303**, 249 (1999); S. De Feyter, E. W.-G. Diau, and A. H. Zewail, *Angew. Chem. Int. Ed. Engl.,* **39**, 260 (2000); E. W.-G. Diau, J. Casanova, J. D. Roberts, and A. H. Zewail, *Proc. Natl. Acad. Sci., USA* **97**, 1376 (2000).
72. E. W.-G. Diau and A. H. Zewail, to be published.
73. D. Zhong, E. W.-G. Diau, T. M. Bernhardt, S. De Feyter, J. D. Roberts, and A. H. Zewail, *Chem. Phys. Lett.* **298**, 129 (1998); M. Chachisvilis and A. H. Zewail, *J. Phys. Chem. A* **103**, 7408 (1999); E. W.-G. Diau, S. De Feyter, and A. H. Zewail, *J. Chem. Phys.* **110**, 9785 (1999).
74. Q. Liu, J.-K. Wang, and A. H. Zewail, *Nature* **364**, 427 (1993); J.-K. Wang, Q. Liu, and A. H. Zewail, *J. Phys. Chem.* **99**, 11309 (1995); Q. Liu, J.-K. Wang, and A. H. Zewail, *J. Phys. Chem.* **99**, 11321 (1995); J. J. Breen, D. M. Willberg, M. Gutmann, and A. H. Zewail, *J. Chem. Phys.* **93**, 9180 (1990); M. Gutmann, D. M. Willberg, and A. H. Zewail, *J. Chem. Phys.* **97**, 8037 (1992); J. T. Su and A. H. Zewail, *J. Phys. Chem.* **102**, 4082 (1998); J. J. Breen, L. W. Peng, D. M. Willberg, A. Heikal, P. Cong, and A. H. Zewail, *J. Chem. Phys.* **92**, 805 (1990); S. K. Kim, J.-K. Wang, and A. H. Zewail, *Chem. Phys. Lett.* **228**, 369 (1994); S. K. Kim, J. J. Breen, D. M. Willberg, L. W. Peng, A. Heikal, J. A. Syage, and A. H. Zewail, *J. Phys. Chem.* **99**, 7421 (1995); A. A. Heikal, S. H. Chong, J. S. Baskin, and A. H. Zewail, *Chem. Phys. Lett.* **242**, 380 (1995).
75. A. H. Zewail, M. Dantus, R. M. Bowman, and A. Mokhtari, *J. Photochem. Photobiol. A: Chem.* **62/3**, 301 (1992); C. Lienau, J. C. Williamson, and A. H. Zewail, *Chem. Phys. Lett.* **213**, 289 (1993); C. Lienau and A. H. Zewail, *Chem. Phys. Lett.* **222**, 224 (1994); C. Lienau and A. H. Zewail, *J. Phys. Chem.* **100**, 18629 (1996); A. Materny, C. Lienau, and A. H. Zewail, *J. Phys. Chem.* **100**, 18650 (1996); Q. Liu, C. Wan, and A. H. Zewail, *J. Phys. Chem.* **100**, 18666 (1996); C. Wan, M. Gupta, J. S. Baskin, Z. H. Kim, and A. H. Zewail, *J. Chem. Phys.* **106**, 4353 (1997); J. S. Baskin, M. Gupta, M. Chachisvilis, and A. H. Zewail, *Chem. Phys. Lett.* **275**, 437 (1997); J. S. Baskin, M. Chachisvilis, M. Gupta, and A. H. Zewail, *J. Phys. Chem. A* **102**, 4158 (1998); A. Douhal, T. Fiebig, M. Chachisvilis, and A. H. Zewail, *J. Phys. Chem. A* **102**, 1657 (1998); M. Chachisvilis, I. Garcia Ochoa, A. Douhal, and A. H. Zewail, *Chem. Phys. Lett.* **293**, 153 (1998); A. Vierheilig, T. Chen, P. Waltner, W. Kiefer, A. Materny, and A. H. Zewail, *Chem. Phys. Lett.* **312**, 349 (1999); C. Wan, M. Gupta, and A. H. Zewail, *Chem. Phys. Lett.* **256**, 279 (1996); M. Chachisvilis and A. H. Zewail, *J. Phys. Chem. A* **103**, 7408 (1999); T. Fiebig, M. Chachisvilis, M. M. Manger, I. Garcia Ochoa, A. de La Hoz Ayuso, A. Douhal, and A. H. Zewail, *J. Phys. Chem A* **103**, 7419 (1999).
76. J. C. Williamson and A. H. Zewail, *Proc. Natl. Acad. Sci., USA* **88**, 5021 (1991); J. C. Williamson, M. Dantus, S. B. Kim, and A. H. Zewail, *Chem. Phys. Lett.* **196**, 529 (1992); J. C. Williamson and A. H. Zewail, *Chem. Phys. Lett.* **209**, 10 (1993); J. C. Williamson and A. H. Zewail, *J. Phys. Chem.* **98**, 2766 (1994); M. Dantus, S. B. Kim, J. C. Williamson, and A. H. Zewail, *J. Phys. Chem.* **98**, 2782 (1994); J. C. Williamson, J. Cao, H. Ihee, H. Frey, and A. H. Zewail, *Nature* **386**, 159 (1997); H. Ihee, J. Cao, and A. H. Zewail, *Chem. Phys. Lett.* **281**, 10 (1997); J. Cao, H. Ihee, and A. H. Zewail, *Chem. Phys. Lett.* **290**, 1 (1998); J. Cao, H. Ihee, and A. H. Zewail, *Proc. Natl. Acad. Sci., USA* **96**, 338 (1999).
77. W. S. Warren and A. H. Zewail, *J. Chem. Phys.* **78**, 2279 (1983); ibid **78**, 2298 (1983); J. J. Gerdy, M. Dantus, R. M. Bowman, and A. H. Zewail, *Chem. Phys. Lett.* **171**, 1 (1990); J. L. Herek, A. Materny, and A. H. Zewail, *Chem. Phys. Lett.* **228**, 15 (1994); E. D. Potter, J. L. Herek, S. Pedersen, Q. Liu, and A. H. Zewail, *Nature* **355**, 66 (1992); E. W.-G. Diau, J. L. Herek, Z. H. Kim, and A. H. Zewail, *Science* **279**, 847 (1998).
78. D. P. Millar, R. J. Robbins, and A. H. Zewail, *Proc. Natl. Acad. Sci., USA* **77**, 5593 (1980); D. P. Millar, R. J. Robbins, and A. H. Zewail, *J. Chem. Phys.* **74**, 4200 (1981); D. P. Millar, R. J. Robbins, and A. H. Zewail, *J. Chem. Phys.* **76**, 2080 (1982).

79. T. Fiebig, C. Wan, S. O. Kelley, J. K. Barton, and A. H. Zewail, *Proc. Natl. Acad. Sci., USA* **96**, 1187 (1999); C. Wan, T. Fiebig, S. O. Kelley, C. R. Treadway, J. K. Barton, and A. H. Zewail, *Proc. Natl. Acad. Sci., USA* **96**, 6014 (1999); The protein work involves D. Zhong of this laboratory. The work on the aminopurine system involves T. Fiebig and C. Wan of this laboratory in collaboration with Olav Schiemann and Jackie Barton.
80. B. Steiger, J. S. Baskin, F. C. Anson, and A. H. Zewail, *Angew. Chem. Int. Ed. Engl.* **39**, 257 (2000); A. Douhal, S. K. Kim, and A. H. Zewail, *Nature* **378**, 260 (1995); T. Fiebig, M. Chachisvilis, M. M. Manger, I. Garcia Ochoa, A. de La Hoz Ayuso, A. Douhal, and A. H. Zewail, *J. Phys. Chem. A* **103**, 7419 (1999); B. A. Leland, A. D. Joran, P. M. Felker, J. J. Hopfield, A. H. Zewail, and P. B. Dervan, *J. Phys. Chem.* **89**, 5571 (1985); A. D. Joran, B. A. Leland, P. M. Felker, A. H. Zewail, J. J. Hopfield, and P. B. Dervan, *Nature* **327**, 508 (1987).

BIBLIOGRAPHY: THE CALTECH RESEARCH

Over the past 20 years at Caltech my group and I have published some 300 scientific papers. The following list includes *some reviews, feature articles,* and *a few books*:

Books

(B1) A. H. Zewail, *Femtochemistry – Ultrafast Dynamics of the Chemical Bond, Vols. I and II,* World Scientific, New Jersey, Singapore (1994).
(B2) A. H. Zewail (Ed.), *The Chemical Bond: Structure and Dynamics,* Academic Press, Boston (1992).

Book Chapters

(B3) J. A. Syage and A. H. Zewail, *Molecular Clusters: Real-Time Dynamics and Reactivity,* in: *Molecular Clusters,* J. M. Bowman and Z. Bacic (Eds.), JAI Press (1998).
(B4) A. H. Zewail, *Femtochemistry: Dynamics with Atomic Resolution,* in: *Femtochemistry & Femtobiology,* V. Sundström (Ed.), World Scientific, Singapore (1997).
(B5) C. Wittig and A. H. Zewail, *Dynamics of Ground State Bimolecular Reactions, Chemical Reactions in Clusters,* E. R. Bernstein (Ed.), Oxford University, New York, p. 64 (1996).
(B6) A. H. Zewail, *Femtochemistry: Concepts and Applications,* in: *Femtosecond Chemistry,* J. Manz and L. Wöste (Eds.), VCH Publishers, Inc., New York, p. 15 (1995).
(B7) P. M. Felker and A. H. Zewail, *Molecular Structures from Ultrafast Coherence Spectroscopy,* in: *Femtosecond Chemistry,* J. Manz and L. Wöste (Eds.), VCH Publishers, Inc., New York, p. 193 (1995).
(B8) A. H. Zewail, *Ultrafast Dynamics of the Chemical Bond – Femtochemistry,* in: *Ultrafast Processes in Chemistry and Photobiology, Chemistry for the 21st Century,* M. A. El-Sayed, I. Tanaka, and Y. N. Molin (Eds.) IUPAC, Blackwell Scientific Publishers, Oxford, p.1 (1995).
(B9) K. E. Jones and A. H. Zewail, *Molecular Mechanisms for Dephasing: Toward a Unified Treatment of Gases, Solids, and Liquids,* in: *Advances in Laser Chemistry,* Vol. 3, Springer Series in Chemical Physics, A. H. Zewail (Ed.), Springer-Verlag, New York, p. 258 (1978).

Reviews

(B10) A. H. Zewail, *Femtochemistry: Recent Progress in Studies of Dynamics and Control of Reactions and Their Transition States, J. Phys. Chem., (Centennial Issue)* **100**, 12701 (1996).
(B11) A. Douhal, F. Lahmani and A. H. Zewail, *Proton-transfer Reaction Dynamics, Chem. Phys. (Special Issue)* **207**, 477 (1996).
(B12) J. C. Polanyi and A. H. Zewail, *Direct Observation of The Transition State, Accounts of Chemical Research (Holy-Grail Special Issue)* **28**, 119 (1995).
(B13) A. H. Zewail, *Coherence – A Powerful Concept in the Studies of Structures and Dynamics, Laser Physics* **5**, 417 (1995).

(B14) A. H. Zewail, *Femtochemistry,* (Feature Article) *J. Phys. Chem.* **97**, 12427 (1993).
(B15) A. H. Zewail, M. Dantus, R. M. Bowman, and A. Mokhtari, *Femtochemistry: Recent Advances and Extension to High-Pressures, J. Photochem. Photobiol. A: Chem.* **62/3**, 301 (1992).
(B16) A. H. Zewail and R. B. Bernstein, *Real-Time Laser Femtochemistry: Viewing the Transition States from Reagents to Products, Chem. & Eng. News,* Vol. 66 November 7, pp. 24–43 (1988) – Feature Article/Special Report; in: *The Chemical Bond: Structure and Dynamics,* A. H. Zewail (Ed.), Academic Press, Boston, p. 223 (1992), with an update section.
(B17) A. H. Zewail, *Femtosecond Transition-State Dynamics,* in: *Structure and Dynamics of Reactive Transition States, Faraday Discuss. Chem. Soc.* **91**, 207 (1991).
(B18) A. H. Zewail, *Femtochemistry: The Role of Alignment and Orientation, J. Chem. Soc., Faraday Trans.* **2 85**, 1221 (1989).
(B19) A. H. Zewail, *Laser Femtochemistry, Science* **242**, 1645 (1988).
(B20) A. H. Zewail, *IVR: Its Coherent and Incoherent Dynamics, Ber. Bunsenges. Phys. Chem.* **89**, 264 (1985).
(B21) N. Bloembergen and A. H. Zewail, *Energy Redistribution in Isolated Molecules and the Question of Mode-Selective Laser Chemistry Revisited (Feature Article), J. Phys. Chem.* **88**, 5459 (1984).
(B22) W. S. Warren and A. H. Zewail, *Phase Coherence in Multiple Pulse Optical Spectroscopy,* in: *Photochemistry and Photobiology: Proceedings of the International Conference,* Vols. I and II, Alexandria, Egypt, January 5–10, 1983, A. H. Zewail, (Ed.) Harwood Academic Publishers, Chur, Switzerland, (1983); Laser Chem. **2** (1,6), 37 (1983).
(B23) A. H. Zewail, *Optical Molecular Dephasing: Principles of and Probings by Coherent Laser Spectroscopy, Acc. Chem. Res.* **13**, 360 (1980);
(B24) A. H. Zewail, *Laser Selective Chemistry – Is it Possible?, Phys. Today* **33**, 2 (1980).
(B25) J. S. Baskin and A. H. Zewail, *Freezing Time – In a Femtosecond, Science Spectra,* Issue 14, p. 62 (1998).
(B26) A. H. Zewail, *What is Chemistry? 100 Years After J. J. Thomson's Discovery, Cambridge Review* **118**, No. 2330, 65 (1997).
(B27) A. H. Zewail, *Discoveries at Atomic Resolution (Small is Beautiful), Nature* (London) **361**, 215 (1993).
(B28) A. H. Zewail, *The Birth of Molecules, Scientific American* **263**, 76 (1990). Also available in other languages: Italian, Japanese, French, Spanish, German, Russian, Chinese, Arabic, Hungarian and Indian.

BIBLIOGRAPHY: SOME GENERAL REFERENCES

Books

(B29) **[Nobel Symposium]** *Femtochemistry & Femtobiology,* V. Sundström (Ed.), World Scientific, Singapore (1997).
(B30) **[Solvay Conference]** *Chemical Reactions and Their Control on the Femtosecond Time Scale,* P. Gaspard & I. Burghardt (Eds.), *Adv. Chem. Phys.* **101**, Wiley, New York (1997).
(B31) **[Lausanne Conference]** *Femtochemistry,* M. Chergui (Ed.), World Scientific, Singapore (1996).
(B32) **[Berlin Conference]** *Femtosecond Chemistry,* Volumes 1 and 2, J. Manz and L. Wöste (Eds.), VCH, Weinheim (1995).
(B33) **[Amsterdam Conference]** *Femtosecond Reaction Dynamics,* D. A. Wiersma, (Ed.), Royal Netherlands Academy of Arts and Sciences, North Holland, Amsterdam, (1994).
(B34) *Ultrafast Processes in Chemistry and Biology – Chemistry for the 21st Century,* M. A. El-Sayed, I. Tanaka, and Y. N. Molin (Eds.), IUPAC, Blackwell Scientific , Oxford, (1994).
(B35) *Density Matrix Method and Femtosecond Processes,* S. H. Lin, R. Alden, R. Islampour, H. Ma and A. A. Villaeys, World Scientific, Singapore (1991).
(B36) *Principles of Nonlinear Optical Spectroscopy,* S. Mukamel, Oxford Univ. Press, Oxford (1995).
(B37) *Ultrafast Dynamics of Chemical Systems,* J. D. Simon (Ed.), Kluwer, Boston (1994).
(B38) *Femtosecond Real-Time Spectroscopy of Small Molecules & Clusters,* E. Schreiber, Springer, New York (1998).

Special Issues

(B39) *Femtochemistry – Ten Years of*, A. W. Castleman, Jr. and V. Sundström (Eds.), *J. Phys. Chemistry*, June Issue (1998).

(B40) *Femtochemistry*, J. Manz & A. W. Castleman, Jr. (Eds.), *J. Phys. Chemistry*, December Issue *97* (1993).

(B41) *Ultrafast Lasers in Chemistry*, *Is. J. Chem*, M. Shapiro and S. Rosenwaks, **34**, No. 1 (1994).

Reviews

(B42) *Chemistry in Microtime*, G. Porter, in: *The Chemical Bond: Structures and Dynamics*, A. H. Zewail (Ed.), Academic Press, Inc., Boston, p.113 (1992).

(B43) *Molecular Photophysics*, G. Beddard, in: *Reports on Progress in Physics* **56**, 63 (1993).

(B44) *Time-Resolved Vibrational Spectroscopy in the Impulsive Limit*, L. Dhar, J. A. Rogers, and K. A. Nelson, *Chem. Rev.* **94**, 157 (1994).

(B45) *The Dynamics of Wave Packets of Highly-Excited States of Atoms and Molecules*, I. Sh. Averbukh and N. F. Perel'man, *Sov. Phys. Usp.* **34**, 572 (1991).

(B46) *Hot Electron Femtochemistry at Surfaces*, J. W. Gadzuk, in [B32], Vol. II, p. 603.

(B47) *Dynamics of Nonthermal Reactions: Surface Femtosecond Chemistry*, R. R. Cavanagh, D. S. King, J. C. Stephenson, and T. F. Heinz, *J. Phys. Chem.* **97**, 786 (1993).

(B48) *Femtochemistry*, J. G. Thorne and G. S. Beddard, *Chemistry & Industry*, June 20 Issue, p. 456 (1994).

(B49) *Wave Packet Dynamics: New Physics and Chemistry in Femto-time*, B. M. Garraway and K-A Suominen, *Rep. Prog. Phys.* **58**, 365 (1995).

(B50) *Femtosecond Broadband Absorption Spectroscopy of Fragments Formed in the Photodissociation of Gas-phase Molecules*, J. H. Glownia, R. E. Walkup, D. R. Gnass, M. Kaschke, J. A. Misewich, and P. P. Sorokin, in [B32], Vol. I, p. 131.

(B51) "Theory of Ultrafast Nonadiabatic Excited-State Processes and their Spectroscopic Detection in Real Time", W. Domcke and G. Stock, *Advances in Chem. Phys.* **100**, 1 (1997)

(B52) "Femtosecond Surface Science", J. A. Misewich, T. F. Heinz, P. Weigand and A. Kalamarides, in: *Advanced Series in Physical Chemistry*, **vol. 5**, H.-L. Dai and W. Ho, World Scientific, Singapore (1995).

(B53) *Femtosecond Time-Resolved Photochemistry of Molecules and Metal Clusters*, T. Baumert, R. Thalweiser, V. Weiss, and G. Gerber, in [B32], Vol. II, p. 397.

Overviews

(B54) R. Hoffmann, "Pulse, Pump and Probe", *American Scientist* **87**, 308 (1999)

(B55) Y. Tanimura, K. Yamashita and P. A. Anfinrud, "Femtochemistry", *Proc. Natl. Acad. Sci., USA* **96**, 8823 (1999); P. Anfinrud, R. de Vivie-Riedle, and V. Engel, **ibid.**, 8328 (1999).

(B56) *Designing the Molecular World*, P. Ball, Princeton University Press, Princeton (1994).

(B57) *Taming the Atom*, H. C. von Baeyer, Random House, New York (1992).

(B58) *Atoms, Electrons, and Change*, P. W. Atkins, W. H. Freeman and Co., New York (1991).

(B59) V. K. Jain, "The World's Fastest Camera", *The World and I, The Washington Times Pub. Co.*, October Issue, p. 156 (1995).

(B60) B. Nordén, www.nobel.se/announcement-99/chemistry99.html

(B61) M. Nauenberg, C. Stroud and J. Yeazell, "The Classical Limit of an Atom", *Scientific American*, June Issue, p. 44 (1994).

(B62) Douglas L. Smith, "Coherent Thinking", *Engineering and Science, LXII*, Number 4, p. 6 (1999).

Chemistry 2000

ALAN J. HEEGER, ALAN G. MacDIARMID and HIDEKI SHIRAKAWA

"for the discovery and development of conductive polymers"

THE NOBEL PRIZE IN CHEMISTRY

Speech by Professor Bengt Nordén of the Royal Swedish Academy of Sciences.
Translation of the Swedish text.

Your Majesties, Your Royal Highnesses, Ladies and Gentlemen,

Chemistry! We all associate chemistry with test tubes, stinking laboratories and explosions – Alfred Nobel's dynamite was born in such an environment. Perhaps the development of new knowledge in chemistry, more than any other science, has been characterized as a sparkling interplay between theory on one hand, the safe and predictable, and, on the other hand, the explosive and surprising reality. When we by chance discover something that may become valuable, we talk about "serendipity" – after the tale about the three princes of Serendip, who traveled widely and had the gift of drawing far-reaching conclusions from whatever they encountered. This year's Nobel Prize in Chemistry is being awarded to three scientists, whose unexpected discovery gave birth to a research area of great importance.

But let us go back to the beginning. In Japan, in 1973, a group of scientists were studying the polymerization of acetylene into plastics – acetylene was the gas that the Swedish engineer Gustaf Dalén once tamed to bring light in the dark for sailors in the form of blinking buoys (1912 Nobel Prize in Physics). Polymerization is the process by which many small molecules react to form a long chain – a polymer. Professors Ziegler and Natta were awarded the 1963 Nobel Prize in Chemistry for a technique for polymerizing ethylene or propylene into plastics; the Japanese scientists used the same catalyst for polymerizing acetylene. One day a visiting researcher in the laboratory, the story goes, added more catalyst than written in the recipe: actually one thousand times too much! Imagine the surprise among your invited dinner guests if, rather than using a few drops of Tabasco in the soup, you had added the whole bottle! The result was a surprise also to the scientists. Instead of the expected black polyacetylene powder that normally was obtained, and that was of no use, a beautifully lustrous silver colored film resulted.

It was, however, only its appearance that was metallic. The material did not conduct electricity. The breakthrough was not made until four years later in collaboration between physicist Alan Heeger and chemists Alan MacDiarmid and Hideki Shirakawa, continuing the experiments with the silver colored film. They tried to oxidize the film using iodine vapor, and – Bingo! The conductivity of the plastic increased by as much as ten million-fold; it had become conductive like a metal, comparable to copper. This was a surprising discovery, to the researchers as well as to others – we are all used to plastics, in contrast to metals, being insulators, which is why we cover electrical cords in plastic.

The discoverers started pondering what had happened. In order to conduct electricity the plastic would somehow have had to mimic metals, making their electrons easily mobile. Polyacetylene can be seen as beads on a string made up of carbon atoms linked by chemical bonds, alternatingly between single and double bonds. It is the electrons of the double bonds that give rise to the electrical conductivity. But this only happens after oxidizing the polymer chain a little here and there, for example using iodine. And why is that? The iodine removes one electron from a carbon atom, thus creating a hole in the electronic structure into which an electron from a neighboring atom can jump, whereupon a new hole is formed and so on. A hole, i.e. lack of electron, corresponds to a positive charge, and the movement of the hole along the chain gives rise to a current.

The exciting idea of being able to combine the flexibility and low weight of plastics with the electric properties of metals has stimulated scientists all over the world, resulting in a novel research field bordering physics and chemistry. Various theoretical models and new conductive, but also semi-conductive, polymers followed during the 1980s in the wake of the first discoveries. Today we can see several possible applications. How about electrically luminous plastic that may be used for manufacturing mobile phone displays or the flat television screens of the future? Or the opposite – instead using light to generate electric current: solar-cell plastics that can be unfolded over large areas to produce environmentally friendly electricity. Finally, lightweight rechargeable batteries may be necessary if we are to replace the combustion engines in today's cars with environmentally friendly electric motors – another application where electrical polymers might find use.

In parallel with the development of conducting polymers, there is an ongoing development of what we might call "molecular electronics," where the very molecules perform the same tasks as the integrated circuits we just heard about in the Nobel Prize in Physics, with the difference that these could be made incomparably smaller. In laboratories around the world, scientists are working hard to develop molecules for future electronics. And among test tubes and flasks, and in the interplay between theory and experiment, we may some day again be astonished by something unexpected and fantastic. But this is a different story, and perhaps a different Nobel Prize…

Professors Heeger, MacDiarmid and Shirakawa,

You are being rewarded for your pioneering scientific work on electrically conductive polymers. Your serendipitous discovery of how polyacetylene could be made electrically conductive has led to the prolific development, pursued by yourself and by others, of a research field of great theoretical and experimental importance. It has inspired chemists and physicists all over the world to important collaborations, and has had, and will no doubt continue to have, consequences of great benefit to mankind.

Dr. Heeger. May I convey to you my warmest congratulations on behalf of the

Royal Swedish Academy of Sciences and ask you to come forward to receive the Nobel Prize in Chemistry for the year 2000 from the hands of His Majesty the King.

Dr. MacDiarmid. May I convey to you my warmest congratulations on behalf of the Royal Swedish Academy of Sciences and ask you to come forward to receive the Nobel Prize in Chemistry for the year 2000 from the hands of His Majesty the King.

Dr. Shirakawa. May I convey to you my warmest congratulations on behalf of the Royal Swedish Academy of Sciences and ask you to come forward to receive the Nobel Prize in Chemistry for the year 2000 from the hands of His Majesty the King.

ALAN J. HEEGER

I was born on a bitter cold morning (20° F below zero) in Sioux City (Iowa) on January 22, 1936. I was told that when my father went out in the cold that morning to go to the hospital to visit his wife and newborn first son, his car would not start. Despite advice to the contrary, he walked to the hospital; his ears were frostbitten on the way.

The Heeger family came to Sioux City (Iowa) from Russia as Jewish immigrants in 1904 when my father was a small boy (age 4). My mother was born in Omaha (Nebraska); she was a first generation child of Jewish immigrants. My mother and father were married in the midst of the Great Depression.

My early years were spent in Akron (Iowa), a small midwestern town of 1000 people, approximately 35 miles from Sioux City. I went to elementary school in Akron. My brother, Gerald, was born in Akron. My father was the manager and, subsequently the owner, of a general store that served the local farming community. I have a strong memory of the day I was told that my father had a weak heart and that he had to go to the hospital. He died when I was nine years old on the same day that Franklin Roosevelt died; it was his 45th birthday.

After my father's death, we moved to Omaha, so my mother could be closer to her family. She raised us as a single parent in a house that we shared with her sister and her sister's children.

One of my earliest memories (long before my father died), is of my mother telling me of the importance of getting a university education. When she graduated from high school, she received a scholarship to go on to university but went to work instead; she was needed by her parents to help support the family. It was always clear to me that it was my responsibility to go to university; prior to my generation no one on either side of my family had an education that went beyond high school. I and my brother were the first in our family to receive the PhD degree.

My high school years were fun and frustrating, typical of the teen years. The most important accomplishment was meeting my wife, Ruth. I have loved her for nearly fifty years, and she remains my best friend.

I was always a good student, but I do not remember science being especially easy. On the contrary, I recall that in high school, physics was somewhat mysterious. I was impatient to get on with my education, to get on with more important things, and therefore completed high school one year early.

My undergraduate years at the University of Nebraska were a special time in my life; the combination of partying and intellectual awakening that is what the undergraduate years are supposed to be. I went to the University with the goal of becoming an engineer; I had no concept that one could pursue science as a career. After one semester, I was convinced that engineering was not for me, and I completed my undergraduate studies with a dual major in Physics and Mathematics. The highlight was a course (in my senior year) in Modern Physics taught by Theodore Jorgensen. Professor Jorgensen introduced me to quantum physics and twentieth century science. I was honored by the University of Nebraska in 1998 with a Doctor of Science (h.c.) and had the pleasure of giving a Physics colloquium at that time. Ted Jorgensen came to the lecture; he was 92 and working hard on revising his book on the Physics of Golf.

Again, I was impatient to get on with "real physics". I started the path toward my PhD in Physics at UC Berkeley while working part time for Lockheed Space and Missile Division in Palo Alto, CA. On Monday, Wednesday and Friday, I would wake up early and drive the Bayshore Freeway to Berkeley to attend classes. After sitting in class all morning, I had lunch and then got back on the freeway to return to work in Palo Alto. Naturally, after such a morning I fell asleep at the wheel almost every trip. Thus, it was not a terribly difficult decision; Ruth and I moved into student housing at Berkeley, and I started research on a full time basis.

When I started at Berkeley, my goal was to do a theoretical thesis under Charles Kittel. Thus, when the decision was made to go for my degree on a full-time basis, I went first to Kittel and asked if I could work for him. Kittel had just returned from a trip to Moscow where he met Landau, and he told me that Landau required that a prospective student had to pass a rigorous examination before he would agree to take the student into his research group. Kittel indicated that I should take the PhD qualifier and come back to him after I had done so. When I came back to discuss my future with him, Kittel told me that he would take me on. He said, however, that although I could do a thesis under his direction in solid state theory, he did not think I would be a first-rate theorist. He recommended instead that I consider working with someone who does experimental work in close interaction with theory. This was perhaps the best advice that anyone ever gave me – and I followed his advice. I joined the research group of Alan Portis.

I remember with clarity my first day in the laboratory. I was doing "original research"; at last I was involved with real physics. After only one day of carrying out magnetic measurements on an insulating antiferromagnet, $KMnF_3$, I wrote a theory of antiferroelectric antiferromagnets and presented it to Portis with great pride. He was patient with me then and again a few days later when I apologized and told him my theory was nonsense. Through my interactions with Portis (I recall spending many hours talking with him in his office), I learned how to think about physics; more important, I began to learn about good taste in the choice of problems.

After completing my degree, I went directly to join the Physics Department at the University of Pennsylvania where I remained for over twenty years. It was an exciting period for condensed matter physics at PENN. Eli Burstein had made major progress in building the solid state group; he convinced

Robert Schrieffer to come to Penn, and he and Schrieffer attracted an outstanding group of young people. Beginning with my experimental studies of magnetic impurities in metals and the Kondo Effect, I learned many-body physics from Schrieffer.

Anthony Garito introduced me to tetracyanoquinodimethane (TCNQ); I brought him into my research group for post-doctoral research. We worked together from 1970 through 1975 on the metal-physics of TTF-TCNQ and on the discovery of the Peierls instability in quasi-one-dimensional π-stacked molecular crystals. Although the direct observation of the incommensurate Peierls distortion with wave number $q = 2k_F$ proved that we were on the right track, this was a time of controversy and stress.

In 1975, the first papers on the novel metallic polymer, poly(sulfur-nitride), $(SN)_x$ appeared in the literature. I was intrigued by this unusual quasi-1d metal and wanted to get into the game. I learned that Alan MacDiarmid, a professor in the Chemistry Department at PENN, had a background in sulfur-nitride chemistry, and I made an appointment to see him with the goal of convincing him to collaborate with me and to synthesize $(SN)_x$. I recall that we met late in the afternoon of an autumn day. After quite a long discussion during which I made little progress toward my goal, I realized that while I was saying "$(SN)_x$", he was hearing "$(Sn)_x$". Needless to say, he was not impressed with my enthusiasm for $(Sn)_x$ being a metal; any chemist knew that *tin* was a metal!

Once MacDiarmid and I got past this initial language problem, a true collaboration began. We realized that it was a long reach across the Chemistry-Physics boundary, and we were determined to learn from one another. Although we collaborated during the week, we typically met on Saturday mornings with no agenda; just to try to learn from one another. At that time, I was fascinated with the metal-insulator transition as envisioned by Mott. I recall that I tried to convey my interest in this problem to MacDiarmid by asking him to consider a linear chain of hydrogen atoms as a model system. He balked right away; a linear chain of hydrogen atoms did not exist. After discussion, we focused in on the abstraction of a chain of π-bonded -CH- units as an example of a system that would have one unpaired electron per repeat unit. Shortly thereafter, MacDiarmid went to Japan for a visit. MacDiarmid is a very visual person. He loved the golden color of films and crystals of $(SN)_x$, and he showed samples and photos of this golden material during his lectures. After one such lecture, a Japanese scientist came up to him during the coffee break and told MacDiarmid that he, too, had some shiny films. Thus, MacDiarmid was introduced to Hidekei Shirakawa and to polyacteylene.

When MacDiarmid returned from Japan, he told me with great excitement about $(CH)_x$. With the help of a small addition to an ONR grant from the Program Officer, Kenneth Wynne, we were able to bring Hideki Shirakawa to PENN as a Visiting Scientist. The initial discovery of the remarkable increase in electrical conductivity of $(CH)_x$ and the identification of that increase as resulting from a transition from insulator (semiconductor) to metal followed in a very short time.

The soliton in polyacetylene was born with the observation of an electron spin resonance (esr) signal in the pure material where there should not have been one. Building on the earlier work by Michael Rice on phase-solitons, I realized that if one drew a domain wall between the two identical forms with opposite bond alternation, one would have an unpaired spin and postulated that the origin of the esr signal might be a bond-alternation domain wall. Curt Fincher, then a graduate student in my research group, had recently discovered the doping-enhanced infrared vibrational modes which became a signature of the doping. In a luncheon seminar before the solid state group at PENN, I argued that these doping-induced IR modes might arise from the enhanced electric field at IR frequencies that would result if a charged bond-alternation domain wall were to move back and forth driven by the external field of the incident IR radiation. Schrieffer listened closely and made some comments about "kinks" at the end of my talk. A few days later he showed me how the mid-gap state would arise from the formation of such a bond-alternation domain wall and how that mid-gap state would have a reversed spin/charge relation relative to that of fermions. Wu-Pei Su then worked this out in detail, and the SSH papers were written.

I was drawn to Santa Barbara by the promise of a singular opportunity to build a special Physics Department, by the promise of continuing my close collaboration with Bob Schrieffer, by the opportunity to work with Fred Wudl, and – frankly – by the lure of this beautiful place. Wudl, then a synthetic chemist at Bell Laboratories, and I were recruited to UC Santa Barbara together and enjoyed a close and productive collaboration over a period of 15 years.

Daniel Moses and I have worked together for twenty years, initially at PENN and then at UCSB. Dan dragged me into ultra-fast pulsed laser spectroscopy and into fast-transient photoconductivity as probes of the excited states of semiconducting polymers. Dan continues in his efforts to resolve the remaining fundamental scientific issues in the field of semiconducting polymers with creativity and with determination.

In 1986, in the process of building the Macromolecular division of our newly formed Materials Department, we convinced Paul Smith to leave DuPont Central Research and come to UCSB. Whereas I and Alan MacDiarmid and most of the early players in the conducting polymer field were amateurs in the field of polymer science, Paul was a professional. He quickly hammered into my head the importance of making conducting polymers processible, and he had the annoying habit of asking me embarrassing questions such as "What is the intrinsic electrical conductivity of a conducting polymer?". Anything I know about the processing and mechanical properties of polymers, I learned from Paul.

In 1990, Paul Smith and I decided that conducting polymers as materials had developed to a level of maturity that commercial products were possible. With this as a goal, we founded UNIAX Corporation. Fortunately, on a trip to China in 1986, I met Yong Cao and immediately realized that he was a remarkable scientist. I was able to bring him to Santa Barbara in 1987. Initially, he worked with Paul and with me at UCSB. When we founded UNIAX, Yong Cao was the first employee. His creativity, determination and scientific strength were critical to our scientific progress and to the success of UNIAX. During the 1990's, UNIAX played a leading role in developing the science and technology of conducting polymers with many important contributions.

The twenty-five years since the discovery of conducting polymers have taken me on a great ride; always on the frontier and always with the challenge of exciting discoveries. In 1990, the discovery of polymer LEDs by Richard Friend and colleagues at Cambridge gave the field a boost with the promise of important technology and with the excitement of an entirely new set of phenomena to study. In 1992, while doing post-doctoral research in my group at UCSB, Serdar Saricifici discovered ultrafast photo-induced electron transfer from semiconducting polymers to acceptors such as C_{60}. This discovery resulted in the development of polymer photodetectors and photovoltaic cells that offer promise for use in a variety of applications. In 1996, the discovery of amplified spontaneous emission and lasing (simultaneously by our group, by Richard Friend's group at Cambridge and by Valy Vardeny's group at Utah) opened yet another potentially important direction. And it goes on.

None of this could have been accomplished without the hard work, dedication and creativity of the students and post-docs with whom I have had the pleasure of working over the past forty years. I thank them all.

I have enjoyed the life of a scientist while sharing both the exciting days and the disappointments with Ruth. She has filled my life with love and surrounded me with beauty. She has also gallantly put up with my eccentricities for more than forty years. We have succeeded in starting an academic dynasty; our two sons, Peter and David are both academics. Peter is a professor and medical doctor who is doing research on immunology at Case Western Reserve University. David is a professor and neuroscientist at Stanford University where he studies human vision. I have had the great pleasure of collaborating and publishing articles (as co-author) with both of my sons. Now I am looking forward to the emergence of my four grandchildren, Brett, Jordan, Julia and Alice, as the next generation of the Heeger family. Of all the congratulations that I have received as a result of the Nobel Prize, I took greatest pleasure from their pride in their grandfather.

SEMICONDUCTING AND METALLIC POLYMERS: THE FOURTH GENERATION OF POLYMERIC MATERIALS

Nobel Lecture, December 8, 2000

by

ALAN J. HEEGER

Department of Physics, Materials Department, Institute for Polymers and Organic Solids, University of California at Santa Barbara, Santa Barbara, CA 93106, USA.

I. INTRODUCTION

In 1976, Alan MacDiarmid, Hideki Shirakawa and I, together with a talented group of graduate students and post-doctoral researchers discovered conducting polymers and the ability to dope these polymers over the full range from insulator to metal[1,2]. This was particularly exciting because it created a new field of research on the boundary between chemistry and condensed matter physics, and because it created a number of opportunities:

Conducting polymers opened the way to progress in understanding the fundamental chemistry and physics of π-bonded macromolecules;

Conducting polymers provided an opportunity to address questions which had been of fundamental interest to quantum chemistry for decades:

Is there bond alternation in long chain polyenes?

What is the relative importance of the electron-elecron and the electron-lattice interactions in π-bonded macromolecules?

Conducting polymers provided an opportunity to address fundamental issues of importance to condensed matter physics as well, including, for example, the metal-insulator transition as envisioned by Neville Mott and Philip Anderson and the instability of one-dimensional metals discovered by Rudolph Peierls (the "Peierls Instability").

Finally – and perhaps most important – conducting polymers offered the promise of achieving a new generation of polymers: Materials which exhibit the electrical and optical properties of metals or semiconductors *and* which retain the attractive mechanical properties and processing advantages of polymers.

Thus, when asked to explain the importance of the discovery of conducting polymers, I offer two basic answers:

(i) They did not (could not?) exist.

(ii) They offer a unique combination of properties not available from any other known materials.

The first expresses an intellectual challenge; the second expresses a promise for utiltity in a wide variety of applications.

Conducting polymers are the most recent generation of polymers[3]. Polymeric materials in the form of wood, bone, skin and fibers have been used by man since prehistoric time. Although organic chemistry as a science dates back to the eighteenth century, polymer science on a molecular basis is a development of the twentieth century. Hermann Staudinger developed the concept of macromolecules during the 1920's. Staudinger's proposal was openly opposed by leading scientists, but the data eventually confirmed the existence of macromolecules. Staudinger was awarded the Nobel Prize in Chemistry in 1953 "for his discoveries in the field of macromolecular chemistry". While carrying out basic research on polymerization reactions at the DuPont company, Wallace Carothers invented nylon in 1935. Carothers' research showed the great industrial potential of synthetic polymers; a potential which became reality in a remarkably short time. Today, synthetic polymers are used in larger quantities than any other class of materials (larger by volume and larger by weight, even though such polymers have densities close to unity). Polymer synthesis in the 1950's was dominated by Karl Ziegler and Giulio Natta whose discoveries of polymerizarion catalysts were of great importance for the development of the modern "plastics" industry. Ziegler and Natta were awarded the Nobel Prize in Chemistry in 1963 "for their discoveries in the field of the chemistry and technology of high polymers". Paul Flory was the next "giant" of polymer chemistry; he created modern polymer science through his experimental and theoretical studies of macromolecules. His insights and deep understanding of macromolecular phenomena are summarized in his book, *Principles of Polymer Chemistry*, published in1953, and still useful (and used) today. Flory was awarded the Nobel Prize in Chemistry in 1974 "for his fundamental achievements, both theoretical and experimental, in the physical chemistry of macromolecules".

Because the saturated polymers studied by Staudinger, Flory, Ziegler and Natta are insulators, they were viewed as uninteresting from the point of view of electronic materials. Although this is true for saturated polymers (in which all of the four valence electrons of carbon are used up in covalent bonds), in conjugated polymers the electronic configuration is fundamentally different. In conjugated polymers, the chemical bonding leads to one unpaired electron (the π–elecron) per carbon atom. Moreover, π-bonding, in which the carbon orbitals are in the sp^2p_z configuration and in which the orbitals of successive carbon atoms along the backbone overlap, leads to electron delocalization along the backbone of the polymer. This electronic delocalization provides the "highway" for charge mobility along the backbone of the polymer chain.

As a result, therefore, the electronic structure in conducting polymers is determined by the chain symmetry (i.e. the number and kind of atoms within the repeat unit), with the result that such polymers can exhibit semiconducting or even metallic properties. In his lecture at the Nobel Symposium (NS-81) in 1991, Professor Bengt Rånby designated electrically conducting polymers as the *"fourth generation of polymeric materials"*[3].

The classic example is polyacetylene, $(\text{-CH})_n$, in which each carbon is σ-bonded to only two neighboring carbons and one hydrogen atom with one π-electron on each carbon. If the carbon-carbon bond lengths were equal, the chemical formula, $(\text{-CH})_n$ with one unpaired electron per formula unit would imply a metallic state. Alternatively, if the electron-electron interactions were too strong, $(\text{-CH})_n$ would be an antiferromagnetic Mott insulator. The easy conversion to the metallic state on doping[1,2] together with a variety of studies of the neutral polymer have eliminated the antiferromagnetic Mott insulator as a possibility.

In real polyacetylene, the structure is dimerized as a result of the Peierls Instability with two carbon atoms in the repeat unit, $(\text{-CH=CH})_n$. Thus, the π-band is divided into π- and π*-bands. Since each band can hold two electrons per atom (spin up and spin down), the π-band is filled and the π*-band is empty. The energy difference between the highest occupied state in the π-band and the lowest unoccupied state in the π*-band is the π-π* energy gap, E_g. The bond-alternated structure of polyacetylene is characteristic of conjugated polymers (see Fig.1). Consequently, since there are no partially filled bands conjugated polymers are typically semiconductors. Because E_g depends upon the molecular structure of the repeat unit, synthetic chemists are provided with the opportunity and the challenge to control the energy gap by design at the molecular level.

Although initially built upon the foundations of quantum chemistry and condensed matter physics, it soon became clear that entirely new concepts were involved in the science of conducting polymers. The discovery of nonlinear excitations in this class of polymers, solitons in systems in which the ground state is degenerate and confined soliton pairs (polarons and bipolarons) in systems in which the ground state degeneracy has been lifted by the molecular structure, opened entirely new directions for the study of the interconnection of chemical and electronic structure. The spin-charge separation characteristic of solitons and the reversal of the spin-charge relationship (relative to that expected for electrons as fermions) in polyacetylene challenged the foundations of quantum physics. The study of solitons in polyacetylene[4], stimulated by the Su-Schrieffer-Heeger (SSH) papers [5,6] dominated the first half of the 1980's.

The opportunity to synthesize new conducting polymers with improved/desired properties began to attract the attention of synthetic chemists in the 1980's. Although it would be an overstatement to claim that chemists can now control the energy gap of semiconducting polymers through molecular design, we certainly have come a long way toward that goal.

Reversible "doping" of conducting polymers, with associated control of the electrical conductivity over the full range from insulator to metal, can be accomplished either by chemical doping or by electrochemical doping. Concurrent with the doping, the electrochemical potential (the Fermi level) is moved either by a redox reaction or an acid-base reaction into a region of energy where there is a high density of electronic states; charge neutrality is maintained by the introduction of counter-ions. Metallic polymers are, there-

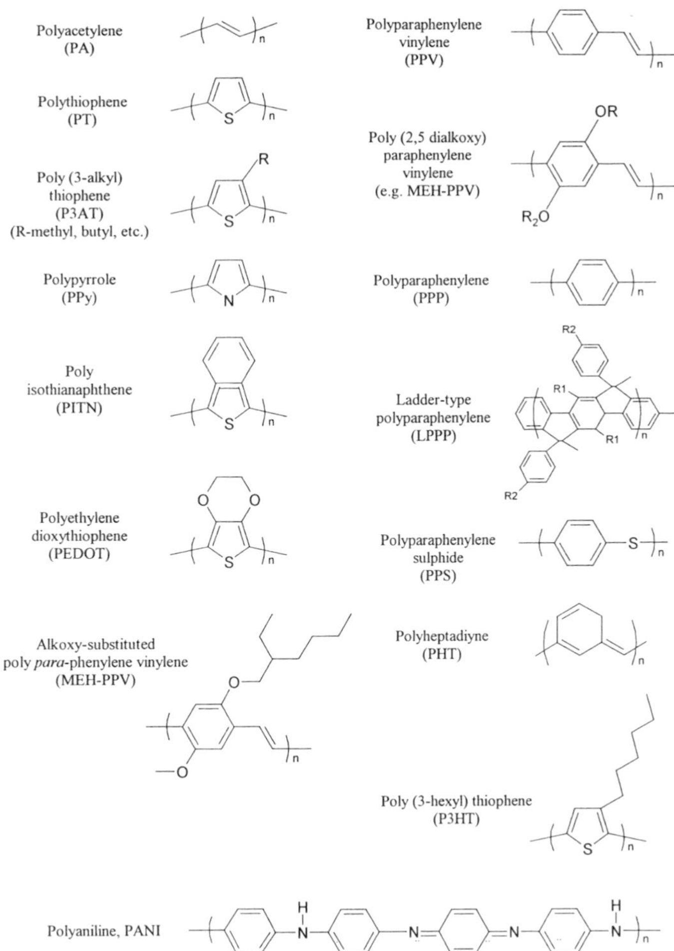

Figure 1. Molecular structures of a few conjugated polymers; note the bond-alternated structures.

fore, salts. The electrical conductivity results from the existence of charge carriers (through doping) and from the ability of those charge carriers to move along the π-bonded "highway". Consequently, doped conjugated polymers are good conductors for two reasons:

(i) Doping introduces carriers into the electronic structure. Since every repeat unit is a potential redox site, conjugated polymers can be doped n-type (reduced) or p-type (oxidized) to a relatively high density of charge carriers[7].

(ii) The attraction of an electron in one repeat unit to the nuclei in the neighboring units leads to carrier delocalization along the polymer chain and to charge carrier mobility, which is extended into three dimensions through interchain electron transfer.

Disorder, however, limits the carrier mobility and, in the metallic state, limits the electrical conductivity. Indeed, research directed toward conjugated polymers with improved structural order and hence higher mobility is a focus of current activity in the field.

Fig. 2 shows the early results on the conductivity of polyacetylene as a function of the doping level; even in these early studies the conductivity increased by more than a factor of 10^7 to a level approaching that of a metal[2].

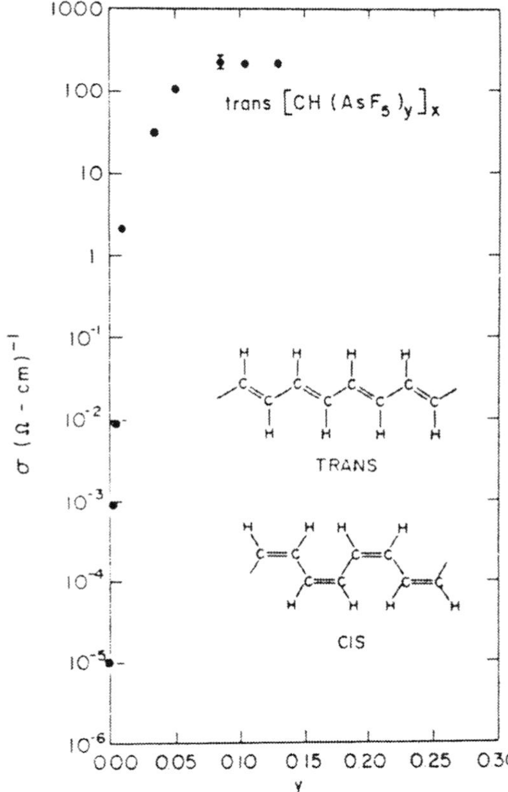

Figure 2. Electrical conductivty of *trans*-$(CH)_x$ as a function of (AsF_5) dopant concentration. The *trans* and *cis* polymer structures are shown in the inset.

The electrochemical doping of conducting polymers was discovered by the MacDiarmid-Heeger collaboration in 1980 and opened yet another scientific direction[8]. The electrochemistry of conducting polymers has developed into a field of its own with applicaions that range from polymer batteries and electrochromic windows to light-emitting electrochemical cells[9].

Although I have emphasized the processing advantages of polymers, even as late as 1990, there were no known examples of stable metallic polymers which could be processed *in the metallic form* (a requirement for broad use in industrial products). This major outstanding problem was first solved with polyaniline, PANI. PANI has been investigated extensively for over 100 years and attracted interest as a conducting material for several important reasons; the monomer is inexpensive, the polymerization reaction is straightforward and proceeds with high yield, and PANI has excellent stability. As shown by Alan MacDiarmid and his collaborators in the mid-80's, polyaniline can be rendered conducting through two independent routes; oxidation (either chemically or electrochemically) of the leuco-emeraldine base or protonation of the emeraldine base through acid-base chemistry[10]. Because the insertion of counter-ions is involved in both routes; conducting polyaniline is a salt (a polycation with one anion per repeat unit).

Processing high molecular weight polyaniline into useful objects and devices proved to be a difficult problem. Yong Cao, Paul Smith and I made important progress in 1991 by using functionalized protonic acids to both convert PANI to the metallic form and, simultaneously, render the the resulting PANI-complex soluble in common organic solvents[11]. The functionalized counter-ion acts like a "surfactant" in that the charged head-group is ionically bound to the oppositely charged protonated PANI chain, and the "tail" is chosen to be compatible with non-polar or weakly polar organic liquids (in the case of solutions) or the host polymer (in the case of blends). The processibility of PANI induced by the "surfactant" counter-ions has enabled the fabrication of conducting polymer blends with a variety of host polymers[12]. Since the blends are melt-processible as well, the counter-ion induced processibility of of polyaniline provides a route to conducting polymer blends for use in industrial products. The "surfactant" counterions offer an unexpected advantage; they lead to the formation of a self-assembled network morphology in the PANI polyblends[13]. Because of these interpenetrating networks, the threshold for the onset of electrical conductivity in blends with traditional insulating host polymers is reduced to volume fractions well below 1%[14]. The PANI network is sufficiently robust that it remains connected and conducting even after the removal of the host polymer, thus opening another new direction; the fabrication of novel electrodes for use in electronic devices. The use of "surfactant" counter-ions was introduced with the goal of making PANI processible in the conducting form. The self-assembly of phase-separated networks was an unexpected – but very welcome – bonus.

Conducting polymers were initially attractive because of the fundamental interest in the doping and the doping-induced metal-insulator transition. However, the chemistry and physics of these polymers in their non-doped semiconducting state are of great interest because they provide a route to "plastic electronic" devices. Although polymer diodes were fabricated and characterized in the 1980's[15], the discovery of light emitting diodes (LEDs) by Richard Friend and colleagues at Cambridge in 1990 [16] provided the stimulus for a major push in this direction. The polymer light-emitting diode is, however, only one of a larger class of devices in the emerging class of "plastic" optoelectronic devices, including lasers, high sensitivity plastic photodiodes (and photodiode arrays) and photovoltaic cells, ultrafast image processors (optical computers), thin-film transistors and all-polymer integrated circuits; in each case these sophisticated electronic components are fabricated from semiconducting and metallic polymers. All of these have a common structure: They are thin film devices in which the active layers are fabricated by casting the semiconducting and/or metallic polymers from solution (e.g. spin-casting).

II. DOPING

Charge injection onto conjugated, semiconducting macromolecular chains, "doping", leads to the wide variety of interesting and important phenomena which define the field. As summarized in Fig. 3, reversible charge injection by "doping" can be accomplished in a number of ways.

A. Chemical Doping by Charge Transfer

The initial discovery of the ability to dope conjugated polymers involved charge transfer redox chemistry; oxidation (p-type doping) or reduction (n-type doping) [1,2,7], as illustrated with the following examples:

a. p-type

$$(\pi\text{-polymer})_n + 3/2ny(I_2) \rightarrow [(\pi\text{-polymer})^{+y}(I_3^-)_y]_n \tag{1}$$

b. n-type

$$\begin{aligned}&(\pi\text{-polymer})_n + [Na^+(\text{Napthtalide})^{\cdot}]_y \rightarrow \\ &[(Na^+)_y(\pi\text{-polymer})^{-y}]_n + (\text{Naphth})^o\end{aligned} \tag{2}$$

When the doping level is sufficiently high, the electronic structure evolves to that of a metal.

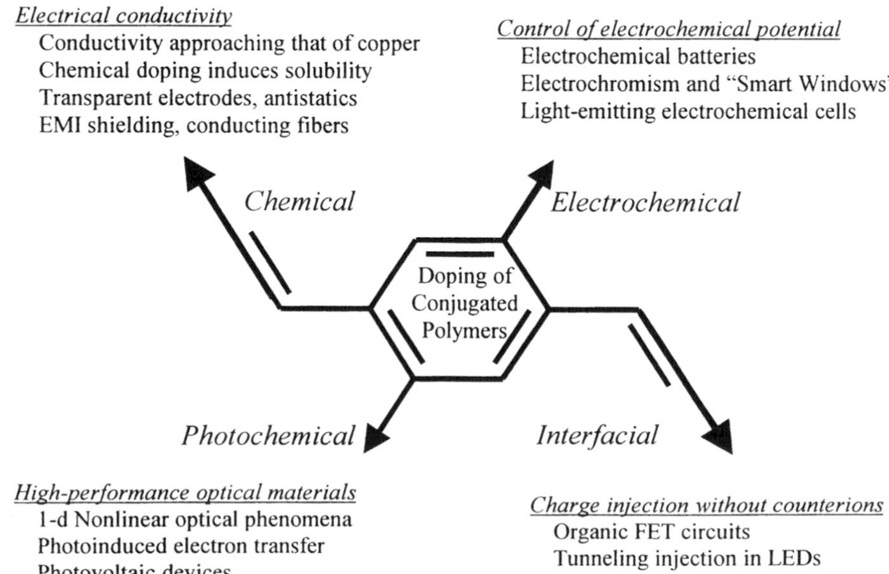

Figure 3. Doping mechanisms and related applications.

B. Electrochemical Doping

Although chemical (charge transfer) doping is an efficient and straightforward process, it is typically difficult to control. Complete doping to the highest concentrations yields reasonably high quality materials. However, attempts to obtain intermediate doping levels often result in inhomogeneous doping. Electrochemical doping was invented to solve this problem[8]. In electrochemical doping, the electrode supplies the redox charge to the conducting polymer, while ions diffuse into (or out of) the polymer structure from the nearby electrolyte to compensate the electronic charge. The doping level is determined by the voltage between the conducting polymer and the counter-electrode; at electrochemical equilibrium the doping level is precisely defined by that voltage. Thus, doping at any level can be achieved by setting the electrochemical cell at a fixed applied voltage and simply waiting as long as necessary for the system to come to electrochemical equilibrium (as indicated by the current through the cell going to zero). Electrochemical doping is illustrated by the following examples:

a. p-type

$$(\pi\text{-polymer})_n + [Li^+(BF_4^-)]_{sol'n} \rightarrow [(\pi\text{-polymer})^{+y}(BF_4^-)_y]_n + Li_{(electrode)} \quad (3)$$

b. n-type

$$(\pi\text{-polymer})_n + Li_{(electrode)} \rightarrow [(Li^+)_y(\pi\text{-polymer})^{-y}]_n + [Li^+(BF_4^-)]^{sol'n} \quad (4)$$

C. Doping of Polyaniline by Acid-Base Chemistry

Polyaniline provides the prototypical example of a chemically distinct doping mechanism[10]. Protonation by acid-base chemistry leads to an internal redox reaction and the conversion from semiconductor (the emeraldine base) to metal (the emeraldine salt). The doping mechanism is shown schematically in Fig. 4. The chemical structure of the semiconducting emeraldine base form of polyaniline is that of an alternating copolymer. Upon protonation of the emeraldine base to the emeraldine salt, the proton induced spin unpairing mechanism leads to a structural change with one unpaired spin per repeat unit, *but with no change in the number of electrons*[17,18]. The result is a half-filled band and, potentially, a metallic state where there is a positive charge in each repeat unit (from protonation) and an associated counterion (e.g. Cl^-, HSO_4^-, $DBSA^-$ etc). This remarkable conversion from semiconductor to metal has been well-described, but it is not well understood from the view of basic theory. There are no calculations which show that the metallic (emeraldine salt) final state is lower in energy than the semiconductor and no detailed understanding of the rearrangement reactions sketched in Fig. 4.

D. Photodoping

The semiconducting polymer is locally oxidized and (nearby) reduced by photo-absorption and charge separation (electron-hole pair creation and separation into "free" carriers):

$$(\pi\text{-polymer})_n + h\nu \rightarrow [\{\pi\text{-polymer}\}^{+y} + \{\pi\text{-polymer}\}^{-y}]_n \quad (5)$$

where y is the number of electron-hole pairs (dependent upon the pump rate in competition with the recombination rate). The branching ratio between free carriers and bound excitons (and the closely related issue of the magnitude of the exciton binding energy) is a subject of continuing discussion [19]

Following photoexcitation from the ground state ($1A_g$ in the notation of molecular spectroscopy) to the lowest energy state with proper symmetry ($1B_u$), recombination to the ground state can be either radiative (luminescence) or nonradiative. Some families of conjugated polymers exhibit high luminescence quantum efficiencies (for example, PPV and PPP and their soluble derivatives); others do not (for example, polyacetylene and the polythiophenes). A number of mechanisms have been identified that lead to low quantum efficiencies for photoluminescence. Rapid bond relaxation in the excited state and the formation of solitons with states at mid-gap prevent radiative recombination in polyacetylene[4]. The existence of an A_g state or a triplet excited state below the $1B_u$ state will favor nonradiative recombination (in both cases, direct radiative transitions to the ground state are forbidden). Interchain interactions in the excited state ("excimers") also lead to nonradiative channels for decay [20–22].

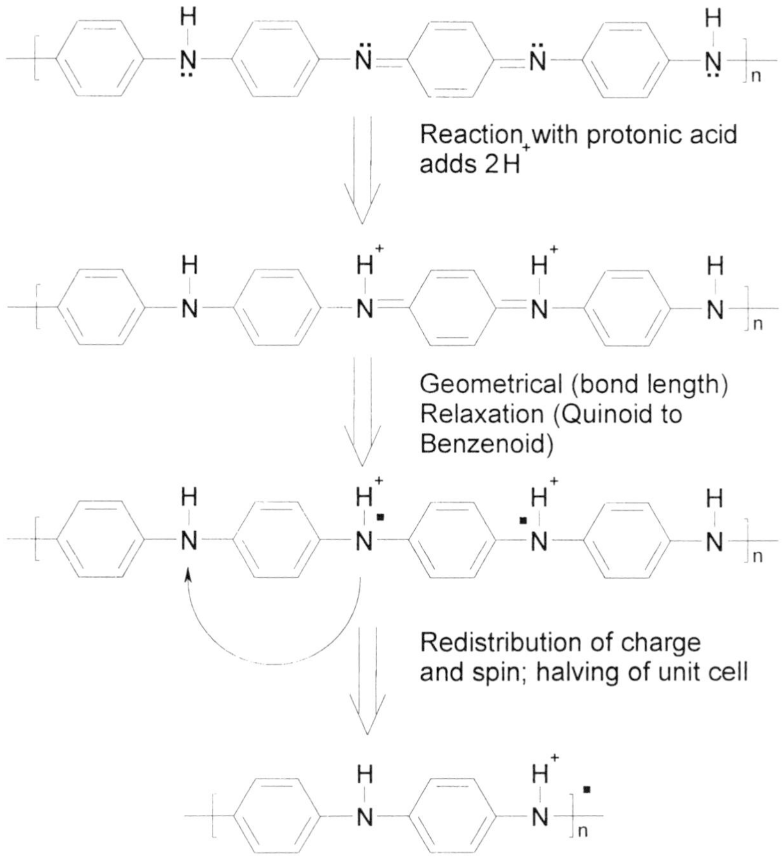

Figure 4. Protonation induced spin unpairing in polyaniline; conversion from insulator to metal with no change in the number of electrons.

E. Charge Injection at a metal-semiconducting polymer (MS) interface
Electrons and holes can be injected from metallic contacts into the π^*- and π-bands, respectively:

a. Hole injection into an otherwise filled π-band

$$(\pi\text{-polymer})_n - y(e^-) \rightarrow (\pi\text{-polymer})_n^{+y} \qquad (5)$$

b. Electron injection into an empty π^*-band

$$(\pi\text{-polymer})_n + y(e^-) \rightarrow (\pi\text{-polymer})_n^{-y} \qquad (6)$$

In the case of charge injection at an MS interface, the polymer is oxidized or reduced (electrons are added to the π^* band or removed from the π-band). However, the polymer is not doped in the sense of chemical or electrochemical doping, for there are no counter-ions. This distinction becomes particularly clear when comparing charge injection in the polymer light-emitting diode[16,23] (where there are no ions) with that in the polymer light-emitting electrochemical cell where electrochemical doping with associated redistribution of ions provides the mechanism for charge injection[9].

As indicated in Fig. 3, each of the methods of charge-injection doping leads to unique and important phenomena. In the case of chemical and/or electrochemical doping, the induced electrical conductivity is permanent, until the carriers are chemically compensated or until the carriers are purposely removed by "undoping". In the case of photo-excitation, the photoconductivity is transient and lasts only until the excitations are either trapped or decay back to the ground state. In the case of charge injection at a metal-semiconductor (MS) interface, electrons reside in the π^*-band and/or holes reside in the π-band only as long as the biasing voltage is applied.

Because of the self-localization associated with the formation of solitons, polarons and bipolarons, charge injection leads to the formation of localized structural distortions and electronic states in the energy gap[4]. In the case of "photo-doping", the redistribution of oscillator strength associated with the sub-gap infrared absorption and the corresponding bleaching of the interband (π-π^*) transition provide a route to nonlinear optical (NLO) response. Real occupation of low energy excited states[4,24] and virtual occupation of higher energy excited states (in the context of perturbation theory)[25] lead to, respectively, resonant and nonresonant NLO response.

By charge-injection doping at an MS interface, the polymer semiconductor can be used as the active element in thin film diodes[15] and field effect transistors (FETs)[26–28]. Tomozawa *et al.*[15] demonstrated the first example of an electronic device component fabricated by casting the active polymer directly from solution; even these early diodes exhibited excellent current-voltage characteristics. Dual carrier injection in metal/polymer/metal structures provides the basis for polymer light-emitting diodes (LEDs)[23]. In polymer LEDs, electrons and holes are injected from the cathode and anode, respectively, into the undoped semiconducting polymer; light is emitted when the injected electrons and holes meet in the bulk of the polymer and recombine with the emission of radiation[16].

Thus, as summarized in Fig. 3, doping is a common feature of conducting polymers; doping leads to a remarkably wide range of electronic phenomena.

III. NOVEL PROPERTIES GENERATE NEW TECHNOLOGY

As emphasized in the Introduction, conducting polymers exhibit novel properties; properties not typically available in other materials. I focus on a few of these as illustrative examples:

(a) Semiconducting and metallic polymers which are soluble in and processible from common solvents;
(b) Transparent conductors;
(c) Semiconductors in which the Fermi energy can be controlled and shifted over a relatively wide range.

In each case, the property is related to a fundamental feature of the chemistry and/or physics of the class of conducting polymers. These novel properties enable a number of applications including polymer LEDs, conducting polymers as electrochromic materials, polymer photodetectors, and polymer photovoltaic cells.

A. Semiconducting and metallic polymers which are soluble in and processible from common solvents

Because the interchain electron transfer interactions of conjugated polymers are relatively strong compared with the Van der Waals and hydrogen bonding interchain interactions typical of saturated polymers, conducting polymers tend to be insoluble and infusible. Thus, there was serious doubt in the early years following the discovery that π-conjugated polymers could be doped to the metallic state as to whether or not processing methods could be developed. Significant progress has been made using four basic approaches:

1. Side-chain functionalization; principally used for processing semiconducting polymers from solution in organic solvents or from water;
2. Precursor route chemistry, principally used for processing polyacetylene and PPV into thin films;
3. Counter-ion induced processing, principally used for processing polyaniline in the metallic form from organic solvents;
4. Aqueous colloidal dispersions created by template synthesis, principally used for processing polyaniline and poly(ethylenedioxythiophene), PEDOT.

The addition of moderately long side chains onto the monomer units resulted in derivatives of polythiophene, the poly(3-alkylthiophenes), or P3ATs; see Fig. 1. Since the side chains decrease the interchain coupling and increase the entropy, these derivatives can be processed either from solution or from the melt. Similarly, side chain functionalization of poly(phenylene vinylene), PPV, (see Fig.1) has progressed to the point where a variety of semiconducting polymers and copolymers are available with energy gaps that span the visible spectrum[29]). Water solubility was achieved by incorporating polar groups such as $(CH_2)_nSO_3^-M^+$ into the side chains (so-called 'self-doped' polymers)[30,31].

The ability to fabricate optical quality thin films by spin-casting from solution has proven to be an enabling step in the development of plastic electronic devices such as diodes, photodiodes, LEDs, LECs, optocouplers, and thin film transistors[32].

The counter-ion induced processibility of "metallic" polyaniline utilizes bifunctional counter-ions such a dodecylbenzenesulfonate to render the polymer soluble[11]. The charge on the SO_3^- head group forms an ionic bond with the positive charge (proton) on the PANI chain; the hydrocarbon tail "likes" organic solvents. Processing PANI in the conducting form resulted in materials with improved homogeneity and crystallinity, and with correspondingly improved electrical conductivities[33]; The solubility of "metallic" PANI in organic solvents has also enabled the fabrication of conducting blends of PANI with a variety of insulating host polymers[11,12]. These polymer blends exhibit a remarkably low percolation threshold as a result of the spontaneous formation of an interpenetrating network morphology[13].

Template-guided synthesis of conducting polymers was first reported by S.C. Yang and colleagues[34]. The molecular template, in most cases, polyacids such as polystyrene sulfonic acid, binds the monomer, for example aniline, to form molecular complexes which are dispersed in water as colloidal particles. Upon polymerization, the aniline monomers form polyaniline and remain attached to the template to form the template-polyaniline complex. By judicious choice of the template molecule and the polymerization conditions, stable sub-micron size colloidal particles of polyaniline-template aggregate can be formed during polymerization. The stabilization against coagulation arises from the Coulomb repulsion between particles, which is a result of the surface charge provided by the extra sulfonic acid groups in polystyrene sulfonic acid. Very stable dispersions of polyaniline-polystyrene sulfonic acid complexes can be made with particle sizes less than 1 μm[34]. Transparent films with a specific resistivity of 1-10 Ω-cm can be cast from such dispersions.

Poly(ethylenedioxythiophene)-polystyrene sulfonate (PEDOT-PSS) can be prepared as a stable dispersion in water[34,35]. Films of PEDOT-PSS are semi-transparent and can be spin-cast with a surface resistance of approximately 500 Ω/square and with 75% transmission.

B. Transparent metallic polymers

Metals reflect light at frequencies below the plasma frequency, ω_p, defined as

$$\omega_p^2 = 4\pi Ne^2/m^* \tag{7}$$

where N is the number of electrons per unit volume, e is the electron charge, and m* is the effective mass (the "optical mass") of the electrons in the solid. At frequencies above the plasma frequency, metals are transparent[36]. For conventional metals (Na, Cu, Ag etc.), N is of order 10^{23} per unit volume. As a result, the plasma frequency is in the ultraviolet; therefore, conventional metals appear shiny and "metallic-looking" in the spectral range over which the human eye is sensitive.

Metallic polymers have a lower density of electrons; both the length of the repeat unit along the chain and the interchain spacing are relatively large compared to the interatomic distances in conventional metals. Typically, therefore, for metallic polymers N is of order 2–5 × 10^{21} cm^{-3}. Thus, for metallic polymers, the plasma frequency is at approximately 1 eV[37,38]. The reflectance of high quality, metallic polypyrrole (doped with PF_6) is shown in Fig. 5. Metallic polymers exhibit high reflectance (and thus look "shiny") in the infrared, but they are semitransparent in the visible part of the spectrum. The residual absorption above the plasma frequency arises from interband (π-π*) transitions between the partially filled π-band and the lowest energy π*-band.

Optical quality thin films of metallic polymers are useful, therefore, as transparent electrodes[39]. For example, polyaniline[40], polypyrrole[41] and PEDOT[42] have been used as transparent hole-injecting electrodes in polymer LEDs (the initial demonstration of mechanically flexible polymer LEDs utilized PANI as the anode[40]). Transparent conducting films can be used for a variety of purposes; for example, as antistatic coatings, as electrodes in liquid crystal display cells or in polymer LEDs, or for fabricating electrochromic windows.

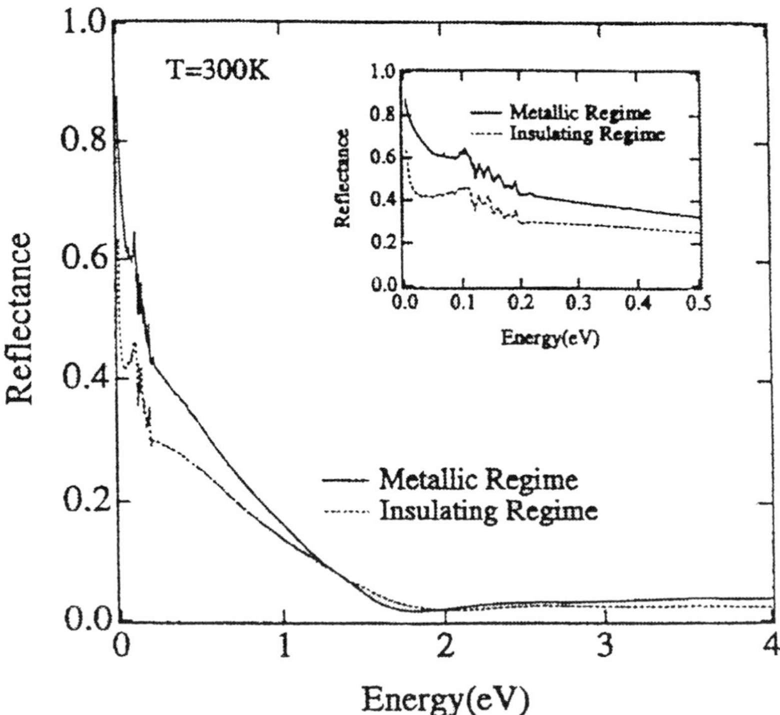

Figure 5. Reflectance spectra of PPy-PF_6 near the metal-insulator transition; inset shows the low energy spectra with expanded scale.

C. Semiconductors in which the Fermi energy can be shifted across the energy gap

In comparison with traditional inorganic semiconductors, semiconducting polymers cannot be considered materials with ultra-high purity. As a result, although many device concepts have been demonstrated using semiconducting polymers as the active materials[32], there was early skepticism that these novel semiconductors could be used in commercial applications.

In some ways, however, semiconducting polymers are more robust than their inorganic counterparts. In particular, whereas pinning of the Fermi energy by surface states is a major problem in conventional semiconductors, the Fermi energy can be controlled and shifted all the way across the energy gap in conjugated polymers. The absence of Fermi level pinning and the ability to shift the chemical potential all the way across the energy gap are fundamentally important; these novel features of semiconducting polymers underlie the operation of polymer LEDs, polymer LECs, and polymer photodiodes.

One might have expected chemical reactions between the metal electrode and the polymer to lead to interface states which pin the Fermi level, as in inorganic semiconductors. Experiments have shown, however, that these interfacial interactions do not lead to pinning of the Fermi level. For example, electroabsorption measurements were used to determine the built-in electric field in metal-semiconductor-metal (MSM) structures of the kind used for polymer LEDs[43]. The results indicated that the maximum internal field is nearly equal to the single particle energy gap; the built-in field directly tracked the difference in work functions of the two meal electrodes, as originally proposed by Parker[23]. Thus, the Fermi level is not pinned by surface states. The absence of Fermi level pinning in semiconducting polymers is a major advantage: Conceptually, it greatly simplifies the device physics, and technologically, it greatly simplifies the device fabrication.

IV. SEMICONDUCTING POLYMERS: ELECTRONIC STRUCTURE AND BOND RELAXATION IN EXCITED STATES

A. Band Structure, Electron-Lattice Interaction, Electron-Electron Interaction and Disorder

Although the linear and nonlinear optical properties of conducting polymers have been investigated for over a decade, there is still controversy over the description of the elementary excitations. Are the lowest energy elementary excitations mobile charge carriers (charged polarons) either injected at the contacts or created directly via inter-band photoexcitation, or are the lowest energy excitations bound neutral excitons[19]? The answer is of obvious importance from the perspective of our basic understanding of the physics of conducting polymers. The answer is also important for applications based on these materials.

The central issue relates to the strength of the electron-electron interactions relative to the bandwidth, relative to the electron-phonon interaction and relative to the strength of the mean disorder potential. Strong electron-electron interactions (electron-hole attraction) lead to the creation of localized and strongly correlated negative and positive polaron pair; neutral polaron excitons. Well screened electrons and holes with associated lattice distortions (charged polarons) on the other hand, are more appropriately described using a band picture supplemented by the electron-phonon interaction. Molecular solids such as anthracene[44] are examples of the former, where the absorption is dominated by excitonic features; whereas inorganic semiconductors such as Si and GaAs are examples of the latter (where rigid band theory is a good approximation).

The electronic structure of conjugated polymers was described by SSH[5,6] in terms of a quasi-one-dimensional tight binding model in which the π-electrons are coupled to distortions in the polymer backbone by the electron-phonon interaction. In the SSH model, photoexcitation across the π-π^* band gap creates the self-localized, nonlinear excitations of conducting polymers; solitons (in degenerate ground state systems), polarons, and bipolarons[4]. Direct photogeneration of solitons and polarons is enabled by the Franck-Condon overlap between the uniform chain in the ground state and the distorted chain in the excited state[45,46]. When the ground state is non-degenerate, as in the PPVs, charged polaron pairs can either separate as mobile charged polarons or form bound polaron-excitons; i.e. neutral bipolarons bound by a combination their Coulomb attraction and their shared distortion. Photoluminescence can be described in terms of the radiative decay of polaron-excitons.

Conjugated polymers are π-bonded *macromolecules*, molecules in which the fundamental monomer unit is repeated many, many times. Thus, N, the Staudinger index as in $(CH)_N$, is large. Since the end-points are not important when N is large, the π-electron transfer integral (denoted as "β" in molecular orbital theory[47], and "t" in tight-binding theory) tends to delocalize the electronic wavefunctions over the entire macromolecular chain. This tendency toward delocalization is limited by disorder (which tends to localize the wavefunctions) and by the Coulomb interaction, which binds electrons when transferred to a nearby repeat unit to the positive charge left behind; i.e. to the "hole".

In principle, disorder can be controlled. Chain extended and chain aligned samples can be prepared. Indeed, by utilizing the method of gel-processing of blends of conjugated polymers in polyethylene[48], a high degree of structural order has been attained[45,46]. Thus, as a starting point, it is useful to consider idealized samples in which the macromolecular chains are chain extended and chain aligned.

The relative importance of the Coulomb interaction versus the band structure is a classic problem of the field. In tight-binding theory, the π-electron band structure extends over a band width, $W = 2zt$ where z is the number of nearest neighbors. Thus, for linear polymers with $z = 2$ and $t \approx 2.5$ eV, $W \approx 10$ eV[4]. Since the size of the monomer, typically 5-10 Å along the chain axis, is the smallest length in the problem, the monomer length is the effective

"Bohr radius". Thus, on general grounds, one expects the electron-hole binding energy to be reduced from 13.5 eV (the electron-proton Coulomb binding energy in the H-atom where the Bohr radius is 0.53Å) by a factor of 10–20 simply because of the change in length scale. Dielectric screening provides an additional reduction factor of ε, where $\varepsilon \approx 3$ (typical of conjugated polymers) is the dielectric constant for an electric field along the chain. Thus, the binding energy is expected to be no more than a few tenths of an eV. Since the typical band widths and band gaps are all in the eV range, one can start with a one-electron band approach and treat the Coulomb energy as a perturbation. In this description, the electron-hole bound states are Wannier excitons which are delocalized over a number of repeat units. There are obvious cases where this argument breaks down. For example, one finds that in structures containing benzene rings, specific sub-bands have bandwidth near zero (nodes in the wavefunction reduce the effective transfer integral to zero). When electrons and holes occupy such narrow bands, the corresponding excitons are easily localized by the Coulomb interaction onto a single monomer. Thus, in general, one can expect both "Wannier -like" excitons and "Frenkel-like" excitons for electrons and holes originating from different bands in the same polymer.

B. Electronic Structure of Polyacetylene

Trans-polyacetylene, trans-$(CH)_x$, was the first highly conducting organic polymer[1,2]. The simple chemical structure, -CH- units repeated (see Fig. 6a), implies that each carbon contributes a single p_z electron to the π-band. As a result, the π-band would be half-filled. Thus, based upon this structure, an individual chain of neutral polyacetylene would be a metal; since the electrons in this idealized metal could move only along the chain, polyacetylene would be a one-dimensional (1d) metal. However, experimental studies show clearly that neutral polyacetylene is a semiconductor with an energy gap of approximately 1.5 eV. Rudolf Peierls showed many years ago[49] that 1d metals are unstable with respect to a structural distortion which opens an energy gap at the Fermi level, thus rendering them semiconductors. In the Peierls instability, the periodicity ($\Lambda = \pi/k_F$) of the distortion is determined by the magnitude of the Fermi wave-vector (k_F). Since $k_F = \pi/2a$ for the half-filled band of trans-$(CH)_x$, the Peierls distortion doubles the unit cell, converting *trans*-polyacetylene into *trans*-$(\text{-HC = CH-})_x$, see Fig. 6b where, schematically, the dimerization is drawn as alternating single bonds and double bonds. The π-band of $(CH)_x$ is split into two sub-bands, a fully occupied π-band (the valence band in semiconductor terminology) and an empty π^*-band (the conduction band), each with a wide bandwidth (~5 eV) and significant dispersion. The resulting band structure and associated density of states, shown in Fig. 7, results from the opening of the bandgap that originates from the doubling of the unit cell as a result of the bond alternation caused by the Peierls instability of the 1d metal.

Shortly after the initial discovery of doping and the metal-insulator transition in polyacetylene, the SSH description of the electronic structure was pro-

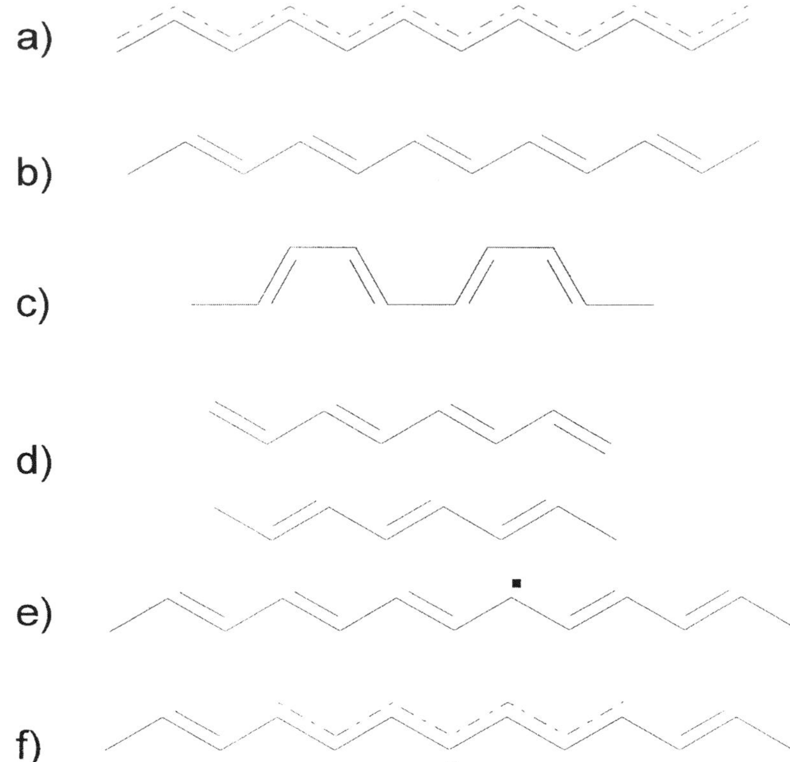

Figure 6. Polyacetylene a) undimerized structure, b) dimerized structure due to the Peierls instability, c) *cis*-polyacetylene, d) degenerate A and B phases in *trans*-polyacetylene, e) soliton in *trans*-polyacetylene.

posed[5,6]. The construction of the remarkably successful SSH Hamiltonian was based on two assumptions: (a) The π-electronic structure can be treated in the tight-binding approximation with a transfer integral $t \approx 2.5$ eV, and (b) The chain of carbon atoms is coupled to the local electron density through the length of the chemical bonds.

$$t_{n,\,n+1} = t_o + \alpha(u_{n+1} - u_n) \tag{8}$$

where $t_{n,n+1}$ is the bond-length dependent hopping integral from site n to n+1, and u_n is the displacement from equilibrium of the nth carbon atom. The first assumption defines the lowest order hopping integral, t_o, in the tight-binding term that forms the basis of the Hamiltonian. The second assumption provides the first-order correction to the hopping integral. This term couples the electronic states to the molecular geometry, giving the electron-phonon (el-ph) interaction where α is the el-ph coupling constant. The bond-length dependent hopping integral is physically correct, as indicated by bond alternation observed in polyacetylene[50,51]. The precise form of Eqn. 8, in which the dependence of the hopping integral on the C-C distance is linearized for small deviations about t_o, is the first term in a Taylor expansion.

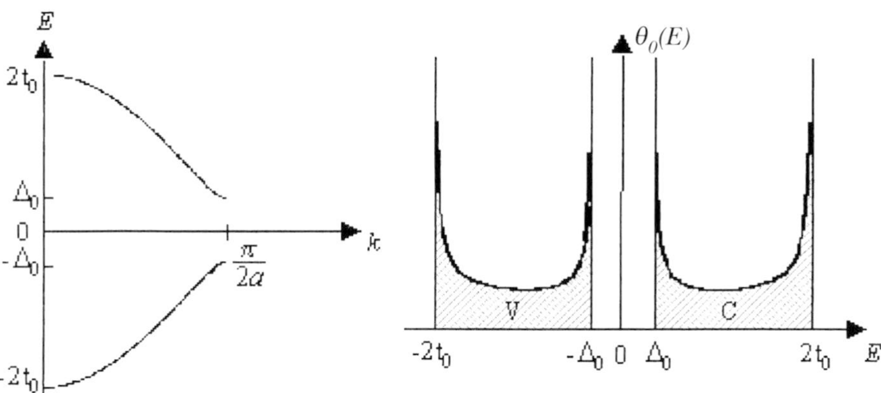

Figure 7. Electronic structure of semiconducting polyacetylene. Left: Band structure. Right: Density of states. The energy opens at k = π/2a as a result of the Peierls distortion.

The resulting SSH Hamiltonian is then written as the sum of three terms [5,6]:

$$H_{SSH} = \sum_{n,\sigma}\left[-t_0 + \alpha(u_{n+1} - u_n)\right]\left(c^+_{n+1,\sigma}c_{n,\sigma} + c^+_{n,\sigma}c_{n+1,\sigma}\right) + \sum_n \frac{p_n^2}{2m} + \frac{1}{2}K\sum_n (u_{n+1} - u_n)^2 \tag{9}$$

where p_n are the nuclear momenta, u_n are the displacements from equilibrium, m is the carbon mass, and K is an effective spring constant. The $c^+_{n,\sigma}$ and $c_{n,\sigma}$ are the fermion creation and annihilation operators for site n and spin σ. The last two terms are, respectively, a harmonic "spring constant" term which represents the increase in potential energy that results from displacement from the uniform bond lengths in $(CH)_x$ and a kinetic energy term for the nuclear motion.

The spontaneous symmetry breaking due to the Peierls instability implies that for the ground state of a pristine chain, the total energy is minimized for $|u_n| > 0$. Thus, to describe the bond alternation in the ground state,

$$u_n -> < u_n > = (-1)^n u_o \tag{10}$$

With this mean-field approximation, the value u_o which minimizes the energy of the system can be calculated as a function of the other parameters in the Hamiltonian[4–6]. Qualitatively, however, one sees that u_o and $-u_o$ both minimize the energy for trans-polyacetylene since the bonds all make the same angle with respect to the chain axis. Hence, the energy as a function of u has a double minimum at $\pm u_o$, as shown in Fig. 8. The corresponding structures of the two degenerate ground states are shown in Fig. 6d. The double minimum of Fig. 8 implies that nonlinear excitations, solitons, will be important. In polymers with a nondegenerate ground state, the degeneracy indicated in Fig. 8 is lifted; for a nondegenerate ground state polymer, the absolute minimum energy occurs at a single value of u.

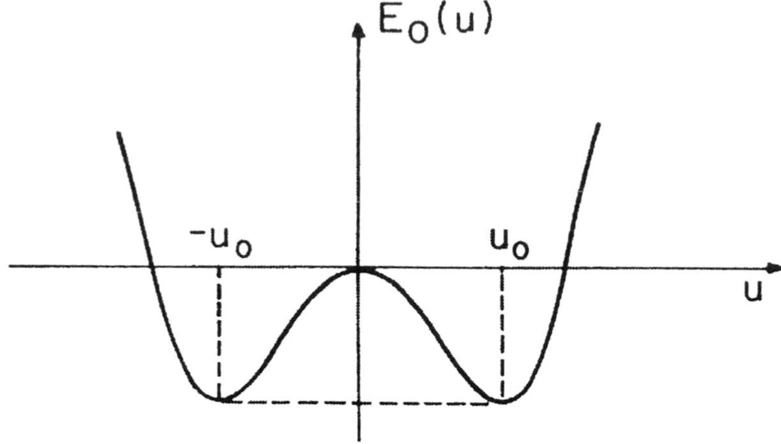

Figure 8. Total energy of the dimerized polyacetylene chain; note the double minimum associated with the spontaneous symmetry breaking and the two-fold degenerate ground state.

C. The electronic structure of poly(phenylene vinylene)

Poly(phenylene vinylene), PPV, and its soluble derivatives have emerged as the prototypical luminescent semiconducting polymers. Since PPV has a non-degenerate ground state, structural relaxation in the excited state leads to the formation of polarons, bipolarons, and neutral excitons. Prior to treating the structural relaxation in the excited state, however, one needs to develop a satisfactory description of the electronic excited states.

In PPV, with eight carbons in the main-chain repeat unit, the π-band is split into eight sub-bands. Since each band can hold precisely 2 electrons per repeat unit, the four π-sub-bands with the lowest energy are filled and the four π^*-sub-bands are empty. Thus, I begin with a description at the one-electron level; i.e. from the point of view of band theory.

Brazovskii and Kirova and Bishop[52,53] treated the band structure of PPV with a tight-binding Hamiltonian, using standard values for the hopping integrals. The el-ph coupling, Eqn. 8, can be included after defining the basic band structure of the semiconductor. From this simple ansatz, basis states which reflect the intrinsic symmetry of the phenyl ring and the dimer were calculated and used to expand the wavefunction. The basis states hybridize as a result of ring-to-dimer hopping, thus yielding the $6 + 2 = 8$ subbands in the π-system of PPV.

The resulting band structure, Fig. 9, has a number of important features. Six of the subbands (three occupied and three unoccupied), labeled D and D* in Fig. 9, are broad, and the corresponding wavefunctions are delocalized along the chain. The other two bands labeled L and L* (one occupied and one unoccupied), are narrow and the corresponding wavefunctions have amplitudes which are localized on the ring. The L and L* subbands derive from the two ring states whose wavefunctions have nodes at the *para* linkage sites; as a result, these states do not participate in hopping and the subbands do not acquire the resulting dispersion. On the contrary, the delocalized D and

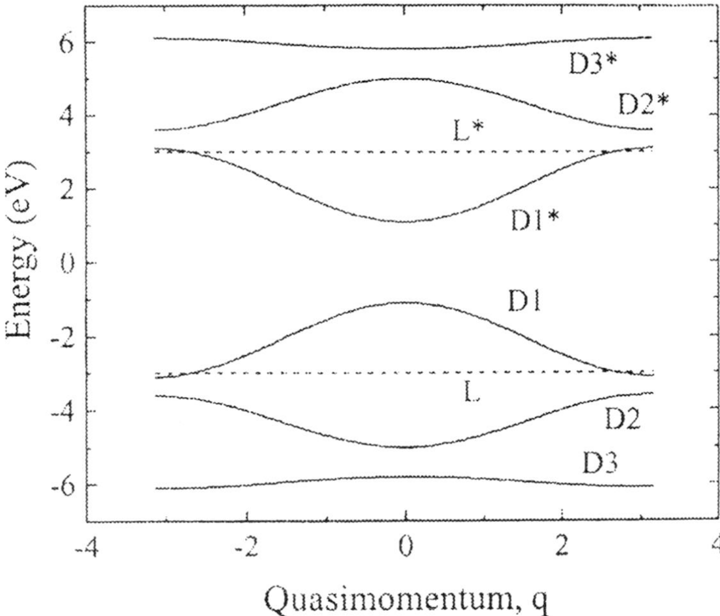

Figure 9. Electronic band structure of PPV.

D* subbands have relatively high dispersion (~1.7 eV), indicating good delocalization over both ring and dimer states.

Starting from this simple yet physically robust model of the band structure, the discussion of the electronic structure can be extended to include the effect of the Coulomb interaction. The delocalized nature of D1 and D1* implies that, for electron correlation effects involving these subbands, it is appropriate to compute an effective mass from the dispersion curves and then compute the corresponding one-dimensional (1d) hydrogenic levels. From the k^2-dependence of the dispersion near the zone center, the effective masses in D1 and D1* are $m^* = 0.067\ m_e$.

Using the well-known result for the binding energy of a hydrogenic-like state in 1d, the exciton associated with the lowest energy electronic transition (D1 - D1*) has a binding energy and effective radii given by:

$$E_b = E_b{}^* ln^2(a_b/a_\perp) \qquad a_b = a_b{}^*[ln(a_b/a_\perp)]^{-1}, \tag{11a}$$

where

$$E_b{}^* = \frac{m^* e^4}{\varepsilon^2 \hbar^2} \qquad a_b{}^* = \frac{\hbar^2 \varepsilon}{m^* e^2}, \tag{11b}$$

ε is the dielectric susceptibility, $\hbar$ is Planck's constant, m^* is the electron effective mass at the zone center in *k*-space (for PPV, $m^* \approx 0.067\ m_e$), e is the electron charge and $\alpha_\perp \approx 2$ Å (the "width" of the chain). From Eqn. 11, $E_b \approx 0.1$–0.2 eV (depending on the value assumed for $\alpha_\perp$) and a radius of approximately 30 Å.

In addition to this weakly-bound Wannier-Mott exciton, there are two other excitons in this model. The degenerate D1-L1* and L1-D1* transitions are treated by assuming an immobile (massive) carrier in the L or L* subband and a mobile carrier in the D or D* subband. The resulting bound state has a binding energy of ~0.8 eV, and is a more tightly bound (but still somewhat delocalized) exciton. Finally, the L-L* transition forms a very tightly bound Frenkel exciton, localized on the phenyl ring.

With the electronic excitations established, they can be compared to the data obtained from optical absorption measurements. An unambiguous test of the agreement between the proposed electronic structure (Fig. 9) and experiment requires data from chain extended and chain oriented samples of macromolecular PPVs. With macromolecular samples, chain end boundary conditions are not important; with oriented polymers, the polarization of the various absorptions with respect to the chain axis can be determined. Recent polarized absorption studies of highy oriented PPV and PPP indicate that the one-electron model is a good starting point[54,55].

D. Solitons, Polarons and Polaron Excitons: The Elementary Excitations of Conducting Polymers

Although bond relaxation in the excited state is implicitly allowed through the bond-length dependent hopping integral in the SSH model, the effect of bond relaxation, and the formation of solitons, polarons and bipolarons, has not been explicit in the previous section. These important concepts are summarized in the following paragraphs. In the related experimental studies, trans-polyacetylene and polythiophene were used as the model systems for the degenerate ground state polymer and the nondegenerate ground state polymer, respectively[4].

a. Solitons

Charge storage on the polymer chain leads to structural relaxation, which in turn localizes the charge. The simplest example of the dramatic effect of this structural relaxation is the soliton in trans-polyacetylene. The soliton is a domain boundary between the two possible degenerate ground state configurations of trans-$(\text{-CH} = \text{CH-})_N$, the "A" Phase and the "B" Phase. For simplicity, we often draw the chemical structure of the soliton as an abrupt change from A Phase to B Phase, as shown in Fig. 6e. In agreement with the predictions of the SSH model [4], however, the experimental evidence indicates that the structural relaxation in the vicinity of the domain boundary extends over approximately fourteen carbon atoms, as illustrated in Fig. 6f. The corresponding spin and charge distributions are similarly delocalized.

The value of a picture like Fig. 6e quickly becomes evident when one tries to understand the quantum numbers of the soliton and the reversed charge-spin relationship that characterizes the solitons of polyacetylene (see Fig. 10c). Since the non-bonding, or mid-gap, state formed by the chain relaxation can then be mapped to a specific atomic site, the resulting distribution of charge and spin can be easily addressed; if the state is unoccupied (doubly occupied), the carbon atom at the boundary is left with a positive (negative) charge, but there are no unpaired spins; the charged soliton is positively (negatively) charged and spinless. Single occupation of the soliton state neutralizes the electronic charge of the carbon nucleus, while introducing an unpaired spin onto the chain. The localized electronic state associated with the soliton is a nonbonding state at an energy which lies at the middle of the π-π^* gap, between the bonding and antibonding levels of the perfect chain. On the other hand, the defect is both topological and mobile due to the translational symmetry of the chain. Such a topological and mobile defect is historically referred to as a soliton[4–6]. The term "soliton" (S) refers simultaneously to all three types of solitons; the quantum numbers for spin and charge enter only when referring to a specific type (for example, neutral solitons found as defects from the synthesis of undoped material or charged solitons created by photoexcitation), and even then, the spin is only implicit. Another feature of the soliton terminology is the natural definition of an "anti-soliton" (AS) as a reverse boundary from B Phase back to A Phase in Figs. 6e and 6f. The anti-soliton allows for conservation of soliton number upon doping or upon photoexcitation.

b. Polarons and Bipolarons

In cases such as poly(thiophene), PPV, PPP and cis-polyacetylene where the two possible bond alternation patterns are not energetically degenerate, confined soliton-antisoliton pairs, polarons and bipolarons, are the stable nonlinear excitation and the stable charge storage states[56,57]. A polaron can be thought of as a bound state of a charged soliton and a neutral soliton whose mid-gap energy states hybridize to form bonding and anti-bonding levels. The neutral soliton contributes no charge and a single spin, as noted above, and the charged soliton carries charge of ±e and no spin; the resulting polaron then has the usual charge-spin relationship of fermions; $q = \pm e$ and $s = 1/2$. The polaron is illustrated schematically for PPP in Fig. 10a and 10b. The positive (negative) polaron is a radical cation (anion), a quasiparticle consisting of a single electronic charge dressed with a local geometrical relaxation of the bond lengths. Similarly, a bipolaron is a bound state of two charged solitons of like charge (or two polarons whose neutral solitons annihilate each other) with two corresponding midgap levels, as illustrated in Fig. 11a and 11b. Since each charged soliton carries a single electronic charge and no spin, the bipolaron has charge ± 2e and zero spin. The positive (negative) bipolaron is a spinless dication (dianion); a doubly charged bound state of two polarons bound together by the overlap of a common lattice distortion (enhanced geometrical relaxation of the bond lengths).

c. Solitons and Polarons in Conjugated Polymers: Experimental Results

The models outlined above for solitons, polarons and bipolarons make concrete predictions of experimentally observable phenomena, and indeed theoretical progress would not have been possible without concurrent experimental investigation. Optical probes of the mid-gap electronic states (see Fig. 10c) and magnetic measurements of spin concentrations and spin distributions contributed greatly to the refinement and verification of theoretical predictions. For a detailed review of the experimental results, the reader is referred to the Review of Modern Physics article on this subject[4].

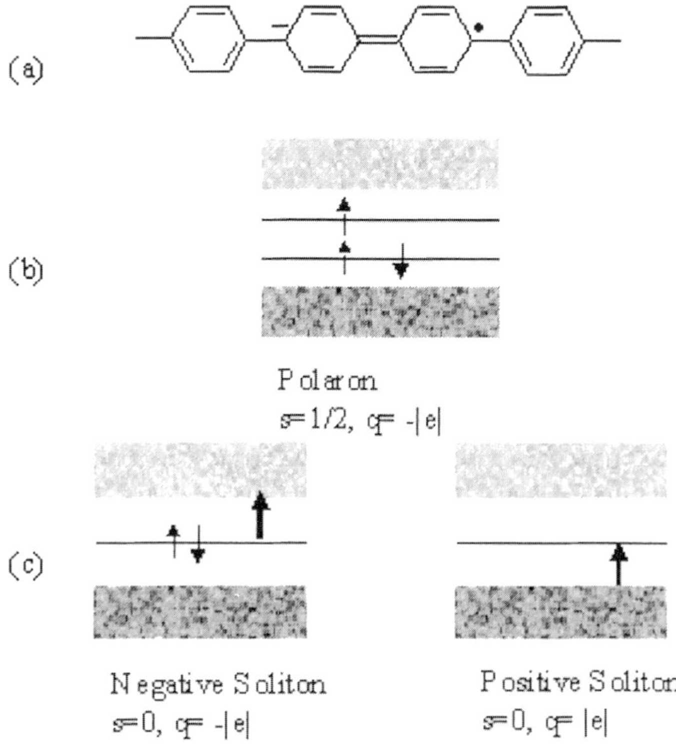

Figure 10. a. Schematic picture of a polaron in PPP; b. Band diagram of an electron polaron; for a hole polaron, the lower gap state is single occupied and the upper gap state is empty; c. Band diagrams for positive and negative solitons with associated electronic transitions.

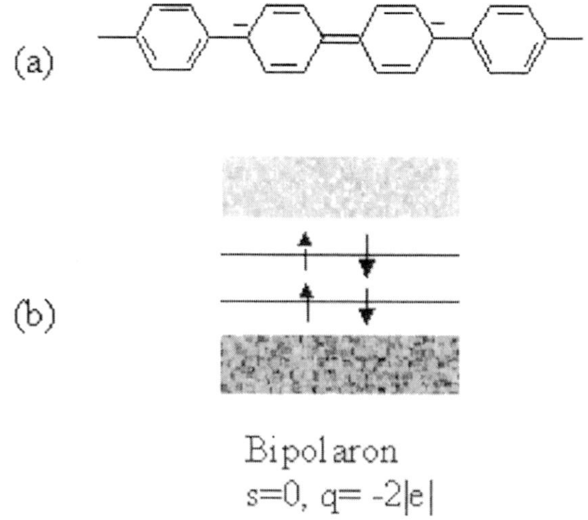

Figure 11. Bipolarons in polymer with non-degenerate ground state. a. Schematic picture of a negative bipolaron in PPP; b. Band diagram for a negative bipolaron. For a positive bipolaron, both gap states are unoccupied.

V. PHOTOINDUCED ELECTRON TRANSFER

The discovery of photoinduced electron transfer in composites of conducting polymers (as donors, D) and buckminsterfullerene, C_{60}, and its derivatives (as acceptors, A) opened a number of new opportunities for semiconducting polymers[58,59]. A schematic description of the photoinduced electron transfer process is displayed in Fig. 12.

A broad range of studies has been carried out to fully characterize this photoinduced charge transfer, culminating with a study of the dynamics of photoinduced electron transfer from semiconducting polymers to C_{60} using fs time resolved measurements[60,61]. These studies have demonstrated that the charge transfer occurs within 50 femtoseconds after photo-excitation. Since the charge transfer rate is more than 1000 times faster than any competing process (the luminescence lifetime is greater than 300 picoseconds), the quantum efficiency for charge separation approaches UNITY! Moreover, the lifetime of the charge transferred state is metastable[62,63].

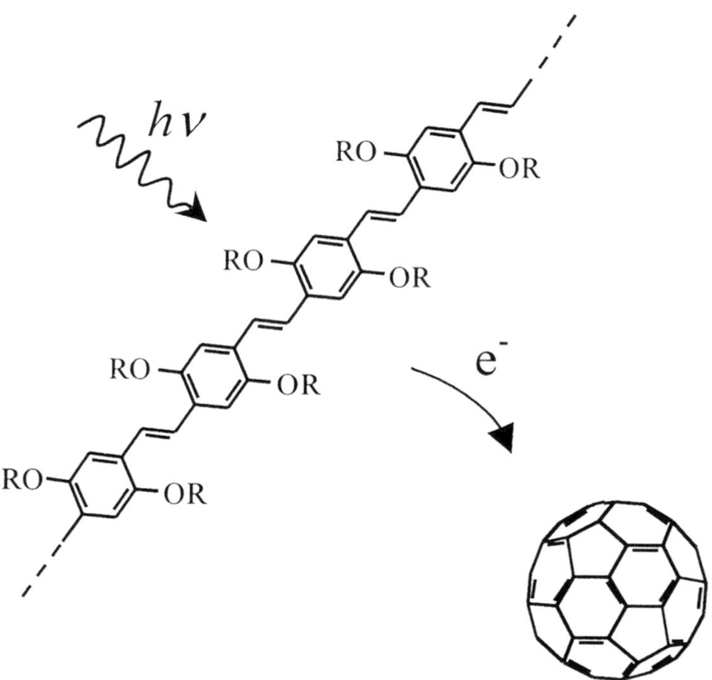

Figure 12. Illustration of photoinduced electron transfer from a conjugated semiconducting polymer to C_{60}.

Semiconducting polymers are electron donors upon photoexcitation (electrons promoted to the antibonding π* band). The idea of using this property in conjunction with a molecular electron acceptor to achieve long-lived charge separation is based on the stability of the photoinduced nonlinear excitations (such as polarons) on the conjugated polymer backbone. Buckminsterfullerene, C_{60}, is an excellent electron acceptor capable of taking on as many as six electrons[64]; C_{60}, therefore, forms charge transfer salts with a variety of strong donors. It is reasonable, therefore to consider electron transfer from photo-excited semiconducting polymer to C_{60}. Fig. 13 shows a schematic energy level diagram of the photoinduced electron (or hole) transfer process, which can be described in terms of the following steps:

Step 1: $D + A \rightarrow {}^{1,3}D^* + A$, (excitation on D);
Step 2: ${}^{1,3}D^* + A \rightarrow {}^{1,3}(D - A)^*$, (excitation delocalized on D-A complex);
Step 3: ${}^{1,3}(D - A)^* \rightarrow {}^{1,3}(D^{\delta+} - A^{\delta-})^*$, (charge transfer initiated);
Step 4: ${}^{1,3}(D^{\delta+} - A^{\delta-})^* \rightarrow {}^{1,3}(D^{+\bullet} - A^{-\bullet})$, (ion radical pair formed);
Step 5: ${}^{1,3}(D^{+\bullet} - A^{-\bullet}) \rightarrow D^{+\bullet} + A^{-\bullet}$, (charge separation);

where the donor (D) and acceptor (A) units are either covalently bound (intramolecular), or spatially close but not covalently bonded (intermolecular); 1 and 3 denote singlet or triplet excited states, respectively.

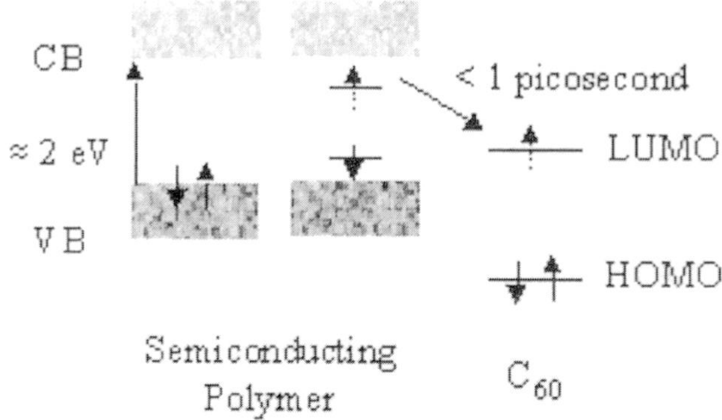

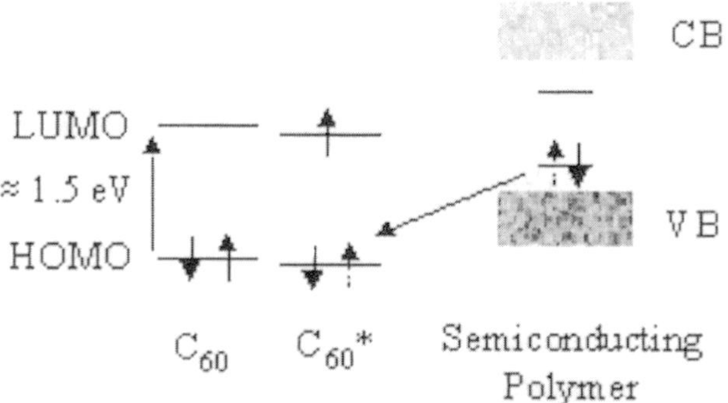

Figure 13: Energy band diagram for photoinduced electron transfer and photoinduced hole transfer between semiconducting polymers and C_{60}.

Since the partial charge transfer at Step 3 is strongly dependent on the surrounding medium there is a continuous range for the transfer rate, $0 < \delta < 1$. At Step 4, $\delta = 1$, *i.e.* a whole electron is transferred. At each step, the D-A system can relax back to the ground state either by releasing energy to the "lattice" (in the form of heat) or through light emission (provided the radiative transition is allowed). The electron transfer step (step 4) describes the formation of an ion radical pair; this does not occur unless $I_{D*} - A_A - U_C < 0$, where I_{D*} is the ionization potential of the excited state (D*) of the donor, A_A is the electron affinity of the acceptor, and U_C is the Coulomb energy of the separated radicals (including polarization effects). Stabilization of the charge separation (step 5) can be enabled by carrier delocalization on the D^+ (and/or A^-) species and by structural relaxation. The symmetrical process of hole transfer from the photoexcited acceptor to the donor is described in an analogous way, and can be driven by photoabsorption in spectral regions where the acceptor can be photoexcited but not the donor.

Even though the photoinduced electron transfer reaction is energetically favorable, energy must be conserved in the process. In semiconducting polymers, the excess energy is readily taken up by promoting the hole to a higher energy state in the π^*-band. Ultimately, therefore, the ultrafast nature of the photoinduced electron-transfer process results from the delocalized nature of the π-electrons. Once the photoexcited electron is transferred to an acceptor unit, the resulting cation radical (positive polaron) species on the conjugated plymer backbone is relatively stable. Thus, photoinduced electron transfer from the conjugated polymer donor onto an acceptor moiety can be viewed as *"photodoping"* (se Fig. 3). The forward-to-reverse asymmetry of the photoinduced charge separation in the conducting polymer/C_{60} system is nevertheless remarkable; the asymmetry is orders of magnitude greater than that observed in the photosynthesis of green plants.

Definitive evidence of charge transfer and charge separation was obtained from light-induced ESR (LESR) experiments[58,63,65]. Upon illuminating the conjugated polymer/C_{60} composites with light with $h\nu > E_{\pi-\pi^*}$ where $E_{\pi-\pi^*}$ is the energy gap of the conjugated polymer (donor), two photoinduced ESR signals can be resolved; one at $g = 2.00$ and the other at $g = 1.99$ [58,63]. The higher g-value line is assigned to the conjugated polymer cation (polaron) and the lower g-value line to C_{60}^- anion. The assignment of the lower g-value line to C_{60}^- is unambiguous, for this low g-value was measured earlier[64]; the higher g-value is typical of conjugated polymers. The LESR signal vanishes above 200K; this rules out permanent photochemical changes as the origin of the ESR signal and demonstrates the reversibility of the photoinduced electron transfer. The integrated LESR intensity shows two peaks with equivalent intensities. The temperature dependence of the LESR signal intensity shows Arrhenius behavior with activation energy of approx. 15 meV. This result suggests a phonon assisted back relaxation mechanism of the photoinduced charge separated state[63].

In MEH-PPV/C_{60} composites, the strong photoluminescence of MEH-PPV is quenched by a factor in excess of 10^3, and the luminescence decay time is reduced from $\tau_o = 550$ ps to $\tau_{rad} << 60$ ps (the instrumental resolution) indicating that charge transfer has cut-off the radiative decay[58,62,63]. An estimate of the transfer rate, $1/\tau_{ct}$, can be obtained from the quenching of the photoluminescence;

$$1/\tau_{ct} \approx (1/\tau_{rad}) I_o / I_{comp} \tag{12}$$

where $1/\tau_o$ and $1/\tau_{rad}$ are the radiative decay rates, I_o and I_{comp} are the integrated photoluminescence intensities of MEH-PPV and the MEH-PPV/C_{60} composite, respectively. The data imply, therefore, that $1/\tau_{ct} > 10^{12}$; i.e. electron transfer occurs on the sub-picosecond time scale. The ultrafast charge transfer process was subsequently time resolved[60,61]; the data directly confirm that charge transfer occurs in less than one hundred femtoseconds.

The photoinduced electron transfer process serves to sensitize the photoconductivity of the semiconducting polymer[66]. Time-resolved transient photocurrent measurements indicate that the addition of as little as 1% of C_{60} into the semiconducting polymer results in an increase of initial photocurrent by an order of magnitude. This increase of the photocarrier generation efficiency is accompanied by an increase in lifetime of the photocarriers upon adding C_{60}. Thus, the ultrafast photoinduced electron transfer from the semiconducting polymer onto C_{60} not only enhances the charge carrier generation in the host polymer but also serves to prevent recombination by separating the charges and stabilizing the charge separation.

Conjugated polymers with higher electron affinities (e.g. cyano-substituted PPV) function as the acceptor component with MEH-PPV as the donor [67,68]. Through control of the morphology of the phase separation into an interpenetrating bicontinuous network of D and A phases, high interfacial area was achieved within a bulk material: a "bulk D/A heterojunction" that yields efficient photoinduced charge separation[67–70]. These all-polymer phase separated blends were successfully used in fabricating solar cells with efficiencies that approach those fabricated from amorphous silicon[69].

VI. METALLIC POLYMERS AND THE METAL-INSULATOR TRANSITION

A. Metallic Polymers with High Performance Electrical and Mechanical Properties

An early (and continuing) goal of the field of conducting polymers was the creation of materials with high electrical conductivity *and* with the excellent mechanical properties of polymers. This goal has been achieved; more importantly the conditions for realizing this combination of properties are understood and can be applied to new materials.

Electrical conductivity results from the existence of charge carriers and the ability of those carriers to move. In principle, broad π-electron bandwidths (often several eV) can lead to relatively high carrier mobilities[4]. As a result of the same intra-chain π-bonding and the relatively strong inter-chain electron transfer interaction, the mechanical properties (Young's modulus and tensile strength) of conjugated polymers are potentially superior to those of saturated polymers. Thus, metallic polymers offer the promise of truly high performance; high conductivity plus superior mechanical properties.

This combination of high electrical conductivity and outstanding mechanical properties has been demonstrated for doped polyacetylene[70–75]. Unfortunately, since doped polyacetylene is not a stable material, the achievement of *stable* high performance conducting polymers remains an important goal.

B. Metal-Insulator Transition in Doped Conducting Polymers

a. The Role of Disorder

Ioffe and Regel[76] argued that as the extent of disorder increased in a metallic system, there was a limit to metallic behavior; when the mean free path becomes equal to the inter-atomic spacing, coherent metallic transport would not be possible. Thus, the Ioffe-Regel criterion is defined as

$$k_F l \approx 1; \tag{13}$$

where k_F is the Fermi wave number and l is the mean free path. The metallic regime corresponds to $k_F l >> 1$. Based on the Ioffe-Regel criterion, Mott proposed[77,78] that a metal-insulator (M-I) transition must occur when the disorder is sufficiently large that $k_F l < 1$. In recognition of Anderson's early work on disorder induced localization, Mott called this M-I transition the "Anderson transition"[79]. In the limit where $k_F l << 1$ (i.e. where the strength of the random disorder potential is large compared to the bandwidth), all states become localized and the system is called a "Fermi glass"[80]. A Fermi glass is an insulator with a continuous density of localized states occupied according to Fermi statistics. Although there is no energy gap, the behavior is that of an insulator because the states at the Fermi energy are spatially localized.

The scaling theory of localization demonstrated that the disorder-induced M-I transition was a true phase transition with a well defined critical point[81]. MacMillan[82] and Larkin and Khmelnitskii[83], showed that near the critical regime of Anderson localization a power law temperature dependence is to be expected for the conductivity.

The M-I transition in conducting polymers is particularly interesting; critical behavior has been observed over a relatively wide temperature range in a number of systems, including polyacetylene, polypyrrole, poly(p-phenylene-vinylene), and polyaniline[33]. In each case, the metallic, critical and insulating regimes near the M-I transition have been identified. The critical regime is tunable in conducting polymers by varying the extent of disorder (i.e. by studying samples with different $\rho_r \equiv \rho(1.4K) / \rho(300K)$), or by applying external pressure and/or magnetic fields. The transitions from metallic to critical behavior and from critical to insulating behavior have been induced with a magnetic field, and from insulating to critical and then to metallic behavior with increasing external pressure[33].

In the metallic regime, the zero temperature conductivity remains finite with magnitude that depends on the extent of the disorder. Metallic behavior has been demonstrated for conducting polymers with $\sigma(T)$ remaining constant as T approaches zero[33]. Well into the metallic regime where the mean free path extends over many repeat units, the residual resistivity will become small, as in a typical metal. However, this truly metallic regime, with $k_F l >> 1$ has not yet been achieved.

b. Infrared Reflectance Studies of the Metallic State and the Metal-Insulator Transition
Infrared (IR) reflectance measurements have played an important role in clarifying the metal physics of conducting polymers. High precision reflectance measurements were carried out over a wide spectral range on a series of PPy-PF_6 samples in the insulating, critical, and metallic regimes near the M-I transition[84]. Since the reflectance in the infrared (IR) is sensitive to the charge dynamics of carriers near the Fermi energy (E_F), such a systematic reflectance study can provide information on the electronic states near E_F and how those states evolve as the system passes through the M-I transition. The data demonstrate that metallic PPy-PF_6 is a 'disordered metal' and that the M-I transition is driven by disorder; similar results were obtained for polyaniline[37,85].

Fig. 14 shows $\sigma(\omega)$ for a series of PPy-PF_6 samples (A-F) as obtained from Kramers-Kronig transformation of $R(\omega)$ at room temperature. For the most metallic samples, $R(\omega)$ exhibits metal-like features; a free carrier plasma resonance as indicated by a minimum in $R(\omega)$ near 1 eV and high $R(\omega)$ in the far IR. As PPy-PF_6 goes from the metallic to the insulating regime via the critical regime (Samples A through F), $\sigma(\omega)$ is gradually suppressed in the IR. In the insulating regime (F), $\sigma(\omega)$ remains well below that of the metallic sample (A) throughout the IR. The $\sigma(\omega)$ spectra are in excellent correspondence with the transport results (this is especially clear at low frequencies in Fig. 14); the better the quality of the sample, as defined by the higher $\sigma_{dc}(300K)$, the higher $R(\omega)$ in the IR[84].

In the far-IR (below 100 cm^{-1}), the Hagen-Rubens (H-R) approximation provides an excellent fit to $R(\omega)$[36],

$$R_{H\text{-}R}(\omega) = 1 - (2\omega/\pi\sigma_{H\text{-}R})^{1/2}, \qquad (14)$$

where $\sigma_{H\text{-}R}$ is the ω-independent conductivity. The $\sigma_{H\text{-}R}$ values obtained from the Hagen-Rubens fits are in remarkably good agreement with the measured values of $\sigma_{dc}(300K)$. The excellent fits and the agreement between $\sigma_{H\text{-}R}$ and $\sigma_{dc}(300K)$ imply a weak ω-dependence in the corresponding optical conductivity, $\sigma(\omega)$, for $\omega < 100$ cm^{-1}.

The $\sigma(\omega)$ data are fully consistent with the "localization-modified Drude model" (LMD)[86]

$$\sigma_{LD}(\omega) = \sigma_{Drude}\{1 - C[1 - (3\tau\omega)^{1/2}]/(k_F l)^2\} \qquad (15)$$

where k_F is the Fermi wave-number and l is the mean free path. In this model, the zero frequency limit determines the constant C (precisely at the M-I Transition, C=1) while a fit to $\sigma(\omega)$ determines $k_F l$. The suppressed $\sigma(\omega)$ as ω approaches zero arises from weak localization induced by disorder[84]. The

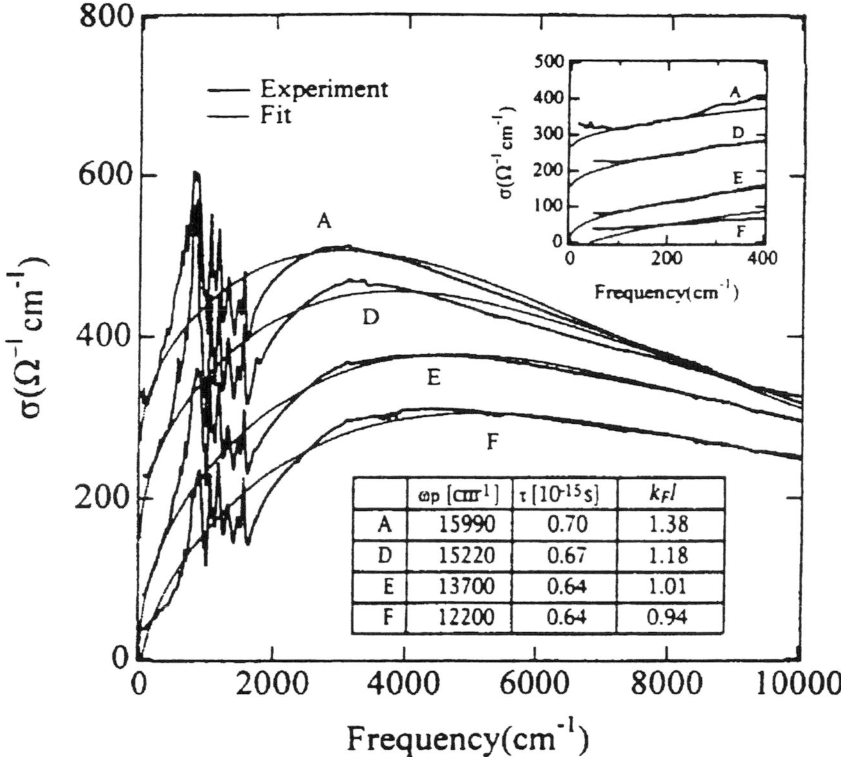

Figure 14. Frequency dependent conductivity, $\sigma(\omega)$, for Ppy-PF_6. The inset shows the far-IR data in greater detail.

$\sigma(\omega)$ data from the various regimes were fit with the functional dependence predicted by the LMD model; Fig. 14 illustrates the excellent agreement of the fits to the data with the parameters summarized in the inset (except for the phonon features around 400–2000 cm^{-1} which are not included in the LMD model). There are small deviations for $\omega < 100\ cm^{-1}$ (more important in the less metallic samples), below which phonon-assisted hopping makes a measurable contribution to $\sigma(\omega)$ and to the dc conductivity.

The parameters obtained from the fits are reasonable. The screened plasma frequency, $\Omega_p = \omega_p/(\varepsilon_\infty)^{1/2} = 1.5 \times 104\ cm^{-1}$, is in good agreement with the frequency of the minimum in $R(\omega)$, and τ is typical of disordered metals ($\tau \sim 10^{-14}$–10^{-15} s). The quantity k_Fl is of particular interest for it characterizes the extent of disorder and is often considered as an order parameter in localization theory[87,88]. For all four samples represented in Fig. 14, $k_Fl \approx 1$, implying that all are close to the M-I transition. As the disorder increases and the system moves from the metallic regime (sample A with $k_Fl = 1.38$) to the critical regime (sample E with $k_Fl = 1.01$), k_Fl approaches the Ioffe-Regel limit[76], precisely as would be expected. In the insulating regime (sample F) $k_Fl = 0.94 < 1$, consistent with localization of the electronic states at E_F.

Thus, the IR reflectance data obtained for metallic polymers indicate that

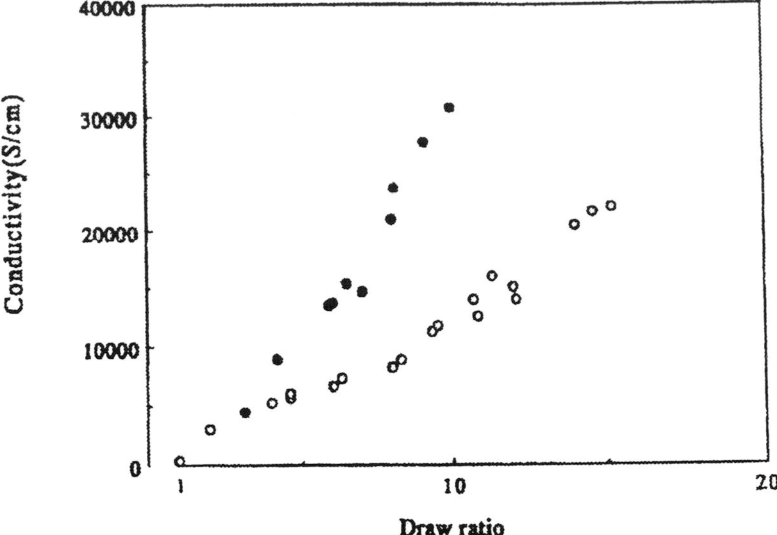

Figure 15. Electrical conductivity of iodine-doped polyacetylene (parallel to the draw axis) vs. draw ratio; solid points, thin films of thickness 3–5 μm; open circles, films of thickness 25–30 μm.

they are disordered metals near the disorder-induced M-I transition. There is remarkable consistency between the conclusion obtained from transport studies and from IR reflectance measurements.

C. Striving Toward More Perfect Materials - Chain Extension, Chain Alignment and Inter-chain Order

Experimental studies have established that for conducting polymers, the electrical properties and the mechanical properties improve together, in a correlated manner, as the degree of chain extension and chain alignment are improved. Polyacetylene remains the prototype example.

Fig.15 shows the correlation between the electrical conductivity (σ) and the draw ratio (λ) for iodine doped polyacetylene films; the conductivity increases approximately linearly with the draw ratio[73]. The slope of σ versus λ is approximately a factor of two larger for the thinner films (evidently the details of polymerization and/or doping result in more homogeneous, higher quality material for the thinner films).

X-ray diffraction studies of these drawn films exhibited a high degree of structural order that improves with the draw ratio; the structural coherence length perpendicular to the draw direction increases by about a factor of two as the films are drawn, from 10 nm at $\lambda = 4$ to 20 nm at $\lambda = 15$[73]. Consistent with the chain orientation and the improved structural order, the anisotropy ($\sigma_{||}/\sigma_{\perp}$) in the electrical conductivity increased with the draw ratio, approaching 250 as $\lambda \rightarrow 15$ (although $\sigma_{||}$ increased dramatically as a function of λ, $\sigma_{\perp}$ remains essentially constant). These data set a lower limit on the intrinsic anisotropy and demonstrate that heavily doped polyacetylene is a highly anisotropic metal with relatively weak interchain coupling.

Does the weak interchain coupling imply poor mechanical properties? The answer is "No"; for films with $\lambda = 15$, Young's modulus reaches 50 GPa and the tensile strength approaches 1 GPa, mechanical properties that are characteristic of high performance materials. More importantly, the data demonstrate a direct correlation between the electrical conductivity and the mechanical properties. The linear relationship implies that the increase in both the conductivity and the modulus (or tensile strength) with draw ratio result from increased uniaxial orientation, improved lateral packing and enhanced interchain interaction. The correlations between the electrical conductivity and the mechanical properties observed for polyacetylene are general. This correlation has been demonstrated in every case studied[31], and can be understood as a general feature of conducting polymers.

Although the electrical conductivity is enhanced by the relatively high mobility associated with intra-chain transport, one must have the possibility of inter-chain charge transfer to avoid the localization inherent to systems with a 1d electronic structure[77]. The electrical conductivity becomes three-dimensional (and thereby truly metallic) only if there is high probability that an electron will have diffused to a neighboring chain prior to traveling between defects on a single chain. For well-ordered crystalline material in which the chains have precise phase order, the interchain diffusion is a coherent process. In this limit, the condition for extended anisotropic transport is the following[89]:

$$L/a \gg (t_o/t_{3d}) \tag{16}$$

where L is the coherence length, a is the length of the chain repeat unit, t_o is the intra-chain π-electron transfer integral, and t_{3d} is the inter-chain π-electron transfer integral.

A precisely analogous argument can be constructed for achieving the intrinsic strength of a polymer material; i.e. the strength of the main-chain covalent bonds. If E_o is the energy required to break the covalent main-chain bond and E_{3d} is the weaker inter-chain bonding energy (from Van der Waals forces and hydrogen bonding), then the requirement is coherence over a length L such that[90]

$$L/a \gg E_o / E_{3d}. \tag{17}$$

In this limit, the large number, L/a, of weak interchain bonds add coherently such that the polymer fails by breaking of the covalent bond. The direct analogy between Eqns. 15 and 16 is evident. When the interchain structural order is "nematic" (i.e. without precise phase order), the conditions become $L/a \gg (t_o/t_{3d})^2$ and $L/a \gg (E_o/E_{3d})^2$, respectively[89]. Thus, the achievement of high performance materials with nematic order requires even greater longer range intra-chain coherence.

In fact, for conjugated polymers, E_o results from a combination of σ and π bonds (the latter being equal to t_o) and E_{3d} is dominated by the interchain transfer integral, t_{3d}. Thus, the inequalities imply that, quite generally, the conductivity and the mechanical properties will improve in a correlated manner as the degree of chain alignment is increased. This prediction is in excellent agreement with data obtained from studies of the poly(3-alkylthiophenes), the poly(phenylene vinylenes), poly(thienylene vinylene) and polyacetylene[31].

Assuming that the linear correlation persists to even higher draw ratios, the extrapolated modulus of 300 GPa for polyacetylene would imply that the electrical conductivity of perfectly oriented polyacetylene would be approximately 2×10^5S/cm[91]. However, this extrapolated value might still be limited by structural defects (e.g. sp^3defects, etc.) rather than by intrinsic phonon scattering.

A theoretical estimate of the intrinsic conductivity, limited by phonon scattering, yields the following expression[89]:

$$\sigma_{ll} = (e^2/2ha)na^3(2\pi M\omega_o t_o{}^2/\alpha^2 h)\exp[h\upsilon_{ph}/k_B T] \qquad (18)$$

where σ_{ll} is the conductivity parallel to the chain, h is Plancks's constant, α is the strength of the electron-phonon interaction, M is the mass of the repeat unit of length α, and t_o is the transfer integral (2-3 eV). Using parameters obtained from experiment, Eqn. 17 predicts $\sigma_{ll} = 2 \times 10^6$ S/cm at room temperature[89], more than an order of magnitude greater than that achieved to date. The exponential dependence upon T arises from the fact that backscattering must involve a high energy phonon with wave number near the zone boundary. The high conductivity values and the exponential increase in conductivity on lowering the temperature predicted by Eqn. 17 indicates that the achievement of metallic polymers in which the macromolecules are chain extended and chain aligned is a major opportunity.

D. The Possibility of Superconductivity in Metallic Polymers

Superconductivity has not yet been observed in doped conjugated polymers. Might one expect superconductivity? If so what are the materials requirements?

There is every reason to be optimistic that superconductivity will be discovered in doped conjugated polymers:

(i) They are metals;

(ii) The coupling of the electronic structure to the molecular structure is well known. Upon doping, the bond lengths change such that charge is stored in solitons, polarons and bipolarons. Thus, the electron-phonon interaction which is responsible for superconductivity in conventional metals, leads to important effects in metallic polymers. In this context, doubly charged bipolarons can be thought of as analogous to real-space Cooper pairs.

As shown above, however, currently available metallic polymers are barely metallic; their electronic properties are dominated by disorder with mean free paths close to the Ioffe-Regel criterion for disorder induced localization.

It is interesting to compare the transport properties of available metallic polymers with those of the underdoped High T_c superconductors at doping levels where $k_F l \sim 1$. For example, when $YBa_2Cu_3O_{7-\delta}$ is underdoped to values of δ such that the resistivity *increases* as the temperature is lowered with $\rho_r \equiv \rho(1.4K)/\rho(300K) \approx 2$ (i.e. with temperature dependence similar to that found in the best metallic polymers), the disorder associated with the random occupation of the oxygen sites quenches the superconductivity.

The phase diagrams of $MBa_2Cu_3O_{7-\delta}$ (where M = Y, Dy, etc.) have been thoroughly studied[92]. For large δ, $MBa_2Cu_3O_{7-\delta}$ are antiferromagnetic insulators; increasing the oxygen concentration introduces carriers into the CuO_2 planes and results in a transition from the antiferromagnetic insulating phase to the metallic and superconducting phase. The changes in resistivity which occur during this evolution from insulator to metal (and superconductor) have been followed in detail in a continuous set of resistivity measurements carried out on ultrathin films[93,94]. Thus, the prototypical High-T_c "superconductors" show behavior characteristic of the disorder-induced metal-insulator transition. Near the metal-insulator transition, the ρ vs T curves look remarkably similar to those obtained from "metallic" polyaniline, polypyrrole, and PPV. At doping levels near the metal-insulator transition there is no sign whatever of the fact that these copper-oxide systems exhibit superconductivity with the highest known transition temperatures.

The point is quite clear. One cannot expect to make any progress toward superconductivity in metallic polymers until there is a major improvement in materials quality to the point where the mean free path is much longer than the characteristic monomer repeat units. When this has been achieved, $k_F l >> 1$ and the resistivity will be truly metal-like, decreasing as the temperature decreases. The unresolved question is how to realize the required improvements in structural order? This remains as a major challenge to the field.

Although the required structural order is lacking in materials that are currently available, there is evidence that in metallic polyaniline, the electronic states near the Fermi energy interact strongly and selectively with a specific optical phonon mode[95]. The 1598 cm^{-1} Raman-active vibrational mode (A_g symmetry) exhibits a distinct resonance enhancement associated with the mid-IR absorption in metallic polyaniline. Since the mid-IR oscillator strength results from the intraband free-carrier Drude absorption, the resonant enhancement suggests that the symmetry of the electronic wavefunctions near E_F matches the vibrational pattern of the 1598 cm^{-1} normal mode. Quantum chemical calculations of the electronic wavefunctions and analysis of the normal modes verified that this is in fact true[95]. Since the coupling of the electronic states near E_F to lattice vibrations is known to lead to superconductivity in metals, the demonstration of resonant coupling to a specific vibrational mode provides some optimism; at least this is the kind of feature that one would like to see.

The story of superconductivity in metallic polymers has not yet begun. If and when superconductivity is discovered in this class of materials, that discovery will create a new research opportunity that will be both exciting and potentially important.

VII. APPLICATIONS

Solid state photonic devices are a class of devices in which the quantum of light, the photon, plays a role. They function by utilizing the electro-optical and/or opto-electronic effects in solid state materials. Because the interband optical transition (absorption and/or emission) is involved in photonic phenomena and because photon energies from near infrared (IR) to near ultraviolet (UV) are of interest, the relevant materials are semiconductors with band gaps in the range from 1 to 3 eV. Typical inorganic semiconductors used for photonic devices are Si, Ge, GaAs, GaP, GaN and SiC. Photonic devices are often classified into three categories: light sources (light emitting diodes, diode lasers etc.), photodetectors (photoconductors, photodiodes etc.) and energy conversion devices (photovoltaic cells). All three are important. Because photonic devices are utilized in a wide range of applications, they continue to provide a focus for research laboratories all over the world.

Most of the photonic phenomena known in conventional inorganic semiconductors have been observed in these semiconducting polymers[32]. The dream of using such materials in high performance "plastic" photonic devices is rapidly becoming reality: High performance photonic devices fabricated from conjugated polymers have been demonstrated, including light emitting diodes, light emitting electrochemical cells, photovoltaic cells, photodetectors, and optocouplers; i.e. all the categories which characterize the field of photonic devices. These polymer-based devices have reached performance levels comparable to or even better than their inorganic counterparts.

This Nobel Lecture is, perhaps, not the proper place to discuss in detail the many current and anticipated applications of semiconducting and metallic polymers; there is simply neither time in the presentation nor space in the written document to do justice to the subject. Suffice it to say here that I am convinced that we are on the verge of a revolution in "Plastic Electronics". I refer those interested to recent reviews that focus on these important developments[97].

VIII. CONCLUDING COMMENTS

I am greatly honored to receive the Nobel Prize in Chemistry for "the discovery and development of conducting polymers". The science of semiconducting and metallic polymers is inherently interdisciplinary; it falls at the intersection of Chemistry and Physics. In creating and expanding this "fourth generation"of polymers, we attempted to understand nature with sufficient depth that we could achieve materials with novel and unique properties; properties that are not otherwise available. This was (and is) an elegant and somewhat dangerous exercise; elegant because it required the synthesis of knowledge from the two disciplines, and dangerous because one is always pushing beyond the knowledge and experience of his background. I started out as a physicist; however, I am what I have become. I have evolved, with the help of many colleagues in the international scientific community, into an interdisciplinary scientist. That my work and that of my colleagues Alan MacDiarmid and Hideki Shirakawa has had sufficient impact on Chemistry to be recognized by the Nobel Prize gives me, therefore, particular satisfaction.

REFERENCES

1. H. Shirakawa, E.J. Louis, A.G. MacDiarmid, C.K. Chiang and A.J. Heeger, Chem. Commun. 578, (1977).
2. C.K. Chiang, C.R. Fincher, Jr., Y.W. Park, A.J. Heeger, H. Shirakawa and E.J. Louis, Phys. Rev. Lett. 39, 1098 (1977).
3. B. Rånby in Chapter 3 of *Conjugated Polymers and Related Materials: The Interconnection of Chemical and Electronic Structures,* Edited by W.R. Salaneck, I. Lundström and B. Rånby, Oxford University Press, Oxford, (1993).
4. A.J. Heeger, S. Kivelson, J.R. Schrieffer and W.P. Su, Rev. Mod. Phys.60, 781 (1988).
5. W.P. Su, J.R. Schrieffer and A.J. Heeger, Phys. Rev. Lett. 42, 1698(1979).
6. W.P. Su, J.R. Schrieffer and A.J. Heeger, Phys. Rev. B22, 2099 (1980).
7. C.K. Chiang, S.C. Gau, C.R. Fincher, Jr., Y.W. Park, A.G. MacDiarmid, Appl. Phys. Lett. 33,18 (1978).
8. P.J.Nigrey, A.G. MacDiarmid and A.J. Heeger, Chem. Commun. 96, 594 (1979).
9. Q. Pei, G. Yu, C. Zhang, Y. Yang and A.J. Heeger, Science, 269, 1086 (1995).
10. W.R. Salaneck, I. Lundstrom, W.S. Haung and A.G. MacDiarmid, Synth. Met.13, 291 (1986).
11. Y. Cao, P. Smith and A.J.Heeger, Synth. Met. 48, 91 (1992).
12. Y. Cao, P. Smith and A.J. Heeger, U.S. Patent 5,232,631.
13. C.Y. Yang, Y. Cao, P. Smith and A.J. Heeger, Synth. Met. 53, 293 (1992).
14. M. Reghu, Y. Cao, D. Moses, and A.J. Heeger, Phys. Rev. B497,1758 (1993).
15. H. Tomozawa, D. Braun, S. Philips, A.J. Heeger and H. Kroemer, Synth. Met.22, 63 (1987); H. Tomozawa, D. Braun, S.D. Philips, R. Worland, and A.J. Heeger, Synth. Met. 28, C687 (1989).
16. J.H. Burroughes, D.D.C. Bradley, A.R. Brown, R.N. Marks, R.H. Friend, P.L. Burns and A.B. Holmes, Nature, 347, 539 (1990).
17. A.G. MacDiarmid and A.J.Epstein in "Conjugated Polymeric Materials: Opportunities in Electronics, Optical Electronics and Molecular Electronics", Ed. by J.L. Bredas and R.R. Chance (Kluwer Academic Publishers, Dordrecht, 1990), p.53.
18. F. Wudl, R.O. Angus, F.L.Lu, P.M. Allemand, DJ. Vachon, M. Nowak, Z.X. Liu,and A.J. Heeger, J. Amer. Chem. Soc. 109, 3677 (1987).
19. N.S. Sariciftci, "Primary Photoexcitations in Conjugated Polymers" (World Scientific Publishing Co., Singapore, 1998).
20. S.A. Jenekhe, and J.A. Osaheni, Science, 265, 765 (1994).
21. J. Cornil, D.A. do Santos, X. Crispin, R. Silbey and J.L. Bredas, J. Amer. Chem. Soc. 120, 1289 (1998).
22. L.J. Rothberg, M. Yan, F. Papadimitrakopoulos, M.E. Galvin, E.W. Kwock and T.M. Miller, Synth. Met. 80, 41 (1996).
23. I.D. Parker, J. Appl.Phys. 75, 1656 (1994).
24. E.S. Maniloff, D. Vacar, D.W. McBranch, H. Wang, B.R. Mattes, J. Gao and A.J. Heeger, Opt. Commun., 141, 243 (1997).
25. L.Y. Chiang, A.F. Garito and D.J. Sandman, Proceedings of the Materials Research Society, Vol. 247 (Materials Research Society, Pittsburgh, 1992).
26. F. Garnier, R. Hajlaoui, A. Yasser and P. Svrivastava, Science, 265, 1684(1994).
27. J.H. Burroughes, C.A. Jones and R.H. Friend, Nature, 335, 137 (1988).
28. C.J. Drury, C.M.J. Mutsaers, C.M. Hart, M. Matters, and D.M. de Leeuw, Appl. Phys. Lett. 73, 108 (1998).
29. F. Hide, M. Diaz-Garcia. B. Schwartz, M. Andersson, Q.Pei and A.J. Heeger, Science, 73, 1833 (1996).

30. M. Kobayashi, N. Colaneri, M. Boysel, F. Wudl, and A.J. Heeger, J. Chem. Phys., 83, 5717 (1985).
31. A.J. Heeger and P. Smith in "Conjugated Polymers", Ed. by J.L. Bredas and R. Silbey (Kluwer Acad. Publishers, Dordrecht, 1991), p. 141.
32. G. Yu and A.J. Heeger in "The Physics of Semiconductors", Ed. by M. Schleffer and R. Zimmerman (World Scientific Publishing Co., Singapore, 1996), Vol. 1.
33. R. Menon, C.O. Yoon, D. Moses and A.J. Heeger in "Handbook of Conducting Polymers" 2nd Edition, Ed. by T.A. Skotheim, R.L. Elsenbaumer and J.R. Reynolds (Marcel Dekker, Inc., New York, 1998), p. 85.
34. L. Sun and S.C. Yang, Polymer Preprints,33, 379 (1992).G. Heywang and F. Jonas, Adv. Mater. 4, 116 (1992).
35. J. Friedrich and K. Werner, U.S. Patent 5,300,575.
36. C. Kittel, "Introduction to Solid State Physics", (John Wiley and Sons, Inc. New York, 1986).
37. K.Lee, A.J. Heeger and Y. Cao, Phys. Rev. B48, 14884 (1993).
38. K. Lee, R. Menon, C.O. Yoon and A.J. Heeger, Phys. Rev. B52, 4779 (1995).
39. Y. Cao, G.M. Treacy, P. Smith and A.J. Heeger, Appl. Phys. Lett. 60, 2711 (1992).
40. G. Gustafsson, Y. Cao, G.M. Treacy, F. Klavetter, N. Colaneri, and A.J. Heeger, Nature, 357, 477 (1992).
41. J. Gao, A.J. Heeger, J.Y. Lee and C.Y. Kim, Synth. Met. 82, 221 (1996).
42. Y. Cao, G. Yu, C. Zhang, R. Menon and A.J. Heeger, Synth. Met. 87, 171 (1997).
43. I.H. Campbell, T.W. Hagler, D.L. Smith and J.P. Ferraris, Phys. Rev. Lett., 76, 1900 (1996).
44. M. Pope and C.E. Swenberg, "Electronic Prosesses in Organic Crystals" (Clarendon Press, Oxford, 1982).
45. T.W. Hagler and A.J. Heeger, Chem. Phys. Lett., 189, 333 (1992).
46. T.W. Hagler,K. Pakbaz,, K. Voss, and A.J. Heeger, Pjys. Rev. B44, 8652 (1991).
47. A. Streitweiser, "Molecular Orbital Theory for Organic Chemists" (John Wiley and Sons, New York, 1961).
48. P. Smith, P.J. Lemstra, J.P.L. Pijpers and A.M. Kile, Colloid and Polym. Sci. , 259, 1070 (1981).
49. R.E. Peierls, "Quantum Theory of Solids" (Clarendon Press, Oxford, 1955).
50. C.R. Fincher,C.E. Chen, A.J. Heeger, A.G. MacDiarmid, and J. B. Hastings, Phys. Rev. Lett.48, 100 (1982).
51. C.S. Yannoni and T.C. Clarke, Phys. Rev. Lett. 52 1191 (1983).
52. S. Brazovskii, N. Kirova and A.R. Bishop, Opt. Mater. 9, 465 (1998).
53. N. Kirova, S. Brazovskii, and A.R. Bishop, Synth. Met. 100, 29 (1999).
54. E. K. Miller, D. Yoshida, C.Y. Yang and A.J. Heeger, Phys. Rev. B59, 4661 (1999).
55. E.K. Miller, C.Y. Yang and A.J. Heeger, Phys. Rev. B62, 6889 (2000).
56. K. Fesser, A.R. Bishop and D. Campbell, Phys. Rev. B27, 4804 (1983).
57. S.A. Brazovski and N. Kirova, JETP, 33, 4 (1981).
58. N.S. Sariciftci, L. Smilowitz, A.J. Heeger and F. Wudl, Science, 258, 1474 (1992).
59. S. Morita, A.A. Zakhidov and Y. Yoshino, Sol. State Commun., 82, 249 (1992).
60. B. Kraabel, D. McBranch, N.S. Sariciftci, D. Moses, and A.J. Heeger, Mol. Cryst. and Liq. Cryst., 256, 733 (1994); B. Kraabel, J.C. Hummelen, D. Vacar, D. Moses, N.S. Sariciftci and A.J. Heeger, J. Chem. Phys. 104, 4267 (1996).
61. G. Lanzani, C. Zenz, G. Cerullo, W. Graupner, G. Leising, U. Scherf, S. De Silvestri, Synth. Met. 111-112, 493 (2000).
62. L. Smilowitz, N.S. Sariciftci, R. Wu, C. Gettinger, A.J. Heeger and F. Wudl, Phys. Rev. B47, 13835 (1993).
63. N.S. Sariciftci and A.J. Heeger, Int. J. of Mod. Phys. B8, 237 (1994).
64. P.M. Allemand, G. Srdanov, A. Koch and F. Wudl, J. Amer. Chem. Soc. 113, 1050 (1991).
65. K. Lee, R. Jansson, N.S. Sariciftci and A.J. Heeger, Phys. Rev. B49, 5781 (1994).

66. C.H. Lee, G. Yu, D. Moses, K. Pakbaz, C. Zhang, N.S. Sariciftci, A.J. Heeger and F. Wudl, Phys. Rev. B48, 15425 (1993).
67. G. Yu and A.J. Heeger, J. Appl. Phys., 78, 4510 (1995).
68. J.J.M. Halls, C.M. Walsh, N.C. Greenham, E.A. Marseglia, R.H. Friend, S.C. Moratti and A.B. Holmes, Nature,376, 498 (1995).
69. G. Yu, J. Gao, J.C. Hummelen, F. Wudl and A.J. Heeger, Science, 270, 1789 (1995).
70. C.Y. Yang and A.J. Heeger, Synth. Met. 83, 85 (1996).
71. K. Akagi, M. Suezaki, H. Shirakawa, H. Kyotani, M. Shimamura, and Y. Tanabe, Synth. Met. 28, D1 (1989).
72. J. Tsukomoto, Jpn. J. of Appl. Phys. 29, 1 (1990).
73. Y. Cao, P. Smith and A.J. Heeger, Polymer, 32, 1210 (1991).
74. T. Schimmel, W. Reiss, J.Gmeiner, M. Schworer, H. Naarmann, and N. Theophilou, Sol. State. Commun. 65, 1311 (1998).
75. S. Tokito, P. Smith and A.J. Heeger, Polymer, 32, 464 (1991).
76. A.F. Ioffe and A.R. Regel, Prog. Semicond. 4, 237 (1960).
77. N.F. Mott and E.A. Davis, "Electronic Processes in Noncrystalline Matrials (Oxford Univ. Press, Oxford, 1979).
78. N.F. Mott, "Metal-Insulator Transition" (Taylor & Francis, London, 1990).
79. P.W. Anderson, Phys. Rev. B109, 1492 (1958).
80. P.W. Anderson, Comm. on Sol. State Phys., 2, 193 (1970).
81. E. Abrahams, P.W. Anderson, D.C. Licciardello and T.V. Ramakrishnan, Phys. Rev. Lett., 42, 695(1979).
82. W.L. McMillan, Phys. Rev. B24, 2739 (1981).
83. A.I. Larkin and D.E. khmelnitskii, Sov. Phys. JETP, 56, 647 (1982).
84. K. Lee, E.K. Miller, A.N. Aleshin,R. Menon, A.J. Heeger, J.H. Kim, C.O. Yoon and H. Lee, Adv. Mater. 10, 456 (1998).
85. K. Lee, A.J. Heeger and Y. Cao, Synth. Met. 69, 261 (1995).
86. N.F. Mott and M. Kaveh, Adv. Phys. 34, 329 (1985).
87. T.G. Castner, in "Hopping Transport in Solids", Ed. byM. Pollak and B.I. Shlovskii (Elsevier Science, Amsterdam, 1991).
88. P.A. Lee and T.V. Ramakrishnan, Rev. Mod. Phys., 57, 287 (1985).
89. S. Kivelson and A.J. Heeger, Synth. Met. 22,371, (1987).
90. A.J. Heeger, Faraday Discuss., Chem. Soc. 88, 1 (1989).
91. S. Tokito, P. Smith and A.J. Heeger, Polymer, 32, 464 (1991).
92. J.M. Tranquada in "Earlier and Recent Aspects of Superconductivity", Ed. by J.G. Bednorz and K.A. Muller (Springer-Verlag, Berlin, 1990).
93. T. Wang, K.M. Beauchamp, D.D. Berkley, B.R. Johnson, J.X. Liu, J. Zhang, and A.M. Goldman, Phys. Rev. B43, 8623 (1991).
94. G. Yu, C.H. Lee, A.J. Heeger, N. Herron, E.M. McCarron, L. Cong, G.C. Spalding, C.A. Nordman, and A.M. Goldman, Phys. Rev. B45, 4964 (1992).
95. N.S. Sariciftci, A.J. Heeger, V. Krasevec, P. Venturini, D. Mihailovic, Y. Cao, J. Libert and J. L. Bredas, Synth. Met. 62, 107 (1994).
96. M.D. McGehee, E.K. Miller, D. Moses and A.J. Heeger, in "Advances in Synthetic Metals: Twenty Years of Progress in Science and Technology", Ed. by P. Bernier, S. Lefrant and G. Bidan (Elsevier Science S.A., Lausanne, 1999).
97. M.D. McGehee and A.J. Heeger, Adv. Mater, *22,* 1655 (2000).

Erratum

Page 405, line 19 should read as:

energy state of the π^*-band[ref]. Ultimately, therefore, the ultrafast nature of the

where ref. refers to M. J. Rice and Yu. N. Gartstein, Phys. Rev. B53, 10764 (1996).

Alan G. MacDiarmid

ALAN G. MACDIARMID

I was born a Kiwi (a New Zealander) in Masterton, New Zealand on April 14, 1927, and still am a Kiwi by New Zealand law, although I became a naturalized United States citizen many years ago in order to have the right to vote in US elections and, hence, voice my political opinions in a meaningful way. My father, an engineer, was unemployed for four years during the Great Depression which hit New Zealand rather severely in the early 1930s. Since jobs were believed to be more plentiful in the vicinity of Wellington, the capital city of New Zealand, located at the bottom of the North Island, we moved to Lower Hutt a few miles from Wellington. There my two older brothers and my elder sister were able to find jobs while I and my younger sister were still at primary school.

My mother and father set the stage for nurturing a warm, loving united, mutually supportive family who always pulled together and also helped others outside the family in need when necessary. Although we did not have too much food, my mother was always inviting other, less fortunate people to meals. On such occasions, my older brothers and sister would frequently remind me and my younger sister at meals not to ask for more food by saying to us out loud at the table, "FHB," which meant, "Family Hold Back," i.e., don't eat too much! We had no phone or refrigerator. In one of the houses we lived in Lower Hutt, our hot water came from water pipes embedded in the brick at the back of the open fireplace in the living room. This resulted in our weekly bath night – where the younger children used the bath water from the older children, to which we were allowed to add more hot water if any still remained! For most of my time at primary school, I went to school barefooted, like most of the other kids. The soles of our feet literally became leather!

Even though I have been away from New Zealand for about 50 years, my brothers and sisters and I (my parents passed on several years ago) are still very closely connected to each other. Throughout the decades we have telephoned each other about every ten days and we all keep up to date with what we are each doing. Shortly after learning of my being a recipient of the Nobel Prize I was speaking to one of my brothers in New Zealand by phone and I said how lucky I was to have been raised in a poor family which was also a close loving family. The fact that we were poor made us self reliant and conscious of the value of money. The fact that we were closely knit taught us the important aspects of interpersonal relationships. Everyone expects "the important things" in life that such as birthday and Christmas presents, but it is the "little unimportant" actions which actually are the real important things. These put the flesh on the skeleton of any relationship. Several hundred of

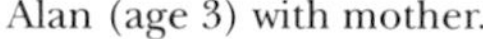

Alan (age 3) with mother.

Alan (age 9) with sister Alice.

these each week – the unimportant, the unexpected, the unnecessary, "the little things", are the things that really count. We are lucky to have been brought up in this environment, but there is a statement on the wall of my study at home in suburban Philadelphia which reads, "I am a very lucky person and the harder I work the luckier I seem to be"!

It is my home life while growing up through high school, which I consider to have been the single most important factor in any success which I may have had in life. As my parents always said, "...an 'A' grade in a class is not a sign of success." Success is knowing that you have done your best and have exploited your God-given or gene-given abilities to the next maximum extent. More than this, no one can do.

For a period in grade school, I attended a two-room school in Keri Keri (town population, 600) where most of my school chums were Maori boys and girls from whom I learned so much. During much of my time at grade school I had an early morning, pre-school job delivering milk on my bicycle for Mr. Bradley, who had a few cows in a nearby paddock. My mother was superb – she would get up with me while it was still dark to make me hot tea to send me on my way. When I started high school it was necessary to give up my milk route. Instead, I delivered the "Evening Post" newspaper on my bicycle after school.

When my father retired (on a very small pension) and moved away from Wellington, it was necessary for me to leave Hutt Valley High School after only three years at the age of 16 and take a low-paying, part-time job as "lab boy"/janitor in the chemistry department at Victoria University College, as it was then known. The total student population was 1200; the Chemistry

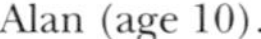
Alan (age 10).

Alan (age 10) with sister Alice in high school uniform.

Department had a faculty of 2! I boarded with friends of my parents and, as a part-time student, took only two courses – one in chemistry and one in mathematics. During this time I became a resident at Weir House, the University dormitory for men. This I found to be one of the most enjoyable and maturing times of my life where I made many good friends from the other ninety residents, with some of whom I still keep in close contact. I remained a part-time student throughout my B.Sc. and M.Sc. studies at Victoria University College. After completing my B.Sc. degree I graduated to the position of demonstrator. Since the age of 17 I have supported myself financially, assisted later only by scholarships and fellowships for which I am most grateful.

My interest in chemistry was kindled when I was about ten years old at which time I found one of my father's old chemistry text books dating back to the late 1800's when he was studying engineering. I spent hours pouring over the pages in complete confusion but with burning curiosity! Some clarification of a type occurred when I rode my bicycle to the public library in Lower Hutt and entered the children's section. There, on the right hand side on the bottom shelf, in the new books section, was a book with a bright blue cover. It was called, "The Boy Chemist." I took it out and continually renewed it by borrowing it for over a year and carried out most of the experiments in it. One of my duties as lab boy, when I was not washing dirty labware or sweeping floors, was to prepare demonstration chemicals for Mr. A.D. "Bobbie" Monro, the lecturer in first-year chemistry. On one occasion he asked me to prepare some S_4N_4–beautiful bright orange crystals. When it became time for me to start my M.Sc. thesis, I asked Mr. Monro if I could look at some of its chemistry. He agreed. This resulted in my first publication in 1949. Its derivatives

Alan (age 12) with bicycle.

were highly colored. Color continued to be one of the driving forces in my future career in chemistry. I love color. Little did I know that thirty years later this was going to be a key factor which would shape my professional life.

In 1950, I had the good fortune to receive a Fullbright fellowship from the U.S. State Department to do a Ph.D. at the University of Wisconsin in the USA where I studied under Professor Norris F. Hall, majoring in Inorganic Chemistry, studying the rate of exchange in ^{14}C-tagged complex metal cyanides. It was at the University of Wisconsin that I became president of the International Club – the largest student organization on campus and had the crucial chance meeting of my life when I met my future wife, Marian Mathieu, at an International Club dance. During this time I was elected by the Department of Chemistry to the position of Knapp Research Fellow and had the privilege of living rent free in the beautiful old ex-governor's mansion on the shores of Lake Mendota.

When I was still at the University of Wisconsin I was successful in obtaining a New Zealand Shell graduate scholarship to study silicon hydrides at Cambridge University, England under the directorship of Professor H.J. Emeléus. It was there that Marian and I were married in the chapel at my college, Sidney Sussex College.

After a brief appointment as a junior faculty member at Queens College of the University of St. Andrews, Scotland, I accepted a junior position on the faculty of the Department of Chemistry at the University of Pennsylvania where I have been for the past 45 years and became father of three daughters and a son and grandparent of nine lovely boys and girls. I grew to love teaching and the stimulation of young fresh inquiring minds. I am still fully en-

Alan (age 12) with parents and one brother and sisters.

gaged in teaching as well as research and indeed have requested to teach a section of first-year chemistry at Penn later this year.

I had the good fortune to meet my future friend and colleague, Professor Alan J. Heeger, Professor of Physics at the University of Pennsylvania. On one occasion he came to my office and informed me that Mort Labes, Professor of Chemistry at Temple University, had published a paper on a highly conducting material. I asked Heeger its formula and he replied, "sss-nnn-ex". Being an inorganic chemist, I wrote down on a piece of paper, "$(Sn)_x$" and said, "Of course you expect it to be conducting, it's a metal!" To which Heeger replied on paper, "No, not $(Sn)_x$, but $(SN)_x$! This was the beginning of our each learning each other's scientific language. I told him that I had made the precursor to $(SN)_x$, i.e. S_4N_4 during my M.Sc. thesis work in New Zealand. He asked me if I could make some $(SN)_x$ – as golden crystals. We were ultimately successful, and co-published many papers together, on this conducting polymer.

When I was a Visiting Professor at Kyoto University in Japan, lecturing on molecular silicon compounds, I visited Tokyo Institute of Technology in 1975 and described our work on $(SN)_x$. Hideki Shirakawa and I met over a cup of green tea after a lecture I gave and as I was showing a sample of our golden $(SN)_x$, he showed me a sample of his silvery $(CH)_x$.

I asked him how he had made this silvery film of polyacetylene and he replied that this occurred because of a misunderstanding between the Japanese language and that of a foreign student who had just joined his group. Shirakawa had been polymerizing ordinary acetylene welding gas using a Ziegler-Natta catalyst and had been obtaining a rather uninteresting black-brown powder. He told the new student to repeat this work using a con-

Alan (age 15) in high school uniform.

Alan (with dog) with parents and brothers and sisters.

centration of the catalyst which was milli-molar. A few days later the student came back and said that the stirring bar would not go around in the flask. Shirakawa went to the laboratory and, sure enough, instead of the black-brown powder, there were lumps of silvery-pinkish jelly floating around. Shirakawa asked what the student had done and the student replied that he had done exactly as Shirakawa had told him; he had made the catalyst with a concentration of "x-molar"– in other words, he had made the catalyst 1000 times more concentrated than Shirakawa had told him! Shirakawa was most intrigued by this observation, since as all good chemists know, a catalyst should only increase the rate of a chemical reaction and should not alter the nature of the product. This then started Shirakawa investigating this silvery form of polyacetylene. I asked Shirakawa if he could join me for a year at the University of Pennsylvania since I was already interested in conducting materials such as the golden $(SN)_x$ films. He stated that he could and when he arrived we tried to make the silvery polyacetylene, $(CH)_x$, more pure and, hence, increase its conductivity. However, we found that the purer we made the $(CH)_x$, by elemental analysis, the lower was its conductivity! Since we had found previously that by adding bromine to the golden $(SN)_x$ material, we could increase its conductivity tenfold, we thought that perhaps the impurity in the polyacetylene was acting as a dopant and was actually increasing the conductivity of the polyacetylene, rather than decreasing it. We therefore decided to add some bromine to the silvery $(CH)_x$ films and immediately, within a few minutes at room temperature, the conductivity increased many millions of times. We then collaborated with my colleague, Professor Alan Heeger, who was well-versed in the physics of conducting materials. The rest

Alan (age 21) and Alice.

is history! When Alan left Penn almost 10 years ago, my ongoing collaboration with my good friend Professor Art Epstein (Physics Dept, Ohio State Univ.) continued at an even more rapid pace.

One of the transparencies I showed at the very end of my Nobel Lecture in Stockholm on December 8, 2000 is given below. Every word carries real meaning and emotion from my heart.

I wish to extend my personal thanks to:

- My *(late)* wife, ***Marian,*** for her dedicated support and love during our 36 years of marriage.
- My loving partner ***Gayl Gentile*** for her untiring personal and professional support throughout the past 9 years.
- My mother, ***Ruby*** and father, ***Archibald MacDiarmid*** for providing a loving and solid home foundation on which to base my life.
- My brothers and sisters, ***Colin, Roderick, Sheila, Alice*** for their ceaseless, loving emotional support during the past 73 years!
- To my children, ***Heather, Dawn, Duncan and Gail,*** for their understanding and forbearance in my not spending as much time with them as I might have during their childhood years.
- To my delightful grandchildren who never cease to be a pleasure with their many questions and boundless enthusiasm.

We all owe so much to those who have gone before us – "we stand on the shoulders of giants."

Copies of the very last transparencies given in my Nobel Lecture are reproduced below. They carry a very special message to all of us.

*Seeking the Great **White Bird** of Absolute Truth*

The dependency of any one person's research on the labors of scores of earlier scientific pioneers is illustrated very beautifully by a few sentences of this variation from a book by Olive Schreiner, written at the turn of the century, entitled, "The Story of an African Farm." I would like to share with you this adapted portion.

The story concerns a young hunter who, in his youth, heard about the great white bird of "absolute truth" which lived at the very top of a high mountain far in the east. He had spent all his life seeking it without success – and now he was growing old.

The old thin hands cut the stone ill and jaggedly, for the fingers were stiff and bent. The beauty and strength of the man were gone.

At last, an old, wizened, shrunken face looked out above the rocks. He saw the eternal mountains still rising to the white clouds high above him.

The old hunter folded his tired hands and lay down by the precipice where he had worked away his life.

"I have sought," he said, "for long years I have labored; but I have not found her. By the rough and twisted path hewn by countless others before me, I have slowly and laboriously climbed. I have not rested. I have not repined. And I have not seen her; now my strength is gone. Where I lie down, worn out, other men will stand, young and fresh. By the steps that I, and those before me, have cut, they will climb; by the stairs that we have built, they will mount. They will never know those who made them, their names are forgotten in the mists of time. At the clumsy work they will laugh; when the stones roll, they will curse us; but they will mount, and on our work they will climb, and by our stair! They will find her, and through us!"

The tears rolled from beneath the shriveled eyelids. If truth had appeared above him in the clouds now, he could not have seen her, the mist of death was in his eyes.

... Then slowly from the white sky above, through the still air, came something falling ... falling ... falling. Softly it fluttered down and dropped on to the breast of the dying man. He felt it with his hands ~

~ it was ~

~ a feather.

"SYNTHETIC METALS": A NOVEL ROLE FOR ORGANIC POLYMERS

Nobel Lecture, December 8, 2000

by

ALAN G. MACDIARMID

Department of Chemistry, University of Pennsylvania, Philadelphia, Pennsylvania 19104-6323, USA.

INTRODUCTION

An organic polymer that possesses the electrical, electronic, magnetic, and optical properties of a metal while retaining the mechanical properties, processibility, etc. commonly associated with a conventional polymer, is termed an "intrinsically conducting polymer" (ICP) more commonly known as a "synthetic metal". Its properties are intrinsic to a "doped" form of the polymer. This class of polymer is completely different from "conducting polymers" which are merely a physical mixture of a nonconductive polymer with a conducting material such as a metal or carbon powder distributed throughout the material.

THE CONCEPT OF DOPING

Conjugated organic polymers are either electrical insulators or semiconductors. Those that can have their conductivity increased by several orders of magnitude from the semiconductor regime are generally referred to as 'electronic polymers' and have become of very great scientific and technological importance since 1990 because of their use in light emitting diodes[1]. *Trans*-$(CH)_x$ and the emeraldine base form of polyaniline are used in Figure 1 to illustrate the increases in electrical conductivity of many orders of magnitude which can be obtained by doping. The conductivity attainable by an electronic polymer has very recently been increased an infinite number of times by the discovery of superconductivity in regioregular poly(3-hexylthiophene)[2]. Although this phenomenon was present only in a very thin layer of the polymer in a Field Effect (FET) configuration at a very low temperature (~2 K) it represents an historical quantum leap – superconductivity in an organic polymer!

Prior to the discovery of the novel protonic acid doping of polyaniline, during which the number of electrons associated with the polymer chain remain unchanged,[3] the doping of all conducting polymers had previously been accomplished by redox doping. This involves the partial addition (reduction)

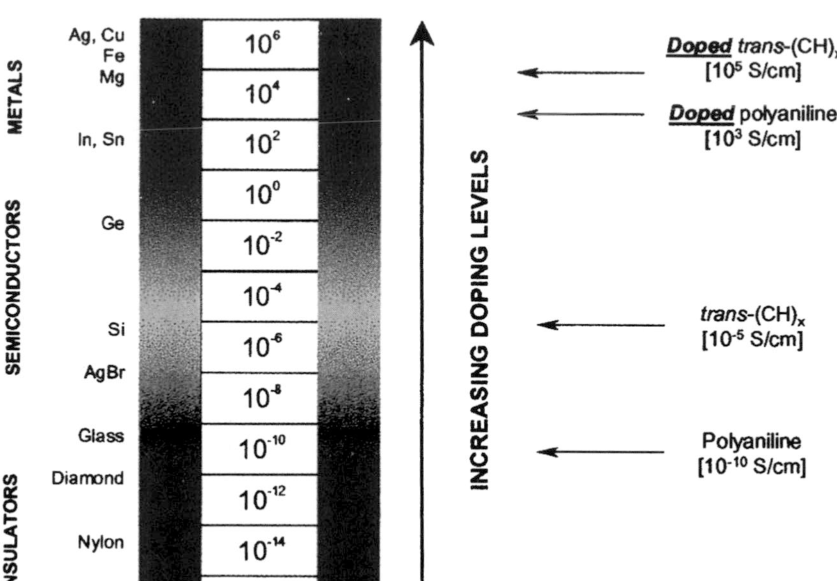

Figure 1. Conductivity of Electronic Polymers

or removal (oxidation) of electrons to or from the pi system of the polymer backbone[4–6].

The concept of doping is the unique, central, underlying, and unifying theme which distinguishes conducting polymers from all other types of polymers[7]. During the doping process, an organic polymer, either an insulator or semiconductor having a small conductivity, typically in the range 10^{-10} to 10^{-5} S/cm, is converted to a polymer which is in the 'metallic' conducting regime (~1 to 10^4 S/cm). The controlled addition of known, usually small ($\leq$ 10 per cent) non-stoichiometric quantities of chemical species results in *dramatic* changes in the electronic, electrical, magnetic, optical, and structural properties of the polymer. Doping is reversible to produce the original polymer with little or no degradation of the polymer backbone. Both doping and undoping processes, involving dopant counterions which stabilize the doped state, may be carried out chemically or electrochemically[6]. Transitory doping by methods which introduce no dopant ions are also known[8].

By controllably adjusting the doping level, a conductivity anywhere between that of the non-doped (insulating or semiconducting) and that of the fully doped (highly conducting) form of the polymer can be easily obtained. Conducting blends of a (doped) conducting polymer with a conventional polymer (insulator), whose conductivity can be adjusted by varying the relative proportions of each polymer, can be made[9]. This permits the optimization of the best properties of each type of polymer.

Since the initial discovery in 1977, that polyacetylene $(CH)_x$, now commonly known as the prototype conducting polymer, could be *p*- or *n*-doped, either chemically or electrochemically to the metallic state[7,10,11], the development of the field of conducting polymers has continued to accelerate at an unexpectedly rapid rate and a variety of other conducting polymers and their derivatives have been discovered[5,6]. This rapid growth rate has been stimulated by the field's fundamental synthetic novelty and importance to a cross-disciplinary section of investigators – chemists, electrochemists, biochemists, experimental and theoretical physicists, and electronic and electrical engineers and to important technological emerging applications of these materials.

In the "doped" state, the backbone of a conducting polymer consists of a delocalized pi system. In the undoped state, the polymer may have a conjugated backbone such as in *trans*-$(CH)_x$ which is retained in a modified form after doping, or it may have a non-conjugated backbone, as in polyaniline (leucoemeraldine base form), which becomes truly conjugated only after *p*-doping, or a non-conjugated structure as in the emeraldine base form of polyaniline which becomes conjugated only after protonic acid doping.

REDOX DOPING

All conducting polymers (and most of their derivatives), e.g. poly-(para-phenylene), 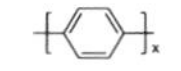 poly(phenylenevinylene), 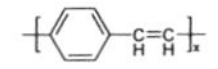 polypyrrole, 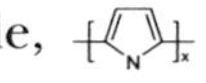 polythiophene, 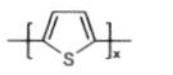 polyfuran, 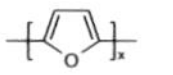 poly(heteroaromatic vinylenes) 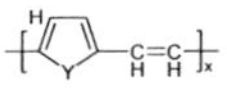 (where Y=NH, NR, S and 0); polyaniline, 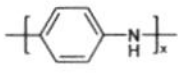 ; etc., undergo either *p*- and/or *n*-redox doping by chemical and/or electrochemical processes during which the number of electrons associated with the polymer backbone changes[5,6]. Selected examples of the different types of doping are presented below.

Chemical and electrochemical p-doping. P-doping, i.e. partial oxidation of the π backbone of an organic polymer, was first discovered by treating *trans*-$(CH)_x$ with an oxidizing agent such as iodine[7], viz.,

$$trans\text{-}[CH]_x + 1.5xy\, I_2 \rightarrow [CH^{+y}(I_3)_y^-]_x \quad (y \leq 0.07)$$

This process was accompanied by an increase in conductivity from ~10^{-5} S/cm to ~10^3 S/cm. If the polymer is stretch-oriented five- to six-fold before doping, conductivities parallel to the direction of stretching up to ~10^5 S/cm can be obtained [5,6].

Approximately 85% of the positive charge is delocalized over 15 CH units (depicted below, for simplicity over only five units) to give a positive soliton, viz.,

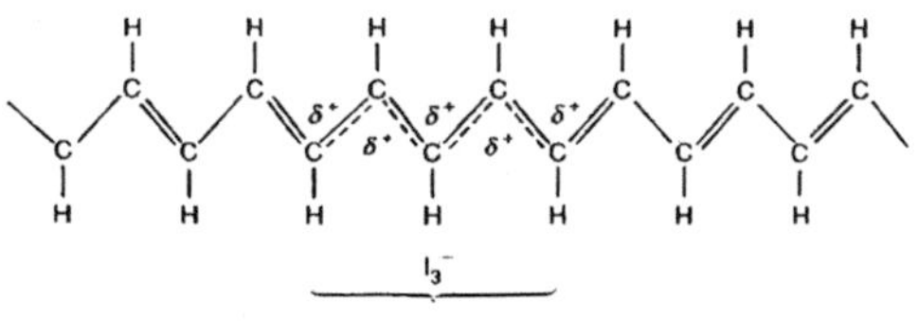

P-doping can also be accomplished by electrochemical anodic oxidation by immersing a *trans*-$(CH)_x$ film in, e.g. a solution of $LiClO_4$ dissolved in propylene carbonate and attaching it to the positive terminal of a DC power source, the negative terminal being attached to an electrode also immersed in the solution[10], viz.,

$$trans\text{-}[CH]_x + (xy)(ClO_4)^- \rightarrow [(CH^{+y}(ClO_4)_y^-]_x + (xy)e^- \quad (y \leq 0.1)$$

Chemical and electrochemical n-doping. N-doping, i.e. partial reduction of the backbone pi system of an organic polymer, was also discovered using *trans*-$(CH)_x$ by treating it with a reducing agent such as liquid sodium amalgam or preferably sodium naphthalide[7], viz.,

$$trans\text{-}[CH]_x + (xy)Na^+(Nphth)^- \rightarrow [Na_y^+(CH)^{-y}]_x + Nphth \quad (y \leq 0.1)$$

The antibonding π^* system is partially populated by this process which is accompanied by an increase in conductivity of ~10^3 S/cm.

N-doping can also be carried out by electrochemical cathodic reduction[11] by immersing a *trans*-$(CH)_x$ film in, e.g. a solution of $LiClO_4$, dissolved in tetrahydrofuran and attaching it to the negative terminal of a DC power source, the positive terminal being attached to an electrode also immersed in the solution, viz.,

$$trans\text{-}[CH]_x + (xy)Li^+ + (xy)e^- \rightarrow [Li_y^+(CH)^{-y}]_x \quad (y \leq 0.1)$$

In all chemical and electrochemical p- and n-doping processes discovered for $(CH)_x$ and for the analogous processes in other conducting polymers, counter 'dopant' ions are introduced which stabilize the charge on the polymer backbone. In each case, spectroscopic signatures, e.g. those of solitons, polarons, bipolarons, etc., are obtained characteristic of the given charged polymer. However, the doping phenomena concept extends considerably beyond that given above to 'doping' processes where no counter dopant ion is involved, i.e. to doping processes in which transitory 'doped' species are produced, which have similar spectroscopic signatures to polymers containing dopant ions. Such type of doping can provide information not obtainable by chemical or electrochemical doping. Examples of such types of redox doping which can be termed 'photo-doping' and 'charge-injection doping' are given below.

DOPING INVOLVING NO DOPANT IONS

Photo-doping. When *trans*-$(CH)_x$ for example, is exposed to radiation of energy greater than its band gap, electrons are promoted across the gap and the polymer undergoes 'photo-doping'. Under appropriate experimental conditions, spectroscopic signatures characteristic of, for example, solitons can be observed[12], viz.,

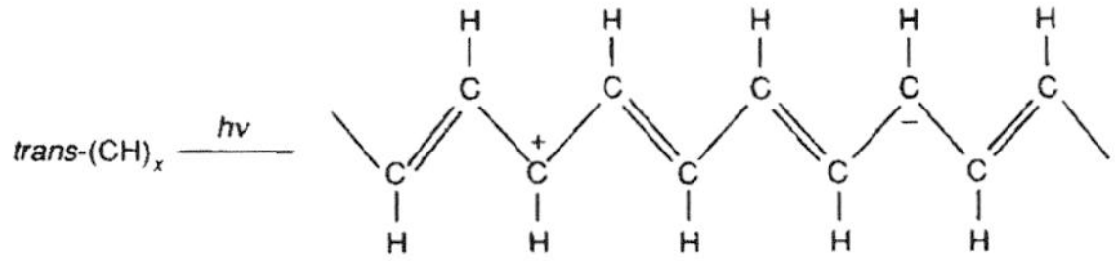

$$trans\text{-}(CH)_x \xrightarrow{h\nu}$$

The positive and negative solitons are here illustrated diagrammatically for simplicity as residing only on one CH unit; they are actually delocalized over ~15 CH units. They disappear rapidly due to recombination of electrons and holes when irradiation is discontinued. If a potential is applied during irradiation, then the electrons and holes separate and photoconductivity is observed.

Charge-injection doping. Charge-injection doping is most conveniently carried out using a metal/insulator/semiconductor (MIS) configuration involving a metal and a conducting polymer separated by a thin layer of a high dielectric strength insulator. It was this approach, which resulted in the observance of superconductivity in a polythiophene derivative, as described previously. Application of an appropriate potential across the structure can give rise, for example, to a surface charge layer, the 'accumulation' layer which has been extensively investigated for conducting polymers[8,13]. The resulting charges in the polymer, e.g. $(CH)_x$ or poly(3-hexylthiophene), are present without any associated dopant ion. The spectroscopic properties of the charged species so formed can therefore be examined in the absence of dopant ion. Using this approach, spectroscopic studies of $(CH)_x$ show the signatures characteristic of solitons and the mid-gap absorption band observed in the chemically and electrochemically doped polymer. However, coulombic interaction between charge on the chain and dopant ion is a very strong interaction and one that can totally alter the energetics of the system.

NON-REDOX DOPING

This type of doping differs from redox doping described above in that the number of electrons associated with the polymer backbone does not change during the doping process.

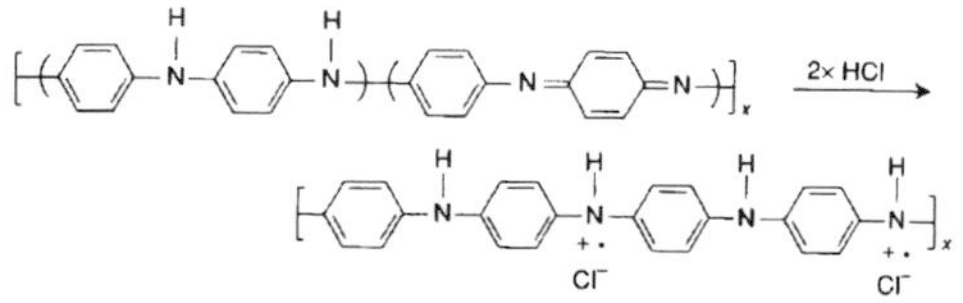

2× HCl

The energy levels are rearranged during doping. The emeraldine base form of polyaniline was the first example of the doping of an organic polymer to a highly conducting regime by a process of this type to produce an environmentally stable polysemiquinone radical cation. This was accomplished by treating emeraldine base with aqueous protonic acids and is accompanied by a nine to ten order of magnitude increase in conductivity (up to ~3 S/cm) to produce the protonated emeraldine base[14–16]. Protonic acid doping has subsequently been extended to systems such as poly(heteroaromatic vinylenes)[17].

THE POLYANILINES

The polyanilines refer to a very important class of electronic/conducting polymers. They can be considered as being derived from a polymer, the base form of which has the generalized composition: $-[(C_6H_4-NH-C_6H_4-NH)_y-(C_6H_4-N=C_6H_4=N)_{1-y}]_x-$

and which consists of alternating reduced, $-C_6H_4-NH-C_6H_4-NH-$ and oxidized, $-C_6H_4-N=C_6H_4=N-$ repeat units[3,14,15]. The average oxidation state can be varied continuously from y = 1 to give the completely reduced polymer, $-(C_6H_4-NH-C_6H_4-NH-C_6H_4-NH-C_6H_4-NH)_x-$ to y = 0.5 to give the 'half-oxidized' polymer, $-[(C_6H_4-NH-C_6H_4-NH)-(C_6H_4-N=C_6H_4=N)]_x-$ to 0 to give the completely oxidized polymer, $-(C_6H_4-N=C_6H_4=N-C_6H_4-N=C_6H_4=N)_x-$. The terms 'leucoemeraldine', 'emeraldine', and 'pernigraniline' refer to the different oxidation states of the polymer where y = 1, 0.5, and 0, respectively, either in the base form, e.g. emeraldine base, or in the protonated salt form, e.g. emeraldine hydrochloride[3,14,15]. In principle, the imine nitrogen atoms can be protonated in whole or in part to give the corresponding salts, the degree of protonation of the polymeric base depending on its oxidation state and on the pH of the aqueous acid. Complete protonation of the imine nitrogen atoms in emeraldine base by aqueous HCl, for example, results in the formation of a delocalized polysemiquinone radical cation[3,15,18] and is accompanied by an increase in conductivity of $\sim 10^{10}$.

The partly protonated emeraldine hydrochloride salt can be synthesized easily either by the chemical or electrochemical oxidative polymerization of aniline[3,14,15]. It can be deprotonated by aqueous ammonium hydroxide to give emeraldine base powder (a semiconductor).

ALLOWED OXIDATION STATES

As can be seen from the generalized formula of polyaniline base, $-[(C_6H_4-NH-C_6H_4-NH)_y-(C_6H_4-N=C_6H_4=N)_{1-y}]_x-$ the polymer could, *in principle*, exist in a continuum of oxidation states ranging from the completely reduced material in the leucoemeraldine oxidation state, y = 1 to the completely oxidized material in the pernigraniline oxidation state, y = 0. However, we have shown[16] that at least in N-methyl-2-pyrrolidinone (NMP) solution in the range y = 0 to y = 0.5 (emeraldine oxidation state) only two chromophores are present, characteristic of y = 1 and y = 0.5 species and that all intermediate oxidation states consist, at the molecular level, only of *mixtures* of the chromophores characteristic of these two states.

Since most of the properties of polyaniline of interest are concerned with the solid state, we have carried out a series of studies in the solid state which show that the same phenomenon is true in the y = 1 to y = 0. 5 oxidation state range and in the y = 0. 5 to y = 0 oxidation state range. Within each of these ranges all intermediate oxidation states consist, at the molecular level, only of mixtures of the chromophores characteristic of the two states defining the beginning and end of each range[19,20].

DOPING

Polyaniline holds a special position amongst conducting polymers in that its most highly conducting doped form can be reached by two completely different processes – protonic acid doping and oxidative doping. Protonic acid doping of emeraldine base units with, for example 1M aqueous HCl results in complete protonation of the imine nitrogen atoms to give the fully protonated emeraldine hydrochloride salt[14,15]:

As shown in Figure 2, protonation is accompanied by a 9 to 10 order of magnitude increase in conductivity reaching a maximum in ~1 M aqueous HCl.

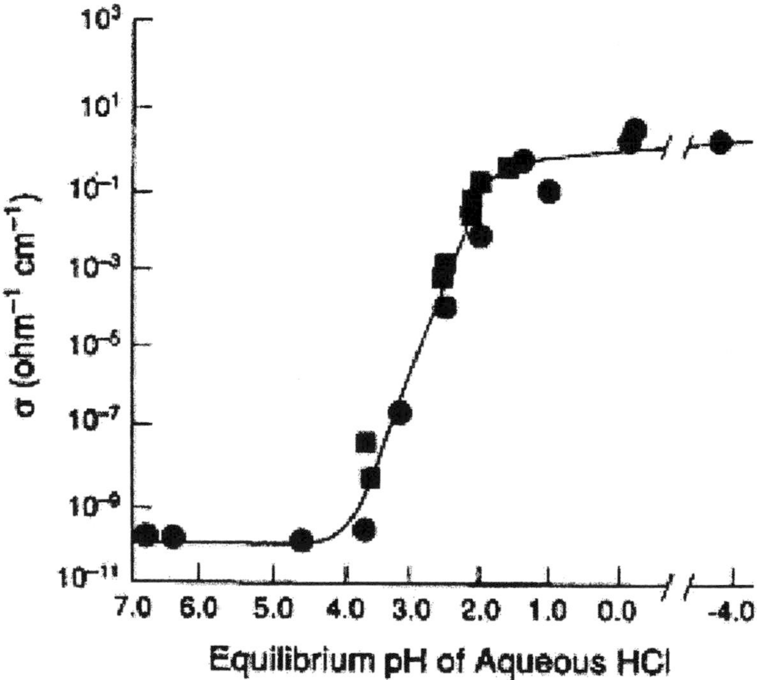

Figure 2. Conductivity of emeraldine base as a function of pH of HCl dopant solution as it undergoes protonic acid doping (● and ■ represent two independent series of experiments)[14,15].

The same doped polymer can be obtained by chemical oxidation (p-doping) of leucoemeraldine base[3]. This actually involves the oxidation of the σ/π system rather than just the π system of the polymer as is usually the case in p-type doping. Its reaction with a solution of chlorine in carbon tetrachloride proceeds to give emeraldine hydrochloride:

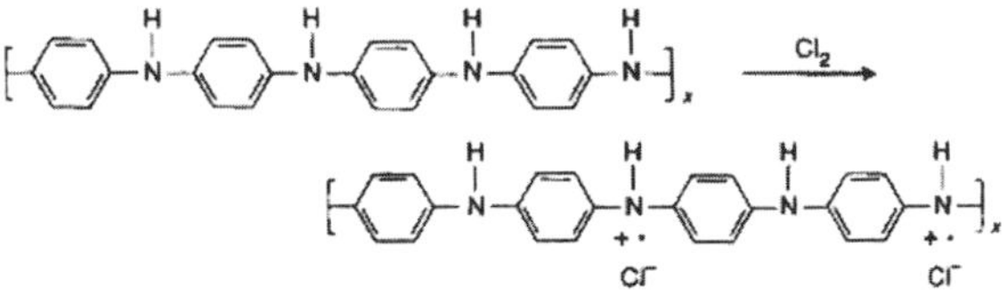

NANOELECTRONICS

The basic purpose of this research is to blend the now well-established field of elecronic/conducting polymers with the new, emerging field of nanoscience, by electrostatic fabrication ("electrospinning") to produce "nanoelectronics" – electronic junctions and devices significantly smaller than the diameter of a human hair (~50,000 nm). It is commonly accepted that a nanomaterial is defined as one consisting of a substance or structure which exhibits at least one dimension of less than 100 nm (0.1 μm)[21].

Our objectives were: (i) to develop a method by which nanofibers (diameter < 100 nm) of organic polymers could be controllably and reproducibly fabricated such that in one given preparation, all fibers would have a diameter < 100 nm *and* (ii) to reproducibly and controllably fabricate, for the first time, nanofibers of electronic polymers (in their semiconducting and metallic regimes) *and/or* their blends in conventional organic polymers for the purpose of ascertaining their applicability in the fabrication of nanoelectronic devices.

We have made substantial progress in achieving these objectives by using a relatively little known, simple, convenient and inexpensive "electrospinning" method[22–27]. We have previously reported[23] fabrication of the first conducting polymer fibers (diameter ~950 nm to 2,100 nm) of polyaniline doped with *d,l* camphorsulfonic acid (PAn.HCSA) as a blend in polyethylene oxide (PEO). We were surprised to find that an electronic polymer, such as polyaniline, which might have been expected to be more susceptible to degradation than most conventional organic polymers, survived, without observable chemical or physical change, following the 25,000 V electrospinning fabrication process in air at room temperature.

Electrospinning. The electrospinning technique involves a simple, rapid, inexpensive, electrostatic, non-mechanical method in which a polymer solution in a variety of different possible common solvents, including water, is placed in a hypodermic syringe or in a glass pipette, at a fixed distance (5–30 cm) from a metal cathode[24]. The positive (anode) terminal of a variable high voltage transformer is attached to the metal tip of the hypodermic syringe or to a wire inserted into the polymer solution in the glass pipette, the negative terminal being attached to the metal cathode. The tip of the syringe can be placed vertically over the cathode or at any other convenient angle to it. When the voltage applied between the anode and cathode reaches a critical value, ~14,000 V at a ~20 cm separation, the charge overcomes the surface tension of the deformed drop of the polymer solution on the tip of the syringe and a jet is produced. Since the polymer molecules all bear the same (positive) charge, they repel each other while traveling in air during a few milliseconds from the anode to cathode and become separated[25]. At the same time, evaporation of the solvent molecules occurs rapidly. Evaporation of solvent is also enhanced because the similarly-charged (positive) solvent molecules repel each other. Under appropriate conditions, dry, meters-long fibers accumulate on the surface of the cathode resulting in a non-woven mesh of nano- to micron diameter fibers depending on experimental parameters (Figure 3).

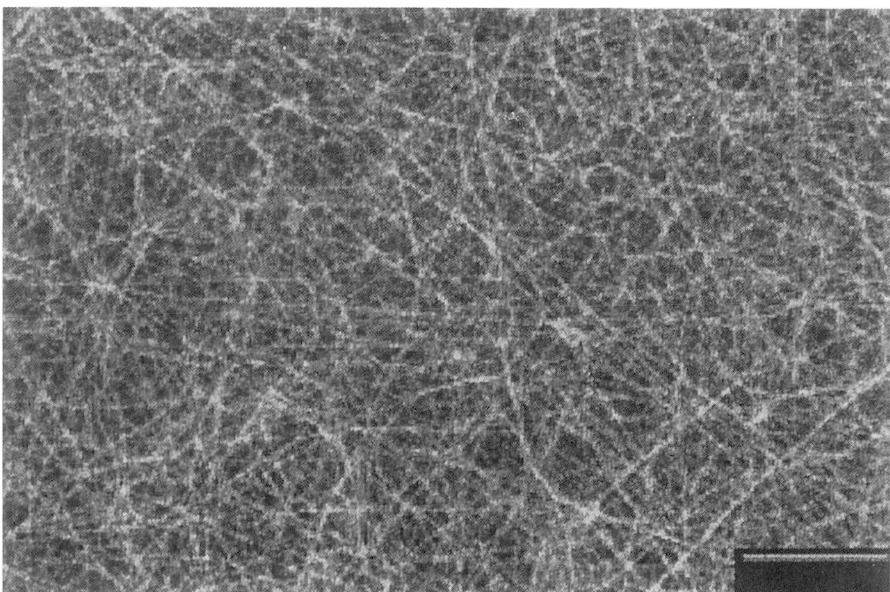

Figure 3. 50 wt% Nanofiber blend of PAn.HCSA fabricated from 2 wt% PAn.HCSA and 2 wt% PEO from chloroform solution at 25,000 V (anode/cathode separation, 25 cm). Scale bar: 100,000 nm.

Nanofiber Fabrication. Since the submicron fibers (500–1,600 nm) obtained in our initial work[23] were not classifiable as true "nanofibers," our immediate objective was to break the "nanotechnology barrier" and to consistently and reproducibly fabricate true nanofibers (diameter <100 nm) of an organic polymer. This was accomplished (see Figure. 4) using an 8 wt% solution of polystyrene (Mw 212,400) in tetrahydrofuran at a potential of 20,000 V between the anode and cathode which were separated by 30 cm. The fibers were collected as a mat on an aluminum target and were found to have diameter characteristics: average, 43.1 nm; maximum, 55.0 nm; minimum, 26.9 nm. Other studies involving polystyrene gave fibers whose diameters were consistently < 100 nm; average, 30.5 nm; maximum, 44.8 nm; minimum, 16.0 nm. It might also be noted that the above 16 nm fiber is only ~30 polystyrene molecules wide. It is also of interest to note that a 16 nm fiber such as the one mentioned above lies well within the ~ 4–30 nm diameter range of multiwalled carbon nanotubes[26].

Electronic Polymer Fibers. By using a previously observed method for producing polyaniline fibers[27] we have prepared highly-conducting sulfuric acid-

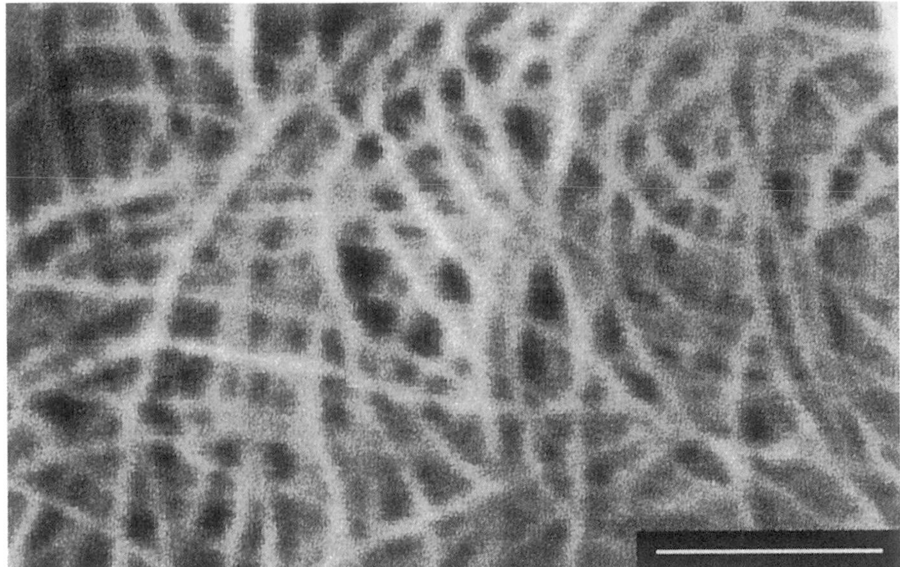

Figure 4. Electrospun fibers of polystyrene (see text). Scale bar: 1000 nm. The extended length of the fibers is clearly visible.

doped polyaniline fibers (diameters: average, 139 nm; maximum, 275 nm; minimum, 96nm) by placing a ~ 20 wt% solution of polyaniline in 98% sulfuric acid in a glass pipette with the tip ~ 3 cm above the surface of a copper cathode immersed in pure water at 5,000 V potential difference. The fibers collect in or on the surface of the water. The conductivity of a single fiber was ~ 0.1 S/cm, as expected since partial fiber de-doping occurred in the water cathode. The diameter and length of the fibers appear (Figure 5) to be sen-

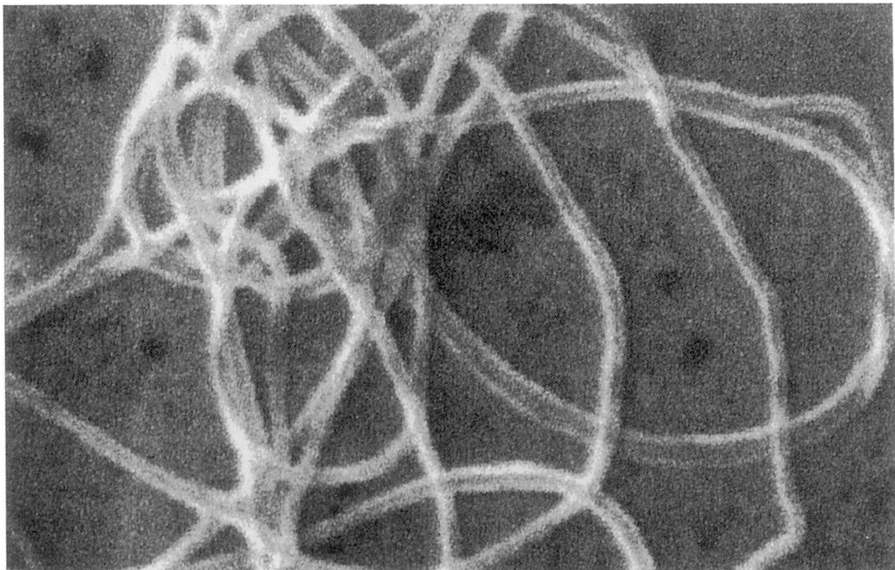

Figure 5. 100% polyaniline fiber with an average diameter of 139 nm.

sitive to the nature of the polyaniline used. No great difficulty is foreseen in producing fibers < 100 nm diameter.

It is relatively easy to prepare conducting blends of PAn.HCSA in a variety of different conventional polymers such as polyethylene oxide, polystyrene, polyacrylonitrile, etc. For example, ~ 20 wt% blends of PAn.HCSA in polystyrene (Mw 114,200) are obtained by electrospinning a chloroform solution; fiber diameter characteristics: average, 85.8 nm; maximum, 100.0 nm; minimum 72.0 nm. These fibers are sufficiently electrically conductive that their SEMs may be recorded without the necessity of applying a gold coating.

Separate, individual nanofibers can be collected and examined if so desired. An appropriate substrate – glass slide, silicon wafer or loop of copper wire, etc. – is held between the anode and cathode at a position close to the cathode for a few seconds to collect individual fibers (see Figure 6).

Figure 6. Polystyrene fibers collected on a bent copper wire (magnification 33x) and subsequently coated with a thin layer of polypyrrole by *in situ* deposition from aqueous solution. Scale bar: 1 mm.

Current/voltage (I/V) curves are given in Figure 7 for a single 419 nm diameter fiber (Fiber 1) and for a ~600nm diameter fiber (Fiber 2) of a blend of 50 wt% PAn.HCSA and polyethylene oxide collected on a silicon wafer coated with a thin layer of SiO_2. Two gold electrodes separated by 60.3 μm are deposited on the fiber after its deposition on the substrate.

Nanofibers as Substrates. The large surface to volume ratio offered by nanofibers makes them excellent, potentially useful substrates for the fabrication of coaxial nanofibers consisting of superimposed layers of different materials. Catalysts and electronically active materials can be deposited on them by chemical, electrochemical, solvent, chemical vapor, or other means, for use in nanoelectronic junctions and devices.

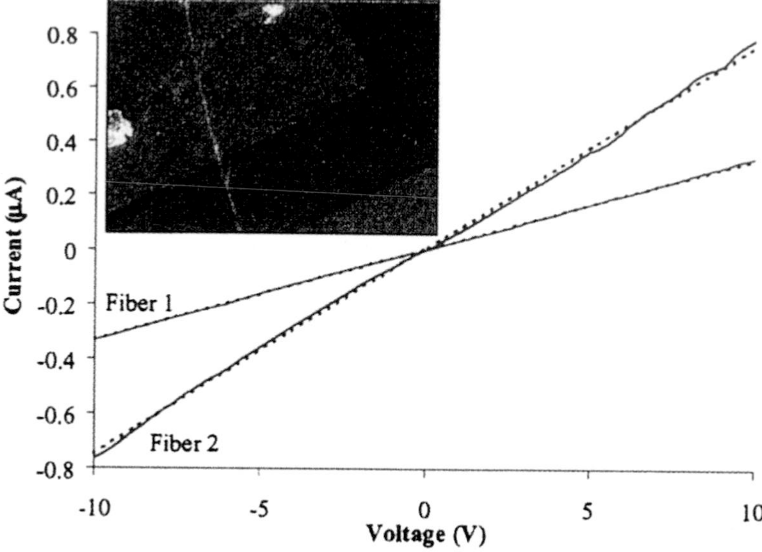

Figure 7. Current/voltage curves of 50 wt% PAn.HCSA/PEO blend nanofiber.

We have found, for example, that polyacrylonitrile nanofibers can be easily and evenly coated with a 20–25 nm layer of conducting polypyrrole (Figure 8) by immersion in an aqueous solution of polymerizing polypyrrole[28]. Analogously, we have found that electroless deposition of metals can also be performed. Polyacrylonitrile fibers, for example, can be evenly coated with gold by electroless deposition[29].

Carbon Nanofibers. As previously reported[30] polyacrylonitrile fibers may be

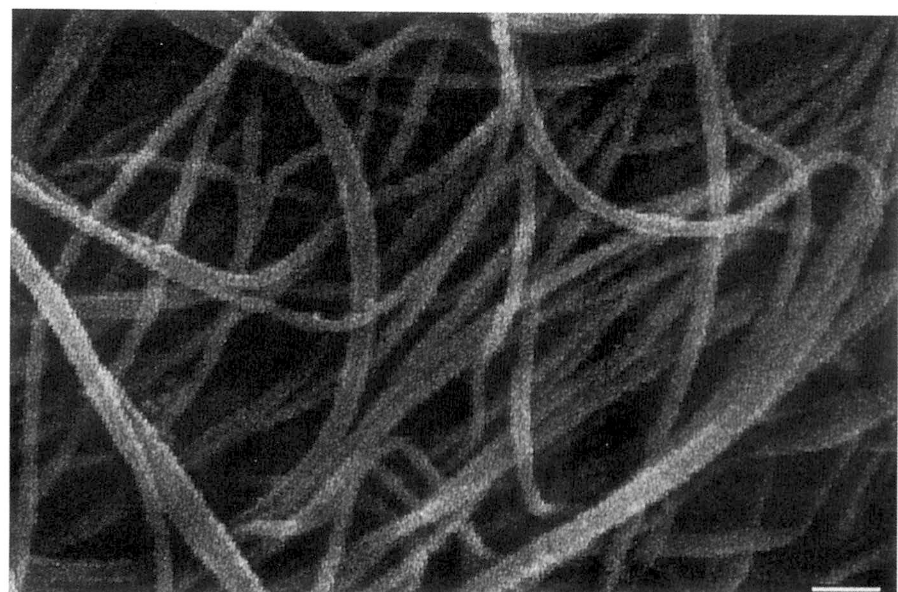

Figure 8. SEM of conducting polypyrrole coated polyacrylonitrile nanofibers. Scale bar: 1000 nm.

thermally converted to carbon nanofibers with some shrinkage. We have similarly converted polyacrylonitrile nanofibers to carbon nanofibers.

In summary, electronic polymers have been used for the past 20 years to produce rectifying diodes by Schottky and p/n junctions, transistors, light-emitting devices, photovoltaic cells, rechargeable batteries, etc[1]. Now, the ability to fabricate nanofibers of electronic polymers which are only a few molecules thick suggests the emergence of a field of nanoelectronics whereby the electronic properties of such nanofibers can be exploited for technological purposes.

LINE PATTERNING OF CONDUCTING POLYMERS[30]

One of the exciting challenges of the first part of this century will be the development of low-cost disposable plastic/paper electronic devices[31–33]. Conventional inorganic conductors, such as metals, and semiconductors, such as silicon, commonly require multiple etching and lithographic steps in fabricating them for use in electronic devices. The number of processing steps and chemical etching steps involved limit the minimum price and therefore their applicability in disposable electronics. On the other hand, conducting polymers combine many advantages of plastics, e.g. flexibility and processing from solution, with the additional advantage of conductivity either in the metallic or semiconducting regimes; however, the lack of simple methods to obtain inexpensive conductive polymer shapes/patterns limit many applications. We here describe a novel, simple and cheap method to prepare patterns of conducting polymers by a process which we term, "Line Patterning".

Line Patterning uses the difference in selected physical and/or chemical properties between a substrate and insulating lines which have been printed on it by a conventional copying or printing process towards a fluid (or vapor) to which they are both simultaneously exposed. The substrate and printed lines react differently or at different rates with the fluid (or vapor) to which they have been exposed. This results in a non-uniform deposition on the substrate as compared to the printed lines. If the fluid contains a conducting polymer, which remains as a film after evaporation of the solvent, a pattern of conducting polymer results. A pattern is first designed on a computer and is then printed on, for example, an overhead transparency using a standard, non-modified office laser printer.

The printed (insulating) lines can be easily removed, if necessary, in a few seconds by ultrasonic treatment in toluene, dissolving the printed lines and leaving a clean pattern of deposited material on the substrate whose shape was originally defined by the now non-existent printed lines. Line Patterning has the following advantages: no photolithography is involved; no printing of conducting polymer is involved; it uses only, e.g., a standard office laser printer, which is not modified in any way; commercially available flexible, transparent plastic or paper substrates can be used; solutions of commercially available conducting or non-conducting polymers can be used from which the polymers may be deposited on sustrates; it is inexpensive; rapid development of customized patterns (within hours) from a computer designed pattern to product is routine.

We have exploited, for example, the observation that a commercial dispersion of poly-3,4-ethylenedioxythiophene (PEDOT, "Baytron P", Bayer Corp.) wets commercial plastic overhead transparency, but not the lines printed on it by a standard office laser printer. A coating of PEDOT can be applied by a roller and after evaporation of the solvent; the printed lines can be easily and cleanly removed by sonication, leaving only the conducting polymer on the transparency.

Two electrodes were prepared in this way, each containing 25 lines/inch. A drop of a standard commercial Polymer Dispersed Liquid Crystal Display (PDLC)[34] mixture containing an optical adhesive and 15 μm spacer spheres was placed on the center of each electrode. The second electrode was placed on top at an angle of 90° to the first. This resulted in a (25 × 25), i.e. 625 pixels/sq. inch matrix. [Figure 9] Exposure to U/V light for a few minutes resulted in polymerization of the mixture to bind the two electrodes together and to produce a free-standing working PDLC liquid crystal display device. When an electrode pattern of 100 lines/inch was used a working 10,000 pixel/sq. inch display was produced.

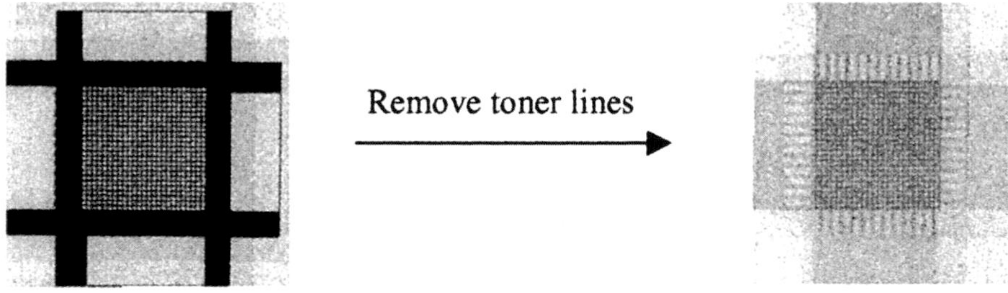

Figure 9. 625 pixel PDLC Liquid Crystal Display.

We have devised a novel way of separating conducting polymer circuits from each other by making use of the height (~4-5 μm) of the printed toner lines, using a standard office printer, above the substrate e.g. on an overhead transparency. This is illustrated [Figure 10 and 11] by a "push button" switch to open and close a simple electrical circuit. A combination of two patterned transparencies where the two adjacent conductive areas are electrically separated from each other by two ~4-5 μm non-conductive printed toner lines is obtained by placing the printed lines on top of each other as shown in Figures 11 and 12. Depression of the areas labeled "PRESS" causes the upper transparency to bend. This electrically connects the conducting PEDOT surfaces. When released, the transparency film returns back to its original position, thus breaking the electrical circuit.

The two-dimensional conducting polymer circuits may be readily converted to three-dimensional circuits by two different methods as shown in Figure 12 simply by (i) stapling two two-dimensional circuits together using a common office desk stapler. The metal staple joins together electrically the con-

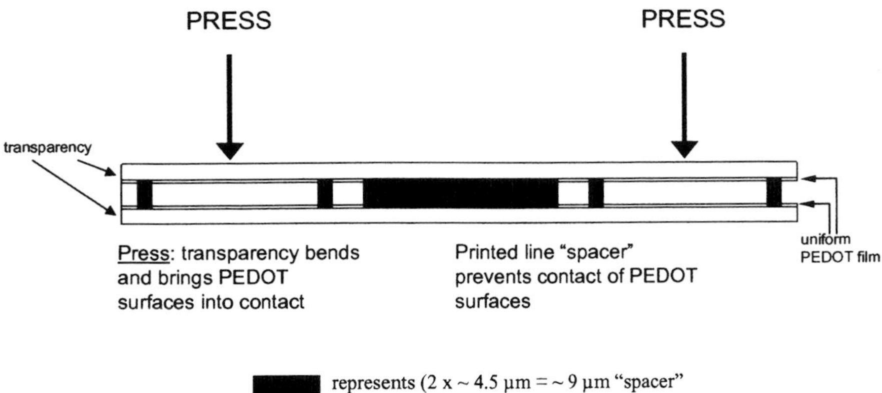

Figure 10. A Simple Electronic Circuit *("push button" switch).*

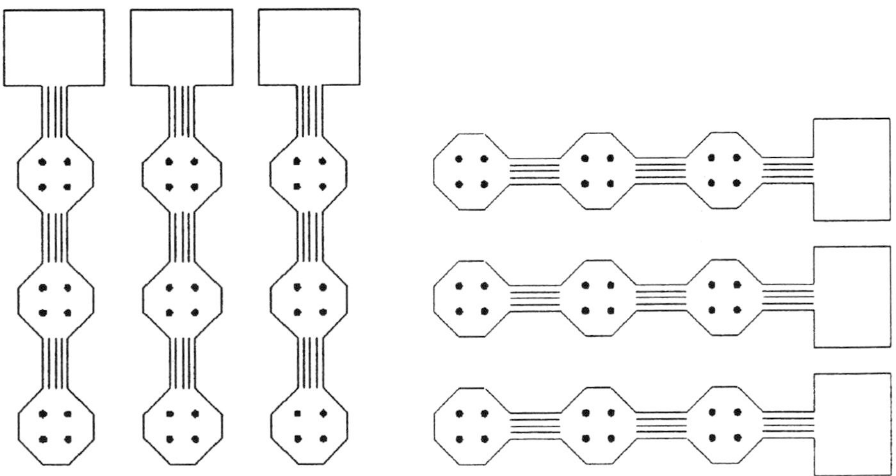

Figure 11. "Push button" switch.

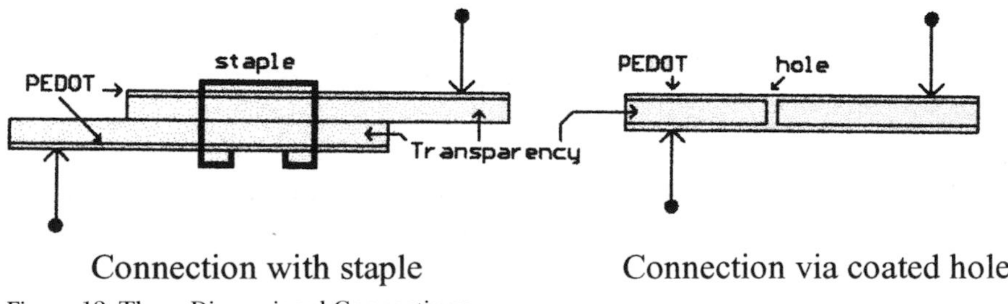

Figure 12. Three-Dimensional Connections.

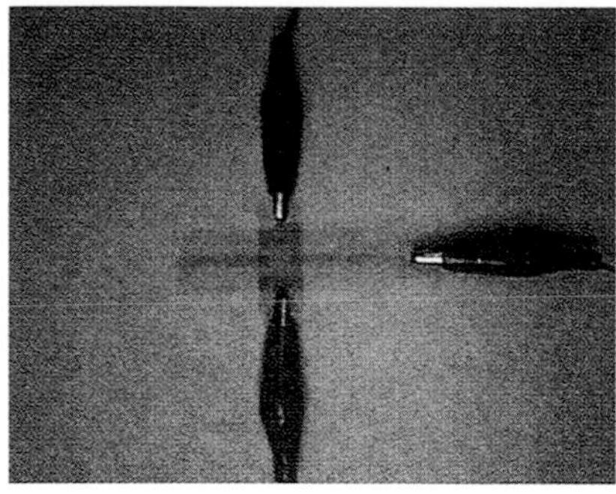

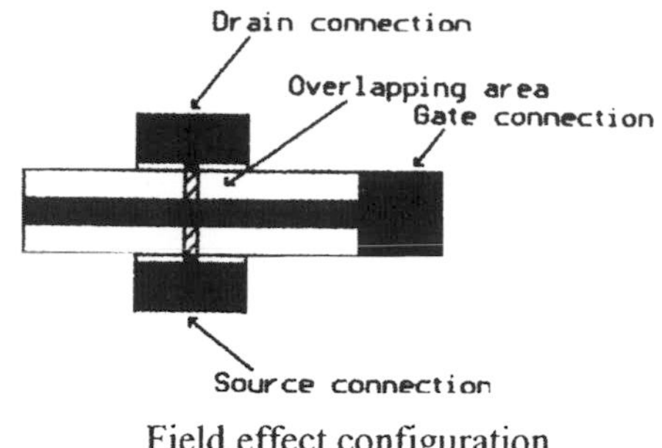

Figure 13. Field-Effect Doped "PEDOT".

ducting polymer areas on two different substrates *or* (ii) making a pinhole through the sheets, as shown, before applying the PEDOT solution. Some of the solution enters the pinhole and joins together electrically the conducting surfaces on the two different circuits.

We have recently observed a curious field effect which thin films of PEDOT exhibit when exposed to a positive gate potential in an FET configuration as shown in Figure 13. A source/drain electrode and a gate electrode are prepared by Line Patterning and are covered by a thin layer of PEDOT as described above. A drop of the optical adhesive containing spacer spheres described above is placed on the source/drain electrode upon which the gate

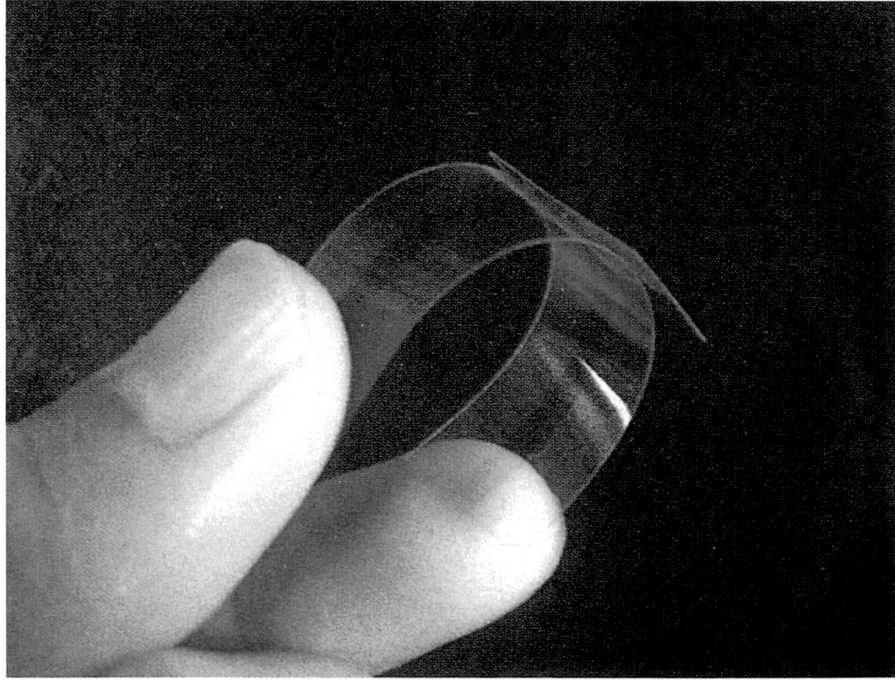

Figure 14. FET-type device.

electrode is then placed at 90°. The two electrodes are manually squeezed together and the optical adhesive is polymerized by exposure to U/V light as was done for the PDLC liquid crystal display described previously. Several thousand of these interconnected transistor-type devices could be readily fabricated per square in. by the Line Patterning process, if it were considered desirable. The free-standing, flexible device shown in Figures 13 and 14 is produced.

The device exhibits the same general reversible features commonly associated with a field effect transistor (FET) as shown in Figure 15.

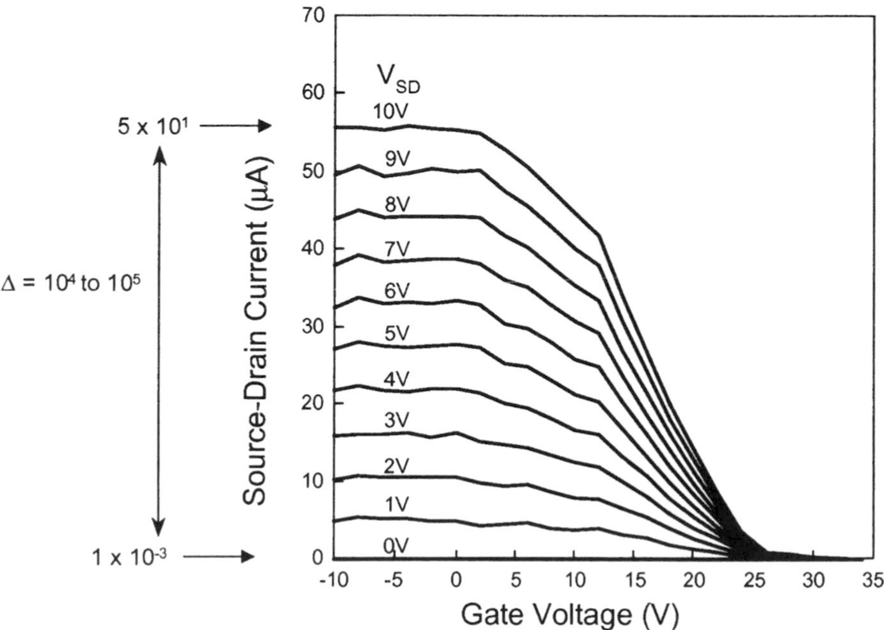

Figure 15. Field Effect Transistor characteristics.

The doped "metallic" PEDOT film ($\sigma \sim 2$ S/cm at room temperature) would not be expected to show a change in conductivity by exposure to a field of this type under the configuration used. We believe this effect presents an entirely new method for ascertaining the nature of highly doped "metallic" conducting polymers. It has frequently been postulated that a doped conducting polymer consists of metallic "islands" surrounded by lowly conducting "beaches" as shown in Figure 16. We postulate that in the effect we have observed only the lowly conducting "beaches" and not the metallic "islands" respond to the applied field. The application of an electric field changes the conductivity of the "beaches" and hence the extent of electrical percolation in the source/drain PEDOT electrode between the metallic "islands", hence changing the bulk conductivity of the material. The response time for our device is much slower than for a conventional field effect transistor. We therefore conjecture that the chief changes in conductivity are probably caused by

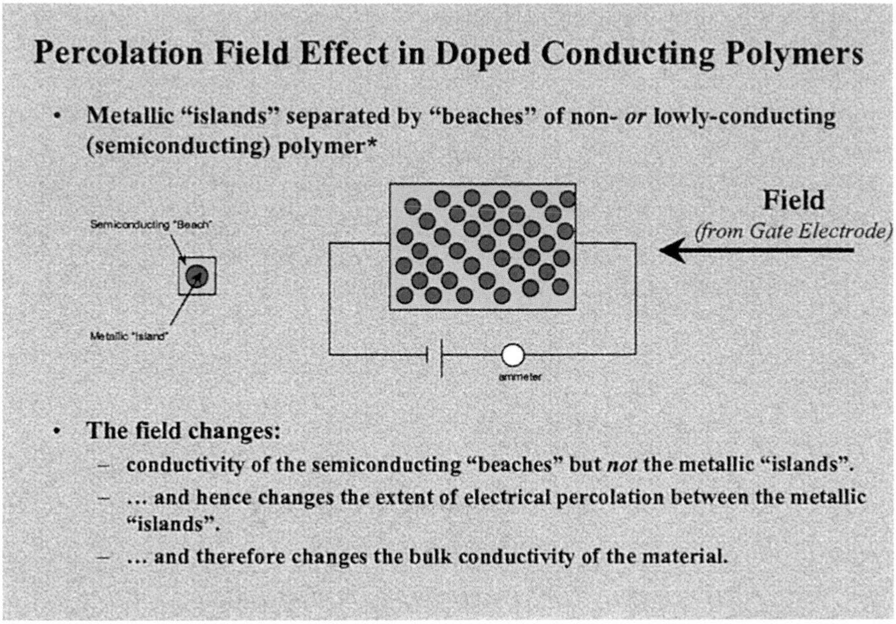

Percolation Field Effect in Doped Conducting Polymers

- **Metallic "islands" separated by "beaches" of non- *or* lowly-conducting (semiconducting) polymer***

Semiconducting "Beach"
Metallic "Island"
ammeter
Field
(from Gate Electrode)

- **The field changes:**
  - **conductivity of the semiconducting "beaches" but *not* the metallic "islands".**
  - **... and hence changes the extent of electrical percolation between the metallic "islands".**
  - **... and therefore changes the bulk conductivity of the material.**

Figure 16. Percolation field effect in doped conducting polymers.

slow diffusion of the dopant anions under influence of the applied field. On removal of the field the system reverts to its original state.

Preliminary studies show that the effect is also present in polyaniline; it may therefore possibly be found in many other conducting polymers and would thus represent a general phenomenon characteristic of all conducting polymers, at least within certain ranges of doping.

SUMMARY

- Polyacetylene, $(CH)_x$, the simplest organic polymer, can be reversibly doped to the metallic regime by partial oxidation or reduction either chemically or electrochemically.
- Polyaniline can be doped to the metallic regime by a simple acid/base protonation.
- A large number of electronic conductive polymers are now known.
- A variety of technological applications of electronic conductive polymers, present and projected, are apparent.

ACKNOWLEDGMENT

This Nobel Prize has world-wide implications since it shows the ever-increasing importance of interdisciplinary research – in this case collaborative research between a polymer chemist, Hideki Shirakawa, Alan Heeger, a physicist, and myself, an organometallic chemist. Each of us had the task of learning the specialized scientific language of the other in order to collectively focus on one specific scientific challenge, an example of where 1+1 + 1 is more than 3!

The prize is also recognition of the good fortune that Alan, Hideki and I had in having each other as such excellent colleagues and also in having such creative colleagues in each of our respective individual research groups – the work stemming from a research group cannot be better than the persons carrying it out. The Prize is a recognition of them and their work and also the work of countless others world-wide during the past 23 years who put the "flesh on the skeleton work" carried out by us in the 1970's at Penn. If it were not for them there would be no prize today in the field.

Research in an experimental science (and also in many other fields) cannot be accomplished without financial support for stipends, apparatus, supplies and the like. A funding organization and project officers within such an organization have tremendous control over the future of science and technology in any given country. In this respect Dr. Kenneth J. Wynne, my contracting officer at the U.S. Office of Naval Research for many years, before his recent retirement, had the scientific intuition and foresight to fund our first work on conducting polymers – the first funding of work of this type anywhere in the world. He funded it because of its scientific interest. The fact that it now has great technological potential was not a consideration at that time.

"Of what use is a beautiful poem"? It gives intellectual stimulation and enjoyment. Similarly with research. If it has some practical use, that is merely "icing on the cake!"

EARLY STUDIES

Polyacetylene, $(CH)_x$: Alan J. Heeger (*formerly,* Physics Dept, Univ. of Penn) Hideki Shirakawa (Tsukuba University) and many undergraduate, graduate students and post doctoral fellows. *Financial Support:* Principally, US Office of Naval Research;* University of Pennsylvania Materials Science Laboratory.

Polyaniline: Arthur J. Epstein (Physics Dept, Ohio State University) and many undergraduate, graduate students and post doctoral fellows. *Financial Support:* Principally, US Office of Naval Research;* University of Pennsylvania Materials Science Laboratory.

RECENT STUDIES

Nanofibers ("Electrospinning"): I.D. Norris, J. Gao, F.K. Ko, W.E. Jones, Jr., A.T. Johnson, Jr. *Financial Support:* US Office of Naval Research;* Army Research Office – MURI.

Line Patterning: D. Hohnholz, H. Okuzaki. *Financial Support:* Subcontract, Kent Displays, Inc. (ONR-SBIR Program); US Office of Naval Research;* Fellowship from Ministry of Education, Science, Culture and Sports, Japan.

* *Dr. K.J. Wynne,* Program Manager, US Office of Naval Research

REFERENCES

1. H.S. Nalwa, (ed.), *Handbook of Organic Conductive Materials and Polymers,* Wiley, New York (1997); T.A. Skotheim, R.L. Elsenbaumer, J.F. Reynolds (eds.), *Handbook of Conducting Polymers,* 2nd Ed., Marcel Dekker, New York (1998).
2. J.H. Schon, A. Dodabalapur, Z. Bao, C. Kloc, O. Schenker, B. Batlogg, *Nature,* **410**, 189 (2001).
3. A.G. MacDiarmid and A.J. Epstein, *Faraday Discuss. Chem. Soc.,* **88**, 317 (1989) and references therein.
4. A.G. MacDiarmid and A.J. Heeger, *Synth.* Met., 1, 101 (1979/80) and references therein.
5. T.A. Skotheim, (ed.), *Handbook of Conducting Polymers, 1 & 2,* Marcel Dekker, New York (1986).
6. M.G. Kanatzidis, *Chemical & Engineering News,* 3 December, 36 (1990).
7. C.K. Chiang, C.R. Fincher, Jr., Y.W. Park, A.J. Heeger, H. Shirakawa, E.J. Louis, A.G. MacDiarmid, *Phys. Rev. Lett.,* **39**, 1098 (1977); C.K. Chiang, M.A. Druy, S.C. Gau, A.J. Heeger, E.J. Louis, A.G. MacDiarmid, *J. Am. Chem. Soc.,* **100**, 1013 (1978).
8. K.E. Ziemelis, A.T. Hussain, D.D.C. Bradley, R.H. Friend, J. Rilhe and G. Wegner, *Phys. Rev. Lett.,* 66, 2231 (1991).
9. V.G. Kulkarni, W.R. Mathew, J.C. Campbell, C.J. Dinkins, and P.J. Durbin, in *49th ANTEC Conference Proceedings,* Society of Plastic Engineers and Plastic Engineering, Montreal, Canada, 5-9 May, p. 663 (1991); L.W. Shacklette, N.F. Colaneri, V.G. Kulkarni and B. Wessling, in *49th ANTEC Conference Proceedings,* Society of Plastic Engineers and Plastic Engineering, Montreal, Canada, 5-9 May, p. 665 (1991).
10. P.J. Nigrey, A.G. MacDiarmid and A.J. Heeger, *J. Chem. Soc. Chem. Commun.,* 594 (1979).
11. D. MacInnes, Jr, M.A. Druy, P.J. Nigrey, D.P. Nairns, A.G. MacDiarmid and A.J. Heeger, *J. Chem. Soc. Chem. Commun.,* 317 (1981).
12. A.J. Heeger, S. Kivelson, J.R. Schrieffer and W.-P. Su, *Rev. Mod. Phys.,* **60**, 781 (1988) and references therein.
13. J.H. Burroughes, C.A. Jones, and R.H.Friend, *Nature,* **335**, 137 (1988); J.H. Burroughes, D.D.C. Bradley, A.R. Brown, R.N. Marks, R.H. Friend, *et al., Nature,* **347,** 539 (1990).
14. J.C. Chiang and A.G. MacDiarmid, *Synth. Met.,* **13,** 193 (1986).
15. A.G. MacDiarmid, J.-C. Chiang, A.F. Richter and A.J. Epstein, *Synth. Met.,* 18, 285 (1987).
16. A.G. MacDiarmid and A.J. Epstein, *Faraday Discuss. Chem. Soc.,* **88**, 317 (1989) and references therein; A.G. MacDiarmid and A.J. Epstein, in *Science and applications of conducting polymers,* ed. W. R. Salaneck, D. T. Clark, and E. J. Samuelsen, Adam Hilger, Bristol, UK, p. 117 (1990).
17. C.C. Han and R.L. Elsenbaumer, *Synth.* Met., **30**, 123 (1989).
18. A.G. MacDiarmid, J.-C. Chiang, A.F. Richter, N.L.D. Somasiri and A.J. Epstein, in *Conducting Polymers,* ed. L. Alcacér, Reidel Publications, Dordrecht, p. 105 (1987).
19. F.L. Lu, F. Wudl, M. Nowak and A.J. Heeger, *J. Am. Chem. Soc.,* **108**, 8311 (1986).
20. Y. Sun, A.G. MacDiarmid and A.J. Epstein, *J. Chem. Soc. Chem. Commun.,* 529 (1990).
21. "Nanotechnology- A Revolution in the Making – Vision for R&D in the Next Decade," report of the Interagency Working Group on Nanoscience, Engineering, and Technology, March 10, 1999.
22. A. Formhals, US Patent No. 1,975,504 (1934).
23. I.D. Norris, M.M. Shaker, F.K. Ko and A.G. MacDiarmid, *Synth. Met.,* **114(2)**, 109 (2000); A.G. MacDiarmid, W.E. Jones, Jr., I.D. Norris, J. Gao, A.T. Johnson, Jr., N.J. Pinto, J. Hone, B. Han, F.K. Ko, H. Okuzaki and M. Llagune, *Synth. Met., 119,* 27 (2001).
24. J. Doshi and D.H. Reneker, *J. Electrost.* **35**, 151 (1995); P.W. Gibson, H.L. Schreuder-Gibson and D. Riven, *AIChE J.* **45**, 190 (1999).

25. D.H. Reneker, A. L. Yarin, H.Fong and S. Koombhongse, *J. Appl. Phys.* **87,** 4531 (2000).
26. S. Iijima, *Nature,* **354,** 56 (1991).
27. D.H. Reneker and I. Chun, *Nanotechnology,* **7(3),** 216 (1996).
28. Z. Huang, P-C. Wang, A.G. MacDiarmid, Y. Xia, and G. M. Whitesides, *Langmuir,* 13 (1997) 6480; R.V. Gregory, W.C. Kimbrell and H.H. Kuhn, *Synth. Met.,* **28(1–2),** C823 (1989).
29. A.M. Sullivan and P.A. Kohl, *J. Electrochem. Soc.,* **142(7),** 2250 (1995).
30. I. Chun, D.H. Reneker, H. Fong, X. Fang, J. Deitzel, N.B. Tan and K. Kearns, *J. Advanced Materials,* **31,** 36 (1999); D. Hohnholz and A.G. MacDiarmid, *Synth. Met., 121,* 1327 (2001).
31. A. Dodabalapur, Z. Bao, A. Makhija, J. G. Laquindanum, V. R. Raju, Y. Feng, H. E. Katz and J. Rogers, *App. Phys. Lett.,* **73,** 142 (1998) and refs. therein.
32. C.J. Drury, C.M.J. Mutsaers, C.M. Hart, M. Matters and D. M. de Leeuw, *App. Phys. Lett.,* **73,** 108 (1998) and refs. therein.
33. H. Okusaki and Y. Osada, *J. Intell. Mater. Syst. Struct.,* **4,** 50 (1993).
34. "Licrystal E7" (Merck Corp., Germany), "NOA-65 Optical Adhesive" (Norland Products, NH USA).

HIDEKI SHIRAKAWA

For the ten years from the third grade of elementary school to the end of high school, I lived in the small city of Takayama, a town of less than sixty thousand, located in the middle of Honshu, Japan. Even though it was far away from Japan's principal cities, Takayama has been called a "little Kyoto" because of the similarity of its landform to Kyoto, the city sits in a basin surrounded by mountains with a river flowing through it, and because of its long-established cultural heritage and tradition. In this small town, rich in natural beauty, I spent my days enthusiastically collecting insects and plants, and making radios. My affinity for science was awakened and grew during in these ten years.

Long after I became a polymer scientist, I occasionally remembered a short composition I had written during my last year in junior high school. At that time students compiled a commemorative collection of compositions describing our future dreams. As I recalled, I wrote something about my wish to be a scientist in the future and to conduct research on plastics useful for ordinary people. I cannot be sure what I wrote exactly because I lost the book of essays during repeated moves afterwards. I had long regretted this loss because I wanted to know more about why and how a junior high school boy decided on a future research career in plastics.

Much to my surprise, I found that the full composition I had lost was printed in every Japanese newspaper the day after the Royal Swedish Academy of Sciences announced its award of the Nobel Prize in Chemistry for 2000 to two friends and myself. After 45 years, I could finally read the complete composition again. I was deeply impressed with the great power of the Nobel Prize.

I was born in Tokyo in August 1936, the third child of Hatsutarou, a medical doctor, and Fuyuno, a daughter of a chief priest of a Buddhist temple. After me, a sister and a brother were born, joining my elder brother, my elder sister and me. After I was born, my family moved many times, following my father's work, but we finally settled in Takayama, my mother's hometown, in 1944 during the confusion toward the end of the war.

My higher education began when I entered Tokyo Institute of Technology in April 1957. In March 1966, I completed my doctoral course and received the degree of Doctor of Engineering. In the same year, I married Chiyoko Shibuya, and we were later blessed with two sons, Chihiro and Yasuki.

There were three specific fields I wished to study at university. One was polymer chemistry, just as I had written in my junior high school composition. The other possibilities were horticulture and electronics. I had decided to major in polymer chemistry only if I successfully passed the entrance examination for Tokyo Institute of Technology. In April 1957, after entering Tokyo Institute of Technology, I mainly studied applied chemistry during my undergraduate career. In Japanese universities, an undergraduate major in a science course has to belong to one of the laboratories in his department during his final year in order to work on a graduation thesis. I was interested in synthesizing new polymers, so I applied to a laboratory conducting synthesis research. But since there were too many applicants who wanted to enter into the laboratory I had chosen, I had to switch to a laboratory working on polymer physics. Initially I was reluctant to work in this field, but actually, I realize that my experiences in this laboratory were of great importance to me when I worked with polyacetylene later on.

I finally began working on polymer synthesis, my original interest, in my graduate program, but I started the work on polyacetylene, the work for which I now share the Nobel Prize, just after I received my doctorate and I became a research associate in April 1966. The initial purpose of this study was to determine the polymerization mechanism of polyacetylene using the Ziegler-Natta catalysts. In the fall of 1967, only a short time after we started this work, we unexpectedly discovered how to synthesize polyacetylene film through an unforeseeable experimental failure.

With the conventional method of polymerization, chemists had obtained polyacetylene in the form of black powder; however, one day, when one a visiting scientist tried to make polyacetylene in the usual way, he only produced some ragged pieces of a film. In order to clarify the reason for the failure, I inspected various polymerization conditions again and again. I finally found that the concentration of the catalyst was the decisive factor for making the film. In any chemical reaction, a very small quantity of the catalyst, about mmol would be sufficient, but the result I got was for a quantity of mol, a thousand times higher than I had intended. It was an extraordinary unit for a catalyst. I might have missed the "m" for "mmol" in my experimental instructions, or the visitor might have misread it. For whatever reason, he had added the catalyst of some molar quantities in the reaction vessel. The catalyst concentration of a thousand-fold higher than I had planned apparently accelerated the rate of the polymerization reaction about a thousand times. Roughly speaking, as soon as acetylene gas was put into the catalyst, the reaction occurred so quickly that the gas was just polymerized on the surface of the catalyst as a thin film.

But we noticed another important factor besides the concentration of the catalyst. Polyacetylene has a property of being insoluble in any solvent, a property which contributed to the formation of the film. Even more surprising, when we observed the film through a transmission electron microscope, we saw that the film was composed of long entangled micro-fibers of polyacetylene. These two properties are essential for the formation of any film, and they were inherent in polyacetylene.

One more important factor contributed to the formation of the film was the Ziegler-Natta catalyst we had used. Most of the Ziegler-Natta catalysts tend to form precipitates which give an inhomogeneous solution. From such an inhomogeneous catalyst, it is very difficult to form polyacetylene film. But the Ziegler-Natta catalyst we had used in our experiment was a unique one. It had good solubility in organic solvents to give a homogeneous solution and it also had high activity to give a high molecular weight and crystalline form of polyacetylene. I could say that nature had prepared us for the way to make polyacetylene film. Later, through the measurements of various absorption spectra of this thin film, we determined the molecular structure of polyacetylene, and thus, we fulfilled the initial purpose of our work.

By chance, this glittering, silvery film, caught the eyes of Professor Alan G. MacDiarmid, one of the co-recipients of the prize, and he invited me to work with him in the U.S.A. In September 1976, I went to the University of Pennsylvania, where Professor Alan J. Heeger, another co-recipient, was also working, and I spent one year there.

I still vividly remember the day of November 23, 1976. With Dr. C. K. Chiang, a postdoctoral fellow who was working under Professor Heeger, I was measuring the electric conductivity of polyacetylene by the four-probe method, adding bromine. At exactly the moment we added bromine, the conductivity jumped so rapidly that he couldn't switch the range of the electrometers. Actually, the conductivity was ten million times higher than before adding bromine. This day marked the first time we observed the doping effect, although it was a pity that the expensive equipment was broken. The discovery of chemical doping is one of the representative results of our collaboration in this period.

After returning to Japan, I continued to work on polyacetylene. What I did first was to shed light on the chemical reaction associated with the doping phenomena. In cooperation with many coworkers, I investigated various spectra of the doped polyacetylene films: infrared absorption, Raman scattering, ultraviolet-visible absorption, the Mössbauer effect, and EXAFS. As a result, we found that the emergence of electrical conductivity on the doped polyacetylene was due to the creation of carbocations or positively charged solitons associated with withdrawing of π electrons from polyacetylene by the dopant when iodine was used as an acceptor dopant.

In November 1979, I moved from Tokyo Institute of Technology to the Institute of Materials Science, University of Tsukuba, where I was appointed Associate Professor. In October 1982, I was promoted to full professor and worked on polyacetylene and other conducting polymers. Since my retirement from University of Tsukuba at the end of March 2000, I have withdrawn from scientific research and other educational activities.

Let me mention two of my major contributions to polyacetylene research during my time at Tsukuba. One is the preparation of oriented films. The significance of polyacetylene being a typical quasi-one dimensional material was recognized very early. In this sense, an oriented film was indispensable to study the intrinsic one-dimensional properties. The polyacetylene films synthesized until then were an isotropic material in which the fibrils were entangled in three-dimensional disorder. I came up with the idea to directly synthesize the uniaxially oriented films by using liquid crystal as a solvent. The same idea was proposed by a scientist from a company. We found that an equimolar mixture of nematic liquid crystals bearing a phenylcyclohexyl moiety was useful for that purpose. We succeeded in simultaneously polymerizing acetylene and synthesizing uniaxially oriented polyacetylene films by orienting the catalyst solution of liquid crystal solvent under flow condition or magnetic field. Further development of this technique enabled us to synthesize helical polyacetylene that consists of clockwise or counterclockwise helical structure of fibrils, by use of chiral nematic liquid crystals. The chiral helicity of the films may be useful for electromagnetic and optical applications.

The other contribution is the synthesis of liquid crystalline conjugated polymers by replacing the hydrogen atom bonded to polyacetylene with a substituent having liquid crystalline nature as the side chain. As these polymers have large substituents, the doping effect is poor. However, these polymers can be modified by introducing various substituents with interesting optical and thermal properties. In addition, they can orient spontaneously in a given range of temperature.

Some other details to be mentioned:

Chair of Master's School in Sciences and Engineering, Graduate School, University of Tsukuba, April 1991–March 1993.

Provost of the 3rd Cluster of Colleges, University of Tsukuba, April 1994–March 1997.

Honors:

The Award of the Society of Polymer Science, Japan (1982), May 1983.

Award for Distinguished Service in Advancement of Polymer Science, the Society of Polymer Science, Japan (1999), May 2000.

Person of Cultural Merits, November 2000.

Order of Culture, November 2000.

THE DISCOVERY OF POLYACETYLENE FILM: THE DAWNING OF AN ERA OF CONDUCTING POLYMERS

Nobel Lecture, December 8, 2000

by

HIDEKI SHIRAKAWA

Emeritus Professor, University of Tsukuba, Tsukuba, Ibaraki 305-8577, Japan.

This lecture is not directly related to our discovery and development of conducting polymers to which the Nobel Prize in Chemistry 2000 was awarded. However, I would like to present my previous work that I had carried out just before we reached the discovery of chemical doping. I do hope my talk will be of use for you, the audience, to deepen your understandings by learning what had happened before and how we did reach the idea of the chemical doping.

I. PROLOGUE

It has been recognized for many years that a very long linear conjugated polyene might have various interesting properties, especially optical, electrical and magnetic properties. The definition of the polyene is that an even number of methyne (=CH–) groups is covalently bonded to form a linear carbon chain bearing one π-electron on each carbon atom. Therefore, the chemical structure of the polyene is best represented by a formula $H(CH{=}CH)_nH$, where n denotes the number of repeating units. Recently *polyacetylene* has become a more popular name instead of polyene because polyacetylene synthesized by the polymerization of acetylene has been used extensively as specimens for various studies.

Pople and Walmsley described in their article in 1962 [1] that "*Although it is not possible to synthesize very long polyenes (polyacetylene) at present, general interest in conjugated polymers with related, but rather more complex, structures makes a full study of the electronic states of this simple polymer worthwhile.*" Although the first polymerization of acetylene was reported in no later than 1958 by Natta and his co-workers [2] who prepared polyacetylene that is structurally identical to the very long conjugated polyene, the work was not accepted widely in the field. Before that time the interests in this compound were limited to theoretical approaches to explain to chemists a red-shift of the absorption maximum (bathochromic effect) and an increase in the absorption coefficient (hyperchromic effect) with increasing the number of repeating units in the conjugation, and to elucidate for physicists bond alternation in connection with an electron-phonon interaction. An accumulation of experimental observations on relatively short polyenes [3–8] coupled with theoretical considerations, such as the free-electron model and simple Hückel molecular orbital treatments, strongly suggested that a difference between the lengths of double and single bonds decreases with increasing the conjugation and that all the bonds tend to be equal their length in an infinitely long polyene. In other words, one would expect that the infinitely long one-dimensional arrangement of π-electrons forms a half-filled band, or that the highest occupied (HO) and the lowest unoccupied (LU) π-electron bands merge with each other, leading to metallic behavior [9, 10]. In the 1950s, however, it became theoretically clear that the polyene with bond alternation is energetically more stable than that with bonds of equal length [1, 11–13]. Since two geometrical isomers, *trans* and *cis*, are possible for each double bond, two isomeric forms, all-*trans* and all-*cis*, are expected as the two extremes of polyacetylene isomers as shown in Scheme 1. Experimentally the carbon-carbon bond lengths in polyacetylene were directly measured by Yannoni and Clarke [14] with use of the nutation NMR spectroscopy, 1.36 and 1.44 Å for the double and single bonds, respectively, in the *trans* form and 1.37 Å for the double bond in the *cis* form.

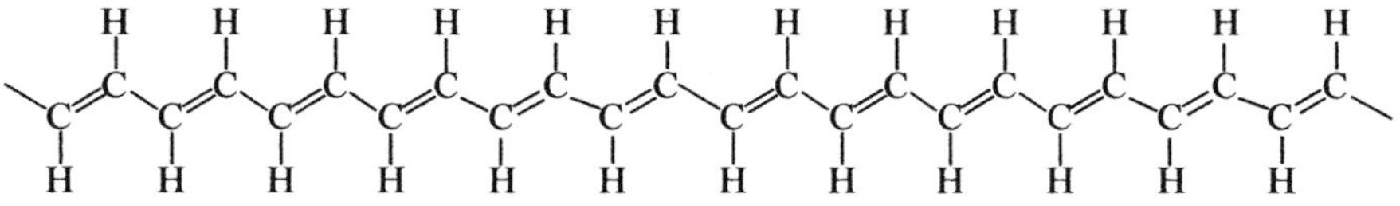

(1)

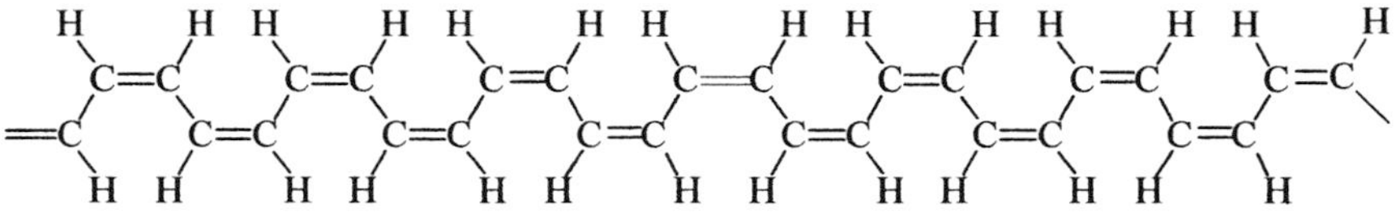

(2)

Scheme 1. All-*trans* (1) and all-*cis* (2) polyacetylene.

Even after the first synthesis by Natta and co-workers [2], polyacetylene remained for some time a material of interest to only a few organic [15, 16] and polymer chemists [17–19] because the product was obtained as insoluble and infusible powders.

II. *TRANS* OR *CIS*?

Among few chemists, S. Ikeda and his co-workers had been studying a mechanism of acetylene polymerization in connection with olefin polymerization by various Ziegler-Natta catalysts. They found that the polymerization yields not only highly polymerized polyacetylene but also benzene, that is, a cyclic trimer of acetylene, and that the ratio of these two products depends upon the species of Ziegler-Natta catalysts used. They also found the formation of alkylbenzenes as a minor by-product of the acetylene polymerization with a catalyst system composed of titanium tetrachloride and trialkylaluminum. In a series of experiments using carbon-14 and deuterium, they noted that a labeled ethyl group is introduced in the ethylbenzene when triethylaluminum labeled with carbon-14 or deuterium is used as the co-catalyst [20]. In another experiments on the oxidation of polyacetylene by alkaline potassium permanganate, they observed the formation of propionic and acetic acids that are derived from the alkyl groups in the trialkylaluminum used as the co-catalyst [20]. From these results, they concluded that polyacetylene and benzene could be formed from the same active site of the catalyst system. Thus, the reaction proceeds by *cis*-opening of the triple bond in acetylene followed by a *cis* insertion into titanium-alkyl bond of the catalyst. This mechanism fits the orbital interaction consideration by Fukui and Inagaki [21], for the role of the catalyst according to which the initially formed configuration of the double bond is *cis* as a result of favored orbital interaction between the inserting acetylene and the active site of the catalyst. Whether cyclic trimerization occurs to give benzene or polymerization proceeds to give polyacetylene is determined by the conformation of the growing chain that takes either *cisoid* or *transoid* structure in the vicinity of the active site of the catalyst [22]. As no *cis* form had been known until then, an important question remained: why is the mechanism capable of yielding only [9] *trans* configuration of the double bonds in polyacetylene? [2, 19].

III. DISCOVERY OF FILM SYNTHESIS

The conventional method of polymerization in the laboratory is such that the catalyst solution is stirred thoroughly by any means to carry out the reaction under a homogeneous condition. The acetylene polymerization has not been an exception. It was customary for polymer chemists who synthesized polyacetylene to bubble acetylene gas into a catalyst solution with stirring. Unfortunately, the product was obtained as an intractable black powder from which it is very difficult to make into samples of a shape suitable for measurement of spectra and physical properties because of its insolubility and infusibility.

Soon after I joined Ikeda's group, we succeeded in synthesizing polyacetylene directly in the form of a thin film [23] by a fortuitous error in 1967. After a series of experiments to reproduce the error, we noticed that we had used a concentration of the Ziegler-Natta catalyst nearly a thousand times greater than that usually used. It is worth noting that the insolubility of polyacetylene contributes to the formation of film. In addition, it was found that the film is composed of entangled micro-fibers called fibrils by a transmission electron microscope observation of an extremely thin film and by scanning electron microscope observation on a surface of a thick film as shown in Figure 1. The fibril diameter is in the range of 20–100 nm depending upon the polymerization conditions. These inherent properties of polyacetylene are absolutely necessary for the film formation even under the higher concentration of the catalyst. One more important factor that should be added is

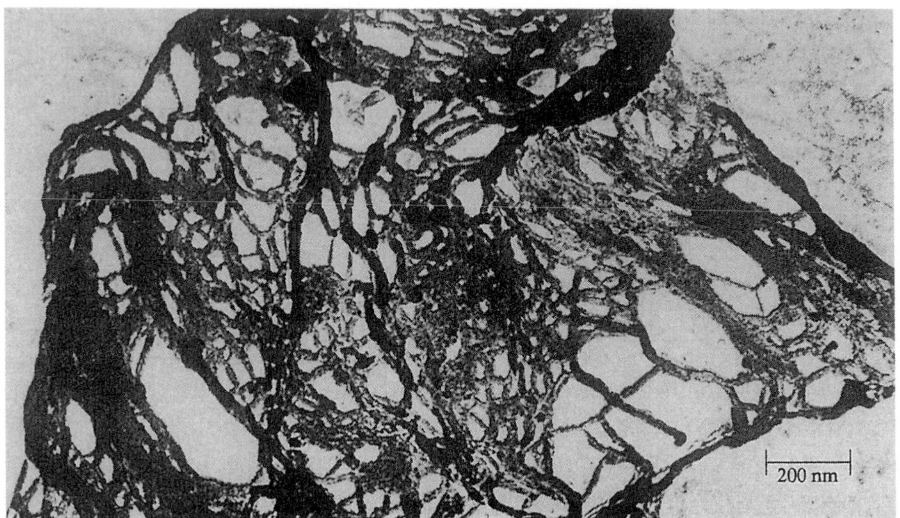

Figure 1a. Transmission electron micrograph of an extremely thin film of polyacetylene.

that the catalyst used at that time, $Ti(O\text{-}n\text{-}C_4H_9)_4\text{-}(C_2H_5)_3Al$, was quite unique from the viewpoint of its good solubility in organic solvents such as hexane or toluene to give a homogeneous solution and its high activity to give exclusively high molecular weight and crystalline polymers. On the contrary, most

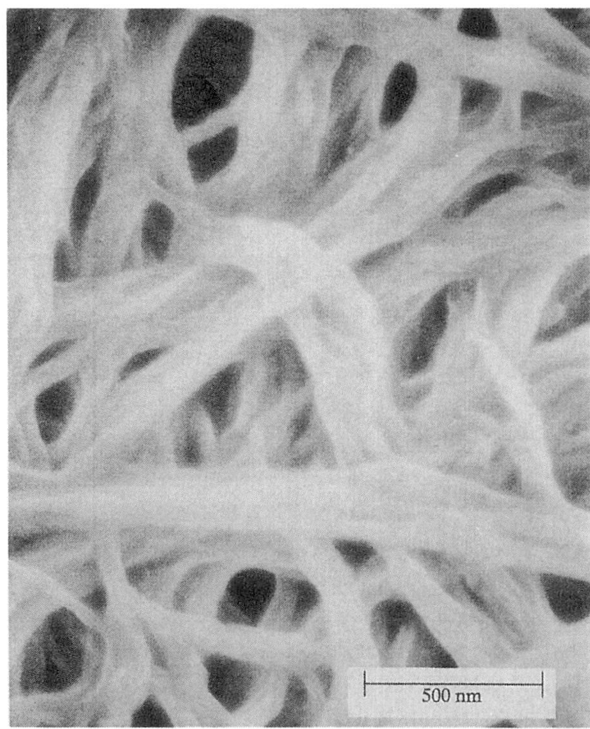

Figure 1b. Scanning electron micrograph of the surface of a thick polyacetylene film.

of Ziegler-Natta catalysts form precipitates to give inhomogeneous solutions when a titanium compound is mixed with alkylaluminum.

The use of the films enabled us to obtain clear infrared spectra as shown in Figure 2 [24] and analysis of these spectra indicated that the configuration of the double bonds strongly depends on the temperature of polymerization. The *trans* contents of polyacetylene prepared by the Ziegler-Natta catalysts decreases with decreasing the polymerization temperature as shown in Table I. A thermal study [25] indicated that the irreversible isomerization of the *cis* form occurs at temperatures higher than 145 °C to give the *trans* one. Thus the *cis* form is thermodynamically less stable than the *trans* one. The observed *cis*-rich polyacetylene synthesized at lower temperatures suggested the *cis*-opening of the triple bond of the acetylene monomer consistent with the *cis*-opening mechanism proposed by Ikeda [22]. In case of the polymerization being carried out at higher temperatures, spontaneous isomerization of the growing *cis* double bonds occurs to give *trans* ones. The *cis*-opening mechanism has been supported by the mutation NMR study [14] and by the infrared study of copolymers of acetylene and acetylene-d_2 [24]. In conclusion, the open problem why only *trans* polyacetylene had been known was solved by the use of films.

Table I. The *trans* contents of polyacetylene prepared at different temperatures

Temperature (°C)	Trans content (%)
150	100.0
100	92.5
50	67.6
18	40.7
0	21.4
–18	4.6
–78	1.9

Catalyst: $Ti(O\text{-}n\text{-}C_4H_9)_4\text{-}(C_2H_5)_3Al$, Ti/Al = 4, [Ti] = 10 mmol/l.

IV ELECTRICAL PROPERTIES OF AS-PREPARED POLYACETYLENE FILMS

The electrical resistivity of as-prepared films with various *cis/trans* contents was measured by the conventional two-probe method under vacuum in a temperature range of –120 to 20 °C [26]. The resistivity and energy gap of *trans*-rich polyacetylene were 1.0×10^4 Ωcm and 0.56 eV, respectively, whereas the values of a *cis*-rich (80 %) one were 2.4×10^8 Ωcm and 0.93 eV, respectively. Hatano and co-workers [17] reported that the resistivity and energy gap measured on compressed pellets of powder polyacetylene synthesized by the same catalyst system are in the range of 1.4×10^4 and 4.2×10^5 Ωcm, and 0.46 eV, respectively, in good agreement with those for *trans*-rich polyacetylene film. In conclusion, it became apparent that the intrinsic electrical properties do not change very much between powder and film.

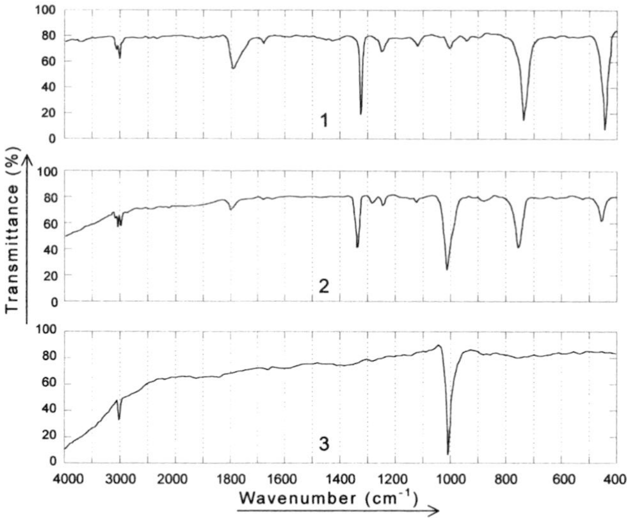

Figure 2. Infrared spectra of polyacetylene synthesized at (1) −78 °C, (2) 20 °C, and (3) 150 °C. After reference 24, Figure 4.

V. HALOGENATION OF THE POLYACETYLENE FILMS

Since no improvement in electrical conductivity was observed in a form of film, we tried to use the polyacetylene films as a source of graphite films as the carbon content of the polymer is as high as 92.3 %. Thermograms of differential thermal analysis of *cis*-rich polymer revealed the existence of two exothermic peaks at 145 and 325 °C and one endothermic peak at 420 °C, which were assigned to *cis-trans* isomerization, hydrogen migration accompanied with crosslinking reaction, and thermal decomposition, respectively [25], as shown in Scheme 2. Thermogravimetric analysis showed that weight

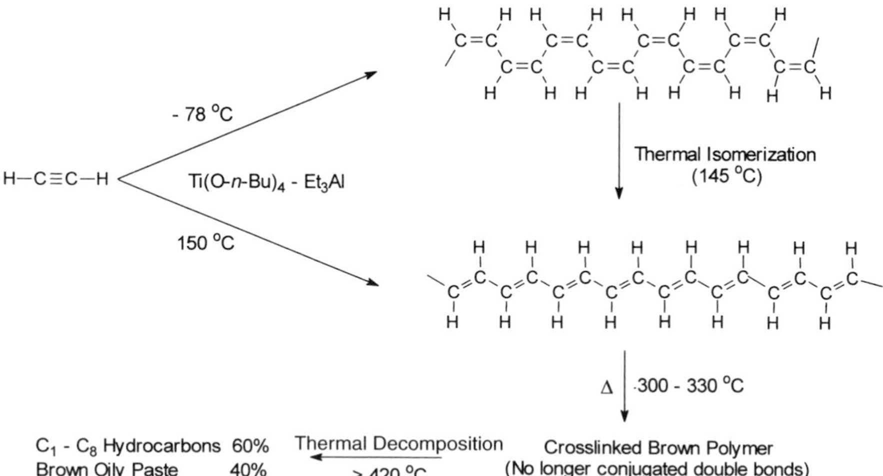

Scheme 2. Thermal characteristics of polyacetylene.

loss reached 63 % at 420 °C. Therefore, pyrolysis of as-prepared polyacetylene films is not suitable for the preparation of graphite films.

Electrophilic addition of halogens such as chlorine and bromine to a carbon-carbon double bond is a well-known reaction and proceeds with good yield and few side reactions. Elimination of hydrogen halide from adjacent carbon atoms in a halogenated hydrocarbon is also a well-known reaction to introduce a carbon-carbon double or triple bond. Since polyacetylene reacts with chlorine and bromine under mild conditions, we thought carbonization might be possible by combining the both reactions. Thus, a polyacetylene film was reacted with chlorine at room temperature to give a chlorinated polyacetylene film, which was subsequently treated with a basic reagent to eliminate hydrogen chloride to give a carbon film in satisfactory yield. However, we found that the carbon film was hardly graphitized even when heated at 2000 °C for several hours.

Fully chlorinated polyacetylene has a chemical structure of $-(CHCl\text{-}CHCl)_n-$, corresponding to a polymer of 1,2-dichloroethylene. It is known, however, 1,2-disubstituted ethylenes hardly ever polymerize. Therefore, the chlorination of polyacetylene is the sole method to synthesize poly(1,2-dichloroethylene). To investigate the chlorination process of the polyacetylene film, *in situ* measurements were planned to obtain the infrared spectra during the chlorination of the polyacetylene film.

It is well known that chlorination of the simplest conjugated polyene, 1,3-butadiene, gives two isomeric products, the 1,2-addition product (3,4-dichloro-1-butene) and the 1,4-addition product (1,4-dichloro-2-butene). Likewise, 1,3,5-hexatriene gives three isomers, 1,2-, 1,4- and 1,6-addition products. The reason why the reaction gives various products is explained by the existence of different resonance structures in the intermediate as shown in Scheme 3. The third isomer is thermodynamically the most stable because the molecule has an inner conjugated diene structure with the largest energy of resonance stabilization.

The motivation of the study was to confirm that the 1,2n-addition of chlorine might occur as the initial step of the reaction as shown in Scheme 4a and 4b. Since the longer inner polyene structure has the larger energy of resonance stabilization, it is anticipated that products of 1,2n-addition with larger n predominate, rather than a 1,2-addition product at the initial step of the reaction or in the partially chlorinated polyacetylene. If such structures could be detected by infrared spectroscopy, one might expect that the positive

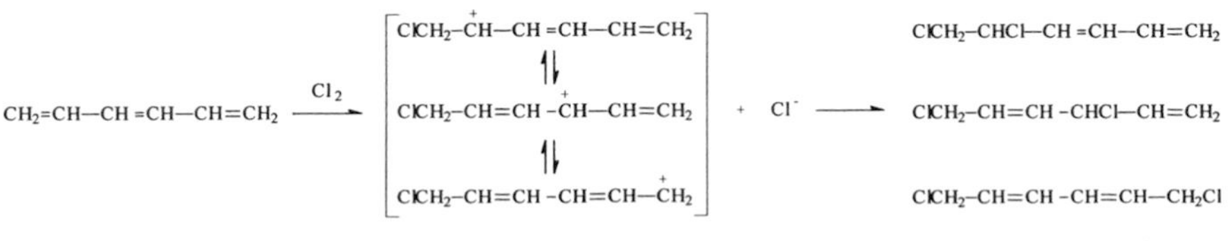

$$CH_2{=}CH{-}CH{=}CH{-}CH{=}CH_2 \xrightarrow{Cl_2} \left[\begin{array}{c} ClCH_2{-}\overset{+}{C}H{-}CH{=}CH{-}CH{=}CH_2 \\ \upharpoonleft\downharpoonright \\ ClCH_2{-}CH{=}CH{-}\overset{+}{C}H{-}CH{=}CH_2 \\ \upharpoonleft\downharpoonright \\ ClCH_2{-}CH{=}CH{-}CH{=}CH{-}\overset{+}{C}H_2 \end{array}\right] + Cl^- \longrightarrow \begin{array}{c} ClCH_2{-}CHCl{-}CH{=}CH{-}CH{=}CH_2 \\ ClCH_2{-}CH{=}CH{-}CHCl{-}CH{=}CH_2 \\ ClCH_2{-}CH{=}CH{-}CH{=}CH{-}CH_2Cl \end{array}$$

Scheme 3. Possible resonance structures in the intermediate during the process of the chlorination of 1,3,5-hexatriene and three chlorinated products.

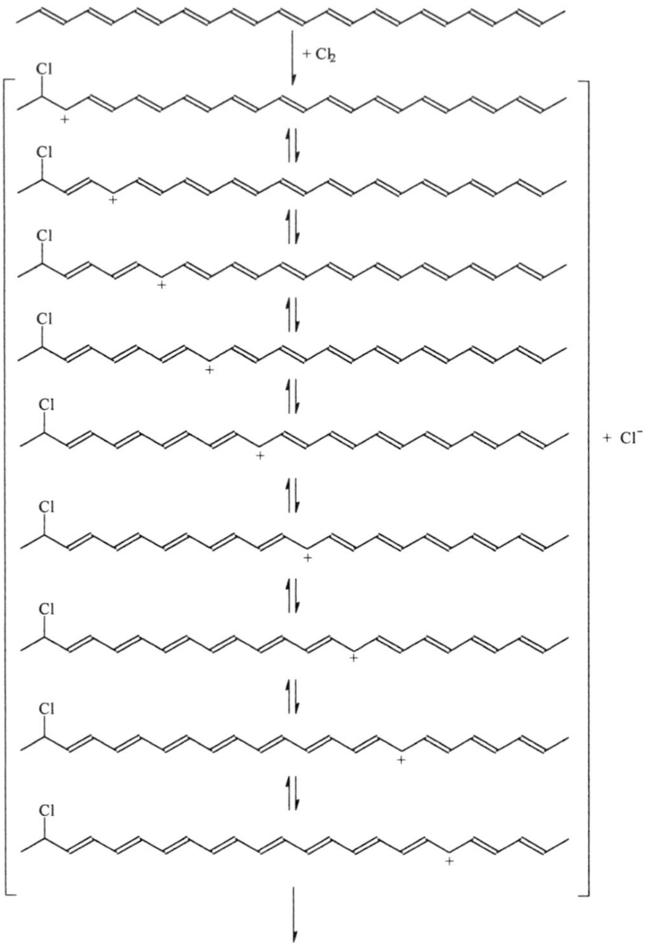

Scheme 4a. Possible resonance structures of the intermediate in the chlorination of polyacetylene.

change on the intermediate carbocation is able to migrate far away from the reaction site through the conjugated polyene structure.

Contrary to this expectation, the infrared spectrophotometer recorded no spectrum but only a 100 % absorption line in the full range of 4000 to 400 cm^{-1} immediately after addition of a trace amount of chlorine to the film. The polyacetylene film changed into an opaque material because of some very strong absorptions. Upon continued reaction with additional chlorine, the spectrum became clearer to give that corresponding to a chlorinated polyacetylene.

VI. EPILOGUE

Later, we know that the very strong absorptions, is the so-called "doping induced infrared band" composed of three bands at 1397, 1288, and 888 cm^{-1} [27–29], which have extremely strong absorption coefficients compared with

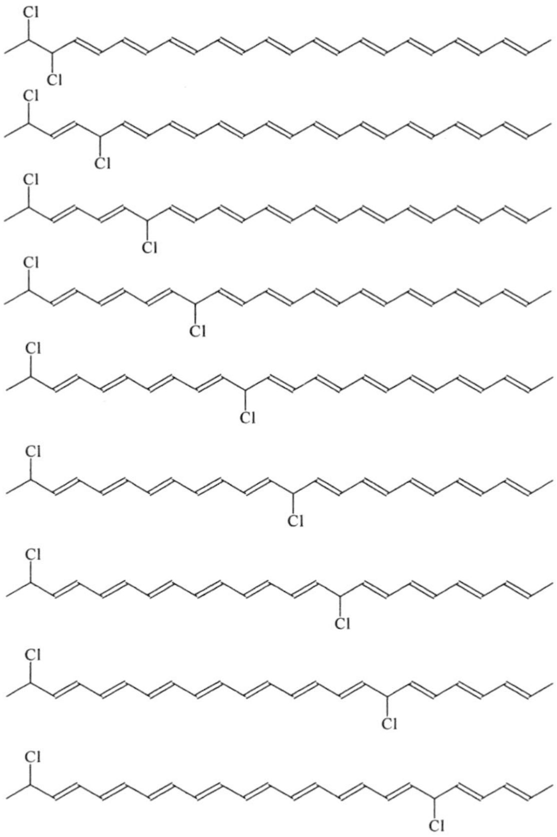

Scheme 4b. Expected chemical structures of partially chlorinated polyacetylene.

those of the as-prepared polymer. The observed isotope shifts of these bands in the spectra of poly(acetylene-$^{13}C_2$) and poly(acetylene-d_2) demonstrated that these bands are of vibration origin [27] in the vicinity of the carbocation or positively charged carbon in the conjugated polyene. At present, the carbocation formed in long conjugated polyenes is widely known as a positively charged soliton [30] that acts as a charge carrier for the electrical conduction. At that time, to my regret, I did not recognize that this carbocation could be a charge carrier and thus polyacetylene could be the first conducting polymer. To open an era of conducting polymers, we had to wait until we intentionally carried out the doping experiment with bromine at University of Pennsylvania, on Tuesday 23rd, 1976 and the later also successively with iodine [31, 32].

ACKNOWLEDGMENT

At the end of my lecture I want to express my heartfelt thanks to the late Professor Sakuji Ikeda who gave me a chance to work on the acetylene polymerization, to Professor Masahiro Hatano who contributed to the semiconducting polymers and encouraged me on my work, to the late Professor Shu

Kambara who organized our polymer research group and supported us in our work even after his retirement, and to Dr. Hyung Chick Pyun with whom I encountered the discovery of polyacetylene film by the fortuitous error. Dr. Takeo Ito was the first graduate student for me and who made a great contribution at the initial stage of this work. My special thanks are to Professors Shiro Maeda, Takehiko Shimanouchi, and Mitsuo Tasumi for helpful discussion on the vibration analyses study. I am indebted to Mr. Shigeru Ando for measurements of solid-state properties and to Professor Yoshio Sakai for electrical conductivity measurements. Throughout this period our research had been partly supported by Grant-in-Aids for Scientific Research from Ministry of Education, Culture and Science of Japan.

REFERENCES

[1] J. A. Pople and S. H. Walmsley, *Mol. Phys.*, **5**, 15 (1962).
[2] G. Natta, G. Mazzanti, and P. Corradini, *Atti Acad. Naz. Lincei, Cl. Sci. Fis. Mat. Nat., Rend.*, (8), 25: 3 (1958).
[3] R. Kuhn, *Angew. Chem.*, **34**, 703 (1937).
[4] J. H. C. Nayler and M. C. Whiting, *J. Chem. Soc.*, 3037 (1965).
[5] F. Bohlmann and H. Manhardt, *Chem. Ber.*, **89**, 1307 (1956).
[6] F. Sondheimer, D. A. Ben-Efraim and R. Wolovsky, *J. Am. Chem. Soc.*, **83**, 1675 (1961).
[7] Y. Takeuchi, A. Yasuhara, S. Akiyama, and M. Nakagawa, *Bull. Chem. Soc. Jpn.*, **46**, 2822 (1973).
[8] K. Knoll and R. R. Schrock, *J. Am. Chem. Soc.*, **111**, 7989 (1989).
[9] N. S. Bayliss, *J. Chem. Phys.*, **16**, 287 (1948), *Quart. Rev.*, **6**, 319 (1952).
[10] H. Kuhn, *Helv. Chim. Acta*, **31**, 1441 (1948).
[11] Y. Ooshika, *J. Phys. Soc. Jpn.*, **12**, 1238 , 1246 (1957).
[12] H. C. Longuet-Higgins and L. Salem, *Proc. Roy. Soc.*, A, **252**, 172 (1959).
[13] M. Tsuji, S. Fuzinaga, and T. Hashino, *Rev. Mod. Phys.*, 32, 425 (1960).
[14] C. S. Yannoni and T. C. Clarke, *Phys. Rev. Lett.*, **51**, 1191 (1983).
[15] L. B. Luttinger, *J. Org. Chem.*, **27**, 1591 (1962).
[16] W. E. Daniels, *J. Org. Chem.*, **29**, 2936 (1964).
[17] M. Hatano, S. Kambara, and S. Okamoto, *J. Polym. Sci.*, **51**, S26 (1961).
[18] W. H. Watson, Jr., W. C. McMordie, Jr., and L. G. Lands, *J. Polym. Sci.*, **55**, 137 (1961).
[19] D. J. Berets and D. S. Smith, *Trans. Faraday Soc.*, **64**, 823 (1968).
[20] S. Ikeda and A. Tamaki, *J. Polym. Sci., Polym. Lett.*, **4**, 605 (1966).
[21] K. Fukui and S. Inagaki, *J. Am. Chem. Soc.*, **96**, 4445 (1975).
[22] S. Ikeda, *Kogyo Kagaku Zasshi*, **70**, 1880 (1967).
[23] T. Ito, H. Shirakawa, and S. Ikeda, *J. Polym. Sci., Polym. Chem. Ed.*, **12**, 11 (1974).
[24] H. Shirakawa and S. Ikeda, *Polym. J.*, **2**, 231 (1971).
[25] T. Ito, H. Shirakawa, and S. Ikeda, *J. Polym. Sci., Polym. Chem. Ed.*, **13**, 1943 (1975).
[26] H. Shirakawa, T. Ito, and S. Ikeda, *Makromol. Chem.*, **179**, 1565 (1978).
[27] I. Harada, Y. Furukawa, M. Tasumi, H. Shirakawa, and S. Ikeda, *J. Chem. Phys.*, **73**, 4746 (1980).
[28] J. F. Rabolt, T. C. Clarke, and G. B. Street, *J. Chem. Phys.*, **71**, 4614 (1979).
[29] C. R. Fincher, Jr., M. Ozaki, A. J. Heeger, and A. G. MacDiarmid, *Phys. Rev. B* **19**, 4140 (1979).
[30] W. P. Su, J. R. Schrieffer, and A. J. Heeger, *Phys. Rev. Lett.*, **42**, 1698 (1979).
[31] H. Shirakawa, E. J. Louis, A. G. MacDiarmid, C. K. Chiang, and A. J. Heeger, *J. Chem. Soc., Chem. Commun.*, 578 (1977).
[32] C. K. Chiang, C. R. Fincher, Jr., Y. W. Park, A. J. Heeger, H. Shirakawa, E. J. Louis, S. C. Gau, and A. G. MacDiarmid, *Phys. Rev. Lett.*, **39**, 1098 (1977).